W9-AOB-314

MATERIALS SCIENCE OF CONCRETE VI

Other Volumes in the *Materials Science of Concrete* series:

Materials Science of Concrete—Special Volume: Ion and Mass Transport in Cement-Based Materials
Edited by R. Doug Hooton, Michael D.A. Thomas, Jacques Marchand, and James J. Beaudoin
©2001, ISBN 1-57498-113-7

Materials Science of Concrete—Special Volume: Role of Calcium Hydroxide in Concrete
Edited by Jan Skalny, Juri Gebauer, and Ivan Odler
© 2001, ISBN 1-57498-128-5

Materials Science of Concrete—Special Volume: Sulfate Attack Mechanisms
Edited by Jacques Marchand and Jan Skalny
© 1999, ISBN 1-57498-074-2

Materials Science of Concrete—Special Volume: The Sidney Diamond Symposium
Edited by Menashi Cohen, Sidney Mindess, and Jan Skalny
© 1998, ISBN 1-57498-072-6

Materials Science of Concrete IV
Edited by Jan Skalny and Sidney Mindess
© 1995, ISBN 0-944904-75-0

Materials Science of Concrete III
Edited by Jan Skalny
© 1992, ISBN 0-944904-55-6

Materials Science of Concrete II
Edited by Jan Skalny and Sidney Mindess
© 1991, ISBN 0-944904-37-8

Materials Science of Concrete I
Edited by Jan Skalny
© 1989, ISBN 0-944904-01-7

MATERIALS SCIENCE OF CONCRETE VI

EDITED BY SIDNEY MINDESS
JAN SKALNY

Published by
The American Ceramic Society
735 Ceramic Place
Westerville, Ohio 43081
www.ceramics.org

The American Ceramic Society
735 Ceramic Place
Westerville, Ohio 43081

Printed in the United States of America.

01 02 03 04 05 5 4 3 2 1

ISBN: 1-57498-069-6

Library of Congress Cataloging-in-Publication Data
A CIP record for this book is available from the Library of Congress.

For more information on ordering titles published by The American Ceramic Society or to request a publications catalog, please call (614) 794-5890 or visit our online bookstore at <www.ceramics.org>.

Contents

Preface

Is there a thing whereof it may be said, See, this is new? It hath been already in the ages which were before us.

There is no remembrance of former things; neither shall there be remembrance of things that are to come with those who shall live afterwards.

— Ecclesiastes 1:10, 11

All too often, this quotation from Ecclesiastes can be used to describe the current state of cement and concrete research. Too much of what we read has already been done before and will almost certainly soon be forgotten. Fortunately, however, this does not apply to the present volume. The 11 chapters that are included cover a wide range of topics: techniques to characterize cement and concrete, early age concrete properties, durability, and modified cements. The authors have done a splendid job of providing information that is timely and mostly new, and that will be referred to for a good many years to come.

As was the case with the previous five volumes in this series, the rapid growth in cement and concrete technology has made it difficult to choose only 11 contributions. Clearly, there are many other topics that should be explored in depth (and that we hope will be covered in future volumes). We hope that the current selection will provide much food for thought for those engaged both in research and in practice. It is more important now than ever to look at the fundamentals underlying the behavior of cement-based construction materials, and this the authors have done.

We would like to thank the contributors for taking the time to prepare these chapters and the publishing staff of The American Ceramic Society for their care in preparing this volume for publication.

Sidney Mindess
Jan Skalny

Phase Composition and Quantitative X-Ray Powder Diffraction Analysis of Portland Cement and Clinker

Quirina I. Roode-Gutzmer and Yunus Ballim
University of the Witwatersrand, Johannesburg, South Africa

"There is an old saying in research: It's okay to sleep with a hypothesis, but you should never marry one. Always be ready to follow a new lead and shift techniques, even if it means giving up a favorite idea."
— J. William Langston

The complexities of quantitative X-ray powder diffraction are discussed in terms of its historic development and how these pertain to the complex phase system of portland cement clinker. The advantages of X-ray diffractometry are compared to other valid techniques, particularly microscopy. The concept of the Rietveld method is rigorously introduced and some results obtained on cement specimens are reviewed.

Introduction

The quantitative phase composition of any material is an indispensable determinant in the relationships between its properties and the material processes that are associated with it. In the case of cement clinker, the phase composition alludes to processes of raw material preparation, burning in the kiln, and subsequent cooling. On the downstream side, the phase composition of clinker and additives such as gypsum governs the cement hydration process, which in turn dictates the ensuing properties of setting, heat of hydration, soundness, load-bearing capacity, and resistance to deterioration or durability in defined environments.

The phase composition of cement and clinker has until recently been estimated by procedures that can, at best, be considered to be semi-quantitative and generally extremely tedious. This has been partly due to the inherent complexity of cement, which has rendered its study by any single analytical technique almost impossible. Of the available analytical techniques, X-ray diffraction (XRD) appears to be one of the few direct ways of characterizing the phase composition of cement. Given its economic and technological importance, together with the complexity of its phase composition, cement has been partly responsible for a number of significant developments in present-day quantitative XRD (QXRD).

The purpose of this paper is first to provide an exposition of the complexity involved in the quantitative aspect of X-ray powder diffraction analysis of cement clinker, and second to show how these complexities may be resolvable by means of a mathematical parameter model, that of the Rietveld method, an account of which is comprehensively provided by Young.[1] The last review of state-of-the-art QXRD analysis of cement and clinker was provided by Struble[2] in 1991, by which time the application of the Rietveld method to cement and clinker had only just begun to make progress. This paper is mainly intended to provide a review of subsequent developments in this field.

X-Ray Powder Diffractometry

In order to appreciate the complexities of phase quantification using X-ray powder diffractometry, the nature of measurements obtained with this technique needs to be fully understood (this aspect is covered in more detail by Klug and Alexander[3]). X-ray powder diffractometry is an operationally definable process that measures, in conventional diffraction geometry, two parameters: the intensity of X-rays diffracted from a certain set of crystal planes of a crystalline compound (phase) contained in a powdered specimen, and the angle at which diffraction takes place with respect to the incident X-rays. The diffractometer should necessarily be properly aligned and the incident X-rays should be monochromatic and have a wavelength of typically between 0.5 and 2.5 Å, depending on which target material is used for the X-ray source.

The validity of these measurements is based on the underlying interaction of X-rays with matter, which is fundamentally expressed by the Bragg equation:

$$2\,d \sin\theta = n\,\lambda \tag{1}$$

where n is the order of reflection ($n = 1$ for our purpose of powder diffraction), λ is the wavelength of the X-ray (e.g., for CuK_α, λ is 1.5418 Å), d is the interplanar spacing between crystal planes *(hkl)* from which diffraction takes place, and θ is the angle of diffraction (which when measured in Bragg-Brentano is expressed as 2θ).

Based on a plot of diffracted intensity as a function of the diffraction angle — a pattern unique to each compound contains peaks having certain angular positions and relative intensities with respect to each other — the phases present in any given crystalline specimen can be identified. Routine identification of phases by powder XRD is accomplished by comparing observed data from a specimen, computationally or manually, with those prepared from pure standard substances contained in a database — the Powder Diffraction File (PDF).[4] The data contained in the PDF are collected, edited, published, and distributed by the International Centre for Diffraction Data.

Portland Cement Clinker Phases

The definition of a mineral or crystalline phase — the former being necessarily of natural origin and the latter referring to both naturally occurring and synthetically prepared compounds — is that it has a certain stoichiometric elemental composition and that the individual atoms are confined to fixed positions relative to one another and repeated at regular intervals in three-dimensional space. It is a phase in the sense that it can be placed on a PTX (pressure, temperature, composition) stability diagram. Thus, the phases that are identified in a clinker by X-ray powder diffractometry must be in agreement with the chemical composition derived from the raw materials used and the thermodynamics of the clinkering process, which includes the high-temperature chemistry in the kiln and the subsequent cooling as the clinker exits the kiln, where kinetic influences are particularly important during cooling.

There are four principal major phases in portland cement clinker: Ca_3SiO_5 (C_3S*), Ca_2SiO_4 (C_2S), $Ca_3Al_2O_6$ (C_3A), and a phase that approximates the composition of Ca_2AlFeO_5 (C_4AF), which refer to the pure compounds tricalcium silicate, dicalcium silicate, tricalcium aluminate, and tetracalcium aluminoferrite, respectively. However, in commercial portland cement clinker variations in the form of these phases occur due to the presence of impurities and the effect of temperature and rate of temperature change during their formation. Under these clinkering conditions, the different forms of the phases are collectively referred to as alite, belite, aluminate, and ferrite, respectively. The quantity in which these phases are present in portland cement clinker is roughly 50–70% alite, 15–30% belite, 5–10% aluminate, and 5–15% ferrite. Minor phases in clinker include free lime (CaO), periclase (MgO), and alkali sulfates.

The crystallography of these phases is complex because of the existence of several possible polymorphs, which occur when different crystallograph-

*Cement chemistry notation: C = CaO, S = SiO_2, A = Al_2O_3, F = Fe_2O_3, and H = H_2O.

ic modifications of the same chemical compound exist. Alite is known to exist in seven polymorphs, belite in five, and aluminate in three, and ferrite is known to have a variable composition in the solid solution series $Ca_2(Al_xFe_{1-x})_2O_5$, where $0 < x < 0.7$. The stability of polymorphs is influenced by relatively low levels of solid solution. For example, the presence of magnesium stabilizes high-temperature monoclinic polymorphs of alite, whereas hydraulic high-temperature polymorphs of belite are stabilized by boron, phosphorous, or strontium and are necessary to prevent it from forming the non-hydraulic γ modification of belite.

A solid solution occurs when there is substitution of one ion for another ion in the structure, usually on an existing lattice site. If ionic or atomic substitution does not occur on existing interstitial sites, they occur as defects to the structure, either on unoccupied sites or interstitial sites (known as Frenkel defects), or they produce vacancies at normally occupied sites to maintain electro-neutrality (known as Schottky defects). The number of defects per unit volume depends on the chemical composition and the thermal and mechanical treatment that the substance has been subjected to. Depending on the type and extent of solid solution, a slight change in unit-cell size of the phase may result, often occurring mainly along well-defined directions. If the solute ions increase above a certain critical concentration, the inherent crystal symmetry becomes altered and a new polymorph is formed. The extent of solid solution also depends largely on the rate of cooling, since rapid cooling favors the retention of solid solution and slow cooling favors exsolution of foreign ions and phase inversion. Taylor[5] has provided detailed descriptions of the chemical composition and polymorphs of cement phases.

Techniques Used in the Analysis of Portland Cement Clinker

Chemical Analysis

A chemical analysis of a clinker or cement provides information normally presented as the analyte oxides, which in a quantitative analysis should total 100 wt% if the analytical method is accurate and all the elements present are analyzed. Typically a portland cement clinker will be composed of 66% CaO, 23% SiO_2, 5% Al_2O_3, 3% Fe_2O_3, and 3% other components. The chemical analysis (obtained by wet chemical or more routinely by X-ray fluorescence spectrometry, XRF) does not provide any indication of the phase composition of the clinker.

Bogue Calculation

The Bogue calculation[6] is a solution of four simultaneous equations using the quantitative elemental composition to obtain estimates of the quantities of the four major phases in portland cement or clinker. Although this calculation has been used as a standard method for decades, it has long been recognized as being inaccurate[7] even when modified by different types of corrections.[8,9] The inaccuracy of this method results from the fact that the phases do not have the composition assumed in the calculation and that equilibrium conditions are not reached during cooling (the latter actually being a cause of the former). The Bogue calculation was, however, devised with the knowledge that it provides only an indication of the potential phase composition. However, if a clinker is produced under conventional pyroprocessing conditions and the compositions of the phases are known exactly in terms of solid-solution, a fairly good estimate of the major phase composition can be attained with the Bogue method. In contrast, a quantitative phase result should predict the pyroprocessing conditions under which the clinker had been prepared. This would satisfy the quality control function of a quantitative phase composition result, which is probably what the Bogue calculation was intended to accomplish.

Optical Microscopy

The technique of optical microscopy is a highly effective means of determining adjustments to raw material composition or plant operating conditions so as to change clinker properties. Ono's method[10] takes into account alite and belite crystallite size, alite birefringence, and belite color, which are indicative of heating rate, time at maximum temperature, maximum temperature, and cooling rate. Although this method has been used to predict strength development,[10] it is not truly quantitative, but rather a very useful qualitative assessment of the clinkering process, which includes raw material preparation, burning in the kiln, and subsequent cooling.

To date, the most accepted accuracy for phase quantification of clinker is accomplished by point counting. This technique is based on the fact that the area of a phase on a polished section of a cement clinker is proportional to the volume fraction of the phase in the material.[11] Error arises from counting statistics, which depends on the number of points, the size of the grid, and the grain size of phases — 4000 points are normally considered to give reasonable precision.[12] Other errors include incorrect identification of phases by the operator and inability to see small inclusions or crystals.

Alternatively, image analysis can be resorted to on digital images of clinker[13,14] so as to potentially avoid the tedium of point counting. This also provides an avenue for automating the analysis. Even though this method would appear to be more objective in terms of avoiding observer errors, it cannot accomplish what the human eye and brain combination can. It is not only variation in etch colors, but crystal modifications, orientations, and shapes that contribute to an observer identifying a particular phase. A further consideration is the loss of information, particularly of resolution, when an optical microscope image is digitized.

Nevertheless, whether point counting or image analysis is used, optical microscopy suffers from limitations introduced by the magnification possible with light microscopy. Aluminate and periclase often occur micro- or cryptocrystalline in clinker, thus being too small to be resolved by optical microscopy.

Scanning Electron Microscopy

Scanning electron microscopy (SEM) of polished sections of clinker (or ground cements dispersed in resin) provides results similar to optical microscopy, but with higher magnification and slightly improved resolution so that very fine-grained material and minute inclusions can be observed. Backscattered electron (BSE) images are generated from the primary high-energy electrons being inelastically scattered back, thereby generating a contrast that is composition differentiating — heavier elements appear brighter than lighter elements. The high-energy electrons also interact with the specimen with the emission of characteristic X-rays; if an energy (or wavelength) dispersive X-ray detector is used, the wavelength emitted is indicative of the element present and its intensity is indicative of the quantity present. In addition, the microstructural features, which include not only the morphology of phases but also the pore structure, can be quantified in an image that is formed by incident secondary electrons of lower energy that collide with the atoms in the specimen. This technique therefore allows quantification of not only the phases, but also the microstructure and exact chemical composition of the phases yielding solid-solution information. The acquisition of this information is further facilitated by the processability of an electronic image that SEM provides. Developments to this technique and the acquisition of excellent results have been obtained by Scrivener[15] and Stutzman.[16,17]

The most important drawback of any microscopic technique, however, is that increased magnification reduces the sampling area and therefore sam-

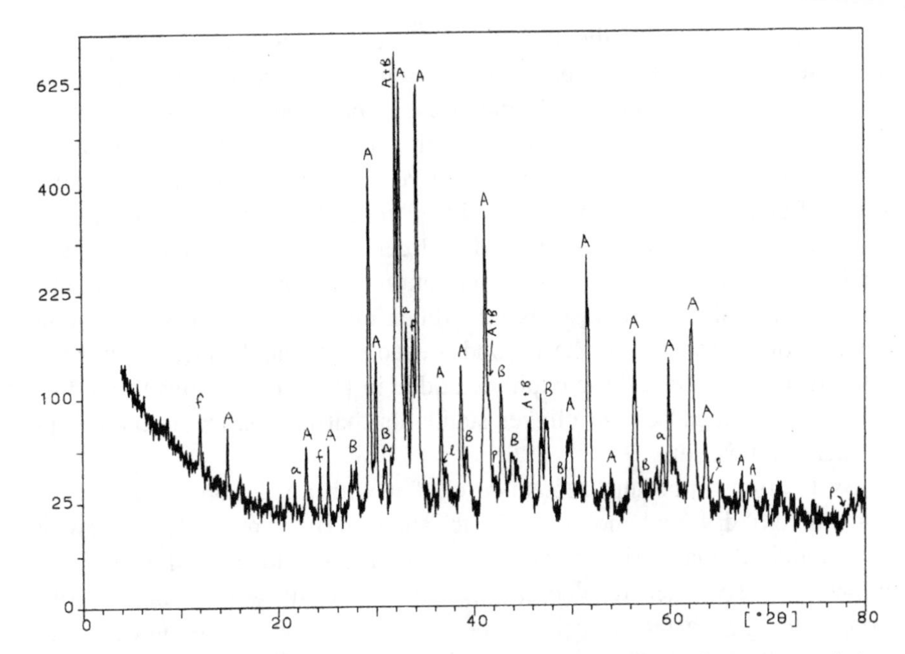

Fig. 1. X-ray powder diffraction pattern of a South African portland cement clinker showing the intense peak concentration at 26–36 °2θ. Phases are labelled A = alite, B = belite, a = aluminate, f = ferrite, l = free lime, p = periclase.

ple representivity. To repeat an example used by Diamond[18]: 10 BSE images of 750× magnification from a 1 cm^2 specimen represent a 0.2% sample of the specimen's microstructure. Therefore, to achieve sample representivity — a crucial requisite for a quantitative result — large numbers of fields must be analyzed and several specimens of the particular sample should be investigated. This requirement makes the application of microscopic techniques tedious.

X-Ray Diffractometry

Portland cements produce a weak-intensity and complex powder XRD pattern where the four major phases all have their strong diffraction lines concentrated in the region 26–36% °2θ (CuK_{α} radiation), giving a severe line overlap problem as shown in Fig. 1. Furthermore, apart from alite dominating the entire powder XRD pattern, it is not at all a simple task to distinguish between the various polymorphs that exist for alite. Sometimes the differences in the fine structure of certain peaks can secure a polymorph identification,[5] but this is not always possible from powder diffraction data.

Attempts to simplify the analysis of clinker by studying each individual phase separately in its various crystallographic modifications are complicated by the extreme difficulty of laboratory preparations yielding each phase in pure isolation from other phases, either from reaction sintering of the reactant oxides or aqueously by selective dissolution. For a quantitative study of phases in commercially produced cements, the method of selective dissolution is the preferred method and can be used in conjunction with many analytical or material characterization techniques. Selective dissolution of some of the clinker phases facilitate more detailed crystallographic analysis of the remaining phases, and if each phase can be effectively separated from the others, these could be added in known quantities to produce synthetic portland cement clinker standards that approximate commercial clinker compositions.

By treating a portland cement or clinker with a solution of salicylic or malic acid in dry methanol, the silicates and free lime are dissolved, leaving the interstitial phases (i.e., ferrite and aluminate) for investigation. The aluminate in the washed residue can then be removed by a saccharose solution, although the ferrite can also be slowly attacked in this process. In a separate experiment, the interstitial phases can be dissolved by a sucrose solution in aqueous KOH (KOSH reagent) so that the silicates and minor phases remain. The amounts of phases selectively isolated by this method depend very critically on controlling the rate of dissolution of the various phases and the reaction time. Gravimetrically, therefore, there is a fairly large range of error associated in establishing a quantitative phase composition. Furthermore, selective dissolution, although very useful, remains an experimentally very laborious technique.

In 1982, the QXRDA methods available that were applied to cements were regarded by Aldridge[19] as unreliable. Even though Gutteridge[20] in 1984 produced good quantitative results, it was achieved at great tedium. Gutteridge employed thorough sample preparation and XRD data collection procedures, made use of more than 30 carefully prepared standards that consisted of various synthetically prepared substituted forms of the four principal cement phases. He also used about 20 more standards to account for minor phases found in cement. In each analysis he made use of an internal standard (rutile) and applied selective dissolution procedures to isolate interstitial and silicate phases.

The process involved in quantifying the phase composition of a portland cement clinker by powder XRD may seem very complex, but the advantages of this technique must be considered against the limitations of other techniques, especially in the light of the significant advances that the Rietveld method has made. As mentioned earlier, the most reliable method of quantifying the phases in a portland cement clinker has been by point counting of a polished section of the specimen under an optical microscope, which is perhaps also the most inexpensive technique. The inabilities to observe minute details and to automate image processing by this technique are, however, alleviated by the use of SEM. Considerably more information about the material is available in a quantitative way, but the SEM technique suffers from being even less bulk representative of the sample than optical microscopy, which is easily resolvable, but at the cost of being laborious. Furthermore, this technique is the most expensive and therefore much less accessible than optical microscopy or X-ray powder diffractometry. The Bogue calculation is of course the least time-consuming and the cheapest, but this method is both indirect and inaccurate and should be restricted for approximate phase composition for clinker production and not used at all for research purposes. Emphasis has been placed on the ability to automate the various techniques under discussion and it is hoped that in the near future accurate quantitative phase compositions can be obtained rapidly and in an online sense for clinker production.

Given these developments, XRD offers an opportunity for a quantitative analysis that is both bulk representative and direct — that is, if phases are to be analyzed, a crystallographic technique would appear to be the most direct. When applied to QXRD, the Rietveld method is particularly aimed at addressing the complexities inherent to both clinker phase composition as well as the interaction phenomena of X-rays with matter. A further advantage of the Rietveld method is that it offers a greater potential for automating quantitative phase analysis than microscopic techniques, thus making it amenable for routine quality control in cement manufacture.

As distinct from clinker, quantifying the phase composition of cement (which is essentially ground clinker admixed with gypsum) becomes even more complicated. Microscopic point counting is best applied on polished sections of crushed pieces or whole nodules of clinker. Analysis of a finely divided powder such as cement by microscopic point counting is rather more complicated and it is necessary to resort to SEM using specialized sample preparation techniques in order to improve the powder sampling statistics. For powder XRD, the use of clinker allows better control of milling so that a narrower particle size distribution (PSD) is obtained in the optimum range of 5–10 μm[3] without sacrificing crystallinity. Cement, although manufactured to have an optimum PSD of 1–40 μm, is generally not optimum for analysis by XRD. In addition, the mechanical attrition achieved in conventional factory milling is harsh and is associated with the production of a considerable amount of heat, which causes crystal defects of various kinds as well as producing amorphous material. Hence, aside from the addition of gypsum, variables introduced by the processing of cement make this a particularly difficult material to analyze for phase composition.

Quantitative X-Ray Powder Diffraction

The idea of phase quantification was already entertained with the first description of the principle of phase identification by Hull[21] in 1919. He claimed that under the same instrumental conditions, a mixture of phases will produce a pattern that equals the superimposed sum of the patterns derived from the separate phases. Although the intensity of each component pattern does depend on the concentration of that phase in the mixture, the relation between intensity and concentration is not simple and is fundamentally not linear, mainly because of the effect of absorption.

Absorption

Absorption is a result of different phases having different mass absorption coefficients, which are calculated from the atomic masses of the elements contained in a formula unit of the phase and the wavelength of the X-rays being absorbed. Absorption essentially leads to the highly absorbing phase being underestimated. Another process that arises from differences in mass absorption coefficients between phases occurs within particles and crystallites and is known as microabsorption. In this process, the diffracted intensity due to absorption is further diminished by increased granularity of the material undergoing diffraction.[22–27] Effects due to microabsorption are usually corrected in quantitative analysis by using the Brindley model,[22] which is incorporated in some Rietveld programs.

Particle and Crystallite Size

Among the specimen-related factors are those of absorption, fluorescence, preferred orientation, and extinction effects. Most of these factors are greatly affected by particle or crystallite size, as was noted in the case of absorption. Effects due to particle size can be experimentally addressed, and must be accounted for in some measure to achieve a quantitative analysis. The crystallite size statistics are greatly improved when the specimen being analyzed is rotated in a plane parallel to its surface, typically one rotation per second, during the XRD data collection. It should be noted that the speed of the data collection should not be in synchronization with the specimen rotation speed, since that would nullify the objective of randomizing the analyte particles.

Crystallinity and Strain

Among the factors that lead to diffraction line-broadening, which directly influences the intensity-concentration relationship, are the degree of crystallinity and strain inherent in crystals. Decrease in the degree of crystallinity results in broad and/or diffuse diffraction lines and in the case of amorphous materials reduces to a "hump" emerging from the background. Strain is introduced in crystal lattices when atoms are displaced from their ideal sites, and can occur in any material to different extents, depending on their atomic arrangement and the kind of treatment the material receives (mechanical, thermal, or otherwise) in sample preparation for analysis. Lattice imperfections can also be introduced into a growing crystal prior to any sample preparation treatment, for example, microtwinning, stacking faults, and so on. Grinding of particularly soft or highly orientated materials result in their lattices becoming plastically deformed, thus causing the crystal lattice to become strained. Line broadening is also crystallite-size dependent, particularly crystallite sizes below about 1 μm.

These factors are important in the sample preparation procedures for QXRDA. To achieve optimum particle sizes while maintaining crystallinity and without introducing undue imperfections in its lattice, it is recommended that the specimen be ground in a liquid medium (the liquid-to-sample ratio should be determined depending on the type of specimen and attrition arrangement). Grinding time should be as short as possible to obtain the desired particle size. If particulate material is very coarse and hard, several short grinding cycles should be used to prevent heating in the mill.

Extinction

Perfect crystals can also present a problem in that X-rays reflected by a set of planes are reflected again in such a way that they destructively interfere and therefore cancel out, resulting in a phenomenon known as extinction. The result is decreased diffracted-beam intensity. This problem occurs typically for minerals like calcite and quartz and can be overcome by introducing lattice imperfections through grinding.

Fluorescence

Another major modifier of the intensity-concentration relationship is that of fluorescence, which results from the X-ray target material being two, three, or four atomic numbers higher than an important element in the specimen. The most common fluorescence problem occurs with iron-containing phases being analyzed by CuK_α radiation. CuK_α radiation has a wavelength of 1.5418 Å, which is slightly shorter than the K absorption edge of iron at 1.7435 Å. The iron atoms in the specimen therefore absorb sufficient energy to eject an electron from its inner K shell and, when the electron returns to its shell, it emits quantized energy known as secondary radiation, which is characteristic of iron (this is essentially the principle used in the technique of XRF analysis). The fluorescent radiation generally becomes part of the background radiation over the entire angular range and can be largely removed as noise by use of a diffracted-beam graphite monochromator. Although the resulting fluorescence can largely be filtered out, it cannot be eliminated and a loss in diffracted intensity results for the phase containing the fluorescent element. In quantitative analysis, therefore, the phase containing the fluorescent element will be underestimated. The only way to eliminate the problem of fluorescence is to use a different X-ray source wavelength.

Preferred Orientation

Preferred orientation arises when there is a strong tendency for the crystallites in a specimen to be orientated in one direction, or one set of directions. The direction along which a mineral tends to orientate is usually determined by the atomic arrangement within the material; for example, gypsum cleaves along the (010) plane where weak hydrogen bonding occurs. For minerals with less obvious cleavage planes, the direction of preferred orientation can be practically assessed by packing the specimen in several different ways into the specimen holder and then observing the diffraction peak with the largest change relative to the others with each repacked specimen. Some mineral phases can have different directions of orientation, depending on how they were crystallized. For example, portlandite, $Ca(OH)_2$, has two different directions in which it preferentially orientates. One is typical for the natural mineral grown under hydrothermal conditions and the other usually occurs with the hydration of lime in ambient conditions typical for hydrating cement.

The degree to which the crystallites preferentially orientate is sensitive to the way the sample is packed. Although the problem of preferred orientation is never really eliminated, its effect can be minimized by ensuring that the particle size is sufficiently small and to pack the sample by side loading or side drifting. This phenomenon can, however, be adequately modeled by use of the March-Dollase function[28] (incorporated in most Rietveld programs), but care should still be exercised to minimize its effect experimentally.

Instrument Geometry

As regards instrument-related factors that influence the intensity-concentration relationship, the diffraction geometry is important. Conventionally the Bragg-Brentano geometry (a detailed account of which is given by Jenkins[29]) is used, where a flat specimen and detector move at angles θ and 2θ, respectively, relative to the static X-ray source. This geometry represents a reasonable compromise between mechanical simplicity and performance. There are, however, three systematic errors inherent in this geometry: axial divergence, flat specimen error, and transparency error.

Axial Divergence

Axial divergence occurs because the X-ray beam from an extended X-ray source (typically a long line focus from a sealed X-ray tube) diverges out of the plane of the focusing circle. The amount of defocusing that results from axial divergence using a flat specimen is reduced by narrowing the divergence angle (usually to 1°), and by using incident and diffracted-beam collimators (Soller slits). Axial divergence introduces asymmetric broadening in the diffracted profile following an approximate $\cot \theta$ dependence, particularly at low diffraction angles.[30] It must also be borne in mind that the diffracted intensity is considerably reduced the more this error is controlled by decreasing the settings of divergence and Soller slits.

Flat Specimen Error

Flat specimen error results because most experimental arrangements use a flat specimen, which is tangential to the focusing circle. To maintain the Bragg-Brentano focusing condition, therefore, the sample should be curved so that it continuously follows the focusing circle. This "out-of-focus" (parafocusing) condition introduces a $\cot \theta$ dependence and results in asymmetric broadening of the diffracted line profile toward low 2θ angles. It is particularly noticeable at low angles, where the irradiated area of the sample is large.

Specimen Transparency

Specimen transparency occurs because the incident X-ray photons penetrate too many atomic layers below the surface of the analyszed specimen. The average diffracting surface therefore lies somewhat below the physical surface of the specimen and therefore farther away from the focusing circle. This is a problem particularly for low absorbing specimens such as organic materials.

Specimen Displacement

Specimen displacement error arises due to practical difficulties in accurately placing the sample on the focusing circle of the goniometer. It is difficult to pack the powdered sample into the holder such that its surface is level with the surface of the holder, the latter being calibrated as the zero level. Specimen displacement typically represents the largest experimental error. It can easily be corrected for, however, with the use of an internal standard in the specimen, usually pure silicon, where the diffracted lines are adjusted in the observed pattern to those expected and calculated from a single crystal of silicon.

Low-Angle Abberations

Although aberrations due to parafocusing and axial divergence are not avoidable in Bragg-Brentano geometry, the asymmetry in the low-angle reflections can be modeled by an adequate Rietveld or profile-modelling method. The error is angle-dependent and only becomes really problematic for materials having their major diffraction peaks at low angles, for example organic and clay materials. In order to study such materials, a calibration should be conducted,[31–33] a feature that is available in some Rietveld programs.

Another important angle-dependent intensity attenuation occurs for microabsorption due to the effect of the Bragg-Brentano geometry. This phenomenon is most severe at low diffraction angles.[27]

Instrument Operating Parameters

Other instrument factors that are important are the type of monochromators used and whether they are for incident or diffracted beams, the type of receiving and divergence slits, and the size of their settings. These factors are critical in conjunction with the target radiation chosen, the kind of X-ray source, the monochromacity of the wavelength, whether filters are used, and the tube voltage and current under which the instrument is operated. These instrument parameters should be chosen to achieve optimum resolution of peaks while maintaining a reasonable diffracted intensity. For quantitative work it is also important to achieve the best signal-to-noise ratio. Cement materials have the problem that the phases are not highly X-ray diffractive and yield a weak intensity pattern with an appreciable background. The signal-to-noise ratio can be improved by operating the X-ray tube source at a high power (the tube output should, however, not exceed 75% of its maximum power output) and to increase the length of time that the sample is radiated (i.e., longer counting times per 2θ step).

Closure

It is clear that the anomalies in the intensity-concentration relationship cannot be completely controlled experimentally. The influencing factors need to be understood and, as far as possible, accounted for in order to achieve reliable QXRD.

Some Highlights of the Development of QXRDA

The following is a summary that is largely drawn from Alexander's review[34] of the history of QXRD between 1936 and 1976, before the advent of the Rietveld method.

The first quantitative phase analysis using X-ray powder diffractometry reported was in 1925 by Navais,[35] who determined the amounts of the mineral mullite in fired ceramics. Measured absolute intensities of diffracted X-rays by components in a binary powder mixture were theoretically addressed by Brentano,[36] Glocker,[37] and Schäfer,[38] but no scheme for its practical application was developed. Thereafter, in 1936 Clark and Reynolds[39] developed a photographic-microphotometric method for the determination of quartz in mine dusts. Their method involved the use of an internal standard to correct for absorption. This method was used extensively by Ballard et al.,[40] who measured peak heights of quartz and a standard (CaF_2 or NiO). The concentration of quartz was then determined with reference to a calibration curve prepared from known mixtures of quartz and internal standard. The problems with this method were the inherent limitation of intensity measurements from photographic densities and the use of peak heights. Furthermore, specimen-related effects such as extinction and microabsorption were neglected.

The reason that a peak-height measurement is not satisfactory for quantitative analysis is that the intensity contained in the width of the peak is not considered. Diffraction line broadening has already been discussed and can arise from differences in crystallinity or crystallite size or from lattice distortions. The area under a peak takes into account the width and to a certain extent the shape of the peak. Determination of peak areas is referred to as integrated intensities, and they are preferred to peak-height intensities. The use of integrated intensities was employed in the analysis of retained austenite in steel.[41–43] An integrated peak-intensity fitting procedure was developed in 1963 by Kennicott.[44]

The difficulties that were apparent in measuring accurate intensities from photographic techniques were essentially ameliorated in 1945 with the advent of the Geiger counter[45] in conjunction with a simpler diffractometer geometry.[4] The counting detection system facilitated precise digital measurements of diffraction intensities, whereas the parafocusing diffraction geometry simplified the theoretical expression of quantitative analysis. With this development, Alexander and Klug[46] in 1948 were able to formulate the basis for quantitative analysis of mixed polycrystalline phases. They showed that for a flat specimen thick enough to yield maximum diffracted intensity, the integrated intensity, I, of reflection i of component J is given by:

$$I_{iJ} = \frac{K_{iJ} x_J}{\rho_J \left[x_J \left(\mu_J - \mu_M \right) + \mu_M \right]} \tag{2}$$

where x_J is the weight fraction of the analyte, ρ_J is the density of the analyte, μ_J is the mass absorption coefficient of component J, μ_M is the mass absorption coefficient of the matrix, K_{iJ} is a constant whose value depends on the diffracting power of the ith reflection of component J and the instrument geometry, and $[x_J (\mu_J - \mu_M) + \mu_M]$ is simply the mass absorption coefficient of the entire specimen, $\bar{\mu}$.

Although they accounted for absorption effects on diffraction intensities, the effects of extinction, microabsorption, and preferred orientation were neglected. The validity of their formulation was, however, demonstrated by their measurements of integrated peak intensities of reference mixtures of analyte quartz *(J)* and an internal standard CaF_2, where the ratio of the quartz/fluorite intensity as a function of the weight fraction of quartz yielded the classic analytical straight line. This method essentially fails as a quantitative method if calibration cannot be accomplished, usually due to lack of availability of pure reference materials or when reflections of different phases overlap. In addition, the presence of amorphous material significantly complicates the analysis.

The addition of an internal standard made it practically possible to compensate for absorption by the specimen. However, Leroux, Lennox and coworkers[47–49] were able to avoid the use of an internal standard by devising a means of directly measuring the total mass absorption coefficient of the specimen ($\bar{\mu}$). For the analysis of small concentrations of analyte without admixing an internal standard, Copeland and Bragg[50] used a "dilution method." A known weight fraction of the analyte, y_J, which is already present in an unknown quantity, x_J, was admixed with the specimen. A strong reflection, k, from another constituent, L, in the specimen, suitable for reference intensity measurement, was used and a straight line graph was obtained of intensity ratio I_{iJ}/I_{kL} as a function of the weight fraction of the analyte, y_J. A negative value of the original weight fraction of the analyte, x_J, is then the intersect of the line with the abscissa. The reason that this approach works successfully is that the matrix, and therefore specimen absorption, remains approximately constant, because the small additions of analyte contribute almost negligibly to the overall specimen absorption.

Copeland and Bragg[50] also treated the general case of an N-component sample producing a diffraction pattern with superimposed "analytical" lines. In order to determine N components, m lines must be measured ($m \geq N$), furnishing sufficient data to permit solution of m simultaneous equations, thus enabling a multicomponent analysis. The accuracy of this method breaks down for significant superposition of peaks — a problem that becomes increasingly complex for large number of components, typically more than three. By diluting $Ca(OH)_2$ in various matrices such as β-C_2S, C_3S, and portland cement, for example, they were able with the use of $Mg(OH)_2$ as an internal standard to achieve relatively accurate amounts of $Ca(OH)_2$ having relative errors within 10%.[51]

Until recently, the main approach to QXRD has been to determine the relative ratios of various intensities (determined from peak heights or integrated peak intensities) and to express these as a function of concentration. This approach was probably used soon after 1919 when Hull[21] first described phase quantification as a possibility with XRD. It was developed as a practical reality by Alexander and Klug[46] in 1948, but its nomenclature as the relative intensity ratio (RIR) method was coined much later by

Chung[52] in 1974. The RIR method was elaborated by Hubbard et al.[53] and Hubbard and Snyder[54] and is essentially defined as the ratio of one peak (with highest intensity and minimal overlap) per phase measured against a peak of an internal standard. When corundum is used as the internal standard, RIR is known as I / I_c — a value that is reported for almost all phases in the PDF, but where I is defined as the main peak of the phase and I_c the main peak intensity for corundum in a 50-50 mixture of corundum and the analyte phase. Although this is not very accurate, mainly due to the use of only one peak per phase, the use of peak heights as opposed to integrated intensities and a range of other factors not considered in the calculation, the reported values of I / I_c are quite useful for achieving rapid estimates in routine XRD analysis.

Chung[52] developed a novel approach to addressing the problem of matrix absorption by introducing a concept of "flushing" the matrix absorption effect. He still used the internal-standard method of Alexander and Klug,[46] which involved the usual procedure of constructing a calibration curve from standards. The matrix absorption effect was then "flushed" by the addition of a considerable amount of a pure substance that is not already present in the sample and whose absorption is exactly known. An equation free of matrix effects can be deduced, where absorption factors are flushed out:

$$X_i = X_f \left(\frac{k_f}{k_i} \right) \left(\frac{I_i}{I_f} \right) \tag{3}$$

where X_i is the weight fraction of component I, X_f is the weight fraction the flushing agent, k is a constant derived from calibration curves where the pure substance I or f is in a matrix of typically corundum ($k_{(I \text{ or } f)} = I_{(I \text{ or } f)} / I_c$), subscript c stands for corundum, and I denotes diffracted intensity by a selected plane of the component denoted by its subscript.

The equation presents a straight line through the origin with slope X_f / K (where $K = k_f / k_i$). Although this method treated the effects of matrix or specimen absorption satisfactorily, typical problems of preferred orientation, microabsorption, extinction, line overlap if separate lines between components cannot be found, particle size, and homogeneity still remained.

The next stage in the development of QXRDA involved attempts to resolve the specimen and instrument-related contributions to a powder diffraction profile, where "profile" refers to the line profile of a single Bragg reflection (peak) and not necessarily the pattern intensity over the entire angular range.[55] It had already been observed as early as 1938 by Jones[56] that a powder diffraction profile is a convolution of a specimen-related function and a function modeling the aberrations introduced by the diffractometer. Specimen-related features are typically crystallite size and strain, which can be assessed using profile fitting or line profile analysis,[57,58] whereas instrument-related features are typically radiation type, axial divergence, and flat specimen error. Profile fitting became particularly useful for quantitative work in that overlapping profiles could be resolved, by a least-squares fitting procedure, into individual component reflections. Integrated intensities, peak positions, and widths could be obtained by this procedure without the use of crystal structure models.[59–62]

Profile fitting, however, was found to be inadequate for severely overlapped reflections. This led to the development of fitting the whole diffraction pattern, as opposed to discrete peaks, over the entire angular range, which is referred to as pattern fitting. This technique of whole-powder pattern fitting was developed in 1981 for neutron powder data by Pawley[63] and in 1986 for X-ray powder data by Toraya.[64]

The next step from pattern fitting over the entire angular range using profile functions that modeled instrument and specimen contributions was the use of structure models from single crystal data of the phases present in the specimen. A diffraction pattern can therefore be calculated from the simultaneous refinement of the crystal structures of the phases present in the specimen, the instrument, and specimen parameters and compared to the observed powder diffraction pattern. The calculated pattern can then be fitted to the observed pattern using a least-squares fit between observed and calculated intensities at every 2θ step in a digital powder diffraction pattern until the best fit is obtained. This is essentially the Rietveld method,[65,66] which was already developed in 1969 but did not find application to quantitative powder diffraction until the 1980s.

The Rietveld Method

The highest form of quantification in any scientific endeavor consists of a mathematical model describing the essential features of the quantitative relationships inherent in vast systems of measurement. In this sense, the Rietveld method represents a mathematical multiparameter model that makes use of the full information content of the powder diffraction pattern. This method was developed by Rietveld[65,66] for the purpose of refining crystal structures from powders using fixed wavelength neutron diffraction for materials that do not produce single crystals. This method immediately proved to be highly successful and led to the publication in 1977 of 172 crystal structures refined by Cheetham and Taylor.[67] In the same year Malmros and Thomas,[68] Young et al.,[55] and Khattak and Cox[69] adapted this method for use with X-ray powder diffraction data. Since then its successful use in quantifying phases from multiphase mixtures has been repeatedly demonstrated.[70–74]

The Basic Rietveld Equation

The observed intensity, y_i, at any arbitrarily chosen point, *I,* in a powder diffraction pattern, may arise from more than one Bragg reflection, either from other phases contained in a multiphase specimen or from symmetrically equivalent crystal planes arising from one phase in the specimen. Each Bragg reflection is diffracted from a plane *(hkl),* where *K* represents the Miller indices, *h, k, l.* The integrated area of a Bragg reflection has an intensity, I_k, that is proportional to the square of the absolute value of the structure factor, $|F_k|^2$. The structure factor is a term that depends on the positions and type of atoms relative to the reflecting planes. Novices in the field of crystallography are encouraged to refer to explanatory literature texts such as Stout and Jensen.[75]

The calculated intensities, y_{ci}, at point *I* in the powder diffraction pattern are therefore a summation of Bragg reflections over all phases *P*, plus the background, and are given by:

$$y_{ci} = \sum_P S_P \sum_P L_K |F_k|^2 \phi(2\theta_i - 2\theta_K) P_K A + y_{bi} \tag{4}$$

where S_P is the scale factor for phase *P; K* represents the Miller indices, *h, k, l,* for a Bragg reflection; L_K contains the Lorentz, polarization, and multiplicity factors; $|F_k|^2$ is the structure factor of the *K*th Bragg reflection; ϕ is the profile shape function; P_K is the preferred orientation function; *A* is an absorption factor; and y_{bi} is the background intensity at the *i*th step.

Phase Quantification

The quantitative phase analysis that the Rietveld method is able to provide relies on a relationship published by Hill and Howard,[71] which requires that the summation be done over all phases present in the mixture:

$$w_P = \frac{S_P M_P V_P}{\sum_P (S_P M_P V_P)} \tag{5}$$

where w_P is the weight fraction of a phase in a mixture of phases, S_P is the Rietveld scaling factor, M_P is the complete unit-cell mass, and V_P is the unit-cell volume of phase *P*.

Although the Rietveld method, when used as a quantitative phase technique, appears to be standardless, it is in fact not. The crystal structure models of the phases serve as the standards, and this can be a very arduous task, particularly with the crystal structure variations that are possible for most of the portland cement clinker phases. Nevertheless, calibration to known mixtures or the addition of an internal standard is not strictly necessary. However, the addition of an internal standard does become necessary when the amount of amorphous material or undefined phases is to be determined.[72]

It is the authors' opinion that an admixed internal standard in portland cement clinkers serves as an internal verification of the phase quantitative results obtained using the Rietveld method. A comparison of these results with point count results obtained on polished sections of clinker nodules should serve as external verification. These verification procedures are necessary in establishing a standardized QXRD-Rietveld method for portland cement clinkers.

Profile Width

A Bragg reflection has a peak position and height, which are fundamentally determined by the crystal structure ($|F_k|^2$), and a width that is dependent on various specimen and instrumental features. Crystallite size and strain broaden the diffraction profile while axial divergence and flat specimen error in parafocusing geometry broaden the profile asymmetrically. Monochromatization limits the breadth of the wavelength profile, but does not introduce any asymmetry.[76] The profile breadth, H_K, is expressed as a full-width-at-half-maximum (FWHM) of the *K*th Bragg reflection[77]:

$$H_K = U \tan^2 \theta_K + V \tan \theta_K + W \tag{6}$$

where *U*, *V*, and *W* are half-width parameters.

Profile Shape Functions

The diffraction profile, apart from being defined by structural features to a position and a relative height, and having a variable width, also contains information on how the tails of the peak decay with distance from the peak position. These decay shapes are critically dependent on the type of radiation source used for diffraction (whether neutron or X-radiation is used) and the shapes may be symmetrical or asymmetrical. Neutron diffraction profiles assume almost exactly a Gaussian peak shape, whereas X-ray diffraction profiles are generally neither Gaussian nor symmetric. The inherent spectral profile of the $K_{\alpha 1}$ line from a Cu target X-ray tube has a width[78] of 0.518×10^{-3} Å[79] and has been shown to be approximately Lorentzian[62] and not completely symmetrical.[80] The step-scan mode of detection also contributes to a more Lorentzian profile shape.

The profile shape function, ϕ, is therefore a convolution of several functions, mostly angle-dependent, that approximate the effects of both instrumental and specimen features of the reflection profiles and is applied over the whole angular range for which diffraction data were collected. It is important to note that the determination of the background, y_{bi}, is critical for correct profile fitting.

Khattak and Cox[69] have shown fundamental problems in representing X-ray diffraction lines with either Gaussian or simple Lorentzian functions. The Voigt function,[76,81,82] a linear convolution of a Gaussian function giving the shape of an X-ray diffraction peak and a Lorentz function giving the correct asymptotic behaviour of the reflection, approximates the observed profile more accurately and over a much larger angular region than the pure Gaussian. The pseudo-Voigt function yields essentially similar results as the Voigt function, but is computationally less demanding, because it conveniently allows the refinement of a mixing parameter, η, determining the fraction of Lorentzian and Gaussian components contained in an observed profile. The equations for Gaussian, Lorentzian, and pseudo-Voigt are expressed, respectively, as follows:

$$\phi = \frac{(4\ln 2)^{1/2}}{(H_K \pi)^{1/2}} \exp \frac{-4\ln 2(2\theta_i - 2\theta_K)^2}{H_K^2} \qquad \text{Gaussian} \quad (7)$$

$$\phi = \frac{2/(H_K \pi)}{1 + 4\left[(2\theta_i - 2\theta_K)^2 / H_K^2\right]} \qquad \text{Lorentzian} \quad (8)$$

$$\phi = \eta L + (1 - n)G \qquad \text{Pseudo-Voigt} \quad (9)$$

where G is Gaussian, L is Lorentzian, and η is the mixing parameter, which is refinable as a linear function.

Other profile shape functions include the Pearson VII function,[83] which varies continuously in shape from a Lorentzian to a Gaussian profile. A survey of profile shape functions was conducted by Young and Wiles,[84] where they concluded that the pseudo-Voigt and Pearson VII functions approximated X-ray diffraction profiles the best, with the pure Gaussian function performing the worst. Howard and Snyder[85] evaluated several profile shape functions and found the Pearson VII function to perform the best. The experience of Roode-Gutzmer (in an unpublished study of several profile shape functions in 1994 and 1995) is that a pseudo-Voigt with approximately 60–70% Lorentzian character works best (and better than a Pearson VII function) for Bragg-Brentano X-ray diffraction patterns acquired in step-scan mode from powdered specimens of portland cement clinkers using CuK_α radiation.

Crystallite-size broadening produces Lorentzian shaped diffraction profiles,[57,86] whereas instrumental broadening and microstrain produce Gaussian profiles.[86] Howard and Snyder[86] suggest, however, that the angular profile of specimen profile broadening may be more reliable than the degree of Gaussian and Lorentzian character to separate size and strain contributions. Although Young and Wiles[84] found that crystallite size broadening is not accurately modeled by a Lorentzian function and that instrumental profile broadening is not well modelled by a Gaussian function, they suggested that analytical functions should rather be based on first principles.

Preferred Orientation

Preferred orientation, P_K, is usually modelled in the Rietveld method with a March-Dollase function[28,87]:

$$P_K = \left(\frac{r^2 \cos^2 \alpha + \sin^2 \alpha}{r} \right)^{-3/2} \qquad (10)$$

where α is the angle between *(hkl)* and the preferred orientation vector, and r is an adjustable parameter.

For platy orientation the direction of preferred orientation is along the cleavage plane *(hkl),* where P_K will assume a positive value. For acicular crystals (i.e., needle- or rod-shaped), orientation essentially occurs in two directions, where the orientation direction is chosen to be perpendicular to both these orientation directions and P_K assumes a negative value. An absolute value of unity for P_K is the theoretical limit for total orientation along (*hkl*) and zero for no preferred orientation.

Other Important Parameters

The effective absorption factor, *A*, depends on instrument geometry and is usually taken to be constant for Bragg-Brentano X-ray powder diffractometers. The Lorentz, polarization, and multiplicity factors at the *K*th Bragg reflection are expressed as L_K. The characteristic radiation from an X-ray tube is considered to be unpolarized and becomes polarized upon diffraction by virtue of the interaction of an electromagnetic wave with the diffracting electrons, where the diffracted beam has decreased amplitude in one of the directional vectors. The degree of polarization depends on the diffraction angle. The Lorentz factor accounts for geometric broadening of reflections as 2θ increases, and the multiplicity factor accounts for differing probabilities that symmetry equivalent planes *(hkl)* will reflect at the same 2θ angle.

Least-Squares Minimization

In the least squares minimization procedure, a set of normal equations derived from the calculated intensities, y_{ci}, are produced with respect to each adjustable parameter. These equations are solvable by inversion of a normal *m* by *m* matrix, where *m* is the number of parameters being refined. An iterative procedure is used where the shifts are calculated and applied to the initial parameters to produce an improved numerical model of the actual pattern. Because the relationships between the adjustable parameters and the intensities are not linear, the starting model must necessarily be close to the correct model, otherwise the nonlinear least-squares procedure could converge to a false minimum. An ad hoc solution to the problem of converging to a false and/or unstable minimum is the introduction of a "damping factor," where the size of the parameter shift can be determined, but this does not always guarantee stability in refinement. This type of nonlinear least-squares algorithm therefore demands a strict refinement strategy. If a parameter is refined at the wrong stage of refinement it can lead to instability. Another problem in refinement strategy is that certain parameters correlate with one another strongly (e.g., the zero point and the unit cell dimensions), since both sets of parameters have the effect of refining the peak position. Changes in certain parameters have similar effects on the calculated intensities, and these parameters should usually not be refined simultaneously.

Refinement Strategy

Initially the refinement strategy for a quantitative phase analysis should first refine instrument zero and the phase scales, which at the outset normally assumes that each phase is present in equal quantities. The instrument zero is essentially the level at which the instrument is calibrated to be the level at which diffraction occurs on the sample surface at 0° 2θ. Technically this should not be a refinable parameter and should be calibrated on the instrument and be a measured input value. It has, however, been found by Roode-Gutzmer for the IUCr-CPD Round Robin (October 1998) that the instrument zero is also sample-holder specific. So if comparative results should be obtained on a set of samples, the exact same sample holder should be consistently used.

Other parameters that should be refined early on are specimen displacement and specimen transparency, both of which have the effect of refining peak positions and can correlate with one another and with the refinement of instrument zero. In the early stages of a refinement it is advisable to consult correlation matrices that are usually produced by least-squares programs. Large correlations would indicate that the parameters should perhaps not be refined simultaneously or that they should be refined with increased damping factors.

The unit-cell dimensions for the major phases can then be refined. Unit-cell dimensions for minor to trace phases, such as free lime and periclase for a portland cement analysis, at this early stage may make the refinement unstable. So essentially at this stage, peak positions have been refined, after which line shapes in terms of widths can be refined, usually in our case with a pseudo-Voigt profile shape function and the half-width Cagliotti parameter W at first. U and V are refined at a later stage only if they are stable, which they usually aren't for minor to trace phases. Peak heights can then be refined in terms of the relative differences found for preferred orientation. At this stage, the procedure can be repeated where peak positions with their new refined shapes and relative heights can be refined, as well as the latter again. Peak asymmetry can then be refined, but should be done only if necessary, for example, when low angle peaks are important in the analysis.

Criteria of Fit

Finally, the fit of the calculated pattern to the observed diffraction data should be assessed. A discussion on the criteria of fit for a Rietveld refinement is summarized by Young.[88] The most often used parameter is a weighted profile residual, R_{wp}:

$$R_{wp} = 100\left[\frac{\sum w_i\left(y_{io} + y_{ic}\right)^2}{\left(\sum w_i y_{io}^2\right)}\right]^{1/2} \tag{11}$$

where y_{io} is the observed intensity at point I, y_{ic} is the calculated intensity at point I, and w_i is the weight assigned to each intensity.

For a perfect refinement the final R_{wp} would equal R_{exp}, the expected profile residual:

$$r_{\exp} = 100\left(\frac{N - P}{\sum w_i y_{io}^2}\right)^{1/2} \tag{12}$$

where N is the number of data points in the pattern and P is the number of parameters refined.

In some quantitative phase analysis programs, the fit of the calculated pattern to the observed pattern is given in terms of the global chi-squared goodness-of-fit factor, χ^2:

$$\chi^2 = \frac{R_{wp}}{R_{\exp}} \tag{13}$$

which should reduce to unity for an exceptional fit. For cement analysis a fit better than a χ^2 value of 5 is not very likely and is acceptable. A goodness-of-fit factor from a Rietveld refinement does not qualify as a verification that the phase composition is quantitatively correct, because one phase can be overestimated at the expense of another without affecting the goodness-of-fit factor at all.

Microabsorption

Absorption contrast (i.e., microabsorption) corrections are not implemented in the profile refinement and are simply correction factors applied after the refinement. Taylor and Matulis[89] use the Brindley model[22] to modify Eq. 5 of Hill and Howard[71] to include the particle absorption factor for each phase:

$$w_P = \frac{S_P M_P V_P}{\tau_P \sum_P \left(S_P M_P V_P / \tau_P \right)} \tag{14}$$

where τ_P is a particle absorption factor for phase P given by[22]:

$$\tau_P = \frac{1}{A_P \int_0^{A_P} \exp\left[-\left(\mu_P - \bar{\mu} \right) x \right] dA_P} \tag{15}$$

where A_P is the particle volume of phase P, μ_P is the mean linear absorption coefficient of phase P, $\bar{\mu}$ is the mean linear absorption coefficient of the solid matrix, and x is the path of the radiation in the particle of phase P.

Values of τ_P are tabulated by Brindley[22] as a function of $(\mu_P - \bar{\mu})R$, where R is the effective particle radius of the phase P and $(\mu_P - \bar{\mu})$ is readily computed when the matrix composition is known. These corrections are sometimes incorporated in Rietveld software and have been demonstrated over the whole composition range of a highly contrasting phase mixture of LiF and $Pb(NO_3)_2$ using CoK_α radiation to yield accurate quantitative results.[89] Without the microabsorption corrections, however, the phase quantitative results were demonstrated to be in significant error.[89]

It is pertinent that the particle diameters of highly contrasting phases be known fairly accurately. These can be estimated by microscopic techniques or from broadening of diffraction peaks according to the Scherrer equation,[3] especially for phases that are highly contrasting and intermixed. It is important, however, that the Scherrer equation estimates crystal thickness and that for microabsorption particle thickness is important, where particles are aggregations of crystals. If it is possible to obtain particle diameters on single phases prior to mixing, these are best obtained by laser particle diffraction techniques. Mean volume diameters calculated by this latter technique have been found by the authors to be very useful for cement analysis. As cement phases are not highly contrasting, it is more important to ensure that the mean particle diameter of the entire specimen is in the optimum range as prescribed by Brindley[22] (preferably where the product of the linear absorption coefficient and the particle diameter does not exceed 0.01) or according to Klug and Alexander[3] (where as a rule of thumb, a mean particle diameter should be 5–10 μm).

Anomalous Dispersion

Another correction factor that improves the accuracy of the quantitative result provided by a Rietveld analysis is that of anomalous dispersion, which has been incorporated by Taylor according to Cromer and Liberman[90] in the most recent versions of SIROQUANT. The effect is a result of the structure factors being calculated using scattering factor polynomial coefficients from international tables that assume that the scattering factors are practically independent of the incident wavelength, which is true except in the case where the radiation wavelength is close to the absorption edge of atoms contained in the specimen. Anomalous dispersion is significant for samples containing cobalt, nickel, and copper when CuK_{α} radiation is used; manganese, iron, and cobalt for CoK_{α} radiation; and chromium, manganese, iron, and cobalt for FeK_{α} radiation. For cement analysis, the presence of iron in ferrite may result in anomalous dispersion, but it is adequately corrected for by this correction procedure. Hence, the choice of the radiation source should be dictated by effects such as fluorescence and not particularly by those of anomalous dispersion.

Rietveld Analysis Software

Most Rietveld analysis programs are freely available, are well distributed, and can usually be obtained via FTP from the Internet. The International Union of Crystallography has a webpage from where connections to other relevant sites are available as well as mirror sites that allow for quicker downloading times. Most of the programs are modifications of the original source code that Rietveld had himself written and distributed.

There are several commercial Rietveld software packages available, of which SIROQUANT[†] is popular for quantitative phase analysis using X-ray powder diffraction data. The authors' experience rests largely with this software package. The computer programs are modular and concise and were written by J.C. Taylor[91] in efficient FORTRAN 77 code for personal computing environments. SIROQUANT has a very user-friendly graphic interface, thus making it useful for industrial applications. The programs run in

[†]Trademark of CSIRO and marketed by Sietronics Pty. Ltd., Australia.

single multiphase mode and refine a multitude of least-squares Rietveld parameters that include phase scales, peak asymmetry, preferred orientation, line widths, instrument zero, unit-cell dimensions, Brindley particle absorption contrast factors, anomalous dispersion corrections, and amorphicity corrections. A choice of the pseudo-Voigt or Pearson VII profile shape functions is available, as well as refining a peak asymmetry factor. The background can be fitted by a linear or cubic spline fitting, although the former has been found to work best, where the points are then user modified. A calibration for the Bragg-Brentano low-angle intensity aberrations can be implemented. Although no features are available to refine specimen transparency and displacement, the former is not important for inorganic materials and the latter can be corrected for by use of an internal standard and by treating the data with a simple angle-dependent relation to shift peak positions to where they should be. This is conveniently done using the DIFFRAC-AT[‡] software developed by SOCABIM. Displacement correction software is also freely available on the Internet.

The calculated patterns in SIROQUANT are derived from crystal structure data contained in a crystal structure database containing data of some 300 common minerals. Minerals can be added easily by the user if needed. Considerable attention has been paid to developing this software for cement analysis and many of the cement phases are therefore contained in this database. Although SIROQUANT has been developed for quantitative phase analysis, it can also be used to refine crystal structures — this feature, if properly used, can allow the determination of exact solid-solution compositions of cement phases. Another significant feature that SIROQUANT offers is the ability to refine and quantify clay type materials for which the material's structure may not be fully known.

Another commercial Rietveld software package for quantitative phase analysis is QUASAR,[§] which offers much the same functionality as SIROQUANT, but reflects a different approach to the Rietveld technique with a different refinement philosophy. Another well-known Rietveld program for quantitative work is RIQAS,[¶] a DOS program quite popularly used in the United States.

For structure refinement and/or solution using non–black box type public programs, the most popular programs are Generalized Structure Analysis

[‡]Trademark of Bruker.

[§]Also developed by CSIRO, but marketed by Philips.

[¶]Marketed by Materials Data Inc.

System (GSAS), written by Larson and von Dreele,[92] and DBWS, written by Wiles, Sakthivel, and Young.[93] GSAS is to be one of the programs used by Stutzman in the development of standardised Rietveld methods for ASTM to be applied to cement clinker analysis. Berliner et al.[94] used GSAS to refine crystal structure models for synthetic portland cement clinker phases. Neubauer et al.[95] used the Rietveld software of Wiles and Young[96] (essentially a precursor to DBWS) modified by Howard to quantify the phases in a synthetic portland cement clinker.

Rietveld Analysis on Portland Cement

There is a large amount of literature demonstrating that the Rietveld method is an accurate means for quantifying phase compositions.[70–74] A serious need has, however, arisen to resolve the problems that are impeding the application of this method as a routine analysis technique. Furthermore, in its application to portland cement clinker, the phase system proves to be particularly complex.

Certain problems were encountered in trying to review published Rietveld analyses on portland cement clinkers. The main problem is intrinsic to the portland cement clinker phase system, and the effect is that the results obtained by most workers using the Rietveld method are not readily verifiable. This is due to the lack of standards available, which is a result of the extreme difficulty in its preparation and the lack of analytical methods that can provide comparable quantitative phase information. Results are often compared to Bogue calculated values, a comparison that may be useful only to show the inaccuracy of the Bogue calculation. A more useful comparison, however, would be to calculate analyte oxides from the phase abundances determined from a Rietveld refinement and to compare these to the analyte oxides obtained from a quantitative chemical analysis (e.g., by XRF), a procedure used by J.C. Taylor, which he called a reverse-Bogue calculation.

Microscopic point counting results, the most useful and accessible, are often not reported with Rietveld results, and if they are, the units in which the phase quantities are expressed as are not mentioned and may not be directly comparable. For example, microscopic point counting values are normally expressed as area or volume percentages, whereas those from a Rietveld analysis are reported as weight percentages. To be correct, comparisons should take into account the density of the phases, the densities being dependent on the crystal structure and solid solution of the phase. Other pertinent information that is often neglected in publications are references to which Rietveld software had been used.

Experimental difficulties occur in cements, where the ferrite phase presents problems regarding fluorescence if CuK_α radiation is used. Ferrite also tends to be underestimated and periclase overestimated if absorption contrast corrections are not applied and if corrections are applied, the particle diameters of these phases should be assessable. Using CoK_α radiation improves peak resolution somewhat and avoids the problem of fluorescence, but anomalous dispersion would then need to be considered for ferrite. The radiation wavelength used in a quantitative phase analysis should therefore be stated in publications.

Certain (detection) limits have been identified in the quantitative phase analysis of clinker using Rietveld on XRD data. These are that lime and periclase, which qualify as minor or trace phases, exhibit only one useful peak each, and these are overlapped. These phases can be more accurately determined by glycol extraction of the lime and point counting of periclase under the microscope, provided the latter is not micro- or cryptocrystalline. Alternatively, the crystal modifications of the phases should allow calculation of incorporated magnesium and the difference between this and the total chemical MgO content should be reconcilable to periclase.

It has also previously been discussed that the addition of an internal standard to portland cement specimens, as an internal verification procedure, is advisable at this stage in the development of an acceptable and routine QXRD-Rietveld method.

The most important problem using the Rietveld method in quantifying the phases in a portland cement accurately is the problem of choosing the correct structure model for alite, the correct structure model choice being even more important since alite is the major phase dominating the XRD powder pattern. An incorrect structure model will not allow adequate deconvolution of overlapped peaks from the other phases.

A short summary of the alite polymorphs will be discussed, but the reader is encouraged to consult Dent Glasser,[97] Regourd,[98] Fayos and Perez-Mendez,[99] Taylor and Aldridge,[100] and Taylor[5] for detailed reviews of C_3S crystal chemistry. Seven polymorphs are known for alite, and their existence is primarily due to differences in orientational disorder displayed by the silica tetrahedra in the structure, which is expected to be dependent on the type and amount of impurities present in C_3S.

Taylor[5] summarizes the C_3S polymorphs as a function of temperature:

$$T_1 \overset{620^\circ C}{\rightleftarrows} T_2 \overset{920^\circ C}{\rightleftarrows} T_3 \overset{980^\circ C}{\rightleftarrows} M_1 \overset{990^\circ C}{\rightleftarrows} M_2 \overset{1060^\circ C}{\rightleftarrows} M_3 \overset{1070^\circ C}{\rightleftarrows} R$$

The high-temperature forms cannot be stabilized to ambient temperature by quenching and therefore require the presence of impurities that substitute for calcium and silicon atoms or holes in the C_3S structure.

Production clinkers are found to contain the M_1 and M_3 polymorphs, usually stabilized by MgO, with M_3 being the most common and more disordered.[101] The monoclinic polymorphs have complicated superstructures,[102] but the overall apparent disorder allows the establishment of an average monoclinic structure[103] comparable to the M_3 substructures of, for example, Nishi and Takéuchi.[102]

The crystallographic modification of C_3S can be identified in a powder diffraction pattern by investigating the fine structure of certain peaks[5] (which is largely improved when the alite phase is isolated from the other cement phases by selective dissolution). Between 51 and 52 °2θ (CuK_α radiation), the M_1 polymorph exhibits a peak that is almost a singlet (this is by virtue of its pseudo-rhombohedral symmetry due to maximum MgO incorporation for this polymorph), whereas the M_3 polymorph exhibits a well-defined doublet, and the T_1 and T_2 forms displaying triplets. These features can also be seen at 32–33 °2θ (CuK_α radiation), but these are more overlapped with other phases.

Of the seven alite polymorphs, Taylor and Aldridge[100] found only two to be remotely adequate for a commercial portland cement clinker: the monoclinic M_3 superlattice[102] and the triclinic form of Golovastikov et al.[104] The triclinic form, even though not found in production clinkers, has the benefit of belonging to a low-crystal symmetry system, which can be refined in a Rietveld refinement to fit a higher-crystal symmetry system. Monoclinic symmetry is a subset of triclinic symmetry and not vice versa. All possible reflections are therefore more likely to be accounted for by the triclinic model than monoclinic ones. The triclinic form of C_3S is therefore often employed in a Rietveld refinement. The disadvantage, however, is that by virtue of the low-symmetry crystal system and a large unit cell containing many atoms, there is an immense number of reflections (typically in the order of 1600),[100] which is computationally very demanding. Taylor and Aldridge[100] dealt with this problem by using the pattern that is measured for alite in the specimen as opposed to calculating one from inadequate structure models. They achieved this by subtracting the alite pattern from the total observed pattern constituting the other portland cement clinker phases, where the quantitative phase composition for the portland cement clinker was accurately known (by microscope point counting). They then indexed the observed reflections according to a rhombohedral symmetry. They found that the use of an observed pattern provided the best phase quantitative results.

While the correct structure choice for alite is difficult, it is extremely important. Also, though it is not a difficult task to identify the various forms in which the other phases appear in a specimen, their correct identification is equally important. Aluminate can appear as either cubic or orthorhombic, but usually appears as a combination in most production clinkers. The monoclinic form occurs only for high alkali contents (typically > 4.6% Na_2O)[105] and has not been found to occur in production clinkers. Apart from being able to distinguish between cubic and orthorhombic forms of aluminate by optical microscopy (by virtue of the cubic form being isotropic and the orthorhombic form being anisotropic and lath-shaped), its characteristic peak at ~33 °2θ is a sharp singlet for the cubic form, which splits into two broader observable peaks for the orthorhombic form (although the lower angle peak is a combination of two Bragg reflections). Modifications to the aluminate structure, $Na_{2x}Ca_{3-x}Al_2O_6$,[105] are largely dependent on alkali content, where the compositional range for cubic is $0 < x < 0.16$, and for orthorhombic is $0.16 < x < 0.20$.

Ferrite varies largely in the amount of iron substituted in solid solution, which can be assessed by the unit-cell dimensions and appears to shrink volumetrically with an increase in x where ferrite is described as a series $Ca_2(Al_xFe_{1-x})_2O_5$.[5] A crystal symmetry change occurs at $x = 0.33$.[106] Mumme[103] and Neubauer et al.[95] established the correct composition for ferrite (and hence the correct structure) by refining the occupancy factors for the octahedral iron site and the tetrahedral aluminum site in the brownmillerite structure of Colville and Geller.[107] This can be accomplished for ferrite in the presence of the other phases (and during a quantitative phase refinement), although the results would improve when the ferrite is isolated from the other phases by selective dissolution.

Belite occurs principally as larnite, β-C_2S, in the majority of production clinkers. The other common polymorph usually contains considerable amounts of magnesium and is mineralogically known as bredigite, α'_L-C_2S, and is usually found for the hydraulic dicalcium silicate phase found in slags. The polymorphs as a function of temperature are summarized by Taylor[5] as follows:

$$\alpha \overset{1425°C}{\rightleftarrows} \alpha'_H \overset{1160°C}{\rightleftarrows} \alpha'_L \underset{690°C}{\overset{630\text{–}680°C}{\rightleftarrows}} \beta \overset{<500°C}{\rightleftarrows} \gamma \overset{780\text{–}860°C}{\rightleftarrows} \alpha'_L$$

The structures of the α'_H, α'_L (where subscripts H and L refer to high and low,[5] presumably referring to the degree of symmetry) and β polymorphs are derived from that of α-C_2S by progressive decrease in symmetry, which arises from changes in the orientations of SiO_4 tetrahedra and small movements in the position of Ca^{2+} ions. The quantities of substituent ions needed to stabilize the higher-temperature polymorphs at ambient temperatures decrease along the sequence from α- to β-C_2S. The structure of the γ polymorph is considerably different from that of the other polymorphs and is associated with a considerable volume change from the polymorph from which it is formed (either from the β or α'_L forms), resulting in the observable phenomenon of "dusting." The γ polymorph is not hydraulic and is therefore an undesirable phase, which is usually avoided by the presence of impurities and quenching. Most production portland cement clinkers contain sufficient stabilizing ions to prevent transformation to the γ modification.

Review of Selected Results

Not many quantitative phase results for cement are available in the literature and, of those, not many satisfy the criteria of being quantitatively verifiable.

Aldridge[19] in 1982 was one of the first to attempt to reconcile microscopic, chemical, and XRD results with one another in a round robin using a thorough statistical approach. At that time, however, no Rietveld-type approaches were applied in quantitative phase analysis of cements. When these techniques became available in cement analysis, these same samples (E1-E6 New Zealand round robin) were reanalyzed by Taylor and Aldridge[108] in 1993 using the Rietveld method in the SIROQUANT software system (Table I). Good agreement exists between results obtained by microscope point counting and using the Rietveld method in SIROQUANT.

In another study by Taylor and Aldridge,[109] three clinker samples (supplied by Gutteridge) from the British Cement Association were analyzed using the SIROQUANT Rietveld software. The results were compared to microscopic point count results (Table II) — these were made available by

Table I. Rietveld results[108] obtained on the New Zealand Standard Cements[19] compared to microscope point counting results (wt%)

Cement sample	Method	Alite	Belite	Aluminate*	Ferrite	χ^2
E1	SQ†	69.0 (10)	14.2 (7)	6.0 (7)	7.2 (6)	3.5
	M‡	71.1 (34)	12.8 (26)	6.7 (13)	9.5 (24)	
E2	SQ	79.2 (9)	7.2 (6)	8.6 (4)	3.3 (6)	3.1
	M	80.7 (34)	7.7 (26)	6.3 (13)	4.9 (24)	
E3	SQ	64.6 (7)	12.2 (7)	7.3 (4)	10.6 (5)	3.4
	M	67.9 (34)	16.8 (26)	5.4 (13)	11.0 (24)	
E4	SQ	49.6 (8)	32.3 (9)	5.2 (5)	8.5 (6)	3.3
	M	50.7 (34)	35.7 (26)	4.0 (13)	10.3 (24)	
E5	SQ	66.6 (8)	8.2 (7)	10.8 (4)	4.8 (4)	3.2
	M	68.7 (34)	14.5 (26)	10.4 (13)	5.2 (24)	
E6	SQ	63.7 (9)	26.6 (7)	6.9 (5)	0.0 (5)	3.5
	M	62.8 (34)	28.6 (26)	8.6 (13)	0.5 (24)	

*Cubic, monoclinic, and orthorhombic forms of C_3A quantified seperately in the Rietveld analysis and summed together. The microscopic point counted value is also a summed value.
†Rietveld results obtained using SIROQUANT.
‡Microscope point counting results obtained in the round robin analysis of Aldridge,[19] averaged over 10, 11, 6, and 7 laboratories for alite, belite, aluminate, and ferrite, respectively.

Table II. Rietveld results[109] obtained on clinker samples from the British Cement Association* compared to microscope point counting results (wt%)

Cement sample	Method	Alite	Belite	Aluminate	Ferrite
S1 PAC	SQ†	58 (1)	23 (1)	14 (1)	5.1 (3)
	M‡	63	19	13	5
S2 PAC	SQ	56 (1)	27 (1)	11 (1)	6.3 (4)
	M	60	21	14	6
S3 PAC	SQ	52 (1)	25 (1)	4.4 (4)	19 (1)
	M	56	24	2	19

*Samples supplied by Gutteridge.
†Rietveld results obtained using SIROQUANT.
‡Micorscope point count results supplied by Gutteridge after Rietveld refinement results were determined.

Gutteridge only after the Rietveld refinement results had been obtained. In both these studies, use was made of a measured alite pattern according to Taylor and Aldridge,[100] as opposed to a calculated one from existing structure models.

Results obtained on the NIST reference materials have proven to be useful. These results are all reported as weight percentages in Table III, with the exception of those obtained by Theisen,[14] which are expressed as volume percentages and could not be converted to weight percentages, by virtue of the interstitial phases (consisting of two phases with largely differing densities) being counted together.

The point count results obtained on the reference materials by the NIST standardization procedure were acquired by D.H. Campbell, a noted and distinguished cement microscopist, and the data satisfy rigorous statistical requirements. These results can therefore be accepted as the most accurate phase composition results that are currently available for these reference materials. They are compared in Table III by point counting results conducted by the FLS laboratory in a study conducted by Theisen[14] to assess the virtues of image analysis as a quantitative phase composition tool. Their results compare fairly well with those obtained by Campbell for NIST.

According to Hoffmänner,[12] 3000 points are sufficient to assess the modal composition of a sample, although up to 4000 points improve the precision of the results. The NIST point counting consistently acquired 3100 points on each polished section, and four polished sections per sample of reference material were analyzed. The results satisfied the requirements not only for precision, but also for accuracy and homogeneity of the sample and/or of sampling. The FLS point count analyses used only two polished sections per sample of reference material and counted fewer points, averaging between 2150 and 3125 points per section. Also, there does not seem to be the same consistency as that employed by NIST. The results obtained by FLS are nevertheless good, but comparisons to Rietveld results should be made by those acquired by NIST on their own reference materials.

The image analysis results of Theisen[14] are comparable with the phase composition reported by NIST. The technique of image analysis demonstrates great potential as an automated means of determining phase abundances from optical microscopy. However, the problems of simulating the human brain's ability to process images and the limitations that are associated with optical microscopy as a technique still remain.

The SEM analyses obtained by Stutzman[110] on these samples appear to have been fairly preliminary in the process of standardizing these reference materials, and the ones listed in Table III are only for interstitial phases and periclase. Better SEM results on these samples may be currently available.

Three sets of Rietveld results are available for these samples, two of which were calculated by Taylor and Aldridge. The two sets of results produced by Taylor and Aldridge appear to differ, but one set does not appear to be better than the other. The published set of data[109] uses similar methodologies to those previously used by Taylor and Aldridge,[108] where the Rietveld software system developed by Taylor,[91] SIROQUANT, was used. Similar refinement strategies were used and the alite pattern used in the refinement was determined from the measured pattern as opposed to calculated ones from structural models. Absorption contrast corrections were applied according to the Brindley model and preferred orientation using the March-Dollase model. The second set of results was presented at a Siemens seminar in 1995 in South Africa, and it is not clear how or why these results are different from those acquired in the first set. It is interesting, however, to note some differences between these two sets of results.

For Reference Material 8486, alite is underestimated in the first set and overestimated in the second. The result for belite in the second set appears to be more accurate and is overestimated in the first set. It is possible that correlation is occurring in the refinement of these two phases. Aluminate is notably overestimated in both sets of results. The overestimation of aluminate for Reference Material 8486 by the Rietveld results actually seems to reflect an underestimation of very fine-grained aluminate in the microscopic analyses as reported by NIST.[111]

For Reference Material 8487, the alite and belite quantities follow the same pattern as that found for Reference Material 8486, possibly confirming the refinement correlation of these two phases. The aluminate in the first set is significantly overestimated, seemingly at the expense of ferrite, which is being significantly underestimated.

For Reference Material 8488, the alite is consistently overestimated for both sets, probably at the expense of belite, which is being significantly underestimated. The aluminate and ferrite quantities are consistently comparable with the NIST-reported phase abundances.

Another set of Rietveld results was acquired on the NIST reference materials by Möller.[112] These results compare very favorably with NIST-reported phase abundances. Unfortunately, the value of these results is not statistically assessable due to the lack of detail on issues such as the

Table III. Various quantitative phase results obtained on NIST portland cement clinker Reference Materials 8486, 8487, and 8488

Reference material number	Material	1987/1989 NIST[111] microscope point counting* (wt%)	1989 Stutzman[110] SEM (wt%)	1997 Theisen FLS microscope point counting† (vol%)	1994 Taylor and Aldridge[109] QXRD Rietveld (wt%)	1995 Möller[112] Krupp Polesius QXRD Rietveld (wt%)	1997 Theisen image analysis (vol%)
8486	Alite	58.47 (1.65)	‡	59.1 ± 1.6	57 (60.0)§	58.0	59.3 ± 2.7
	Belite	23.18 (1.94)		21.7 ± 1.0	25 (22.0)	22.3	23.1 ± 2.9
	Aluminate	1.15 (0.10)	5.6	18.0 ± 1.5¶	3.0 (3.1)	2.4	5.0**
	Ferrite	13.68 (0.63)	13.4		11.0 (11.2)	13.4	12.3
	Free lime	0.18 (0.14)		0.3		0.0	
	Periclase	3.21 (0.72)	5.1	0.9	4.4 (3.8)	4.0	
8487	Alite	73.39 (1.57)		70.6 ± 3.0	72 (74.3)	73.7	72.4 ± 1.2
	Belite	7.75 (1.23)		9.1 ± 2.1	12 (8.5)	9.7	10.9 ± 1.1
	Aluminate	12.09 (0.88)	8.4	17.6 ± 1.4	15 (11.9)	13.6	14.7
	Ferrite	3.27 (0.7)	8.2		0.4 (3.1)	2.2	1.6
	Free lime	2.45 (0.48)		1.8 ± 0.4	1.6 (1.8)	0.9	
	Periclase	0.09 (0.09)	0.0		(0.4)		

Table III, continued

Reference material number	Material	1987/1989 NIST[111] microscope point counting* (wt%)	1989 Stutzman[110] SEM (wt%)	1997 Theisen FLS microscope point counting† (vol%)	1994 Taylor and Aldridge[109] QXRD Rietveld (wt%)	1995 Möller[112] Krupp Polesius QXRD Rietveld (wt%)	1997 Theisen image analysis (vol%)
8488	Alite	64.97 (0.56)		65.9 ± 1.5	67 (67.8)	64.5	65.1 ± 1.1
	Belite	18.51 (0.58)		17.4 ± 2.1	15 (15.1)	19.8	19.5 ± 0.7
	Aluminate	4.34 (1.35)	10.6	16.7 ± 0.9	4.6 (4.6)	3.0	5.7
	Ferrite	12.12 (1.50)	12.4		13 (12.2)	12.7	10.3
	Free lime	0.00 (0.00)		0			
	Periclase	0.05 (0.09)	0.0				

*Microscope point counting was performed by D.H. Campbell, who counted 3100 points per specimen on four polished section specimens of each clinker (therefore, a total of 37 000 points). Values in brackets are observed standard deviations.

†Point counting was done on two polished sections for each clinker; 6350, 4600, and 4300 total points were counted, respectively, for Reference Materials 8486, 8487, and 8488.

‡These results were obtained by Stutzman[110] and referred to by Struble[2] and involved selective dissolution in order to study the interstitial phases, therefore no results for the silicate phases.

§Values in brackets were redetermined by J.C. Taylor and presented in October 1995 at a Siemens Spring Seminar and New Technique Workshop, Johannesburg, South Africa.

¶Aluminate and ferrite counted together as liquid phase for all three reference materials.

**Interstitial phases were distinguished by higher magnification. Free lime was determined chemically and subtracted from the interstitial phases. Aluminate concentrations may appear higher due to periclase being unresolved from the aluminate in the image analysis.[14]

Table IV. Rietveld results[95] obtained on portland cement clinker samples prepared by admixing synthetically prepared cement phases

Cement sample	Method	Alite	Belite	Aluminate	Ferrite
1	Rietveld*	64.8 (0.9)†	20.8 (0.8)	9.6 (0.6)	4.8 (0.6)
	True value‡	65	20	10	5
2	Rietveld	55.3 (1.1)	25.4 (1.2)	9.7 (0.6)	9.6 (0.6)
	True value	55	25	10	10
3	Rietveld	69.9 (0.7)	15.7 (0.8)	5.0 (0.6)	9.4 (0.9)
	True value	70	15	5	10

*Software of Wiles and Young,[96] modified by Howard.
†Standard deviation in brackets.
‡These values are the weight percentages in which the synthetic phases were physically admixed.

Rietveld software employed, the methodology in acquiring the structure model of alite, and whether or not absorption contrast was corrected for.

Neubauer et al.[95] prepared three synthetic portland cement clinkers by admixing synthetic preparations of alite, belite, aluminate, and ferrite. Their Rietveld results, listed in Table IV, reflect the phase composition of the samples accurately. No mention, however, was made regarding the critical use of absorption contrast corrections. The synthetically prepared phases appeared to correspond with those found in production clinkers. The advantage of using synthetically prepared phases was that the structure of each synthetic phase was refined and their correct structure models were used in the quantitative refinement of the synthetic mixtures.

Conclusion

Tremendous progress has been made in the application of the Rietveld method in the quantitative phase analysis of portland cement clinker. Presently, the most satisfactory results have been those of Taylor and Aldridge.[108,109] Since the publication of Taylor's computer programs[91,113] in 1991, the complexities involved in the application of the Rietveld method, especially in the analysis of portland cement clinker, has been consistently addressed in the literature by Taylor and coworkers.[31,100,108,109]

The Rietveld method, by virtue of being a least-squares method, has some inherent limitations, and it has been found by various workers[114,115] that the estimated standard deviations calculated by the Rietveld method are unreliable. According to Rietveld,[116] this should serve as a reminder that this method should not be treated as a black box and that users of this method need to be continually aware of the limitations of least-squares methods in general. Another limitation of least-squares methods is the problem of convergence to false minima or instability of refinement, which necessitates a careful refinement strategy. New developments, however, by Bergman et al.[117,118] have recently begun to address this issue. Bergman et al.[118] have developed a new Rietveld algorithm, BGMN, that essentially eliminates correlation between profile and structure parameters and also overcomes the well-known problems of the March function[28,87] for preferred orientation. In addition, it has become possible to parameterize atomic substitution, which would be particularly useful for the complicated substituted polymorphs of cement phases. This program has already demonstrated excellent quantitative phase analysis results on organic material.

So far, the application of the Rietveld method has been restricted essentially to analyzing cements and clinkers and not so much hydration products of cements. This is understandable in view of the complexity already existing for unhydrated cement, let alone of the added complexity of amorphicity and disordered hydrated structures. Kuzel,[119] however, achieved good quantitative phase results on mixtures of C_3A, gypsum, and ettringite.

Experimental improvements in the way diffraction data are collected can also contribute to the successful development of quantitative phase analysis of cement. Even though position-sensitive detectors have been used since the early 1970s, they have not found routine application. Position-sensitive detectors allow very rapid data collection by collecting the diffraction data over the entire angular range simultaneously. Apart from being able to achieve high intensities for low-diffractive materials such as cement and clinker, the intensity-to-noise ratio required for adequate Rietveld analysis can be achieved in practicable time frames. In addition, phase transformations and kinetics can be studied in situ — this is ideal for observing the hydration progress of cement.

Finally, with the anticipated ASTM standardized Rietveld procedures for cement analysis being developed by Stutzman of NIST, it is strongly recommended that an international round robin be initiated. It is suggested that portland cement clinker reference materials like those that were developed by NIST (e.g., Reference Materials 8486, 8487, and 8488) should be prepared and distributed to willing participants. The Commission for Powder

Diffraction of the IUCr implemented for the first time in 1998 an international round robin on multiphase materials for QXRD analysis using the Rietveld method.

Acknowledgments

We are grateful to Chris Kelaart of Sietronics, Australia, Pieter du Toit of Lafarge, South Africa, and Errol Fernandes of Penn State University for making literature available, to Paul E. Stutzman of NIST and R.A. Young of Georgia Institute of Technology for valuable correspondence, and to R.E. Oberholster and Jens Gutzmer for critically reviewing this manuscript.

References

1. R.A. Young (ed.), *The Rietveld Method.* International Union of Crystallography, Oxford University Press, 1995.
2. L.J. Struble, "Quantitative Phase Analysis of Clinker Using X-Ray Diffraction," *Cement, Concrete, and Aggregates,* **13** [2] 97–102 (1991).
3. H.P. Klug and L.E. Alexander, *X-Ray Diffraction Procedures for Polycrystalline and Amorphous Materials.* John Wiley and Sons, New York, 1974.
4. R. Jenkins and D.K. Smith, "The Powder Diffraction File." IUCr Database Commision Report, August 1987.
5. H.F.W. Taylor, *Cement Chemistry,* second edition. Thomas Telford, 1997.
6. R.H. Bogue, "Calculation of the Compounds in Portland Cement," *Ind. Eng. Chem. Anal. Edn.,* **1**, 192–196 (1929).
7. L.P. Aldridge and R.P. Eardley, "Effects of Analytical Errors on the Bogue Calculation of Compound Composition," *Cem. Technol.,* **4**, 177–182 (1973).
8. A.M. Harisson, H.F.W. Taylor, and N.B. Winter, "Electron-Optical Analyses of the Phases in a Portland Cement Clinker, with Some Observations on the Calculation of Quantitative Phase Composition," *Cem. Conc. Res.,* **15**, 775–780 (1985).
9. H.F.W. Taylor, "Modification of the Bogue Calculation," *Adv. Cem. Res.,* **2**, 73–77 (1989).
10. Y. Ono, "Microscopical Observations of Clinker for the Estimation of Burning Conditions, Grindability and Hydraulic Activity"; pp. 198–210 in *Proceedings of the Third International Conference on Cement Microscopy.* 1981.
11. F. Chayes, *Petrographic Modal Analysis: An Elementary Statistical Appraisal.* Wiley, New York, 1956.
12. F. Hofmänner, "Microstructure of Portland Cement Clinker." Holderbank Management & Consulting Ltd., Rheintaler Druckerei und Verlag, Heerbrugg, Switzerland, 1975.
13. A. Märten, J. Strunge, and D. Knöfel, "Quantitative Structure Analysis of Portland Cement Clinker via Image Analysis — A Suitable Tool for Quality Control"; pp. 44–51 in *Proceedings 16th International Conference on Cement Microscopy.* 1994.
14. K. Theisen, "Quantitative Determination of Clinker Phases and Pore Structure Using Image Analysis," *World Cement Research and Development,* August 1997, pp. 71–76.

15. K.L. Scrivener, "The Microstructure of Anhydrous Cement and Its Effect on Hydration"; pp. 39–46 in *Microstructural Development During Hydration of Cement,* Materials Research Society Symposium Proceedings, Vol. 85. Edited by L.J. Struble and P.W. Brown. Materials Research Society, 1987.
16. P.E. Stutzman, "Cement Clinker Characterization by Scanning Electron Microscopy," *Cement, Concrete, and Aggregates,* **13** [2] 109–114 (1991).
17. P.E. Stutzman, "Applications of Scanning Electron Microscopy in Cement and Concrete Petrography"; pp. 74–90 in *Petrography of Cementitious Materials,* ASTM STP 1215. Edited by S.M. De Hayes and D. Stark. American Society for Testing and Materials, Philadelphia, 1994.
18. S. Diamond, "The Microstructures of Cement Paste in Concrete"; pp. 123–147 in *Proceedings of the 8th International Cement Congress.* 1986.
19. L.P. Aldridge, "Accuracy and Precision of Phase Analysis by Bogue, Microscopic, and X-Ray Diffraction Methods," *Cement and Concrete Research,* **12**, 381–398 (1982).
20. W. Gutteridge, "Quantitative X-Ray Powder Diffraction in the Study of Some Cementive Material"; pp. 11–23 in *Proceedings British Ceramic,* No. 35. Edited by F.P. Glasser. September 1984.
21. A.W. Hull, "A New Method of Chemical Analysis," *J. Am. Chem. Soc.,* **41**, 1168–1175 (1919).
22. G.W. Brindley, "The Effect of Grain or Particle Size on X-Ray Reflections from Mixed Powders and Alloys, Considered in Relation to the Quantitative Determination of Crystalline Substances by X-Ray Methods," *Phil. Mag.,* **36** [7] 347–369 (1945).
23. P.M. de Wolff, "Measurement of Particle Absorption by X-Ray Fluorescence," *Acta. Cryst.,* **9**, 682–683 (1956).
24. R.J. Harrison and A. Paskin, "The Effects of Granularity on the Diffracted Intensity in Powders," *Acta. Cryst.,* **17**, 325–331 (1964).
25. P. Suortti, "Effects of Porosity and Surface Roughness on the X-Ray Intensity Reflected from a Powder Specimen," *J. Appl. Cryst.,* **5**, 325–331 (1972).
26a. A. Taylor, "Influence of Crystal Size on the Absorption Factor as Applied to Debye-Scherrer Diffraction Patterns," *Phil. Mag.,* **35** [7] 215–229 (1944).
26b. A. Taylor, "The Optimum Thickness of Powder Specimens in X-Ray Diffraction Work," *Phil. Mag.,* **35** [7] 632–638 (1944).
27. W. Pitschke, N. Mattern and H. Hermann, "Incorporation of Microabsorption Corrections into Rietveld Analysis," *Powder Diffraction,* **8** [4] 223–228 (1993).
28. W.A. Dollase, "Correction of Intensities for Preferred Orientation in Powder Diffractometry: Application of the March Model," *J. Appl. Cryst.,* **19**, 267–272 (1986).
29. R. Jenkins, "Instruments for the Measurement of X-Ray Powder Patterns"; in *Papers and Abstracts, PDSA-94, An International Workshop on Advanced Powder Diffraction Techniques in Mineral and Materials Processing, South Africa, 24–27 October 1994.*
30. A.J.C. Wilson, *Mathematical Theory of X-Ray Powder Diffractometry,* Philips Technical Library. N.V. Philips, Gloeilampenfabrieken, Eindhoven, 1963. Chapter 4.
31. C.E. Matulis and J.C. Taylor, "Intensity Calibration Curves for Bragg-Brentano Diffractometers," *Powder Diffraction,* **7**, 89–94 (1992).
32. C.E. Matulis and J.C. Taylor, "An Algorithm for Correction of Intensity Aberrations in Bragg-Brentano X-Ray Diffractometer Data: Its Importnace in the Multiphase Full-

Profile Rietveld Quantitation of a Montmorillonite Clay"; pp. 301–307 in *Proceedings of the 41st Annual Conference on Applications of X-Ray Analysis, Denver,* Advances in X-Ray Analysis, Vol. 36. Edited by J.V. Gilfrich, C.R. Hubbard, C.S. Barrett, P.K. Predecki, and D.E. Leyden. Plenum Press, New York, 1992.

33. C.E. Matulis and J.C. Taylor, "A Theoretical Model for the Correction of Intensity Aberrations in Bragg-Brentano X-Ray Diffractometers — Detailed Descriptions of the Algorithm," *J. Appl. Cryst.,* **26**, 351–356 (1993).
34. L.E. Alexander, "Forty Years of Quantitative Diffraction Analysis"; pp. 1–13 in *Proceedings of the 25th Annual Conference on Applications of X-Ray Analysis, Denver,* Advances in X-Ray Analysis, Vol. 20. Edited by H.F. McMurdie, C.S. Barrett, J.B. Newkirk, and C.O. Ruud. Plenum Press, New York, 1976.
35. A.L. Navais, "Quantitative Determination of the Development of Mullite in Fired Clays by an X-Ray Method," *J. Am. Ceram. Soc.,* **8**, 296–302 (1925).
36. J.C.M. Brentano, "The Determination of the Atomic Scattering Power for X-Rays from Powders of Gold, Aluminium for CuK_{α} Radiation," *Phil. Mag.,* **6** [7] 178–191 (1928).
37. R. Glocker, "Principles of Quantitative X-Ray Analysis of the Concentration of Metal Phases in an Alloy or Mixture," *Metallwirtsch.,* **12**, 599–602 (1933).
38. K. Schäfer, "Quantitative X-Ray Analysis of Crystalline Powders," *Z. Kristallogr.,* **99**, 142–152 (1938).
39. G.L. Clark and D.H. Reynolds, "Quantitative Analysis of Mine Dusts," *Ind. Eng. Chem., Anal. Ed.,* **8**, 36–40 (1936).
40. J.W. Ballard, H.I. Oshry, and H.H. Schrenk, "Quantitative Analysis by X-Ray Diffraction. I. Determination of Quartz"; pp. 3520 in *U.S. Bur. Mines Repts. Invest.* June 1940.
41. F.S. Gardner, M. Cohen, and D.P. Antia, "Quantitative Determination of Retained Austenite by X-Rays," *Trans. AIME,* **154**, 306–317 (1943).
42. B.L. Averbach and M. Cohen, "X-Ray Determination of Retained Austenite by Integrated Intensities," *Trans. AIME,* **176**, 401–415 (1948).
43. R.E. Ogilvie, "Retained Austenite by X-Rays," *Norelco Reporter,* **6** [3] 60–61 (1959).
44. P.R. Kennicott, "A Modification of the Busing-Levy Least-Squares Program to Account for Overlapped Data." Rep. No. 63-RL-(3321G). General Electric Research Laboratories, Schenectady, New York, 1963.
45. H. Friedman, "Geiger Counter Spectrometer for Industrial Research," *Electronics,* **18**, 132–137 (April 1945).
46. H.P. Alexander and H.P. Klug, "Basic Aspects of X-Ray Absorption in Quantitative Diffraction Analysis of Powder Mixtures," *Anal. Chem.,* **20**, 886–889 (1948).
47. J. Leroux, D.H. Lennox, and K. Kay, "Direct Quantitative X-Ray Analysis by Diffraction-Absorption Technique," *Anal. Chem.,* **25**, 740–743 (1953).
48. D.H. Lennox, "Monochromatic Diffraction-Absorption Technique for Direct Quantitative X-Ray Analysis," *Anal. Chem.,* **29**, 766–770 (1957).
49. J. Leroux and M. Mahmud, "Influence of Goniometric Arrangement and Absorption in Qualitative and Quantitative Analysis of Powders by X-Ray Diffractometry," *Appl. Spectroc.,* **14**, 131–134 (1960).
50. L.E. Copeland and R.H. Bragg, "Quantitative X-Ray Diffraction Analysis," *Anal. Chem.,* **30**, 196–208 (1958).

51. R.H. Bragg, *Handbook of X-Rays.* McGraw-Hill, New York, 1967. Chapter 12.
52. F.H. Chung, "Quantitative Interpretation of X-Ray Diffraction Patterns of Mixtures. I. Matrix-Flushing Method for Quantitative Multicomponent Analysis," *J. Appl. Cryst.,* **7**, 519–525 (1974).
53. C.R. Hubbard, E.H. Evans, and D.K. Smith, "The Reference Intensity Ratio, I/Ic, for Computer Simulated Powder Patterns," *J. Appl. Cryst.,* **9**, 169–174 (1976).
54. C.R. Hubbard and R.L. Snyder, "RIR-Measurement and Use in Quantitative XRD," *Powder Diffraction,* **3** [2] 74–77 (1988).
55. R.A. Young, P.E. Mackie, and R.B. von Dreele, "Application of the Pattern-Fitting Structure-Refinement Method to X-Ray Powder Diffractometer Patterns," *J. Appl. Cryst.,* **10**, 262–269 (1977).
56. F.W. Jones, "The Measurement of Particle Size by the X-Ray Method," *Proc. R. Soc. London Ser. A,* **166**, 16–43 (1938).
57. F.R.L. Schoening, "Strain and Particle Size Values from X-Ray Line Breadths," *Acta Cryst.,* **18**, 975–976 (1965).
58. R. Delhez, T.H. de Keijser, and E.J. Mittemeijer, "Determination of Crystallite Size and Lattice Distortions through X-Ray Diffraction Line Profile Analysis — Recipes, Methods, and Comments," *Fresenius Z. Anal. Chem.,* **312**, 1–16 (1982).
59. D. Taupin, "Automatic Peak Determination in X-Ray Powder Patterns," *J. Appl. Cryst.,* **6**, 266–273 (1973).
60. W.J. Mortier and M.L. Costenoble, "The Separation of Overlapping Peaks in X-Ray Powder Patterns with the Use of an Experimental Profile," *J. Appl. Cryst.,* **6**, 488–490 (1973).
61. T.C. Huang and W. Parrish, "Accurate and Rapid Reduction of Experimental X-Ray Data," *J. Appl. Phys. Lett.,* **27**, 123–124 (1975).
62. W. Parrish, T.C. Huang, and G.L. Ayers, "Profile Fitting: A Powerful Method of Computer X-Ray Instrumentation and Analysis," *Trans. Am. Crystallogr. Assoc.,* **12**, 55–73 (1976).
63. G.S. Pawley, "Unit-Cell Refinement from Powder Diffraction Scans," *J. Appl. Cryst.,* **14**, 357–361 (1981).
64. H. Toraya, "Whole-Powder-Pattern Fitting Without Reference to a Structural Model: Application to X-Ray Powder Diffractometer Data," *J. Appl. Cryst.,* **19**, 440–447 (1986).
65. H.M. Rietveld, "Line Profiles of Neutron Powder-Diffraction Peaks for Structure Refinement," *Acta. Cryst.,* **22**, 151–152 (1967).
66. H.M. Rietveld, "A Profile Refinement Method for Nuclear and Magnetic Structures," *J. Appl. Cryst.,* **2**, 65–71 (1968).
67. A.K. Cheetham and J.C. Taylor, "Profile Analysis of Powder Neutron Diffraction Data: Its Scope, Limitations, and Applications in Solid State Chemistry," *J. Solid State Chem.,* **21**, 253–375 (1977).
68. G. Malmros and J.O. Thomas, "Least-Squares Structure Refinement Based on Profile Analysis of Powder Film Intensity Data Measured on an Automatic Microdensitometer," *J. Appl. Cryst.,* **10**, 7–11 (1977).
69. C.P. Khattak and D.E. Cox, "Profile Analysis of X-Ray Powder Diffractometer Data: Structural Refinement of $La_{0.75}Sr_{0.25}CrO_3$," *J. Appl. Cryst.,* **10**, 405–411 (1977).

70. P.-E. Werner, S. Salome, G. Malmros, and J.O. Thomas, "Quantitative Analysis of Multicomponent Powders by Full-Profile Refinement of Guinier-Hägg X-Ray Film Data," *J. Appl. Cryst.,* **14**, 149–151 (1979).
71. R.J. Hill and C.J. Howard, "Quantitative Phase Analysis from Neutron Powder Diffraction Data Using the Rietveld Method," *J. Appl. Cryst.,* **20**, 467–474 (1987).
72. D.L. Bish and S.A. Howard, "Quantitative Phase Analysis Using the Rietveld Method," *J. Appl. Cryst.,* **21**, 86–91 (1988).
73. B.H. O'Connor and M.D. Raven, "Application of the Rietveld Refinement Procedure in Assaying Powdered Mixtures," *Powder Diffraction,* **3**, 2–6 (1988).
74. B.H. O'Connor, L. Deyu, B. Jordan, M.D. Raven, and P.G. Fazey, "X-Ray Powder Diffraction Quantitative Phase Analysis by Rietveld Pattern-Fitting — Scope and Limitations"; pp. 269–275 in *Proceedings of the 39th Annual Conference on Application of X-Ray Analysis, Denver,* Advances in X-Ray Analysis, Vol. 33. Edited by C.S. Barrett, J.V. Gilfrich, T.C. Huang, R. Jenkins, and P.K. Predecki. Plenum Press, New York, 1990.
75. G.H. Stout and L.H. Jensen, *X-Ray Structure Determination — A Practical Guide.* Collier MacMillan Limited, London, 1972.
76. D.E. Cox, B.H. Toby, and M.M. Eddy, "Acquisition of Powder Diffraction Data with Synchrotron Radiation," *Austral. J. Phys.,* **41**, 117–131 (1988).
77. G. Cagliotti, A. Paoletti, and F.P. Ricci, "Choice of Collimators for a Crystal Spectrometer for Neutron Diffraction," *Nucl. Instrum. Methods,* **35**, 223–228 (1958).
78. R.L. Snyder, "Analytical Profile Fitting of X-Ray Powder Diffraction Profiles in Rietveld Analysis"; pp. 111–131 in *The Rietveld Method.* Edited by R.A. Young. International Union of Crystallography, Oxford University Press, 1995.
79. H.J. Edwards and J.I. Langford, "A Comparison Between the Variances of the CuK_{α} and FeK_{α} Spectral Distributions," *J. Appl. Cryst.,* **4**, 43–50 (1971).
80. L.K. Frevel, "Profiles for (CuK_{α}) Lines," *Powder Diffraction,* **2** [4] 237–241 (1987).
81. M. Ahtee, L. Unonius, M. Nurmela, and P. Suorti, "A Voigtian Profile Shape Function in Rietveld Refinement," *J. Appl. Cryst.,* **17**, 352–357 (1984).
82. J.I. Langford, "A Rapid Method for Analysing the Breadths of Diffraction and Spectral Lines using the Voigt Function," *J. Appl. Cryst.,* **11**, 10–14 (1978).
83. M.M. Hall, V.G. Veeraghavan, H. Rubin, and P.G. Winchell, "The Approximation of Symmetric X-Ray Peaks by Pearson Type VII Distributions," *J. Appl. Cryst.,* **10**, 66–68 (1977).
84. R.A. Young and D.B. Wiles, "Profile Shape Functions in Rietveld Refinements," *J. Appl. Cryst.,* **15**, 430–438 (1982).
85. S.A. Howard and R.L. Snyder, "An Evaluation of Some Profile Models and the Optimization Procedures Used in Profile Fitting"; pp. 73–80 in *Proceedings of the 32nd Annual Conference on Application of X-Ray Analysis, Denver,* Advances in X-Ray Analysis, Vol. 26. Edited by C.R. Hubbard, C.S. Barrett, P.K. Predecki, and D.E. Leyden. Plenum Press, New York, 1983.
86. S.A. Howard and R.L. Snyder, "The Use of Direct Convolution Products in Profile and Pattern Fitting Algorithms. I. Development of the Algorithms," *J. Appl. Cryst.,* **22**, 238–243 (1989).
87. A. March, "Mathematische Theorie der Regelung nach der Korngestalt bei affiner Deformation," *Z. Kristallogr.,* **81**, 285–297 (1932).

88. R.A. Young, "Introduction to the Rietveld Method"; pp. 1–38 in *The Rietveld Method.* Edited by R.A. Young. International Union of Crystallography, Oxford University Press, 1995.
89. J.C. Taylor and C.E. Matulis, "Absorption Contrast Effects in the Quantitative XRD Analyis of Powders by Full Multi-Phase Profile Refinement," *J. Appl. Cryst.,* **24**, 14–17 (1991).
90. D.T. Cromer and D. Liberman, "Relativistic Calculation of Anomolous Scattering Factors for X-Rays," *J. Chem. Phys.,* **53**, 1891–1898 (1970).
91. J.C. Taylor, "Computer Programs for Standardless Quantitative Analysis of Minerals Using the Full Powder Diffraction Profile," *Powder Diffraction,* **6** [1] 2–9 (1991).
92. A.C. Larson and R.B. von Dreele, "GSAS — Generalized Structure Analysis System." Los Alamos National Laboratory Report LAUR 86–748, 1994.
93. D.B. Wiles, A. Sakthivel, and R.A. Young, "DBWS-9411 — An Upgrade of the DBWS Programs for Rietveld Refinement with PC and Mainframe Computers," *J. Appl. Cryst.,* **8**, 366–367 (1995).
94. R. Berliner, C. Ball, and P.B. West, "Neutron Powder Diffraction Investigation of Model Cement Compounds," *Cem. Conc. Res.,* **27**, 551–575 (1997).
95. J. Neubauer, H.-J. Kuzel, and R. Sieber, "Rietveld Quantitative XRD Analysis of Portland Cement: II Quantification of Synthetic and Technical Portland Cement Clinkers"; pp. 100–111 in *18th International Conference of Cement Microscopy, Houston.* ICMA, 1996.
96. D.B. Wiles and R.A. Young, "A New Computer Program for Rietveld Analysis of X-Ray Powder Diffraction Patterns," *J. Appl. Cryst.,* **14**, 149–151 (1981).
97. L.S. Dent Glasser, "Relationships between Sr_3SiO_5, Cd_3SiO_5, and Ca_3SiO_5," *Acta. Cryst.,* **18**, 455–457 (1965).
98. M. Regourd, "Crystal Chemistry of Portland Cement Phases"; Chapter 3 in *Structure and Performance of Cements.* Edited by P. Barnes. Applied Science, England, 1983.
99. J. Fayos and M. Perez-Mendez, "Crystallo-Chemistry of $Ca_3O(SiO_4)(C_3S)$ Related Phases," *Am. Ceram. Soc. Bull.,* **65**, 1191–1195 (1986).
100. J.C. Taylor and L.P. Aldridge, "Full-Profile Rietveld Quantitative XRD Analysis of Portland Cement: Standard XRD Profiles for the Major Phase Tricalcium Silicate (C_3S: $3CaOSiO_2$)," *Powder Diffraction,* **8** [3] 138–144 (1993).
101. I. Maki and K. Kato, "Phase Identification of Alite in Portland Cement Clinker," *Cem. Conc. Res.,* **12**, 93–100 (1982).
102. F. Nishi and Y. Takéuchi, "Tricalcium Silicate, $Ca_3O(SiO_4)$: The Monoclinic Superstructure," *Z. Kristallogr.,* **172**, 297–314 (1985).
103. W.G. Mumme, "Crystal Structure of Tricalcium Silicate from a Portland Cement Clinker and Its Application to Quantitative XRD Analysis," *N. Jb. Miner. Mh.,* **4**, 145–160 (1995).
104. R. Golovastikov, R.G. Matveeva, and N.V. Belov, "Crystal Structure of the Tricalcium Silicate $3CaO{\cdot}SiO_2 = C_3S$," *Sov. Phys. Cryst.,* **20**, 441–445 (1975).
105. Y. Takéuchi, F. Nishi, and I. Maki, "Crystal-Chemical Characterisation of the Tricalcium Aluminate-Sodium Oxide ($3CaO{\cdot}Al_2O_3{\cdot}Na_2O$) Solid-Solution Series," *Z. Kristallogr.,* **152**, 259–307 (1980).
106. D.K. Smith, "Crystallographic Changes with the Substitution of Aluminium for Iron in Dicalcium Ferrite," *Acta. Cryst.,* **15**, 1146–1152 (1962).

107. A.A. Colville and S. Geller, "The Crystal Structure of Brownmillerite, Ca_2FeAlO_5," *Acta. Cryst.*, **B27**, 2311–2315 (1971).
108. J.C. Taylor and L.P. Aldridge, "Phase Analysis of Portland Cement by Full Profile Standardless Quantitative X-Ray Diffraction — Accuracy and Precision"; pp. 309–314 in *Proceedings of the 40th Conference on Applications of X-Ray Analysis, Denver,* Advances in X-Ray Analysis, Vol. 36. Edited by J.V. Gilfrich, C.R. Hubbard, R. Jenkins, D.K. Smith, T.C. Huang, M.R. James, G.R. Lachance, and P.K. Predecki. Plenum Press, New York, 1993.
109. J.C. Taylor and L.P. Aldridge, "Quantitative X-Ray Diffraction Analysis of Portland Cements"; pp. 141–146 in *International Ceramics Conference, Austceram, Sydney, Australia, July 1994.*
110. P. Stutzman, S. Lenker, H. Kanare, F. Tang, D. Campbell, and L. Struble, "Standard Cement Clinkers for Phase Analysis"; pp. 154–168 in *Proceedings of the 11th International Conference on Cement Microscopy.* ICMA, Duncanville, Texas, 1989.
111. H. Kanare, "Production of Portland Cement Clinker Phase Abundance Standard Reference Materials"; final report, CTL project No. CRA 012-840. Construction Technology Laboratories, Skokie, Illinois, 1987. National Institute of Standards and Technology, Report of Investigation, Reference Materials 8486, 8487, 8488, Portland Cement Clinker, May 22, 1989.
112. H. Möller, "Standardless Quantitative Phase Analysis of Portland Cement Clinkers," *World Cement,* September 1995, pp. 75–84.
113. J.C. Taylor and R.A. Clapp, "New Features and Advanced Applications of SIROQUANT™: A Personal Computer XRD Full Profile Quantitative Analysis Software Package"; pp. 49–55 in *Proceedings of the 41st Conference on Applications of X-Ray Analysis, Denver,* Advances in X-Ray Analysis, Vol. 35. Edited by J.V. Gilfrich, C.R. Hubbard, R. Jenkins, D.K. Smith, T.C. Huang, M.R. James, G.R. Lachance, and P.K. Predecki. Plenum Press, New York, 1992.
114. E. Prince, "Comparison of Profile and Integrated-Intensity Methods in Powder Refinement," *J. Appl. Cryst.,* **14**, 157–159 (1981).
115. M.J. Cooper, "The Analysis of Powder Diffraction Data," *Acta. Cryst.,* **A38**, 264–269 (1982).
116. H.M. Rietveld, "The Early Days: A Retrospective View"; pp. 39–42 in *The Rietveld Method.* Edited by R.A. Young. International Union of Crystallography, Oxford University Press, 1995.
117. J. Bergman, P. Friedel, and R. Kleeberg, "BGMN — A New Fundamental Parameters Based Rietveld Program for Laboratory X-Ray Sources, Its Use in Quantitative Analysis and Structure Investigations," *IUCr-CPD Newsletter,* **20**, Summer 1998.
118. J. Bergman, R. Kleeberg, T. Taut, and A. Haase, "Quantitative Phase Analysis Using a New Rietveld Algorithm — Assisted by Improved Stability and Convergence Behaviour"; in *Proceedings of the 46th Annual Denver Conference on Applications of X-Ray Analysis,* Advances in X-Ray Analysis, Vol. 40 (compact disc). JCPDS–International Centre for Diffraction Data, 1997.
119. H.-J. Kuzel, "Rietveld Quantitative XRD Analysis of Portland Cement: I Theory and Application of the Hydration of C_3A in the Presence of Gypsum"; pp. 87–99 in *18th International Conference of Cement Microscopy, Houston.* ICMA, 1996.

Neutron Diffraction and Neutron Scattering Studies of Cement

R. Berliner
Research Reactor Center, University of Missouri, Columbia, Missouri

"It is not right to know only one thing — one gets stultified by that; one should not rest before one knows the opposite too."
— Vincent van Gogh

Introduction

The heterogeneous nature and chemical complexity of cement pastes makes the characterization of this system difficult. The specific reasons are not hard to find. Cement is made from material that is dug out of the ground and fed to the cement kiln with minimal attempt at purification. Additional complex substances (blast furnace slag, fused microsilica, gypsum, superplasticizers, air entraining agents, accelerators, retarders, and so on) are added to the furnace feedstock, subsequent to the clinkering process or during concrete making, to dispose of unwanted industrial by-products, to control the rate of the cement hydration reactions, to control the paste porosity, or to affect the hardened cement paste morphology. The most important component of a hardened cement paste, calcium silicate hydrate (C-S-H),* is poorly crystalline or amorphous. The stoichiometry of C-S-H, and of other compounds present, is highly variable from place to place within the paste and its composition remains the subject of considerable controversy. A hardened cement paste is a live system: fluids in the cement paste pores (pore solution) continue to react, albeit slowly, with the remaining unhydrated cement and the cement hydration products that form and are embedded in the paste. In response to environmental influences, some of the hydration compounds can become unstable and others precipitate from the pore liquid with beneficial or deleterious effects.

Given the general perception of cement and concrete as rugged, strong materials, it is surprising to the newcomer to learn that the cement paste is, instead, fragile. Dehydration, such as that which occurs in an electron microscope sample stage, can alter irreversibly the morphology of the

*The usual cement chemists notation will be employed here with C = CaO, S = SiO_2, A = Al_2O_3, $\bar{S}$ = SO_4, H = H_2O, D = D_2O.

paste, making microscopic examinations of C-S-H microstructure difficult and sometimes inconclusive.

In spite of the inherent difficulties of the system, considerable understanding of the nature of cement and concrete has been obtained through the use of a wide variety of tools. Studies of the cement pore liquid as a function of time, X-ray diffraction investigations, electron microscope and microprobe studies, NMR experiments, chemical extraction methods, and ultrasonic investigations have all contributed. An overview of the analytical methods applied to cementitious systems can be found in books by Lea[1] and by Taylor.[2] Neutron diffraction and neutron scattering techniques provide another, sometimes unique and sometimes complementary, view of the structure, hydration kinetics, composition, and morphology of cement and cement pastes.

Although the neutron was discovered by Chadwick in the early 1930s,[3] materials science investigations using neutrons awaited the development of suitably intense sources. The powerful research reactors built in the 1960s encouraged the development of a host of neutron beam methods for the investigation of the structure and dynamics of materials. These methods have made substantial contributions to our knowledge of the structure, internal dynamics, thermal vibrations, and magnetic excitations of materials. In many ways, neutron scattering research has made significant contributions to the development of the theory of solids. While the utility of using neutrons as a probe of cementitious materials was recognized nearly three decades ago,[1] until recently, only a few investigations had been reported.

In the last few years, experiments on cementitious systems using neutron diffraction, neutron quasielastic scattering, neutron inelastic scattering, and neutron small angle scattering have been reported. Each of these techniques is based on the unique properties of the neutron in its interaction with matter and each contributes a new view of the state of a dry cement compound or hydrating cementitious systems. The material that follows is intended as an introduction to neutron scattering methods for the nonspecialist. The fundamental properties of the neutron that make it useful for materials science investigations, the production of neutron beams, and the general configuration of some neutron beam instruments will be described. The literature on neutron beam investigations of cement and concrete will be reviewed.

Neutron Production and Properties

Neutrons are spin one-half particles of mass $m = 1.675 \times 10^{-27}$ kg with a magnetic moment of $\mu_n = -1.913\ \mu_N$ where μ_N is the nuclear magneton. They carry no charge and their internal charge distribution dipole moment is smaller than the current detection limit. The deBroglie relation $\vec{p} = h\vec{k}/(2\pi)$ connects the neutron momentum to its wave vector, $\vec{k}$, with $\lambda = 2\pi/|\vec{k}|$. The velocity, $\vec{v}_N = \vec{p}/m_n$, and kinetic energy $E = p^2/(2m_n)$ are calculated from the usual kinematic relations.

It is useful to introduce neutrons by comparison to X-rays because both probes are used for diffraction investigations and more workers are familiar with X-ray methods. X-rays can be obtained with a generator tube or, in the last few years, at a synchrotron X-ray laboratory. Neutrons for neutron beam research arise either from nuclear fission in reactors or from the spallation process. In the latter, accelerator-produced high-energy charged particles produce large numbers of energetic neutrons from collisions with heavy-atom nuclei in the accelerator target.[4]

In the United States, the High Flux Beam Reactor (HFBR, Brookhaven National Laboratory), the High Flux Isotope Reactor (HFIR, Oak Ridge National Laboratory), the Intense Pulsed Neutron Source (IPNS, Argonne National Laboratory), the Los Alamos Neutron Science Center (LANSCE, Los Alamos National Laboratory), the NBSR reactor (National Institute of Science and Technology), and MURR (University of Missouri) maintain active neutron beam research programs and are available to external users by arrangement with local scientific staff or by an open proposal process. Many more neutron beam scientific laboratories are available to European scientists with the major centers in France (Institute Laue Langevin) and England (Rutherford Laboratory). Major neutron beam laboratories exist in Russia (Dubna) and in Japan (JAERI) and several new neutron beam sources are recently completed or under construction in Asia.

In either a reactor or a spallation source, the liberated neutrons are produced at high neutron energies and are moderated by allowing them to approach equilibrium (through multiple collisions) with the atoms of materials at specific temperatures. When the moderator is appropriately designed and at room temperature, thermal neutrons are obtained. Moderators of liquid H_2 or deuterium (at ~20 K) or methane (~150 K) are used to obtain cold neutrons while graphite, suitably isolated, can reach ~2000 K

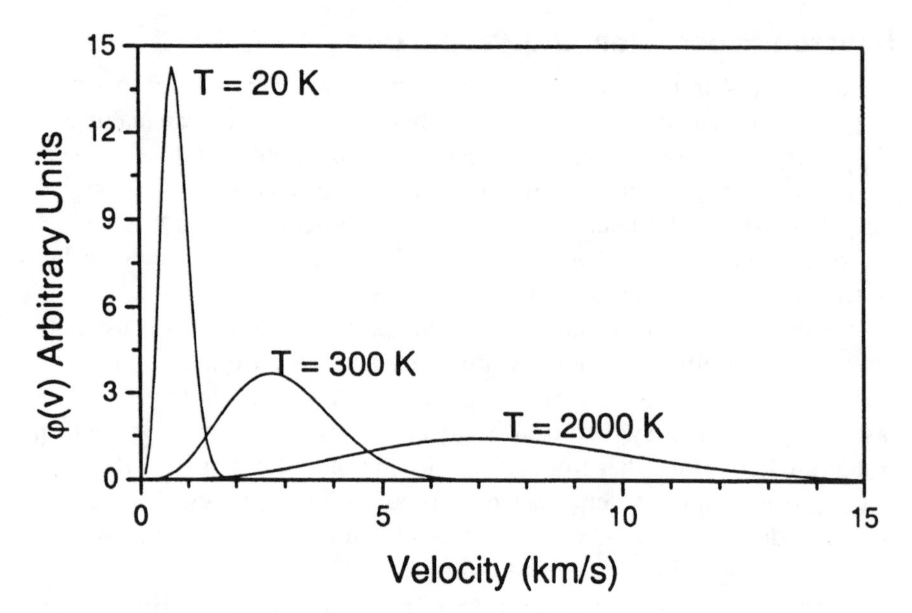

Figure 1. Maxwellian flux distribution, $\phi(v) \propto v^3 \exp[-(mv^2)/(2k_BT)]$ illustrating the intensity distribution as a function of velocity for neutrons from moderators at T = 25, 300, and 2000 K. The curves have been normalized to have the same area. (after Squires, Ref. 7).

by nuclear heating and produces hot neutrons. In each case, neutrons emerge from the moderator at a variety of energies, but changing the moderator temperature effectively serves to shift the peak of the neutron energy spectrum. The terms "cold neutrons" and "hot neutrons" are shorthand for specifying the most probable neutron energy (or wavelength).

The neutron velocity distribution is generally taken as Maxwellian and the consequence of shifting the neutron spectrum with moderators of different temperatures can be seen in Fig. 1. The dramatic increase in the number of low-velocity (long wavelength) neutrons available from the cold moderators greatly enhances the strength of the neutron source for many investigations. The increase in the population of high-energy (short wavelength) neutrons from a hot source or spallation source is important for other classes of experiments. Neutron scattering and neutron diffraction have always been intensity limited as the neutron flux from the most intense neutron source is weak in comparison to the flux of X-rays from common laboratory X-ray tubes and many orders of magnitude weaker than the synchrotron X-ray sources. To compensate for this deficit, careful attention to the optimization of neutron instruments from the source geometry and temperature through beam transport, neutron optics, and detection have played an important role in the development of neutron beam methods.

The numerical relations displayed as Eq. 1 can be used to connect the neutron energy, velocity, wavelength, wave vector, and temperature. Here, the neutron energy is in meV, the wave vector in $Å^{-1}$, the velocity in km/s, and the temperature in K.

$$E = 2.072k^2 = 5.227v^2 = 81.82 / \lambda^2 = 0.08617T$$
$$\lambda = 6.283 / k = 3.956 / v = 9.045 / \left(\sqrt{E}\right) \quad (1)$$

Neutrons interact with nuclei and with magnetic moments. The strength of the neutron-nuclear interaction is dependent on the internal structure of each nuclide and does not vary in a systematic way with atomic number, mass, or charge. In contrast, the strength of X-ray scattering is (almost completely) proportional to Z, the number of electrons in an atom. The interaction of neutrons with nuclei is different even for isotopes of the same element and varies, for those nuclides with nuclear spin, with the relative neutron-nuclear spin orientation. Figure 2 illustrates the strength of neutron scattering (the scattering amplitude, b) as a function of atomic weight and compares the scattering strength with that of X-rays. The irregular nature of the neutron curve is due to the superposition of nuclear resonance scattering, different for each nuclide, onto a smooth contribution due to potential scattering.

Neutron Scattering Theory

There are a number of excellent reviews of neutron scattering theory available. Particularly noteworthy is the classic book by Bacon,[5] and comprehensive theoretical treatments by Lovesey,[6] Squires,[7] and Sears.[8] The application of neutron scattering and neutron diffraction techniques to materials science is the subject of a book edited by Kostorz.[9] An extensive description of experimental techniques is contained in a three volume set of books edited by Price and Skold.[10]

The application of neutron methods to cementitious systems includes diffraction, inelastic scattering, incoherent inelastic scattering, and small angle scattering. A reprise of the formal theory covering all of these areas would be inappropriate in the context of this review. For a comprehensive

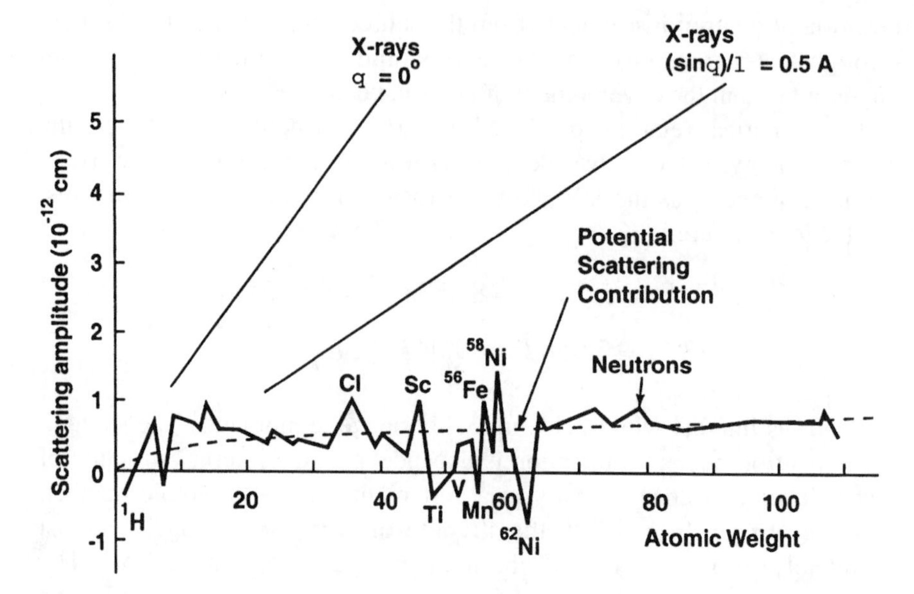

Figure 2. Irregular variation of neutron scattering amplitude with atomic weight due to superposition of resonance scattering on the slowly increasing potential scattering; for comparison the regular increase for X-rays is shown. [From Ref. 5 after *Research (London)*, **7**, 257 (1954)].

description of neutron scattering theory, the reader is referred to the standard works on neutron scattering listed above. Except for a brief account of the fundamental concepts, only the theoretical results and a guide to the interpretation of experiments will be described.

The interaction of neutrons with materials is inherently a quantum mechanical process. Nevertheless, a simple analysis reveals the major features of the interaction. Consider a neutron incident on a single atom fixed to the origin of the coordinate system. The theoretical description begins with the incident neutron represented as a plane wave,

$$\Psi = e^{ikz} \tag{2}$$

where $|\vec{k}| = 2\pi / \lambda$ is the incident neutron wave vector. Part of the incident wave is scattered creating an outgoing spherical wave,

$$\Psi = -(b / r)e^{ikr} \tag{3}$$

where r is measured from the center of the target nucleus. The quantity b is the scattering length which is, in principle, a complex number ($b = \alpha + i\beta$)

and denotes the strength of the scattering. The imaginary part of the scattering length is important only for those few nuclei that exhibit strong absorption and for most purposes, it is sufficient to consider b as a real, scalar quantity.

The complete neutron wavefunction is then the superposition of the incident and scattered waves

$$\Psi = e^{ikz} - (b / r)e^{ikr} \tag{4}$$

The scattering cross section is then given by the ratio of the outgoing neutron current to the incident neutron flux:

$$\sigma = \frac{4\pi r^2 v \left|(b / r)e^{ikz}\right|^2}{v\left|e^{ikz}\right|^2} = 4\pi b^2 \tag{5}$$

where v is the neutron velocity. Values of b are normally obtained by measurement as they depend, in detail, on the properties of the target nucleus, which are insufficiently well known.

The forgoing analysis strictly pertains to a nucleus fixed at the origin and unable to recoil. When the target nucleus is free, the cross section is reduced, $\sigma_{free} = [A / (A / (A + 1)]^2 \sigma_{bound}$, where A is mass of the target nucleus. Additional modifications of the simple formula given in Eq. 5 are required to account for the effect of nuclear spin on the neutron-nuclear interaction (see Ref. 5, p. 32 ff).

In a typical neutron-scattering experiment, a beam of neutrons characterized by their initial energy E_0, energy distribution ΔE_0, central wave vector $\vec{k}_0$, and wave vector distribution $\Delta\vec{k}_0$, falls upon a specimen material. Formally, the number of neutrons scattered per unit time into the solid angle $\Delta\Omega$with final wave vector $\vec{k}$ and energy E is given by:

$$d^2 n = \Phi_0\left(\vec{k}_0, \Delta\vec{k}_0\right) N \left(\frac{d^2\sigma}{d\Omega dE}\right) \Delta\Omega\Delta E \tag{6}$$

where $\phi_0(\vec{k}_0,\Delta\vec{k}_0)$ is the initial neutron flux as a function of the wave vector, N is the number of scattering centers in the specimen, and $d^2\sigma / d\Omega dE$ is the double differential scattering cross section.

Calculation of the cross section is accomplished by completely specifying the initial and final states of the neutron and the scattering system, summing the contributions from all the scattering centers for all final states, and averaging over the initial states of the system with the appropriate statistical weights. For an ensemble of scattering centers where we ignore the effects of spin and where the target system is unchanged by the scattering event (elastic scattering), the cross section can be shown to be:

$$\frac{d\sigma}{d\Omega} = \frac{1}{N}\left|\sum_{R} b_{\vec{R}} \exp\left(i\vec{Q}\cdot\vec{R}\right)\right|^2 \tag{7}$$

Here, $\vec{R}$ is the position vector of any nucleus and $\vec{Q}$ is the wave vector change (momentum transfer) for the neutron. The quantity $B_{\vec{R}}$ is the scattering length for the atom at position $\vec{R}$.

The elements that comprise the specimen material may consist of several different isotopes, each scattering neutrons with different strengths. In some cases, the scattering strength differences between isotopes of the same element can be substantial. In addition, the various isotopes will have different values of nuclear spin with consequent variation of scattering strength. The arrangement of the isotopes for each of the constituent elements in the specimen will almost always be random. Consider a specimen composed of a single element. The average of Eq. 7 over all possible arrangements of the isotopes for that element will then yield:

$$\frac{d\sigma}{\sigma\Omega} = \left\{\overline{|b|^2} - \left|\bar{b}\right|^2\right\} + \left|\bar{b}\right|^2\left|\sum_{R} \exp\left(i\vec{Q}\cdot\vec{R}\right)\right|^2 \tag{8}$$

where $\bar{b} = \Sigma w_i b_i$ with w_i and b_i being the isotopic abundance and scattering length associated with the ith isotope, and similarly, $\overline{|b|^2} = \Sigma w_i b_i^2$.

Equation 8 shows that a specimen consisting of a single element that is composed of several isotopes will have two components to the scattering of neutrons. The first contribution will be proportional the mean square difference of the scattering length for all the atoms in the specimen ($\overline{|b|^2} - |\bar{b}|^2$) and gives rise to what is called incoherent scattering. This term is the result of the uncorrelated variations of the scattering length due to the isotopic disorder. It does not depend on the structure of the specimen material and provides no structural information. The origin of this term can be seen to be completely analogous to the "disorder scattering" observed in alloys.[11] The second term is called coherent scattering. It produces all of the interference effects and is proportional to the square of the average scattering length. The quantity $\sigma_{incoh} = 4\pi(\overline{|b|^2} - |\bar{b}|^2)$ is often called the incoherent or disor-

Table I. Neutron coherent scattering lengths and coherent, incoherent, and absorption cross sections for the most common elemental constituents of cementitious materials. The quantities listed (except for D) are the atomic values (the average over the natural isotopic distribution).[7]

	Z	b_c (10^{-12} cm)	σ_c (10^{-24} cm^2)	σ_i (10^{-24} cm^2)	σ_a (10^{-24} cm^2)
H	1	–3.390(11)	1.7568(10)	80.26(6)	0.3326(7)
D	1	6.671(4)	5.592(7)	2.05(3)	0.000519(7)
C	6	6.6460(12)	5.550(2)	0.001(4)	0.00350(7)
O	8	5.803(4)	4.232(6)	0.000(8)	0.00019(2)
Na	11	3.63(2)	1.66(2)	1.62(3)	0.530(5)
Mg	12	5.375(4)	3.631(5)	0.08(6)	0.063(3)
Al	13	3.449(5)	1.495(4)	0.0082(6)	0.231(3)
Si	14	4.1491(10)	2.1633(10)	0.004(8)	0.171(3)
S	16	2.847(1)	1.0186(7)	0.007(5)	0.53(1)
Cl	17	9.5770(8)	11.562(2)	5.3(5)	33.5(3)
K	19	3.67(2)	1.69(2)	0.27(11)	2.1(1)
Ca	20	4.70(2)	2.78(2)	0.05(3)	0.43(2)
Fe	26	9.45(2)	11.22(5)	0.40(11)	2.56(3)

dered scattering cross section for an element, while $\sigma_{coh} = 4\pi|\bar{b}|^2$ is called the coherent scattering cross section. Table I shows the scattering length and the coherent, incoherent, and absorption cross section for those elements typically found in cementitious materials. Note the magnitude of the incoherent scattering cross section for hydrogen in comparison to all of the other elements. It is also worth pointing out that atom for atom, the coherent scattering amplitude of D is 140% that of Ca. More extensive tables of these quantities are available.[8,12] For those rare experiments with spin-polarized neutrons and oriented nuclei, these formulas must be extended to include the variations of scattering length for different neutron-nuclear spin orientations.

When the atoms in the specimen material are arranged into perfect crystals, Eq. 7 takes on the forms:

$$\left(\frac{d\sigma}{d\Omega}\right)_{\text{coh}} = \frac{(2\pi)^3}{v_0}\sum_{\vec{\tau}}\delta\left(\vec{Q}-\vec{\tau}\right)\left|F_N(\vec{\tau})\right|^2$$

$$\left(\frac{d\sigma}{d\Omega}\right)_{\text{incoh}} = \sum_{d}\frac{\sigma_{\text{incoh},\vec{d}}}{\pi}e^{-2W_d(\vec{Q})} \qquad (9)$$

where the first term describes the Bragg scattering and the second describes the incoherent scattering. The quantity $F_N(\vec{\tau})$ is the familiar structure factor:

$$F_N(\vec{\tau}) = \sum_d b_d \exp\left[i\vec{Q}\cdot\vec{d} - W_d\left(\vec{Q}\right)\right] \tag{10}$$

where the summation is over the vector $\vec{d}$, which is the displacement of the dth atom from the origin of the unit cell. The quantity $W_d(\vec{Q})$ is the Debye-Waller factor and can be shown to be related to the average of the displacement, $\vec{u}(\vec{l},\vec{d})$, of the dth atom in the l-unit cell from its regular (equilibrium) atomic position. To a good approximation, $W_d(\vec{Q}) = 1/2 \langle\{\vec{Q}\cdot\vec{u}(\vec{l},\vec{d})\}\rangle$.

There are aspects of this short summary of neutron nuclear scattering that should be emphasized. First, in contrast to the case for X-rays, the scattering amplitude, b, can sometimes be negative. The sign arises from the quantum mechanical analysis of the scattering process and is related to the sign of the amplitude of the reflected neutron wave function. While measured quantities are always proportional to b^2, the variation in the sign of b can have a substantial impact on the scattering from some isotopic mixtures and ordered alloys. This has a special relevance for the study of cementitious materials because of the large difference that exists in the strength of scattering between hydrogen ($b = -0.374 \times 10^{-12}$ cm) and deuterium ($b = 0.6674 \times 10^{-12}$ cm). Second, neutron-nuclear scattering is independent of scattering angle, unlike the case for X-rays where it decreases with increasing scattering angle.

The elementary properties above are the basis of the utility of neutrons of a probe of materials in general and cementitious materials in particular.

1. The mass of the neutron ensures, through the deBroglie relation, that the wavelength of a thermal (300 K) neutron (~1.77 Å) is commensurate with the spacing between atoms in solids and liquids. Interference effects (diffraction) can therefore provide information about the structure of materials. If the wavelength were substantially larger or smaller, diffraction effects would produce little structural information on the atomic scale.
2. The energy of a thermal neutron (25.8 meV) is on the same order as the internal elementary excitations (atomic and molecular vibrations, magnetic orientations) in solids. Interaction of neutrons with these excitations results in easily measured changes of neutron energy and provides information on the nature of these internal degrees of freedom. In the X-ray case, a photon with a wavelength

of 1.77 Å would have an energy of 7 keV. Energy exchange between this photon and thermal vibrations of a typical solid would change the photon energy by one part in 3×10^5. Measurement of these small photonic energy changes is a formidable experimental problem.

3. The interaction of neutrons with matter is weak and can be accurately treated theoretically using the first Born approximation. Detailed quantitative comparison of neutron experimental results to theoretical calculations enriches the scope of applications of these techniques.
4. Neutrons penetrate deeply into matter because they lack an electric charge and because of their weak interactions with most nuclei. Large sample volumes can thus be studied independently, in most circumstances, of surface or grain effects. For example, typical $1/e$ penetration depths for neutrons in a dry cement are ~3 cm in comparison to the case for Cu-K_α X-rays where the penetration is 30 μm.
5. Specimens examined with neutron methods can be enclosed in special environments (furnace, cryorefrigerator, pressure cell, etc.) with little interference. In the X-ray case, such experiments are almost always substantially more difficult.
6. The neutron magnetic moment and its interaction with magnetic moment distributions provides many opportunities for the study of magnetic materials.

The advantages of neutrons as a probe for condensed matter research are offset, to some extent, by the scarcity of neutron beam sources and instruments. In the United States, neutron beam facilities at the National Laboratories are typically oversubscribed by a factor of 2 or 3. The same situation exists in Europe, where neutron beam laboratories and instruments are proportionally much more numerous.

Neutron Diffraction

Neutron diffraction investigations of cementitious systems can be made on the anhydrous compounds, the hydrating paste, or the hardened cement. In spite of the advantages of the neutron as a probe (in comparison to X-rays), few experiments have been reported, in part because of the lack of familiarity of the technique among workers in the field and in part due to the relative scarcity of neutron beam laboratories.

Neutron diffraction experiments can be performed at a reactor or at a spallation source, but the instruments at each source type are quite different. A schematic diagram of a typical reactor-based diffractometer is shown in Fig. 3. Neutrons from the reactor core pass through a beam tube and strike the monochromator crystal, which selects a wavelength slice from the neutron energy spectrum by crystal diffraction. The monochromatic incident beam bathes the specimen and the angular distribution of neutrons scattered from the specimen is recorded by the detector assembly. This type of instrument is quite analogous to an X-ray diffractometer. The handicap of low neutron source intensities has been overcome, in some respects, by the use of a position-sensitive detector that collects a large portion of the diffraction pattern simultaneously and by focusing neutron monochromator optics, which make more effective use of the neutron source flux.

A spallation source powder diffractometer is shown in Fig. 4.[13] Neutrons are produced at the source by pulses of high-energy charged particles with a repetition rate of (typically) 10–30 Hz. The accelerator target and its surroundings are designed so that the high-energy spallation neutrons are moderated and the result is a neutron pulse that emerges from the region of the target and is transported through a beam tube to the diffractometer. The time of flight of the neutrons is determined by their energy through Eq. 1 and the energy distribution of the neutrons emitted from the moderator results in a distribution of neutron arrival times at the specimen. Detectors are arranged around the specimen position so the detector angle and the time-of-flight (TOF) are sufficient to determine, through Bragg's Law, the d spacing or momentum transfer (Q) for the detected neutrons. The diffraction pattern for pulsed source diffractometers is often presented as intensity vs. TOF or d spacing while reactor-based diffractometer data is almost always shown as intensity vs. scattering angle.

Neutron diffraction investigations of cementitious materials were first performed by Christensen and Lehmann at the Institute Laue Langevin in the mid-1980s.[14] The instrument employed was the D1B diffractometer (Fig. 3), which possesses a position-sensitive detector that spans 80° of the diffraction pattern with a resolution of 0.2° and can collect a spectrum in a few minutes. In the first of their experiments, they studied the hydration of the calcium aluminates C_3A, $C_{12}A_7$, C_5A_3, CA, CA_2, and CA_6 with heavy water as a function of time and temperature. Heavy water (D_2O) is used for

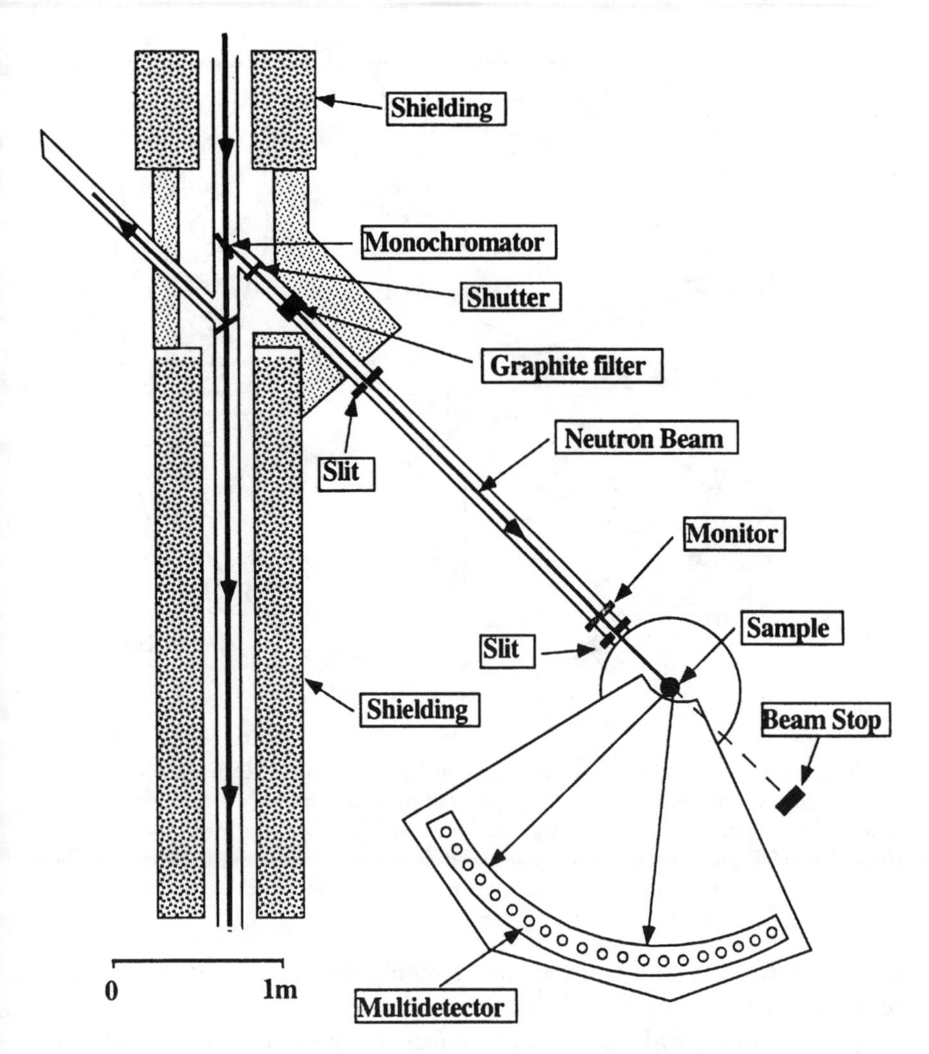

Figure 3. The D1B powder diffractometer at the Institute Laue Langevin. The major elements of the instrument are shown. Neutrons from the reactor core pass through a thermal neutron guide tube and strike the monochromator at the top of the figure. The monochromatic beam bathes the specimen and the scattered neutrons are detected in the multidetector assembly. The multidetector spans 80° of the diffraction pattern at one time with a resolution of 0.2°. (Drawing courtesy of the Institute Laue Langevin.)

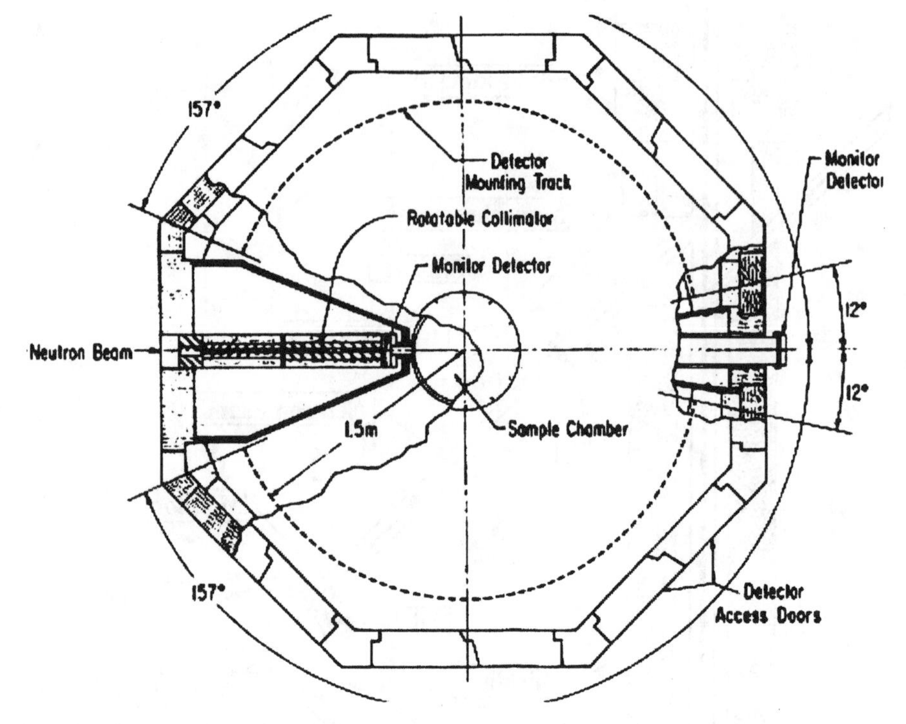

Figure 4. A cross-sectional view of the General Purpose Powder Diffractometer at the Intense Pulsed Neutron Source (Argone National Laboratory). A large number of detector elements, each 30 cm long, nearly fill the space on the detector mounting track. The TOF (in relation to the start of the neutron pulse) and detector number (angular position) are recorded for each neutron scattered from the specimen.[13]

these experiments to avoid the strong incoherent neutron scattering that results from the presence of hydrogen.

X-ray Guinier powder patterns of the reaction products were used to identify the compounds formed. Using the integrated intensity of representative neutron diffraction peaks from the specimen material, they were able to record the time dependence of hydration compound formation and reactant consumption. The hydration compounds C_3AD_6, C_3AD_{12}, and C_3AD_{10} were obtained as reaction products although C_2AD_8, expected from the hydration of C_2A, was not observed. Figure 5 shows the intensity as a function of time at $T = 63°C$ of the hydro-grossular, C_3AD_6, for each of the calcium aluminates: CA_2, CA, C_5A_3, and C_3A. By careful comparison of the consumption of the anhydrous compound with the growth of the hydration product, it was

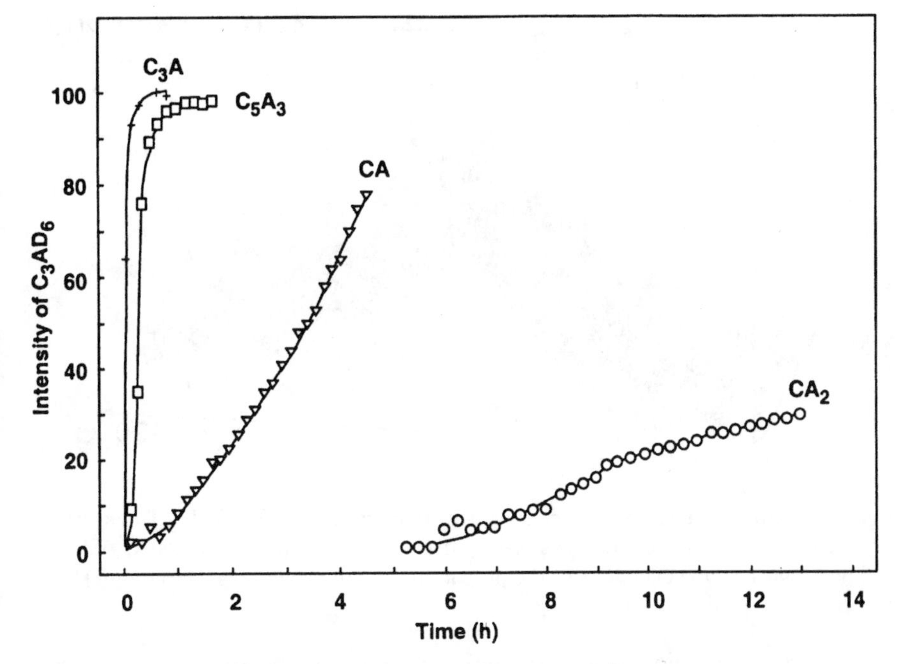

Figure 5. Intensity of C_3AD_6 vs. time for the reaction between D_2O and C_3A, C_5A_3, CA, and CA_2, respectively, at 63°C (Fig. 10 from Ref. 14.).

possible to infer the production of an intermediate amorphous precursor to the formation of the hydrated crystalline reaction products.

Time-resolved neutron powder diffraction was used to study the hydration (with D_2O) of C_3S, β-C_2S, and C_4AF in a second series of experiments.[15] The hydration products of these reactions contained no surprises — the only crystalline phase produced by C_3S or β-C_2S hydration was $Ca(OD)_2$, and C_3S was observed to hydrate faster than β-C_2S. The principal crystalline hydration product of the hydration of C_4AF was C_4AD_6, and the authors argue that the Fe (liberated by the hydrolysis of C_4AF) appears as amorphous Fe_2O_3 .

The $t_{1/2}$ value (the time required for production of 50% of the reaction products) was used to obtain the activation energy for the hydration reactions. An Arrhenius plot of $t_{1/2}$ for the formation of C_3AD_6 from the hydration of C_4AF at various temperatures yielded ε = 58.2 kJ/mol. Similar analyses for C_3A gave ε = 40 kJ/mol, while the result for $C_{12}A_7$, CA, and CA_2 was ε = 105 kJ/mol. Evidently, the compounds with high C/A ratios

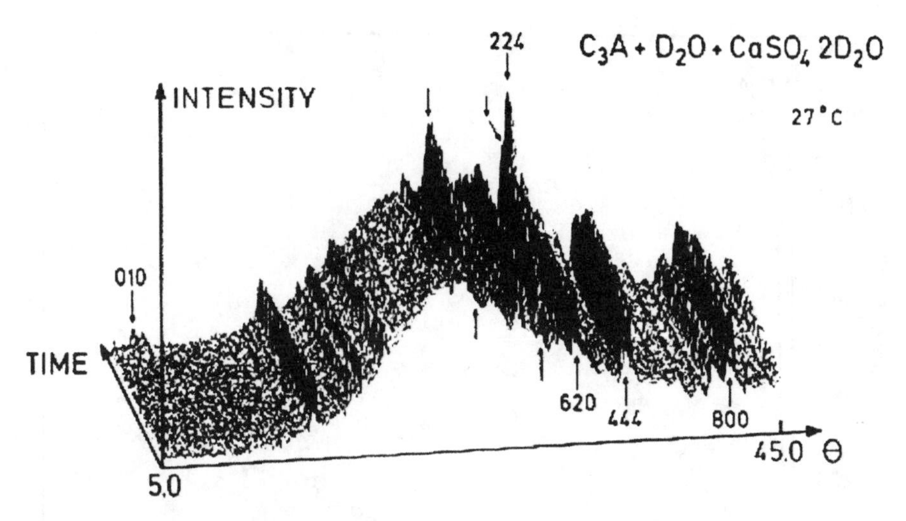

Figure 6. Powder neutron diffraction patterns from a $C_3A \cdot CaSO_4$-D_2O mixture at 27°C recorded at 10 min intervals. Selected Miller indices of C_3A ↑ and $C_4A\bar{S}D_{32}$ ↓ and positions of two reflections of the precursor ↑↓ are shown (Fig. 4 from Ref. 16).

(C_3A and C_4AF) have low activation energies for the formation of crystalline C_3AD_6, while those with low C/A ratios ($C_{12}A_7$, CA, and CA_2) have high activation energies. The authors conclude that the structural properties of Ca^{2+} do not alone satisfactorily explain the differences in reactivity and they speculate that some properties connected with the AlO_4 tetrahedra must play a role.

A third series of time-resolved powder neutron diffraction experiments examined the effect of additives on the hydration reactions of the calcium aluminates.[16] While the results obtained for $Ca(OD)_2$, $CaCO_3$, $CaCl_2$, and SiO_2 additions were much as expected, surprising results were obtained from the experiments where the aluminates were hydrated in the presence of SO_4^{-2}.

The neutron experiments examined the effect of the source of the sulfates [$CaSO_4 \cdot (1/2)D_2O$ and $CaSO_4 \cdot 2D_2O$], variation of the molar ratio (calcium aluminates/ sulfates), temperature (27, 50, and 93°C for the hydration of C_3A), and the form of the aluminate (C_3A, $C_{12}A_7$, and CA). The hydration reactions were followed for different amounts of time from a few hours to 250 days. Figure 6 shows the time sequence of the powder neutron diffraction patterns, illustrating the nature of the raw data from these experiments.

In the conventional view, hydration of C_3A in the presence of sulfates produces ettringite, $Ca_6[Al(OD)_6]_2(SO_4)_3 \cdot xH_2O(C_6A\bar{S}_3H_{32})$, as the first crystalline reaction product. In contrast, the neutron experiments reveal the formation of a heretofore unknown precursor phase. When the hydration is performed at 60°C with a molar ratio (C_3A/sulfates) of 4.26/12.85, the precursor is formed and grows rapidly for several hours. When the molar ratio was increased to 10/1.72 and the hydration was observed at 50°C, the precursor appears as an intermediate phase in the first hour of the hydration. At this ratio and at 27°C, the precursor is the only crystalline reaction product for the first 8 h. At 93°C, the precursor was not observed. Figure 7 shows the time dependence of the production of the precursor, consumption of C_3A, and formation of ettringite at 27°C with a molar ratio of 10.0/1.72 $C_3A/(CaSO_4 \cdot 2D_2O)$ that was observed in this experiment. The disappearance of the precursor phase as ettringite appears strongly suggests that the two phases are linked together. Time-resolved X-ray C_3A-$CaSO_4 \cdot 2D_2O$ hydration experiments at 60°C, designed to reveal the structure of the precursor phase, showed only the production of ettringite, in quantities increasing with time. Finally, the diffraction lines of the precursor phase can be indexed on the ettringite unit cell. Together, these observations make it apparent that ettringite can exist in two modifications, mainly differing in the OD or D_2O occupations or orientations, that are not distinguishable by X-rays.

The utility of neutron powder diffraction measurements has been greatly increased by the application of Rietveld profile refinement methods.[17] In Rietveld refinement, the diffraction pattern data is quantitatively compared to a detailed crystalline atomic model of the material under study.[18]

The neutron intensity diffracted from a powder specimen (neglecting magnetic scattering) can be written as:

$$I(\theta) = A \sum_{hkl} \frac{F_{hkl}^2}{\sin\theta_{hkl} \sin 2\theta_{hkl}} M_{hkl} \left(\frac{2\sqrt{\ln 2}}{H_{hkl}\sqrt{\pi}} \right) \times \exp\left[-(4\ln 2) \left(\frac{2\theta - 2\theta_{hkl}}{H_{hkl}} \right)^2 \right] - Bkg(2\theta) \tag{11}$$

The space group and lattice parameters for the specimen material completely specify the position of the Bragg peaks, $2\theta_{hkl}$, and multiplicity, M_{hkl},

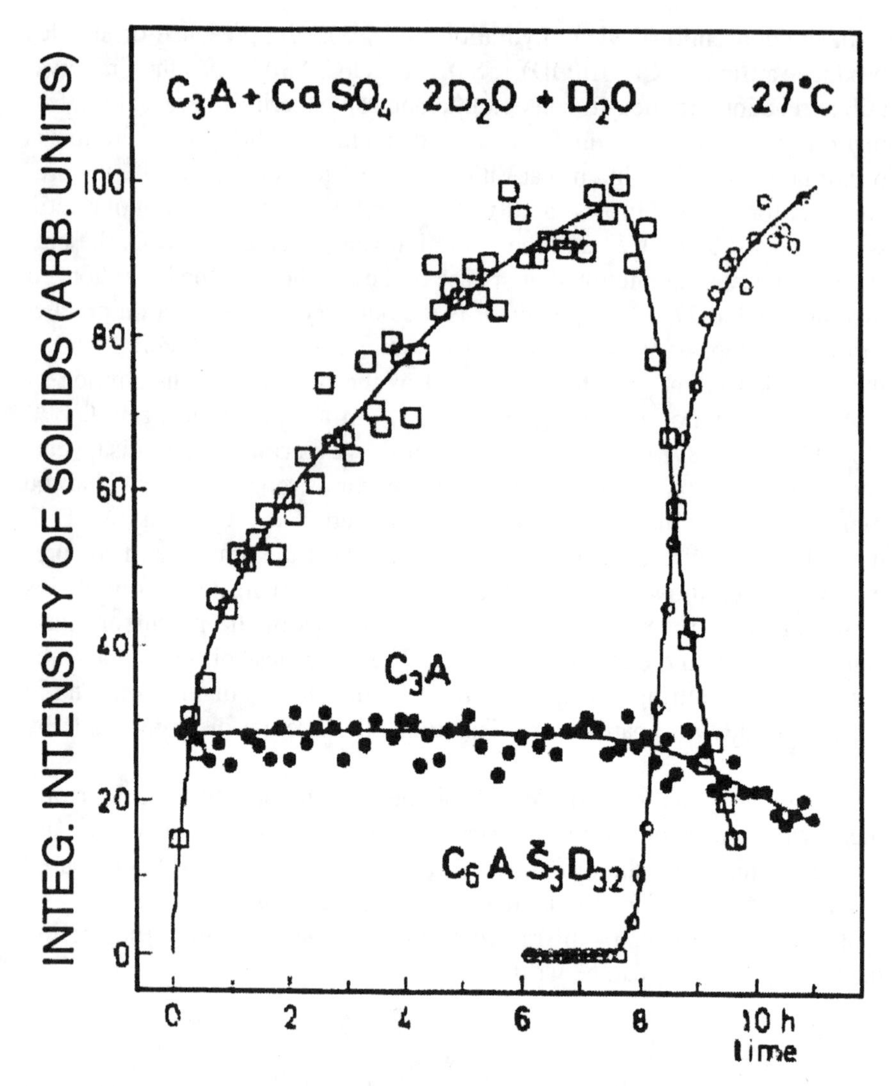

Figure 7. Integrated intensities of solids in the reaction of C_3A-$Ca_SO_4 \cdot 2D_2O$-D_2O at 27°C. The precursor is represented by open squares [Fig. 5(a) from Ref. 16].

for diffraction from the planes with Miller indices (hkl). The quantity 2θ is the scattering angle and F^2_{hkl} is the atomic structure factor. The quantity A is a scale factor that accounts for the incident beam neutron intensity, the detector efficiency, the size of the specimen, and the time spent accumulating the diffraction pattern. The shape of the diffraction peaks is Gaussian with a width H_{hkl} that is determined by the resolution of the spectrometer and by the disorder of the specimen material itself. The quantity is usually parameterized as:

$$H_{hkl} \rightarrow H(2\theta_{hkl}) = U \tan^2 2\theta_{hkl} + V \tan 2\theta_{hkl} + W \tag{12}$$

with U, V, and W as empirical constants dependent on the spectrometer resolution. Additional terms can be added to Eq. 12 to account for internal strain and crystal disorder. The intensity of the diffraction peaks is determined by the structure factor F^2_{hkl}, which is given by Eq. 7 and depends, in detail, on the location, isotopic species, occupation fraction, and thermal vibration amplitudes of the atoms in the crystal unit cell. Additional terms can be added to Eq. 11 to take into account the effect of preferred orientation and absorption on the diffraction peak amplitudes.

When the diffraction data is of suitable quality, the parameters of the model (atom positions, occupations, and so on) can be obtained from a least-squares minimization of the differences between the data and the model as a function of the model parameters. Robust computer implementations of the Rietveld method are widely available[19] and have extended the work of Rietveld to encompass multi-phase analysis and pulsed neutron diffraction data.

Neutron diffraction from synthetic examples of cement clinker compounds was analyzed by Rietveld refinement to determine the suitability of the X-ray single crystal structural models for these materials.[20,21] Monoclinic and triclinic C_3S, β-C_2S, cubic and orthorhombic C_3A, and C_4AF were examined. Figure 8 shows the quality of the agreement between the Rietveld fit, in this case for monoclinic C_3S, and the neutron diffraction data. While good agreement was obtained, the complexity of some of the crystal structures (C_3S in particular) did not allow more than a confirmation of the X-ray structural models.

The cement compound structural models were, in turn, applied[22] to phase analysis of NIST clinker reference materials.[23] Good agreement between the phase composition of the clinkers, determined by several different methods, and the neutron diffraction seven-phase Rietveld refinement (C_3A, β-C_2S, cubic C_3A, orthorhombic C_3A, C_4AF, MgO, and CaO) was obtained. In a multi-phase Rietveld refinement, the relative phase fractions are parameters of the model. The agreement is remarkable in view of the number and complexity of the oxide phases present in a cement clinker.

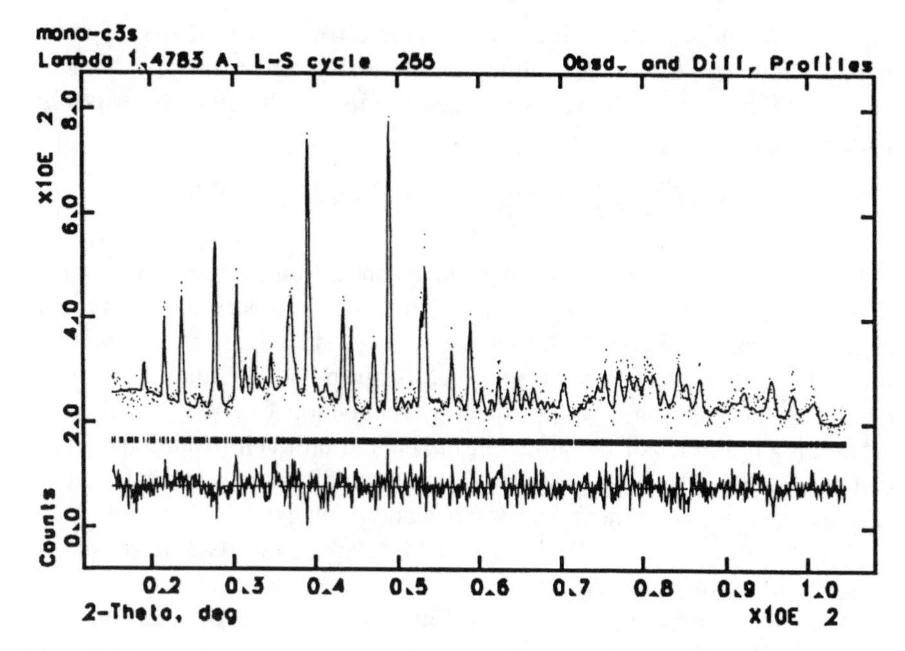

Figure 8. Rietveld refinement results for monoclinic C_3S. The dots are the neutron diffraction data and the solid line is the calculated diffraction pattern for the least-squares optimum fit to the atomic model parameters. The difference between the data and the fit is shown in the lower curve. The location of symmetry-allowed diffraction lines is shown by the short vertical lines (Fig. 1 from Ref. 20).

In-situ neutron diffraction measurements of C_3S specimens hydrated with heavy water (D_2O) were analyzed by two-phase Rietveld refinement [C_3S and $Ca(OD)_2$] to monitor the amount of $Ca(OD)_2$ formed as function of time. Experiments using the HIPD at IPNS and the reactor-based instrument PSD-II at MURR gave similar results. Analyses of the diffraction data by Rietveld refinement indicated that weight fractions of $Ca(OD)_2$ as low as 0.1% could be detected approximately 4–5 h after the start of the reactions (T ~ 22°C).[24]

Neutron diffraction followed by Rietveld refinement provides excellent sensitivity to $Ca(OD)_2$ formation in cement during the early period of (heavy water) hydration and is more sensitive than chemical extraction techniques or X-ray methods. Its use as a monitor of the cement hydration reactions (through the production of $Ca(OD)_2$) is severely compromised, however, by the quite remarkable finding that the reactions proceed 3 or 4 times slower with heavy water than they do with light water and that the calcium silicates are affected more than the calcium aluminates.[25–27] This interesting isotope effect may provide additional avenues for the study of the reactivity of cement compounds.

The basic heavy-atom structure of some cement hydration compounds is known from single crystal X-ray studies.[2] However, very little detailed structural information is available on the location, orientation, and occupation of water for most of the hydrated calcium silicates and aluminates, sulfoalumiantes, and carboaluminates that inhabit ordinary cement pastes. Neutron powder diffraction data from D-substituted hydration compounds analyzed with Rietveld refinement can be used to form structural models of these materials that include the position and orientation of the water molecules within the unit cell. Recently, a detailed structure for the calcium aluminosulfate, ettringite, $Ca_6[Al(OD)_6]_2(SO_4)_3{\cdot}xD_2O$, was reported.[28] As is well known, ettringite appears as one of the hydration products of C_3A in the presence of sulfates. The heavy-ion structure for ettringite was obtained from single crystal X-ray diffraction on a natural mineral specimen.[29]

Polycrystalline ettringite was prepared by precipitation from a solution of $Ca(OH)_2$ and Al_2SO_4 and D-exchanged by alternately soaking the specimen in D_2O and drying it over $ZnCl_2$. The neutron diffraction measurement, conducted at 20 K to suppress the Debye-Waller reduction of the high-angle diffraction peaks, was preformed on the D-substituted material. Ettringite contains four hydroxyl ions and nine crystallographically distinct water molecules in the unit cell. The number of independent atom location variables in the atomic model was minimized by treating each hydroxyl ion and water molecule as a rigid body and refining only the orientation of the water molecule plane and the occupation of the water sites. The neutron diffraction pattern with the Rietveld refinement fit and difference curves is shown in Fig. 9. As can be observed, an excellent correspondence between the model calculation and the data has been obtained. Figure 10 is a view along the c-axis of 2 × 2 × 2 ettringite unit cells. The structure consists of columns of triads of Ca ions separated by $Al(OD)_6$ octahedra along the c-axis. Adjacent to the columns are channels containing triplets of SO_4 ions alternating with one site containing three waters which is only two-thirds occupied. Each of the Ca ions in the columns is coordinated by four water molecules that "paint" the columns with water. The detailed picture of the waters of hydration obtained in this study opens the door to more extensive explorations of hydrated crystal structure and, in particular, to the effect of hydration/dehydration and impurities on the water content and atom site location in cement hydration compounds.

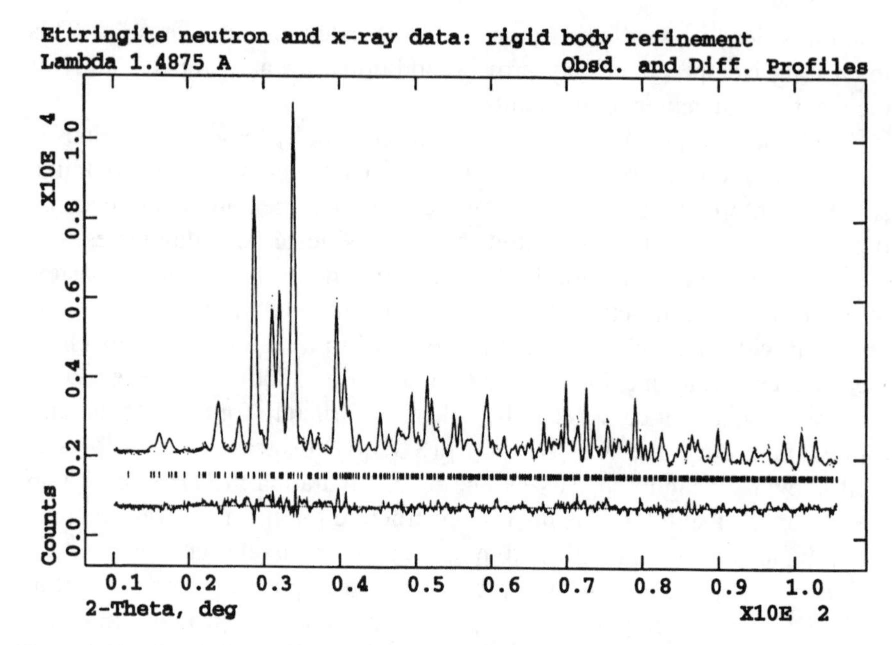

Figure 9. Results of the Rietveld refinement of the ettringite atomic model. The data is shown by the small dots; the calculated intensity pattern, using the D_2O and OD molecule orientations, occupations, and thermal vibrations, is shown as the solid line. The difference between the calculated intensity pattern and the data is shown in the lower part of the figure. The location of allowed reflections is indicated by the short vertical lines (Fig. 1 from Ref. 28).

Neutron Quasielastic and Inelastic Scattering

Thermal neutrons can exchange energy with atoms in materials; these energy changes, easily measured, are the signature of internal diffusive or vibrational atomic motions. Both neutron quasielastic scattering and neutron inelastic scattering have been used as probes of cement pastes. The basic principle behind the application of quasielastic neutron scattering (QNS) to cementitious systems can be easily understood. Hydrogen atoms bound to freely diffusing water molecules in a cement paste can exchange energy with neutrons in "billiard ball" collisions. Hydrogen atoms chemically bound to cement hydration compounds, unable to diffuse or recoil, scatter neutrons elastically. We concentrate on the motions of hydrogen because its large incoherent scattering cross section ($\sigma = 78.9 \times 10^{-24}$ cm^2) ensures that nearly all of the scattering (~99%) from a hydrating cementitious system is due to the hydrogen atoms.

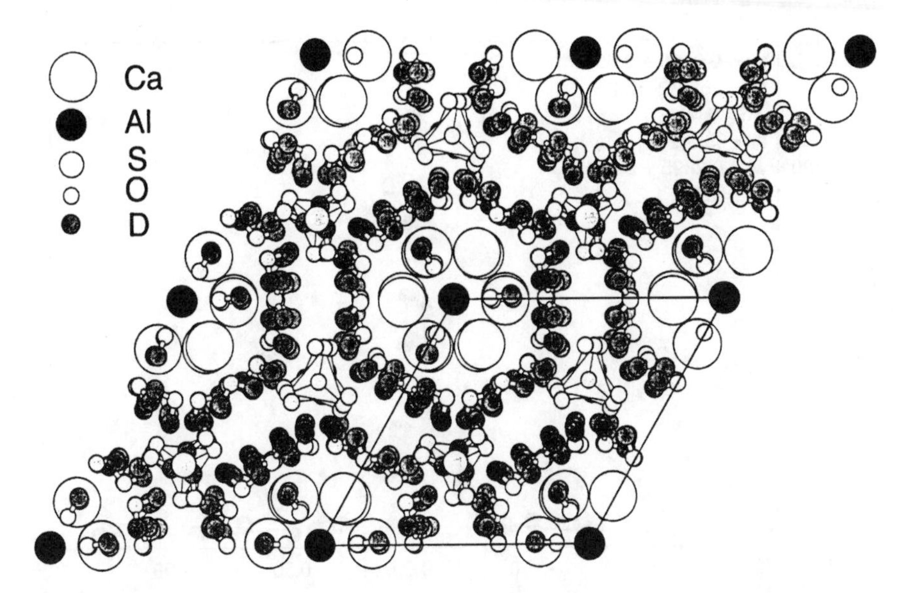

Figure 10. A 2 × 2 × 2 representation of the ettringite unit cell viewed along the c-axis. Triads of Ca ions alternate with $AlOD_6$ octahedra along the columns of atoms that can be seen at the corners of the unit cell. The channels between the columns are filled with sulfate tetrahedra and water. Each Ca ion is coordinated with four waters that are seen to coat the columns with D atoms on the outside (Fig. 2 from Ref. 28).

Figure 11 depicts the energy transfer spectrum for 10 meV incident energy neutrons scattered from a C_3S paste specimen 18.8 h after mixing.[30] The intensity of the scattered neutrons has been decomposed into three parts as in:

$$I = \frac{A_L(w_L / \pi)}{w_L^2 + (x - x_0)^2} + \left(\frac{A_G}{\sqrt{2\pi\sigma}}\right) e^{[(x-x_0)/2W_G]^2} + B \tag{13}$$

Here, $(x–x_0)$ is the scattered neutron energy gain or loss and B is the background intensity. The first term in Eq. 13 is Lorentzian in form and is the result of neutrons exchanging energy with the hydrogen atoms in freely diffusing water molecules. The second component is a Gaussian of width W_G and is the result of scattering from H atoms chemically bound to hydration compounds. The width of the Gaussian is fixed by the energy resolution of the instrument while the width of the Lorentzian (broad in comparison) is dependent on the nature of the diffusive motions of the H atoms. The ratio provides the fraction of the hydrogen in the specimen that is chemically

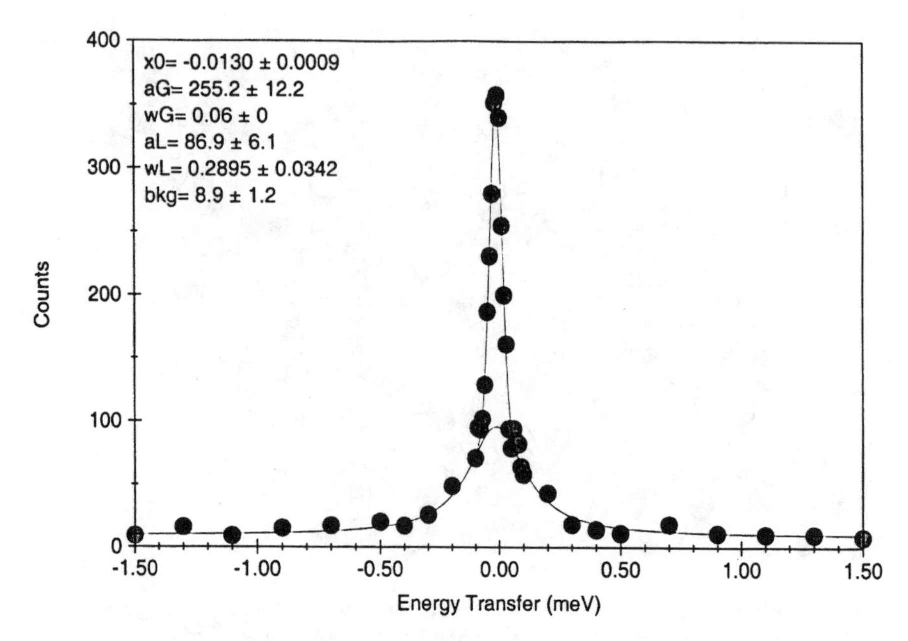

Figure 11. Analysis of QNS data from hydrating cement paste showing the decomposition of the energy transfer spectrum into a Gaussian and a Lorentzian component. The parameters of the least-squares fit are shown [Fig. 2(b) from Ref. 35].

bound and is thus a monitor of the progress of the cement hydration reactions.

The first applications of these methods to cementitious systems were undertaken by the group at NIST and described in their annual reports.[31–33] These experiments demonstrated that quasielastic neutron scattering could be used to monitor the progress of the cement hydration reactions, quantify the fraction of water in a cement paste bound to hydration compounds, probe the environment of the cement pore water, and measure the fraction of frozen water in a cement as a function of temperature.

Different kinds of instruments can be employed for neutron quasielastic and inelastic scattering measurements on cementitious systems. One of these is the classic triple-axis spectrometer, which is shown in Fig. 12. Neutrons from the reactor core pass from the primary shielding and fall upon a monochromator crystal, which is used to select the incident neutron energy, by crystal diffraction, out of the raw reactor neutron energy spectrum. The incident neutron energy can be varied by coordinated rotation of the mono-

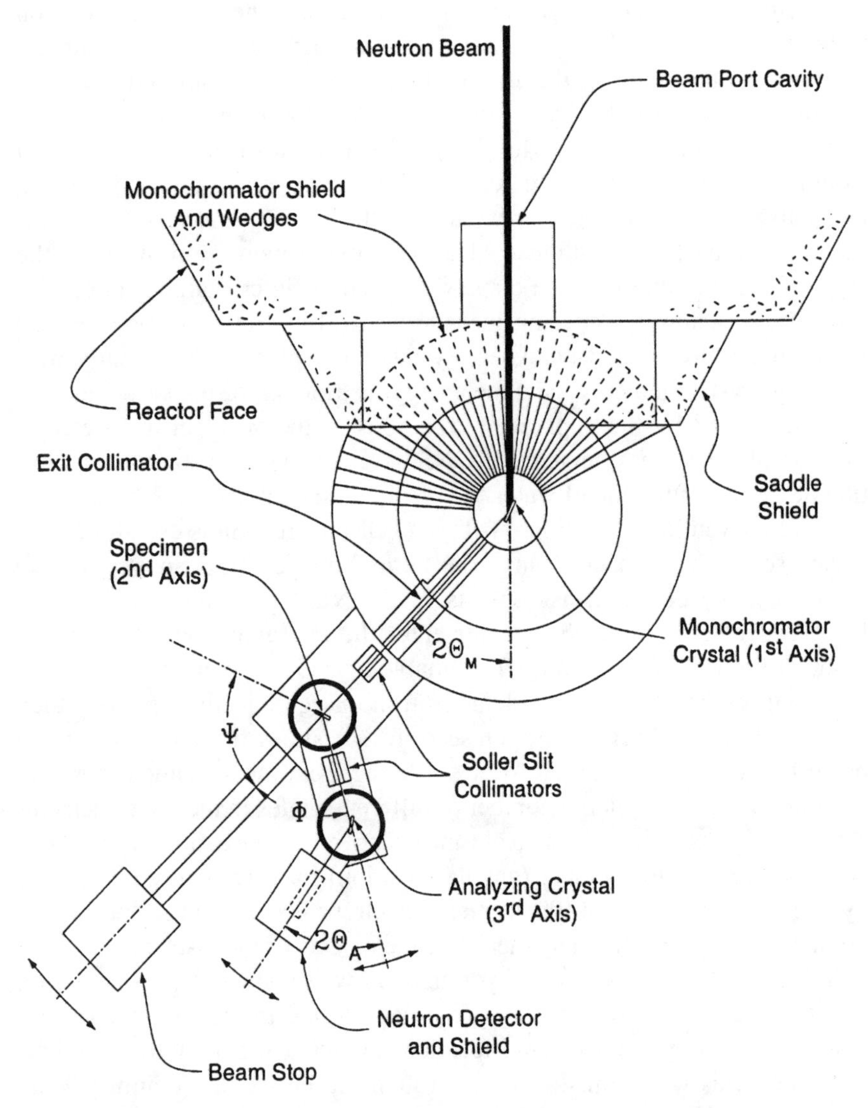

Figure 12. The triple-axis spectrometer. A beam of monochromatic neutrons is selected by Bragg diffraction from the reactor spectrum. The monochromatic beam wavelength is selected by coordinated rotation of the monochromator crystal and shield drum asembly (first axis). The angular distribution of neutrons scattered by the specimen is sampled by rotation of the detector-analyzer arm about the specimen (second axis) and the energy of the scattered neutrons can be analyzed at any angle by analyzer crystal detector rotation using Bragg diffraction (third axis).

chromator crystal and the shield drum (first axis). The specimen rotation table and detector-analyzer arm are rigidly attached to the monochromator drum shield and the angular distribution of neutrons scattered from the specimen can be sampled by rotation of the detector-analyzer arm about the sample axis (second axis). The energy distribution of neutrons scattered from the specimen at any angle can be obtained by Bragg diffraction using the analyzer-crystal and detector rotations (third axis). This versatile instrument, invented by Brockhouse,* has been extensively employed for the study of phonon and magnon dispersion relations in crystalline solids. For quasielastic neutron scattering measurement of hydrating cement pastes, the monochromator and analyzer crystals are replaced by specially bent focusing crystal elements[34] and the energy distribution of the scattered neutrons measured by the appropriate rotations of the analyzer and detector axes. Triple-axis measurements of QNS have been used to study the hydration of C_3S as a function of water-to-cement ratio by the MURR group.[30,35]

More conventionally, a time-of-flight (TOF) instrument is used for quasielastic scattering measurements. The Fermi-chopper TOF Spectrometer,[36] used for many of the measurements by the NIST group,[37–39] is shown in Fig. 13. Neutrons from the reactor enter the instrument via the neutron guide at the left of the figure.† The double crystal monochromator selects, by Bragg diffraction, the incident neutron energy and the neutrons then pass through the Fermi-chopper section that slices the monochromatic beam into a series of pulses. After scattering from the specimen, the neutrons are detected at a large number of different angles in the arc of detector elements. By recording the arrival time (relative to the rotation of the chopper) and the detector number (angular position) for each neutron, the energy and angular spectra for the scattered neutrons is collected at once, without the need for scanning any mechanical elements of the instrument.

The general features of C_3S hydration are well known. An initial period of rapid reaction is followed by a dormant period that ends after several hours. A second period of rapid reaction follows. Ultimately the reactions slow down, as water must diffuse through the layers of calcium silicate hydrate (C-S-H) that coat the remaining cement grains to reach the remain-

*Brockhouse shared the 1996 Nobel Prize in Physics for the invention and application of the triple axis spectrometer.

†Neutrons incident at shallow angles will be reflected from flat surfaces of materials with high scattering length density. This property is exploited in making neutron guides that can transport a neutron beam long distances with little loss.

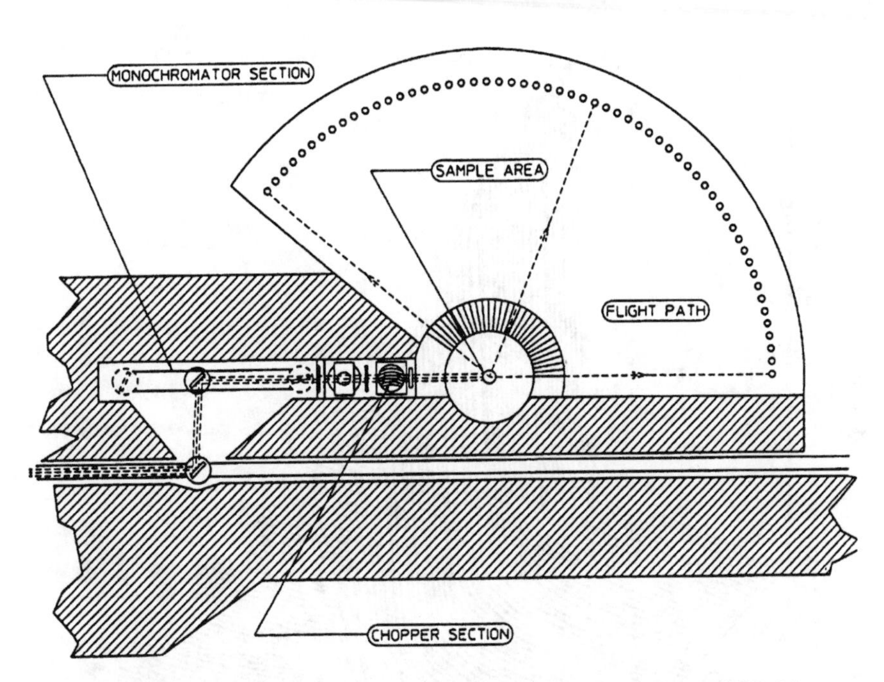

Figure 13. Major elements of the Fermi-chopper spectrometer at the NBSR. Neutrons enter the instrument from the left and a monochromatic beam is selected out of the reactor spectrum by the double-monochromator assembly. The Fermi-chopper slices the beam into short pulses that then strike the specimen. Neutrons scattered from the specimen are collected in the circular arc of detractor elements. By recording the neutron arrival time and detector number, the energy and angular distribution of scattered neutrons is obtained.[36]

ing anhydrous reactant. Figure 14 shows a time sequence of QNS measurements on C_3S hydration at 20°C.[39] The early measurements show the broad Lorentzian peak indicative of the freely diffusing pore water in the C_3S paste. At later times, the measurements illustrate the growth of a sharp Gaussian component to the scattering curves, which is created by the increasing fraction of hydrogen bound to C-S-H and $Ca(OH)_2$. The scattering curves, analyzed as in Fig. 12, provide the free water index (FWI, used in Ref. 39) or the bound water index (BWI = 1−FWI, employed by Ref. 30) as a function of time after mixing. Figure 15 shows the FWI as a function of time for several temperatures, which illustrates the description of the C_3S hydration reaction.[39] The initial burst of reactivity on mixing with water produces a only small amount of solid hydration product and thus a small

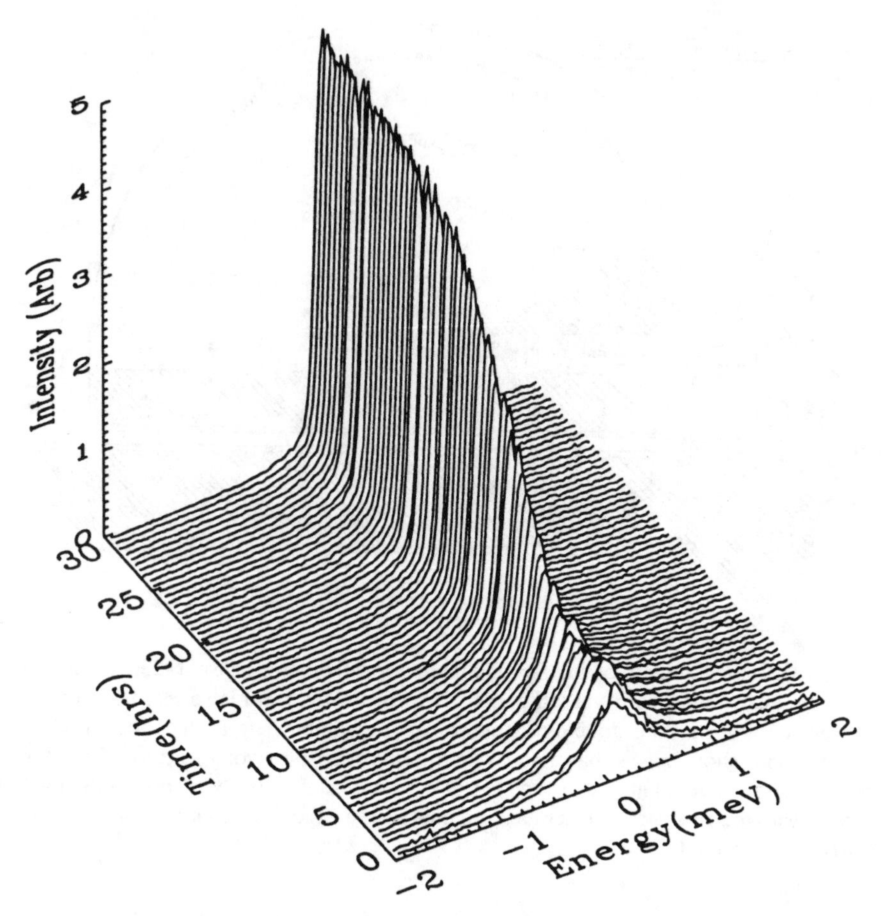

Figure 14. Time sequence of QNS data from a hydrating cement paste. The broad Lorentzian, seen at early times and caused by freely diffusing water in the cement pore liquid, is replaced at later times by the sharp Gaussian feature that is the signature of water bound to cement hydration compounds (Fig. 1 from Ref. 39).

FWI change. What is observed in Fig. 15 is first the dormant period, which extends to longer times at lower temperatures, followed by a rapid drop in the FWI as the hydration reactions proceed. At the sharp change in slope, marked with the arrow at each temperature, the reaction rate, as measured by the FWI, markedly slows, denoting the change to diffusion limited behavior.

The kinetics of the C_3S hydration reactions were analyzed by both the NIST group and the MURR group using a nucleation and growth (Avrami)

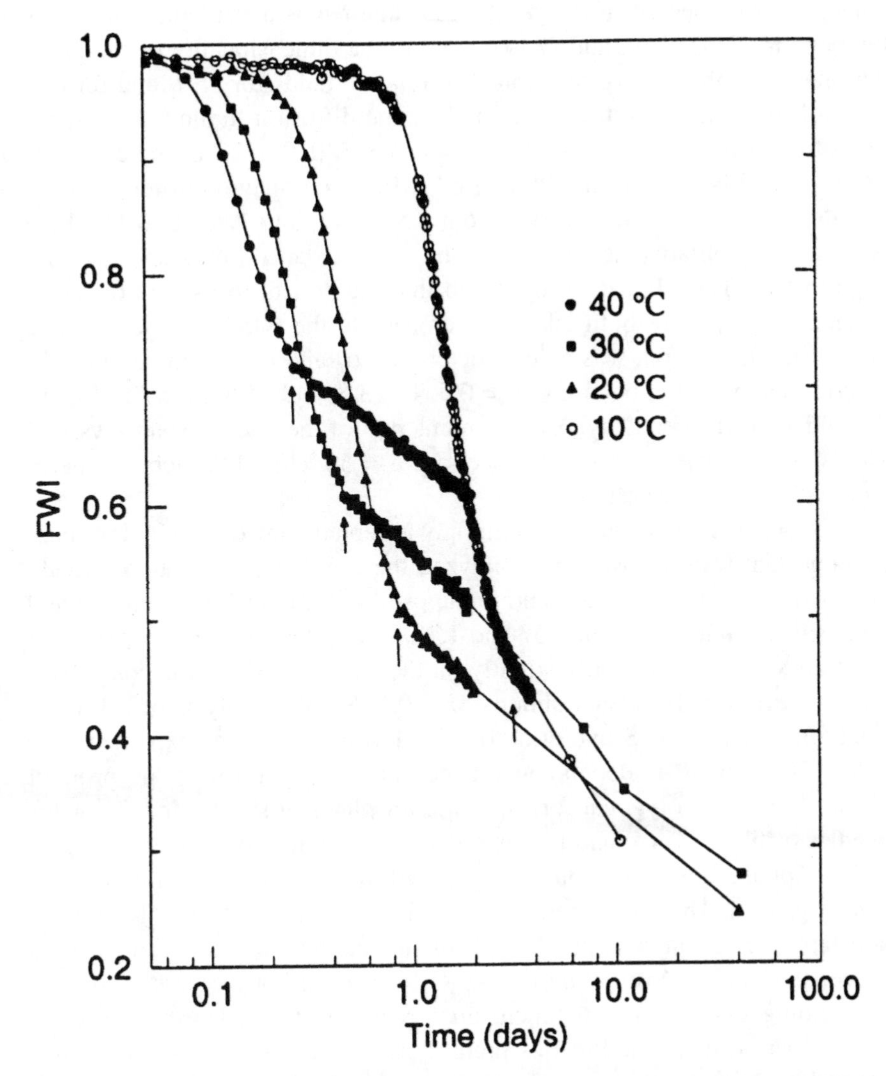

Figure 15. The FWI as a function of time for cement pastes hydrated at T = 10, 20, 30, and 40°C. The arrows in the figure indicate the time when the reaction switches over from a nucleation and growth process to a diffusion-limited one (Fig. 2 from Ref. 39).

model.[40–44] In this model, the amount of C_3S consumed as a function of time can be written as:

$$\alpha - \alpha_0 = 1 - e^{-k(t-t_0)^M} \tag{14}$$

where α is the degree of the reaction at time t, k is a rate constant that is temperature dependent, and M is related to the dimensionality of the product phase and the type of reaction. The relation holds for $t > t_0$ and for $\alpha > \alpha_0$ until the time when the reactions become diffusion limited. As a function of the water-to-cement ratio (w/c = 0.3, 0.5, 0.7) at 20°C, the exponent M was found to vary from 2.06 to 1.84, while as a function of temperature (10–40°C) at w/c = 0.4, M varied from 2.59 at 40°C to 1.88 at 10°C. There is a similar qualitative agreement for the rate constant, k, measured by both experiments. The MURR group found that the best fit to its data occurred when α_0 and t_0 were both taken as zero, while the NIST group, who had better time resolution, was able to determine α_0 and t_0 as a function of the reaction temperature, obtaining t_0 = 0.5, 1.5, 3.5, and 15.5 h for T = 40, 30, 20, and 10°C, respectively. An Ahrennius plot of the rate constant k vs. $1/T$ yields activation energy of the reaction, ε = 30 kJ/mol, which compares well with other measurements.

These results challenge the commonly accepted view of C_3S hydration in an important way. Brown[41] has analyzed the C_3S hydration data of Kondo and Ueda[45] and that of Odler and Schuppstühl[46] with an Avrami model and obtained M values between 0.67 and 1.13 during the acceleratory period of the reaction ,dependent only slightly on the particle size. In the post-acceleratory period, his analysis obtains $M \sim 0.5$. Similar analysis of a Raman scattering study of C_3S hydration obtained $M = 0.85$.[47] The exponent $M = P/S + Q$, where P is dependent on the geometrical form of the product phase: $P = 3$ for polygonal forms, $P = 2$ for plates or sheets, and $P = 1$ for needles or fibers. The quantity S is related to the rate-growth mechanism: $S = 1$ for phase boundary-controlled growth and $S = 2$ for diffusion-controlled growth. The nucleation rate constant is indicated by Q: $Q = 1$ for a constant nucleation rate and $Q = 0$ for nucleation site saturation. Brown argues that the microstructure of C-S-H formed during the early period of hydration is essentially one-dimensional, requiring $P = 1$. During the acceleratory period then, the kinetics predict one-dimensional phase boundary-controlled growth at zero nucleation rate. In the later, post-acceleratory period, Brown sees a shift to diffusion controlled growth ($S = 2$) as the controlling mechanism. The QNS determination of $M \sim 2$ is not consistent with this interpretation. Further consideration of the differences between these different kinds of experiments may well reveal something fundamental about this process.

The diffusion-limited later period of the C_3S hydration reactions was analyzed by both the MURR and the NIST groups using a shrinking-core

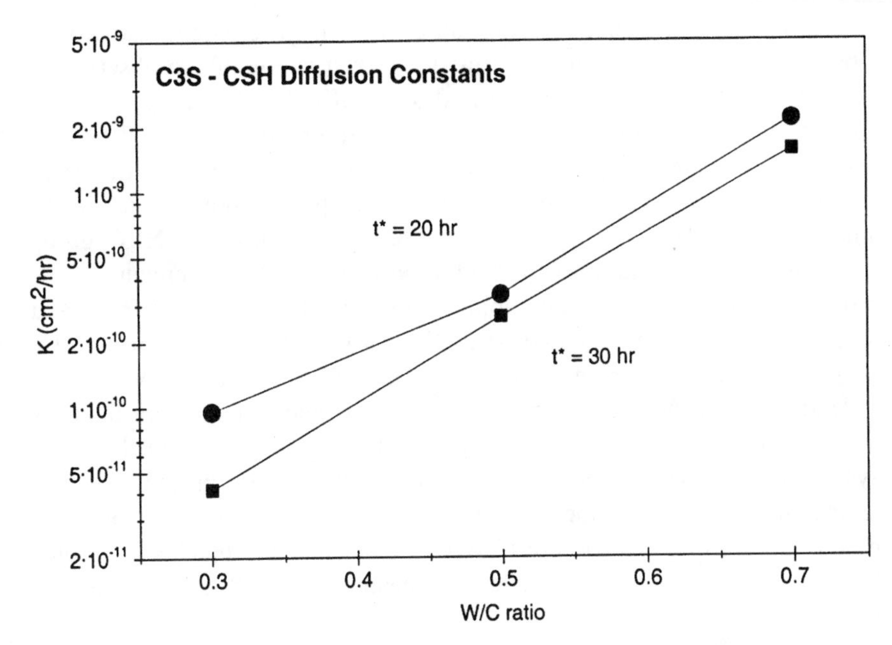

Figure 16. The effective diffusion constant as a function of water-to-cement ratio for the diffusion-limited portion of cement paste hydration. The quantity t^* refers to the time at which diffusion limited hydration is assumed to begin (Fig. 6 from Ref. 35).

model.[48] The effective diffusion constant was found to be essentially independent of temperature: $D^* \sim 2 \times 10^{15}$ m^2/h. The MURR measurements of the effective diffusion coefficient show an approximately exponential dependence on the water-to-cement ratio, which is illustrated in Fig. 16. Similar values of D* were obtained by both experiments where the w/c ratio and temperature conditions overlap.

Because water molecules rigidly bound in ice crystals are similarly unable to diffuse or recoil, QNS can be used to study the freezing of water in hardened cement pastes. Experiments at NIST were able to determine that the fraction of water bound as ice in the pores of a hardened cement paste was independent of the pore size above about 15 nm and was dependent on the w/c ratio.[49] Since cement freeze-thaw resistance is known to be strongly dependent on the w/c ratio, the experiment demonstrates that it is the optimization of the proportion of larger pores that is most important for durability.

While the majority of neutron collisions with chemically bound hydrogen molecules are elastic, some inelastic scattering does occur. In these collisions, neutrons will lose (or gain) an energy characteristic of the atomic vibrational mode of that atom. In this manner, it is possible to observe the $Ca(OH)_2$ that is produced by the hydration of C_3S. The energy transfer spectrum of neutrons scattered from specimens containing $Ca(OH)_2$ exhibits a characteristic peak at 41 meV, which is associated with the oscillation of the hydroxyl ion. The intensity of this peak is proportional to the amount of $Ca(OH)_2$ in the specimen and has been used by the NIST group to extend its examination of the hydration of C_3S.[50] Measurements of the production of $Ca(OH)_2$ were combined with the determination of free water to demonstrate that the sudden reduction in the FWI, shown in Fig. 15, occurs at slightly earlier times than the onset of $Ca(OH)_2$ formation.

Under the reasonable assumption that C-S-H and $Ca(OH)_2$ are the only products formed by the hydration of C_3S, they demonstrate that there is a systematic decrease in the water content of the C-S-H formed at higher temperatures. A measurement of the effective diffusion constant, obtained by monitoring the formation of $Ca(OH)_2$, indicates that the C-S-H product formed at higher temperatures is less permeable to the diffusion of water.

Neutron Small Angle Scattering

A parallel beam of radiation of wavelength transmitted through a slab of material will display scattering about the direction of the incident beam. This effect arises from diffraction of the beam by inhomogenieties in the specimen. As is well known for diffraction effects, there is an inverse relationship between the size of the scattering object and the angle of scattering: objects large in comparison to the radiation wavelength, λ, scatter the beam at small angles to the incident direction while scattering from objects comparable in size to the radiation wavelength will cause scattering at large angles (see Eq. 7). When X-rays are used as the incident radiation, this phenomenon is known as small angle X-ray scattering (SAXS) and the analogous effect for neutrons is called SANS. Extensive development of SAXS theory can be found in books by Guinier and Fournet,[51] Guinier,[52] and Glatter and Kratky.[53] These theoretical developments can generally be taken over directly to the neutron case with a few exceptions where the properties of the neutron lead to new physical phenomena such as magnetic small angle scattering or isotopic contrast. Because of the interest in and importance of the microstructure of cementitious systems, there is a long history of attempts to characterize the internal surface and pore structure of cement pastes. SANS experiments have several advantages for this analysis, but it is only recently that the interpretation of SANS and SAXS data from cementitious systems has been clarified. A critical review of the conflicting results for internal surface area determination in cement pastes, including a discussion of the SANS and SAXS results, has recently become available.[54]

In the discussion of small angle scattering, it is customary to measure the deviation of the scattered beam as a function of the momentum transfer, $Q = (4\pi/\lambda)\sin(\theta_s/2) \approx (2\pi\theta_s)/\lambda$, rather than the scattering angle, θ_s. Because of the inverse relationship between sizes in real space and the angle of scattering (value of the momentum transfer, Q), the spatial resolution of measurements at low Q is limited to details in real space over distances of $d \sim \pi/Q$. If the Q range of small angle scattering measurements is limited to $Q < Q_{max}$, the spatial resolution will be similarly restricted to $d > d_{min} \sim (\pi/Q_{max})$. As long as d_{min} is sufficiently large in comparison to the size of atoms and atomic spacings, which is nearly always the case in small angle scattering experiments, the calculation of small angle scattering intensities can be accomplished by the replacement of the scattering from discrete (atomic) scattering centers with a continuous distribution of scattering strengths, averaged over volumes of the order $(d_{min})^3$.

Development of the formal theory begins with the coherent scattering cross section per atom for neutrons as given by Eq. 7. The quantity b_R, the scattering length per atom, is replaced by a locally averaged scattering length density, ρ_b, and ρ_b itself is replaced by $\rho_b(r) = \bar{\rho}_b + \Delta\rho_b(r)$, to express the scattering length density distribution in terms of deviations from its average value. The cross section, Eq. 7, can then be written:

$$\frac{d\sigma}{d\Omega}(\bar{Q}) = \frac{1}{N}\left|\int_V \left[\rho_b(r) - \bar{\rho}_b\right] e^{i(\bar{Q}\cdot\bar{R})} d^3r\right| \tag{15}$$

Although more complex situations often arise, many small angle scattering experiments can be interpreted in terms of a two-phase model. In this model, the specimen is composed of a matrix material of average scattering length density ρ_{bm} in which are imbedded particles of a different scattering length density ρ_{bp}. Equation 15 then becomes:

$$\frac{d\sigma}{d\Omega}(\bar{Q}) = \frac{1}{N}\left(\rho_{bp} - \rho_{bm}\right)^2 \left|\int_V e^{i(\bar{Q}\cdot\bar{R})} d^3r\right|^2 \tag{16}$$

where $\rho_{bp} = b_p/V_p$ is the scattering length density of the N_p identical particles of scattering length b_p with volume V_p and ρ_{bm} is the scattering length density for the matrix. The integral is over the volume occupied by the specimen.

A single particle form factor can be defined:

$$F_p(\vec{Q}) = \frac{1}{V_p} \int_V e^{i(\vec{Q}\cdot\vec{R})} d^3r \tag{17}$$

where V_p is the particle volume. This provides the normalization $|F_p(0)|^2 = 1$, so that for N_p (identical) particles:

$$\frac{d\sigma}{d\Omega}(\vec{Q}) = \frac{V_p^2 N_p}{N}(\rho_{bp} - \rho_{bm})^2 |F_p(\vec{Q})|^2 \tag{18}$$

In small angle scattering the macroscopic cross-section, $d\Sigma/d\Omega = N(d\sigma/d\Omega)$, where N is the number of atoms in the specimen volume, is often used instead of the cross section per atom.

Equation 18 illustrates the basic features of the SANS from a two-phase system.

1. The strength of the scattering is dependent on $\Delta\rho = \rho_{bp} - \rho_{bm}$, the difference in scattering length density between the particle and the matrix. This quantity, often called the contrast, can be altered in a cementitious system, as will be discussed below, by exchanging H atoms in the specimen for D atoms.
2. If the scattering can be extrapolated reliably to $Q = 0$, $|F_p(0)|^2 = 1$ ensures that:

$$\frac{d\sigma}{d\Omega} = \left[(V_p^2 N_p)/N\right](\rho_{bp} - \rho_{bm})^2 \tag{19}$$

so that the number of particles, N_p, can be obtained if their contrast (composition) and size are known.

3. Guinier has shown that for small values of Qa, where a is a linear dimension of the particles, the scattering cross section can be written:

$$\frac{d\sigma}{d\Omega} = \left[(V_p^2 N_p)/N\right](\rho_{bp} - \rho_{bm})^2 e^{-(Q^2 R_g)/3} \tag{20}$$

where the quantity R_g is the radius of gyration of the particle:

$$R_g^2 = \frac{1}{V_p} \int_{V_p} r^2 d^3 r \tag{21}$$

As is known from elementary mechanics, the radius of gyration for a sphere of radius R is $R_g = \sqrt{0.6}R$. A straight line region in the semi-log plot of I vs. Q^2 is evidence of a Guinier region and provides R_g from the slope of the line. The range of validity of the Guinier approximation is generally taken as $QR_g < 1$, although the exact range varies with the shape of the particles.

4. Porod has shown that for homogeneous particles with sharp boundaries, and for $Q > 1/d_{\min}$ where $d_{\min}$ is the shortest dimension of the particle, the cross section is given by:

$$\frac{d\sigma}{d\Omega}(\vec{Q}) \approx \frac{2\pi A_p}{V_p^2 Q^4} \tag{22}$$

where A_p is the particle surface area.

5. Integration of Eq. 18 over all Q yields a quantity that is called the invariant:

$$\text{Inv} = (2\pi)^2 f_p (1 - f_p) \tag{23}$$

where $f_p = N_p V_p / V$, the volume fraction of the particles.

Combination of these relationships will permit, in principle, the determination of the size, shape, number, and composition of the particles that compose one part of the two-phase model from measurements of the scattered intensity over a sufficiently wide Q range. It must be admitted that many complications exist for the cases where the conditions of the model are relaxed, for example, where particle size or shape distributions, interparticle interference, or diffuse particle boundaries exist.

Most SANS measurements (and a few SAXS laboratories) employ pinhole instruments. Figure 17 illustrates the essential elements of a reactor-based pinhole SANS spectrometer. Neutrons from the reactor core pass through a wavelength filter and are then prepared by a series of apertures so that a tightly collimated beam falls on the specimen. The wavelength filter can be mechanical, such as a helical-vane rotary velocity selector, which

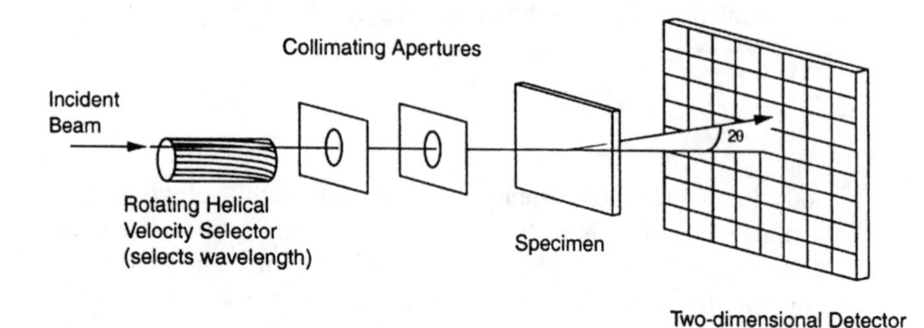

Figure 17. The essential elements of the pinhole SANS instrument are shown. The size of the collimating apertures and the distance between them determines the incident beam collimation. The angular (Q) resolution of the data is dependent on the distance between the sample and the detector and the spatial resolution of the detector. A neutron-absorbing mask in the center of the detector protects it from the transmitted neutron beam.

only transmits neutrons with a particular range of velocities (and thus wavelengths). Alternatively, the wavelength selection can be accomplished by crystal diffraction or by neutron mirror reflection. After passing through the specimen, the transmitted beam, along with the small angle scattered neutrons, is collected in a position sensitive detector. Both the incident- and scattered-beam flight paths are in vacuum or filled with helium gas so that air scattering of the beam does not contaminate the measurement of small angle scattering from the specimen. The length of the scattered beam flight path can be quite large (1–40 m) and the detector assembly is usually moveable so that the data collection angular range (and thus the Q range) can be varied. When the incident beam wavelength is prepared by a velocity selector, the central wavelength and width of the wavelength band can be easily altered (by changing the selector angular velocity) to change the Q range and resolution of the instrument. The center of the position-sensitive detector is protected by a mask that shields it from the transmitted neutron beam. Slight misalignments of the beam or the detector beam-stop can create errors in the scattering data at the very lowest Q values of the measurement. The rapid fall off of the scattered intensity with scattering angle and the presence of stray background radiation in the detector can introduce significant statistical uncertainties into the data at the very largest Q values. In the case of a cement paste, there is no preferred orientation so that the intensity pattern on the face of the detector can be circularly averaged to

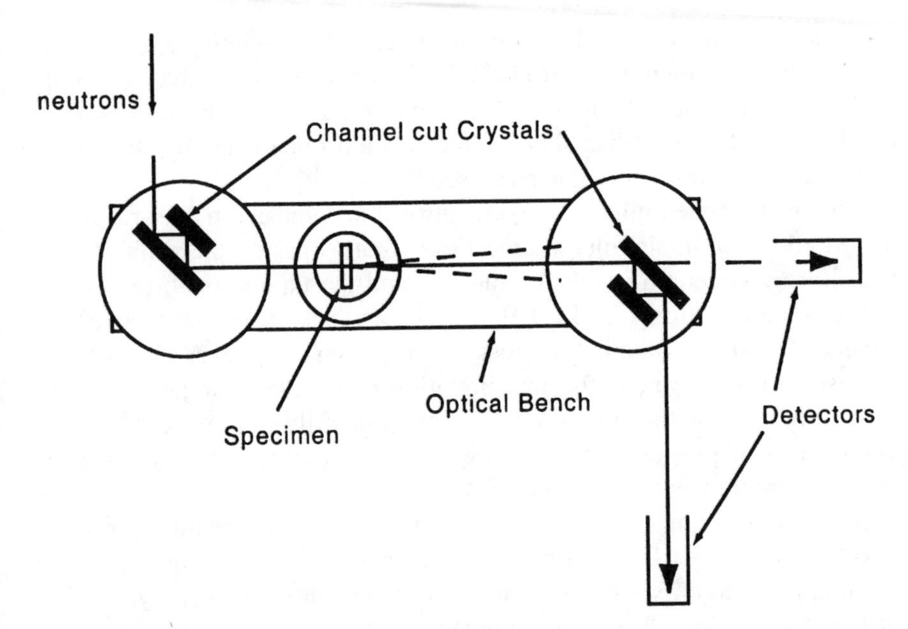

Figure 18. DKS SANS. Neutrons from the reactor are reflected from the first crystal, pass through the specimen, and are reflected again from the second crystal into the detector. Channel cut crystals are used here to decrease the width of the no-sample rocking curves and increase resolution. The small angle scattering is sampled by rotations of the second crystal. The beam transmitted though the second crystal is used to align the rotation of the first crystal element.

obtain I vs. Q as the result of the experiment. The use of a broad incident neutron wavelength band (the Q resolution of the instrument is only weakly dependent on $\delta\lambda$), a large specimen, a position sensitive detector, and rigorous reduction of the instrument background radiation compensates, to some extent, for the (relatively) low intensity of the neutron source.

A different kind of instrument has sometimes been employed for SANS and for SAXS measurements. The basic principle of the DKS (double crystal spectrometer) is shown in Fig. 18.[55] It consists of two crystals oriented so that radiation reflected from the first crystal will be reflected from the second into a detector when the crystals are appropriately arranged. To measure small angle scattering, a specimen is placed between the two crystals and the neutrons scattered into the detector are measured as a function of the rotation of the second crystal. With no specimen in place, the intensity as a function of the rotation of the second crystal (rocking curve) is extremely narrow. With a specimen, the beam of neutrons between the two crystals is broadened by the small angle scattering, giving a much wider rocking curve. The DKS spectrometer is simpler and smaller than a pinhole SANS and can reach much smaller Q values (and thus larger object sizes), but the two crystals act like slits and the data must be treated by slit-deconvolution techniques before it can be used.[56]

The greater penetrating power of neutrons (in comparison to X-rays), the opportunities for in-situ measurement of hydrating pastes, the ability to use very long wavelength incident neutrons, and the unique opportunities to change contrast offered by D_2O-H_2O exchange make SANS attractive as a means to probe the internal microstructure of cement pastes. The issue to be resolved, however, is the interpretation of the small angle scattering data. While the actual measurements are generally straightforward, the complexity of phases, contrasts, object morphologies, and size ranges forms a barrier to simple interpretations.

Small angle neutron scattering was first used to study the microstructure of cements in the early 1980s by Allen and coworkers.[57] They measured the small angle neutron scattering from a set of cement specimens hydrated at different w/c ratios. They also examined the effect of soaking the specimen in heavy water and the change in SANS obtained when the specimen was dried.

In their analysis of the SANS data, they assumed a simplified model of the cement hydration reactions. In this model the hydrated compounds — C_6AFH_{12}, C_4AH_{13}, C_3ACSH_{12}, $C_3S_2H_{2.5}$ — and the hydroxides — $Mg(OH)_2$, $Ca(OH)_2$, NaOH, and KOH — in the appropriate amounts (see Tables 2, 3, and 5 in Ref. 57) were the products of cement hydration. One of the elements of complexity that must be overcome to evaluate SANS data from cements is illustrated in Fig. 19, which shows the scattering length density of these components as a function of D_2O concentration. In the figure, the scattering length of the anhydrous cement compounds is shown as horizontal lines. The difference between the scattering length of D ($b = 6.671 \times 10^{-13}$ cm) and H ($b = -3.74 \times 10^{-13}$ cm) causes the scattering length density of the hydrated compounds to vary linearly with the D_2O concentration. Finally, each of these compounds represents a different volume fraction of the cement paste. The volume-weighted average scattering length density for all the hydrated compounds in this model is shown by the dotted line.‡ The strength of the SANS signal from the specimen in the small-Q Guinier

‡The volume fraction for each of the compounds used to calculate the average scattering length density was obtained from Ref. 57.

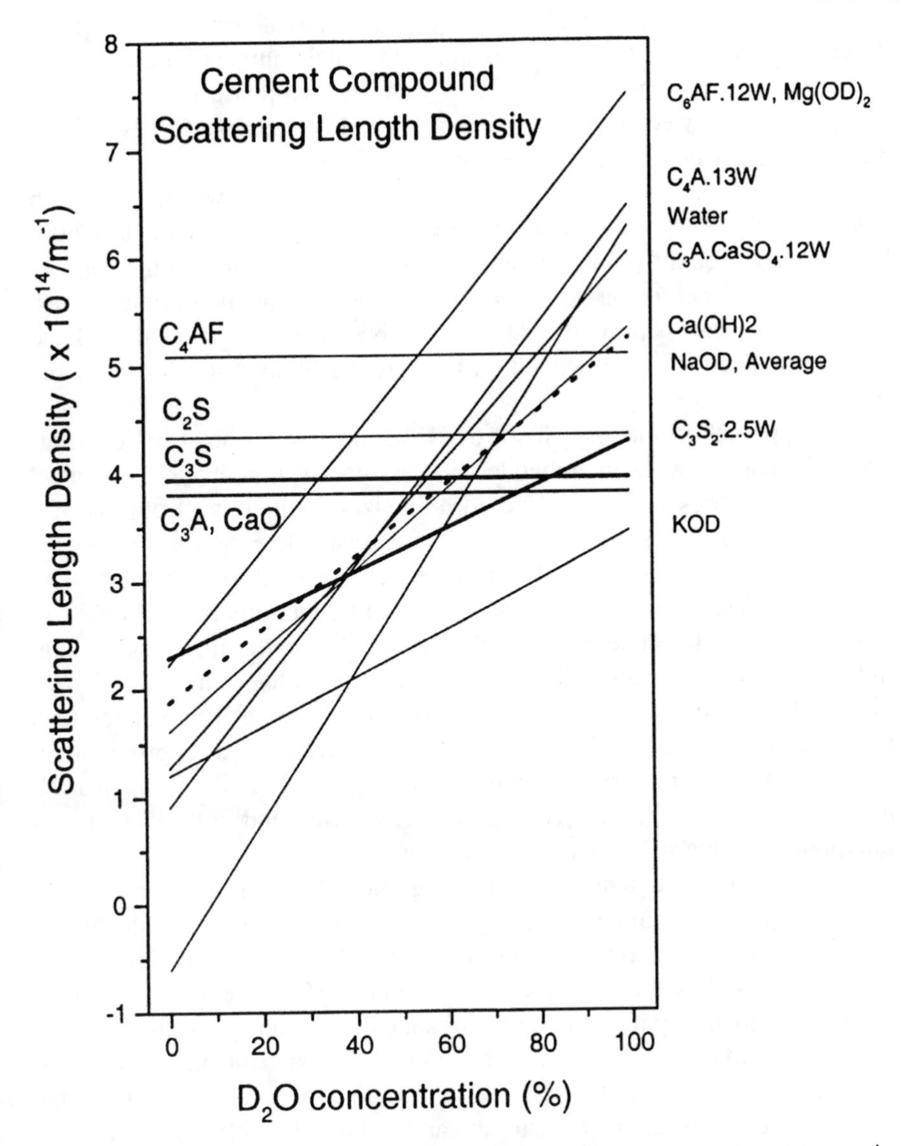

Figure 19. Neutron scattering length density for some cementitious compounds labeled at the right of the figure. The scattering length density for the anhydrous compounds is shown as horizontal lines. The scattering length density for the hydrated compounds varies linearly as a function of the heavy-water concentration. The volume-weighted average for the hydrated compound scattering length density is shown as the dashed line. It is nearly the same as that for NaO(D/H). The densities and volume concentrations have been taken from Ref. 57.

region is directly proportional to the square of the scattering length density difference (the contrast, see Eq. 19) and to the total volume of the scattering particles. If, for example, the scattering in the two-phase model is to be interpreted as between $C_3S_2H_{2.5}$ (the matrix) and pores in the $C_3S_2H_{2.5}$ gel, it matters a great deal whether the pores are filled with light water, heavy water, or air. Note that the contrast between $C_3S_2H_{2.5}$ and water will vanish where the scattering length density curves intersect, at approximately 60% D_2O. If SANS data from a cement paste is to be interpreted on the basis of a two-phase model, one issue is the identification of the appropriate phases. While the contrast is proportional to the SANS intensity, the detailed shape of the SANS curves will depend on the size distribution and shape of the scattering objects.

One approach to the analysis of SANS data for size and shape distribution information has been embedded into a numerical procedure by Vonk.[58] The SANS is presumed to be caused by a size distribution of objects of a known shape. It is a straightforward matter to calculate the form factor for objects of known shape using Eq. 17 (see Ref. 51 or 52 for examples). The parameters describing the object and its size distribution are then varied to minimize the difference between the calculated scattering and the data. Spheres, cylinders, and disks were considered by Allen et al.,[57] and it was ultimately concluded that the SANS from their cement paste specimens was created by a distribution of water-filled spherical pores approximately 5 nm in diameter and a smaller component of pores with diameters near 10 nm. SANS from the specimen that had been dried and not rewetted was interpreted as due to scattering from a much broader distribution of pore sizes with a peak near 5 nm, but extending to much larger diameters. SANS from the specimens immersed in D_2O showed that the heavy water rapidly exchanged with the water in the pores and with the water in the surrounding C-S-H gel. The total porosity was on the order of 1% of the volume. Subsequent experiments by the same group obtained very similar results.[59,60]

Interpretation of more recent SAXS and SANS experiments[61–65] has been strongly influenced by the realization that cement microstructures have many of the characteristics of surface and volume fractals.[66] Evidence for this conclusion comes from electron microscopy of hydrating cement pastes, the development of nucleation and agglomeration models for the growth of C-S-H, and the concordance between the calculated small angle scattering from fractal model structures and that observed from cement pastes.[61] The fractal interpretation of SANS and SAXS data from cementitious materials has changed the identification of the source of the scattering from water filled voids to outer-product C-S-H globules and added considerable detail to the description of the C-S-H gel.

There are many ways of discussing the nature of fractal geometry, but in the simplest terms, fractal objects are self-similar over a wide length range. That is, they appear to be the same when viewed at different magnifications. The fractal description quantifies the irregular geometry of many forms found in nature by allowing them to have non-integer dimensions. As a Euclidean line is characterized by a dimension of 1, a fractal with a dimension $D = 1.6$ would be more irregular than one with $D = 1.4$, and both of these would be more irregular that the Euclidean line. Similar statements can be made about the irregularity of plane surfaces: those fractal forms with $D > 2$ would consist of mountains and valleys that make them less smooth than the Euclidean plane with $D = 2$.

In the Porod approximation, small angle scattering from a monodisperse size distribution of particles with homogeneous scattering length density embedded in a matrix of similarly homogeneous scattering length density should fall off as Q^{-4} for $Q > 1/r_{min}$, the shortest dimension of the particle. Instead, what has often been observed is scattering that decreases as Q^{-x} where $x < 4$. This relationship is often observed over many decades in the magnitude of Q.[62] Calculation of the scattering from model fractal forms has shown that the intensity of small angle scattering should be given by $I \propto Q^{-D}$, where D is the Hausdorf dimension of the fractal.[67]

To clarify the sources of SANS from cement pastes and to develop a tool for the analysis of cement microstructures, Allen and coworkers examined the time-resolved SANS from a series of cement paste specimens from mixing until 28 days after hydration.[63] Using the D11 instrument at the Institute Laue Langevin, it was possible to obtain data over a particularly large Q range (0.0012–0.15 Å^{-1}) with a wavelength of 8 Å and several different detector sample distances (1.78, 4.78, 9.78, and 19.78 m). The scattered intensity was sufficiently large that data could be acquired in a few minutes for the early hydration times. Several different kinds of cement pastes were studied: OPC, $CaCl_2$ accelerated OPC, and OPC with various mineral admixtures. In their model of the cement microstructure, they propose that the SANS is caused by scattering from fractal agglomerates of small C-S-H globules. The measured SANS cross-section data was fit to a semi-empirical functional form that includes effects due to a distribution of sizes for the (assumed) spherical scattering objects and the scale-invariant fractal nature of the agglomerates formed by clusters of these particles:[68]

$$\frac{d\Sigma}{d\Omega} = \left\{ \frac{A}{\left(1 + Q^2 \xi_{max}^2\right)^{(D-1)/2}} \times \frac{\sin\left[(D-1)\arctan\left(Q\xi_{max}\right)\right]}{(D-1)Q\xi_{max}} + B \right\}$$

$$\times \left\{ \frac{3\left[\sin\left(QR_{min}\right) - QR_{min}\cos\left(QR_{min}\right)\right]}{\left(QR_{min}\right)^3} \right\}^2 + \frac{C}{Q^P} \tag{24}$$

In the macroscopic cross section, Eq. 24, the first term in the first bracket models the larger-scale and scale-invariant fractal components of the system. The second (constant) term, B, ensures that the scattering will be dominated by the form factor of the constituent particles (here spheres) at high Q. The second bracket is the form factor [$F(Q)$] for scattering from spherical objects. The Porod scattering is modeled by C/Q^P. The quantity D is the Hausdorf dimension of the fractal objects, R_{min} is the radius of the component particles of the fractal, and ξ_{max} is the correlation length for which the fractal agglomerates remain scale-invariant.

Figure 20 shows a time sequence of SANS intensity measurements from the experiment by Allen and coworkers[63] for w/c = 0.4 OPC as a log-log plot of $Q^4(d\Sigma/d\Omega)$ vs. Q. In this graph, Q^{-4}-Porod scattering shows up as a horizontal line while other powers of Q appear as straight lines. SANS from the anhydrous cement powder exhibits Q^{-4}-Porod scattering from the surfaces of the cement grains with some evidence (from the rise of the curve at high-Q) of 50 nm features. These are probably the result of a small amount C-S-H gel formation on the cement grain surfaces from the moisture in the air. The value of the surface area obtained from this measurement was 690 m^2/kg, a reasonable value for anhydrous cement powder. On mixing with water, there is a slight reduction in the Q^{-4} scattering, which is believed to be due to the softening of the cement grain boundaries. After the end of the induction period, the SANS intensity at high Q increases dramatically, indicating a growing population of 5 nm globules. Analysis of the high-Q Porod scattering yields a rise in the specific surface area of the paste to more than 100 times the value of original surface areas of the clinker grains. At low Q, the SANS intensity decreases with time, caused by the shrinking of the unhydrated cement grains. In the end, the SANS from 28-day-old cement paste can be characterized by a model that contains only the shrunken unhydrated cement grains and the 5 nm cement particles, agglomerated up to a correlation length of 40 nm.[63] Subsequent experiments by the NIST group have extended the fractal interpretation of SANS

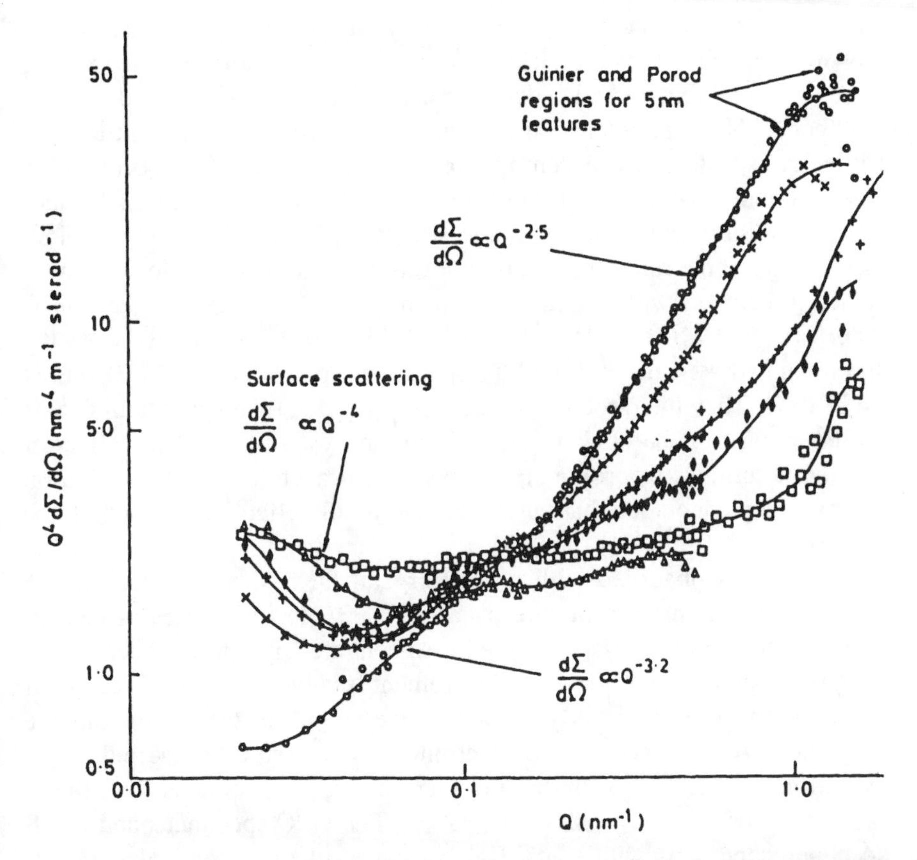

Figure 20. Log-log plots of $Q^4(d\Sigma/d\Omega)$ vs. Q for hydrating 0.4 w/c OPC. Note the change from Q^{-4} Porod scattering for the anhydrous powder to $Q^{-2.5}$ and $\sim Q^{-3}$ power laws for the 28-day hydrated cement. The lines are guides to the eye (Fig. 2 from Ref. 63).

from hydrating cement pastes and applied it to measurements of the effect of silica fume on the development of the volume and surface fractal components of the cement microstructure.[64,65]

An alternative interpretation of SANS from hydrated cement pastes has been proposed. Sabine has put forward an exact solution for the scattering from a distribution of spherical objects of random size that has a particularly simple form.[69] Sabine's function was used to interpret the SANS over a Q range of $0.0056\ \text{Å}^{-1} < Q < 0.21\ \text{Å}^{-1}$, which is similar to that investigated in the experiments by Allen.[63] The result of this experiment was the conclusion that the SANS was due to a globular entity with a diameter of 70 nm. The growth of this feature is correlated with the heat evolution from the reacting paste. A second feature with a diameter near 50 nm is also identified.[70]

Other SANS experiments on cementitious systems have concentrated on particular aspects of the system microstructure. Using a DKS spectrometer, Haussler and coworkers could access very small Q values where the Guinier approximation would hold for large structural features.[71] In this fashion, they measured the radius of gyration, R_g, as a function of time for hydrating cement specimens with w/c ratios between 0.27 and 0.5. At $t = 7$ days, R_g is 900, 1600, and 2100 nm for w/c = 0.27, 0.38, and 0.5, respectively. As the specimen hydrates, the R_g for all the specimens approaches $R_g = 2100$ nm at $t = 28$ days. They interpret this increase as being due to the growth of C-S-H on the surface of the clinker grains. From an examination of the effect of organic solutions (isoproponal, methanol, and acetone) on the hydration process, they concluded that these were not totally effective in stopping the reactions.

The objects observed with perfect-crystal DKS spectrometers are so large that these instruments are usually unsuitable for studies of cement microstructure. If the crystals in the instrument are appropriately bent, however, the angular resolution of the instrument is substantially reduced and it can be used to observe SANS in a Q range similar to that of the pinhole instrument. A bent-crystal DKS spectrometer was used to measure the time evolution of SANS from hydrating OPC and C_3S specimens over the Q range 5×10^{-3} Å^{-1} $< Q < 2.0 \times 10^{-2}$ Å^{-1}.[72] The OPC specimens and a C_3S specimen containing with 1 wt% $CaCl_2$ as an accelerant exhibited a dramatic increase of SANS intensity that began 25–50 h after mixing and cannot easily be understood in terms of the generally accepted time dependence of the cement hydration reactions. Hydration with 100% D_2O, 8% D_2O + 92% H_2O, and 66% D_2O + 34% H_2O was used to examine key regions of scattering contrast (see Fig. 19) as a means of understanding the origin of this scattering. The 66% D_2O + 34% H_2O mixture has a neutron scattering length density very nearly the same as that of anhydrous C_3S and not very different from the C-S-H gel density. This mixture greatly reduces the contrast, and hence the scattering from the interface between C_3S, C-H-S, and the pore water. The 8% D_2O + 92% H_2O solution is known as a null mixture, as it has a neutron scattering length density close to zero — essentially the same as the neutron scattering length density of air. In this manner, it was possible to demonstrate that the dramatic increase of SANS intensity was due to the formation of bubbles within the pore structure, presumably caused by self-desiccation.

The major motivation for SANS investigations of cement pastes has been the characterization of the C-S-H microstructure. As was described earlier, identification of the components of the two-phase model and proper determination of their scattering length density is a crucial step in that effort. If the contrast is misidentified, the density and surface area of the scattering objects cannot be correctly determined. A recent experiment has addressed this issue by using contrast variation. The specimens were 28-day-old OPC cement pastes with w/c = 0.35. In advance of the SANS measurements, each of the specimens was soaked in different D_2O-H_2O mixtures. The strength of the scattering could then be compared with that calculated from different models of the C-S-H gel. Since the microstructure of all of the specimens is the same, the only variable is the neutron contrast. The best fit to the data was to the neutron scattering length density corresponding to 11% RH C-S-H ($C_{1.7}SH_{1.4}$). The authors note that this contrast value is 21% lower than that customarily used ($C_{1.5}SH_{2.5}$) and will increase the previously reported surface areas measured by SANS by 21%.

Summary and Future Developments

The various properties of the neutron described in the first part of this chapter form the basis for the utility of neutron methods, but it is clear that the great attribute of neutrons as a probe for cementitious materials is that they "see" hydrogen. In diffraction experiments, they are sensitive to the position and orientation of (heavy) water molecules and OD ions. After all, the thing that is important about cement is what happens when it is mixed with water, and it is the physical/chemical action of the hydration that leads to the important engineering properties of concrete. Neutron diffraction experiments can now be used to study reaction pathways, hydration/dehydration effects, impurity effects, and thermal and chemical decomposition pathways. The results of these experiments can significantly deepen our understanding of the chemistry of the cement paste system.

Neutrons also "see" the motion/chemical state of hydrogen. Neutrons scattering inelastically from $Ca(OH)_2$ are "marked" by their 41 meV energy transfer, which is characteristic of the OH ion vibration, and their intensity is proportional to the quantity of that compound. QNS experiments on hydrating cement pastes at NIST and at MURR illustrate how neutron energy spectroscopy can differentiate between hydrogen in water molecules freely diffusing in the cement paste and hydrogen bound to the cement hydration compounds. The results of these experiments challenge the results of X-ray and Raman scattering chemical kinetics investigations of cement hydration in a significant way. The identification of C_3S hydration as being "two-dimensional diffusion-controlled growth at zero nucleation rate" in the acceleratory period and "one-dimensional diffusion-controlled growth at zero nucleation rate" in the post-acceleratory period[41] is not consistent with the neutron experiments.[35,39,50]

The experiments on small angle scattering are particularly noteworthy because of the importance of cement microstructure to its engineering properties. The fractal interpretation of SANS data from cements has provided a new language that can be used to characterize the complexity of the microstructure in a meaningful way. These experiments now provide the means to differentiate between different kinds of cement additives on the basis of their microstructural effects.

In spite of the fact that neutron scattering experiments on cementitious systems began nearly 20 years ago, it is only in the last few years that the results of these investigations have begun to make significant additions to our understanding of cement paste. What is really exciting is that the opportunities for the investigation of cementitious systems are by no means exhausted by the experimental efforts described above.

- The description of water in a hydrating system as bound or free oversimplifies its real condition. Higher-resolution QNS measurements of a hydrating paste — experiments that will become possible with improved instrumentation (under construction)[73] — could uncover intermediate conditions of mobility: water trapped in small gel pores or diffusing through a fresh C-S-H gel.
- With the current generation of neutron beam sources and instruments, it is necessary to substitute D for H in order to examine cement hydration using diffraction. A polarized beam and a suitable neutron-polarizing filter combined with a position-sensitive detector would make possible the examination of hydrating hydrogenous (as opposed to deuterated) systems.[§] Experiments employing such devices would allow the direct investigation of the cement hydration isotope effect and reopen diffraction measure-

[§]A discussion of the effect of neutron polarization on the neutron-hydrogen scattering cross-section can be found in Ref. 5.

ments as a tool for monitoring cement hydration kinetics. The practical neutron polarizing filter is currently under development.[74]

- The destructive effect (and economic impact) of water freezing in cement pores is well known. Neutrons may be able to study this problem using the $Ca(OH)_2$ crystals embedded in C-S-H as strain gauges. As the temperature of a cement paste is lowered, pressure from freezing pore water will exert pressure on the crystalline components of the paste.[75] Lattice parameter changes, easily measured by neutron diffraction, are a signature of this internal pressure. Such experiments would be analogous to the use of neutron diffraction measurements to quantify the residual stress in engineering (metallic) materials.[76] These effects might even be profiled as a function of depth within a macroscopic specimen, accessing the different stress environments in a hardened cement paste as a function of distance from the surface.
- It is well known that chemical reactions between a hydrating cement paste and the aggregate that composes concrete can unfavorably impact the strength and durability of concrete. This effect, the alkali-silica reaction (ASR), is the subject of much current research (see Ref. 2, p. 361). It may be possible to study this phenomenon using neutron reflectometry.[77] Neutrons reflected (by mirror reflection) from a flat, polished surface probe the neutron scattering length density of the materials at the mirror interface to a depth of 10–1000 Å. Neutron reflectometry with a synthetic aggregate material in contact with a hydrating cement might be used as a tool to probe the chemical reactions that occur at the interface between the aggregate and the cement.

The experimental investigations of cementitious systems using neutrons that are described here demonstrate a significant contribution to the understanding of cement, cement hydration, and the properties of the hardened cement paste. As the speculations above demonstrate, there remain many opportunities for additional work.

Acknowledgments

The author wishes to thank Andrew J. Allen, Herma G. Buttner, Brigitte Aubert, A. Norland Christensen, S. A. FitzGerald, H. Fjellvag, Ken Herwig, Jim Richardson, Dietmar Schwahn, H. G. Smith, and Terry Udovic for help in locating figures from previous publications and for providing some of the drawings used here. The author gratefully thanks Andrew J. Allen, Hamlin

M. Jennings, and R. A. Livingston for sending reprints and preprints of their work on cementitious systems that have been used to prepare this article. Finally, the author extends his thanks to W. B. Yelon, M. Popovici, F. Ross, and Jan Skalny for reading the text and advancing many helpful suggestions.

References

1. F.M. Lea, *The Chemistry of Cement and Concrete.* Chemical Publishing Company, Inc., New York, 1971.
2. H.F.W. Taylor, *Cement Chemistry,* 2nd edition. Thomas Telford, London, 1997.
3. J. Chadwick, "Possible Existance of a Neutron," *Nature,* **129**, 312 (1932).
4. C.G. Windsor, *Pulsed Neutron Scattering.* Taylor and Francis Ltd., 1981.
5. G. Bacon, *Neutron Diffraction.* Clarendon Press, Oxford, 1975.
6. Stephen W. Lovesey, *Theory of Neutron Scattering from Condensed Matter.* Clarendon Press, Oxford, 1984.
7. G.L. Squires, *Thermal Neutron Scattering.* Cambridge University Press, London, 1977.
8. Varley F. Sears, *Neutron Optics.* Oxford University Press, New York, 1989.
9. G. Kostorz (ed.), *Treatise on Materials Science and Technology,* Vol. 15. Academic Press, New York, 1979.
10. David L. Price and Kurt Skold (eds.), *Neutron Scattering Parts A, B, C,* Methods of Experimental Physics 23. Academic Press, San Diego, 1987.
11. B.E. Warren, *X-Ray Diffraction.* Addison Wesley Publishing, Reading, Massachusetts, 1969. Chapter 12.
12. Varley F. Sears, "Neutron Scattering Lengths and Cross Sections," *Neutron News,* **3**, 26–37 (1992).
13. J.D. Jorgensen, J. Faber, J.M. Carpenter, R.K. Crawford, J.R. Haumann, R.L. Hitterman, R. Kleb, G.E. Ostrowski, F.J. Rotella, and T. G. Worlton, "Electronically Focused Time-of-Flight Powder Diffractometers at the Intense Pulsed Neutron Source," *J. Appl. Cryst.,* **22**, 321–333 (1989).
14. A. Norlund Christensen and M. S. Lehmann, "Rate of Reactions between D_2O and $Ca_xAl_yO_z$," *J. Solid State Chem.,* **51**, 196–204 (1984).
15. A. Norlund Christensen, H. Fjellvag, and M.S. Lehmann, "A Time Resolved Powder Neutron Diffraction Investigation of Reactions of Portland Cement Components with Water," *Acta. Chemica Scandinavica,* **A39**, 593–604 (1985).
16. A. Norlund Christensen, H. Fjellvag, and M.S. Lehmann, "The Effect of Additives on the Reaction of Portland Alumina Cement Components with Water. Time Resolved Powder Neutron Diffraction Investigations," *Acta. Chemica Scandinavica,* **A40**, 126–141 (1986).
17. R.A. Young (ed.), *The Rietveld Method.* Oxford University Press, 1993.
18. H.M. Rietveld, "A Profile Refinement Method for Nuclear and Magnetic Structures," *J. Appl. Cryst.,* **2**, 65–71 (1969).
19. A.C. Larson and R.B. Von Dreele, "GSAS-General Structural Analysis System." Los Alamos National Laboratory Report LAUR 86-748. 1994.
20. R. Berliner, C. Ball, and Presbury B. West, "Neutron Powder Diffraction Investigation of Model Cement Compounds," *Cem. Concr. Res.,* **27**, 551–575 (1997).

21. R. Berliner, C. Ball, and Presbury B. West, "Reply to the Discussion of the Paper 'Neutron Powder Diffraction Investigation of Model Cement compounds,'" *Cem. Concr. Res.,* **28**, 1833–1836 (1998).
22. R. Berliner, C. Ball, and Presbury B. West, "Neutron Powder Diffraction Studies of Portland Cement and Cement Compounds"; pp. 487–492 in *Neutron Scattering in Materials Science II,* Materials Research Society Symposium Proceedings vol. 376. Edited by D.A. Neumann, T.P. Russell, and B.J. Wuensch. 1995.
23. "Report of Investigation, Reference Materials 8486, 8487, 8488." National Institute of Standards and Technology, May 22, 1989.
24. R. Berliner, F. Trouw, and H. Jennings, "Neutron Diffraction Measurement of the Hydration Kinetics of Tricalcium Silicate," *Bull. Am. Phys. Soc.,* **40**, 665 (1995).
25. T.C. King, C.M. Dobson, and S.A. Rodger, "Hydration of tricalcium silicate with D_2O," *J. Mater. Sci. Lett.,* **7**, 861–863 (1988).
26. Takashi Ogura and Seishi Goto, "Hydration of Cement with Heavy Water," *Semento Konkurito Ronbushu,* **46**, 116–121 (1992).
27. Jeff Thomas, private communication, 1996.
28. R. Berliner, "The Structure of Ettringite"; in *Materials Science of Concrete, The Sidney Diamond Symposium.* Edited by Menashi Cohen, Sidney Mindess, and Jan Skalny. American Ceramic Society, Westerville, Ohio, 1998.
29. A.E. Moore and H.F.W. Taylor, "Crystal Structure of Ettringite," *Acta. Cryst.,* **B26**, 386–393 (1970).
30. R. Berliner, M. Popovici, K.W. Herwig, M. Berliner, H.M. Jennings, and J.J. Thomas, "Quasielastic Neutron Scattering Study of the Effect of Water to Cement Ratio on the Hydration Kinetics of Tricalcium Silicate," *Cem. Concr. Res.,* **28**, 231–243 (1998).
31. R.A. Livingston, D.A. Neumann, and J.J. Rush, "Quasielastic Neutron Scattering Study of the Dynamics of Water in Curing Cement"; in *NIST Reactor: Summary of Activities, July 1990 through June 1991.* NIST Technical Note 1292.
32. "Hydrogen in Solids"; in *NIST Reactor: Summary of Activities, October 1992 through September 1993.* NISTIR 5362.
33. "Chemical Physics of Materials"; in *NIST Reactor: Summary of Activities, October 1993 through September 1994.* NISTIR 5594.
34. M. Popovici, K.W. Herwig, R. Berliner, W.B. Yelon, and L. Groza, "High Resolution Neutron Scattering with Bent Monochromators Made of Commercial Silicon Wafers," *J. Neutron Res.,* **7**, 107–117 (1999).
35. R. Berliner, M. Popovici, K. Herwig, H.M. Jennings, and J. Thomas, "Neutron scattering scattering studies of hydrating cement pastes," *Physica,* **B241–243**, 1237–1239 (1998).
36. J.R.D. Copley and T.J. Udovic, *J. Res. Nat. Inst. Stand. and Technol.,* **98**, 71 (1993).
37. R.A. Livingston, D.A. Neumann, Andrew A. Allen, and J.J. Rush, "Application of Neutron Scattering Methods to Cementitious Materials"; pp. 459–469 in *Neutron Scattering in Materials Science II,* Materials Research Symposium Proceedings vol. 376. Edited by Dan A. Neumann, Thomas P. Russell and Bernhardt J. Wuensch. 1995.
38. R.A. Livingston, D.A. Neumann, S.A. FitzGerald, and J.J. Rush, "Quasielastic neutron scattering study of the hydration of tricalcium silicate"; p. 148 in *Neutrons in Research and Industry,* SPIE Proceedings, Vol. 2867. Edited by G. Vourvopoulos. 1997.
39. S.A. FitzGerald, D.A. Neumann, J.J. Rush, D.P. Bentz, and R.A. Livingston, "An in-situ quasielastic neutron scattering study of the hydration reaction in tricalcium silicate," *Chem. Mater.,* **10**, 397 (1998).

40. Paul Wencil Brown, James M. Pommersheim, and Geoffrey Frohnsdorff, "Kinetic Modeling of Hydration Processes"; pp. 245–260 in *Cements Reserarch Progress 1983*. Edited by J. Francis Young. 1983.
41. Paul Wencil Brown, James Pommersheim, and Geoffrey Frohnsdorff, "A Kinetic Model for the Hydration of Tricalcium Silicate," *Cem. Concr. Res.,* **15**, 35–41 (1985).
42. Melvin Avrami, "Kinetics of Phase Change I," *J. Chem. Phys.,* **7**, 1103–1112 (1939).
43. Melvin Avrami, "Kinetics of Phase Change II," *J. Chem. Phys.,* **8**, 212–224 (1940).
44. Melvin Avrami, "Kinetics of Phase Change III," *J. Chem. Phys.,* **9**, 177–184 (1940).
45. R. Kondo and S. Ueda, *V Intl. Sym. Cement II,* Tokyo, 1968. P. 203.
46. I. Odler and J. Schuppstühl, "Early Hydration of Tricalcium Silicate III. Control of the Induction Period," *Cem. Concr. Res.,* **11**, 765–774 (1981).
47. M. Tarrida, M. Madon, B. Le Rolland, and P. Colembet, *Adv. Cem. Bas. Mater.,* **2**, 15–20 (1995).
48. K. Fuji and W. Kondo, "Kinetics of the Hydration of Tricalcium Silicate," *J. Am. Ceram. Soc.,* **57**, 492 (1974).
49. D.L. Gress, T. El-Korchi, R.A. Livingston, D.A. Neumann, and J.J. Rush, "Quantifying freezable water in portland cement paste using quasielastic neutron scattering"; pp. 493-498 in *Neutron Scattering in Materials Science II,* Materials Research Society Proceedings, Vol. 376. Edited by Dan A. Neumann, Thomas P. Russell, and Bernhardt J. Weunsch. 1995.
50. S.A. FitzGerald, D.A. Neumann, J.J. Rush, R.J. Kirkpatrick, X. Cong, and R.A. Livingston, "Inelastic Neutron Scattering Study of the Hydration of Tricalcium Silicate," *J. Mater. Res.,* **14** (1999).
51. A. Guinier and G. Fournet, *Small Angle Scattering of X-Rays.* Trans. by C. B. Walker. Wiley, New York, 1955.
52. A. Guinier, *X-Ray Diffraction.* Trans. by P. Lorrain and D. Lorrain. Freeman, San Francisco, 1963.
53. O. Glatter and O. Kratky, *Small Angle X-Ray Scattering.* Academic Press, London (1982).
54. Jeffrey J. Thomas, Hamlin M. Jennings, and Andrew J. Allen, "The Surface Area of Hardened Cement Paste as Measured by Various Techniques," *Concr. Sci. Eng.,* **1**, (1999).
55. D. Schwahn, G. Meier, and T. Springer, *J. Appl. Cryst.,* **24**, 568–570 (1991).
56. J.S. Lin, C.R. von Bastian, and P.W. Schmidt, "A modified method for slit-length collimation corrections in small-angle X-ray scattering," *J. Appl. Cryst.,* **7**, 439 (1974).
57. A.J. Allen, C.G. Windsor, V. Rainey, D. Pearson, D.D. Double, and N. McN. Alford, "A small angle neutron scattering study of cement porosities," *J. Phys. D: Appl. Phys.,* **15**, 1817–1833 (1982).
58. C.G. Vonk, "On two methods for determination of particle size distribution functions by means of small-angle X-ray scattering," *J. Appl. Cryst.,* **9**, 433 (1976).
59. D. Pearson, A. Allen, C.G. Windsor, N. McN. Alford, and D. Double, "An investigation of the nature of porosity in hardened cement pastes using small angle neutron scattering," *J. Mater. Sci.,* **18**, 430–438 (1983).
60. D. Pearson and A.J. Allen, "A study of ultrafine porosity in hydrated cements using small angle neutron scattering," *J. Mater. Sci.,* **20**, 303–315 (1985).

61. Douglas M. Winslow, "The Fractal Nature of the Surface of Cement Paste," *Cem. Concr. Res.,* **15**, 817–824 (1985).
62. V.M. Castano, P.W. Schmidt, and H.G. Hornis, "Small-angle scattering studies of the pore structure of polymer-modified portland cement pastes," *J. Mater. Res.,* **5**, 1281–1284 (1990).
63. A.J. Allen, R.C. Oberthur, D. Pearson, P. Schofield, and C.R. Wilding, "Development of the fine porosity and gel structure of hydrating cement structures," *Phil. Mag.,* **56**, 263–288 (1987).
64. A.J. Allen, "Time-resolved phenomena in cements, clays and porous rocks," *J. Appl. Cryst.,* **24**, 624–634 (1991).
65. Andrew J. Allen and Richard A. Livingston, "Relationship between differences in silica fume additives and fine scale microstructural evolution in cement based materials," *Adv. Cem. Bas. Mat.,* **8**, 118–131 (1998).
66. B.B. Mandelbrot, *The Fractal Geometry of Nature.* W.H. Freeman, New York, 1983.
67. D.W. Schaefer and K.D. Keefer, "Fractal geometry of silica condensation polymers," *Phys. Rev. Lett.,* **53**, 1383 (1984).
68. S.K. Sinha, T. Freltoft, and J. Kjems, "Observation of power-law correlations in silica-particle aggregates by small-angle neutron scattering"; pp. 97–90 in *Kinetics of Aggregation and Gelation.* Edited by F. Family and D. Landau. North Holland, Amsterdam, 1984.
69. T.M. Sabine, W.K. Bertram, and L.P. Aldridge, "A method for interpreting small angle neutron scattering from quasi-spherical objects"; pp. 499–504 in *Neutron Scattering in Materials Science II,* Materials Research Society Proceedings, Vol. 376. Edited by Dan A. Neumann, Thomas P. Russell, and Bernhardt J. Weunsch. 1995.
70. L.P. Aldridge, W.K. Bertram, T.M. Sabine, J. Bukowski, J.F. Young, and R.K. Keenan, "Small-angle neutron scattering from hydrated cement pastes"; pp. 471–479 in *Neutron Scattering in Materials Science II,* Materials Research Society Proceedings, Vol. 376. Edited by Dan A. Neumann, Thomas P. Russell, and Bernhardt J. Weunsch. 1995.
71. F. Häubler and H. Baumbach, "Structural studies on hydrating cement pastes," *J. Physique IV,* **C8**, 269–372 (1993).
72. R. Berliner, B.J. Heuser, and M. Popovici, "Investigation of portland cement and pure C_3S using 1-D SANS"; pp. 481–486 in *Neutron Scattering in Materials Science II,* Materials Research Society Proceedings, Vol. 376. Edited by Dan A. Neumann, Thomas P. Russell, and Bernhardt J. Weunsch. 1995.
73. P.M. Gehring and D.A. Neumann, "Backscattering spectroscopy at the NIST Center for Neutron Research," *Physica,* **B241–243**, 64–70 (1998).
74. W. Heil, K. Andersen, D. Hofmann, H. Humblot, J. Kulda, E. Lelievre-Berna, O. Scharpf, and F. Taset, "^{3}He neutron spin filter at ILL," *Physica,* **B241–243**, 56–64 (1998).
75. Richard Livingston, private communication.
76. M.T. Hutchings, "Neutron diffraction measurement of residual stress fields: Overview and points for discussion"; in *Mesurement of Residual and Applied Stress Using Neutron Diffraction.* Edited by Michael T. Hutchings and Aaron D. Krawitz. Kluwer Academic Publishers, Dordrecht, 1991.
77. J. Penfold and R.K. Thomas, "The application of the specular reflection of neutrons to the study of surfaces and interfaces," *J. Phys. Condens. Matter,* **2**, 1369–1412 (1990).

Scanning Probe Microscopy: A New View of the Mineral Surface

Christopher Hall
Centre for Materials Science and Engineering, University of Edinburgh, Edinburgh, United Kingdom

Dirk Bosbach*
Institut für Mineralogie, Universität Münster, Münster, Germany

"Ye must have faith. It is a quality which the scientist cannot dispense with."
— Max Plank

We review the use of scanning probe microscopy (and especially atomic force microscopy) as a new research tool in mineral surface chemistry with potential for significant application in cement chemistry. Real-time observations of dissolution and crystal growth processes on the surfaces of minerals such as calcite and gypsum are described. We emphasize how these processes are modified by additives and impurities. The value of well-designed flow experiments is noted and we propose several novel experimental designs for future work on cement materials.

Introduction

Scanning probe microscopy (SPM) is a new tool for materials scientists, and one that has already yielded spectacular information on chemical processes on mineral surfaces. In the last few years some 70 publications have appeared reporting SPM studies of minerals. Only a handful of these come from cement and concrete research groups, but we believe that SPM methods will prove to be valuable as a means of studying the reactivity and surface chemistry of cement materials. In this chapter, we describe these methods and what has been achieved so far.

In SPM, images (Fig. 1) are formed in a manner that has nothing to do with the physics of either light microscopy or electron microscopy. It is well known that in the early 1980s, Binnig and Rohrer working at IBM in Zurich developed the scanning tunneling microscope (STM) and showed extraordinary images of metallic and semiconductor surfaces at atomic resolution.[1–3]

All SPM methods are derived from the original STM concept and all comprise a local probe, a "tip" of nanometer size, which is moved across

*Currently with Institut für Nukleare Entsorgung, Forschungszentrum Karlsruhe, Germany.

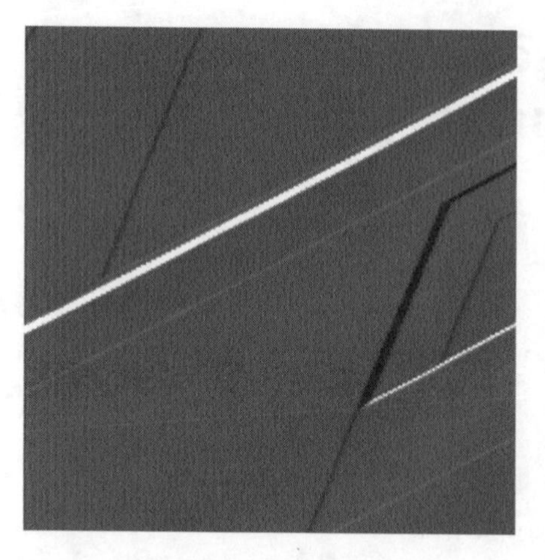 

Figure 1. AFM images of dissolving gypsum. The (010) surface of gypsum imaged under water during active dissolution: a pair of images taken 6 min apart, showing the development of etch features. The field of view is 10 × 10 μm. Single steps are about 0.8 nm high, corresponding to a single layer in the crystal structure (*b*/2). The multiple step running NE to SW is 5 nm in height.[4]

the surface, raster-fashion, to record some local property of that surface. The practical realization of the technique depends above all on the existence of high-performance piezo-drives ("scanners"), which can be moved digitally with extreme precision. A *z* piezo controls the motion of the tip normal to the surface to be imaged and can accurately and rapidly produce displacements as small as 0.01 nm, usually within a feedback loop; *x* and *y* scanners accomplish the rastering parallel to the surface. For atomic and molecular scale imaging, *x* and *y* step sizes can also be set as small as 0.01 nm to collect, say, a 400 × 400 pixel image of an area 4 × 4 nm. However step sizes can be much greater, for instance, 100 nm to image fields as large as 25 × 25 μm. Furthermore, the step rate can be as high as 5 kHz , so that such images can be acquired in a few tens of seconds.

In STM, as originally developed, the local probe is the tip of a metallic wire, sometimes etched after cutting. The local measurement is of the tunneling current produced in response to a potential difference imposed between the tip and the surface. The original experiments were carried out in high vacuum. Mapping the tunneling current provided detailed information on the electronic band structure of the solid substrate. It proved possible to map this with atomic resolution and hence to observe the arrangement of surface atoms (including point defects) and to monitor changes, for

Figure 2. STM image of the (111) surface of the semiconductor iron oxide mineral magnetite. The left image shows two kinds of elementary steps in an image field of 150 × 150 nm. The right image shows atomic resolution of the hexagonal surface lattice with vacancies, image field 6 × 6 nm (P.W. Murray).

example, during oxidation. Figure 2 shows a recent STM image of the surface of the iron oxide mineral magnetite.[5]

Only a few years after STM, atomic force microscopy (AFM), or scanning force microscopy (SFM), was invented. The landmark paper was published in 1986 by Binnig, Quate, and Gerber.[6] In AFM, the local probe is a hard tip, usually of silicon nitride or silicon, and (in the commonest forms of AFM) it is the elevation of the surface that is measured. The tip is integrated into the apex of a flexible and elastic cantilever that bends under the action of the short-range forces acting between tip and surface (Fig. 3). A laser beam reflected from the cantilever detects the deflection and in one common mode of operation the sample z position is continuously adjusted to maintain constant deflection. The tip therefore maps out the surface topography as it is rastered across the sample. Image formation in AFM does not depend on the electrical conductivity of the sample and AFM is equally applicable to conductors and insulators. It is this form of SPM that is therefore most widely applied to minerals. Furthermore, AFM works excellently in water and other transparent liquid media so that we can observe processes such as dissolution and crystal growth directly and in

Figure 3. An AFM imaging probe, scale bar 5 μm (A. Murray, Topometrix Corporation).

real time. For reviews of SPM in mineral surface chemistry, see Refs. 7–9.

Many other SPM methods have been proposed and demonstrated, involving numerous physical interactions between probe and surface. A select few have been incorporated into commercial instruments. One that will probably prove useful in studies of mineral surfaces is the near-field scanning optical microscope (NSOM).[10] The local probe (commonly in the form of an optical fiber drawn down and coated to a submicron aperture) collects light from a small region of the sample surface. A variety of detection and excitation methods can be coupled into the NSOM, including absorption and Raman and fluorescence spectroscopies. In thermal scanning probe microscopy (SThM),[11] the tip is used as a combined heat source/ temperature sensor (peltier junction) and is used to sense the thermal properties of the surface (and also the immediate subsurface, via modulated heating). For polyphasic mineral surfaces (cement clinkers and rocks) this may provide an excellent means of phase identification, since thermal properties of crystalline mineral phases are well defined and characteristic. Finally, scanning electrochemical microscopy[12–14] incorporates the hardware of an STM into an electrochemical cell, in which the imaged surface

Table I. The principal SPM techniques

Method	Acronym	Date
Scanning tunneling microscopy	STM	1982
Atomic force microscopy	AFM	1986
Scanning force microscopy	SFM	1986
Near-field scanning optical microscopy	NSOM	1984
Scanning near-field optical microscopy	SNOM	1984
Magnetic force microscopy	MFM	1987
Thermal probe microscopy	SThM	1988
Scanning electrochemical microscopy	SECM	1989

is the working electrode. The electrical currents or changes in electrical potential that are detected arise from electrolytic and electrodic processes rather than vacuum tunneling. There is considerable potential for the use of SECM to investigate electrochemical (for example, redox) processes on such semiconductor mineral surfaces as cement ferrites in aqueous media.

Table I reviews the principal SPM techniques.

Calcite and Gypsum

Calcite has been a popular subject for AFM investigations of minerals. Calcite is of course a widely distributed rock-forming mineral and is excellent for examination by AFM, since it has perfect crystalline cleavage parallel to the $\{10\bar{1}4\}$ lattice plane. The surface is chemically active in contact with water and aqueous solutions, but rates are conveniently slow. Gypsum also has perfect crystalline cleavage, parallel to double sheets of water molecules in $\{010\}$ planes, and it too has been well studied by AFM. Since calcite and gypsum are both of importance in cement science, we describe the highlights of AFM work on these two minerals to illustrate the capabilities of AFM techniques.

Topography of the Mineral Surface

The simplest way to use an AFM instrument is to obtain topographic images of surfaces of interest. Indeed, most published work on AFM of minerals does only this. There are three reasons why AFM yields new information: (1) there is exquisite z resolution (routinely better than 0.1 nm), even over large x-y fields; (2) the contrast mechanisms are entirely dif-

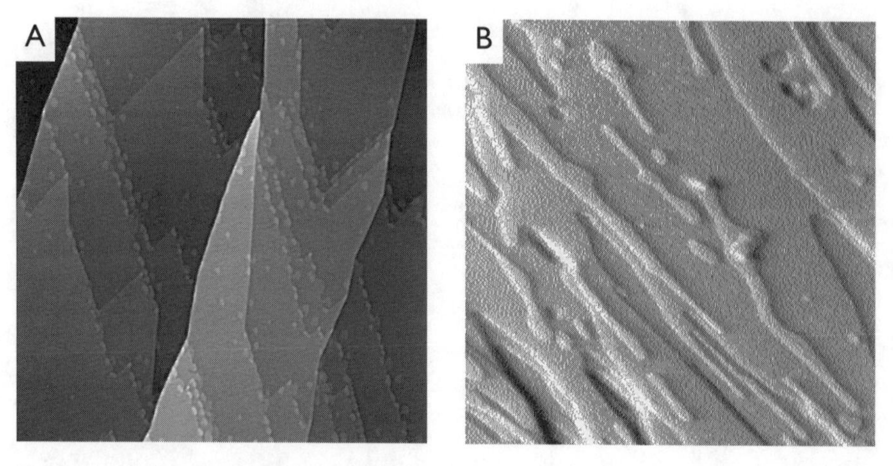

Figure 4. (a) Calcite cleavage surface in air, image field 4.5 × 4.5 μm (D. Bosbach); (b) Gypsum cleavage surface in air 6.6 × 6.6 μm.[15]

ferent from those of electron or light microscopy; and (3) the images can be formed in contact with active environments such as reactive gases, solvents, and above all water and aqueous solutions. In Fig. 4(a) we show an AFM view of a freshly cleaved surface of calcite, imaged in air. The surface clearly consists of large regions of atomically flat material, separated by shallow steps, with straight edges and some crystallographic orientation. The edges of the steps are decorated with small secondary nuclei. This is found with several of the more soluble minerals and probably results from recrystallization of high-energy debris on the surface. A readout of the z piezo motion as we traverse a few of the cleavage steps shows that the step height is about 0.3 nm. This corresponds within the accuracy of the measurement to the c dimension of the calcite unit cell. The cleavage surface of gypsum [Fig 4(b)] has a similar topography, also with elementary cleavage steps about 0.7 nm high, but here rather ragged and without any common orientation. Gypsum is a layer hydrate with hydrogen bonds acting across sheets parallel to (010). These forces are relatively weak and cleavage is energetically easy.

By reducing the x-y scan range and step size, we can make an AFM topographic image of the surface at the atomic scale (see Fig. 5 for gypsum). AFM images of crystalline mineral surfaces commonly show an appearance of this kind: typically displaying a pronounced two-dimensional periodicity, in which both period and angle are found to correspond to

Figure 5. Gypsum (010) cleavage surface in saturated aqueous solution: atomic resolution, image field 18 × 18 nm. Two kink sites are visible on the [001] step edge. Each unit in the surface layer is a sulfate ion.[16]

the known geometry of the crystal lattice. Such is the case for both calcite and gypsum. There has been much discussion of the process by which these atomic-scale images are formed and of how legitimate it is to interpret them as pictures of the atoms or ions that make up the surface. (See Refs. 17–19 for more discussion beyond our scope here.) The received view is that we do not routinely image single individual atoms in AFM, but we do measure a true atomic-scale periodicity generated by mechanical interference across a nanoscale contact area between the tip and the surface lattice. This interference and the AFM signal (e.g., the cantilever deflection) that it generates are sensitive to the surface lattice structure: its crystallographic directions and at least some translational repeat distances. So in that sense we can "see" the surface at unit cell level even in routine AFM. In some cases, it appears that we can also image individual atomic-scale defects, a capability generally regarded as the crucial test of "atomic resolution." In fact, an example is shown in Fig. 5, which shows an isolated kink site in an elementary step on the (010) surface of gypsum. Recent advances in noncontact-mode AFM under UHV conditions confirm that vacancies can be imaged[20,21] and true atom-scale features can be resolved under certain conditions.

We note that the primary AFM measurement is of the voltage applied to the z piezo drive to maintain constant tip deflection. The voltage-displacement calibration of the piezo allows us to convert this to a height map of the surface. Images may be presented as simple grayscale or false-color topographic maps or as shadowed images, mapping the slope of surface features. This display often reveals subtle but real features of the topography particularly well (e.g., Fig. 1).

Dissolution

If the newly cleaved sample of calcite is transferred to an AFM fluid cell and imaged in contact with water, then the microscopic processes of dissolution can be observed directly. Pioneers of this simple experiment[22,23] described a busy microscopic scene, where in the course of a few minutes atomic steps retreated across the field of view and where etch pits appeared and progressively enlarged. The irregular steps of the original cleavage surface were gradually transformed into a more geometrical alignment of features, which was shown to reflect the underlying crystallographic orientation of the mineral lattice. Shallow rhombic etch pits appear in the $(10\bar{1}4)$ surface with sides aligned along the directions $[\bar{4}41]$ and $[48\bar{1}]$ (Fig. 6). These pits rapidly enlarge and merge, so that the primary mechanism of dissolution is the retreat of elementary steps. The earliest of these studies established that the step velocity could be accurately measured and that it depended on crystallographic orientation. Recently, AFM measurements have been used to show how etch pits on calcite grow by preferential detachment of ions from specific sites in step edges.[24–28]

The early work on gypsum[4,16,29] was of the same kind and produced similar conclusions. Step velocities were also found to depend on crystallographic orientation, and it was found that the velocities reflected the same stability as underlay the morphology in crystal growth.[15]

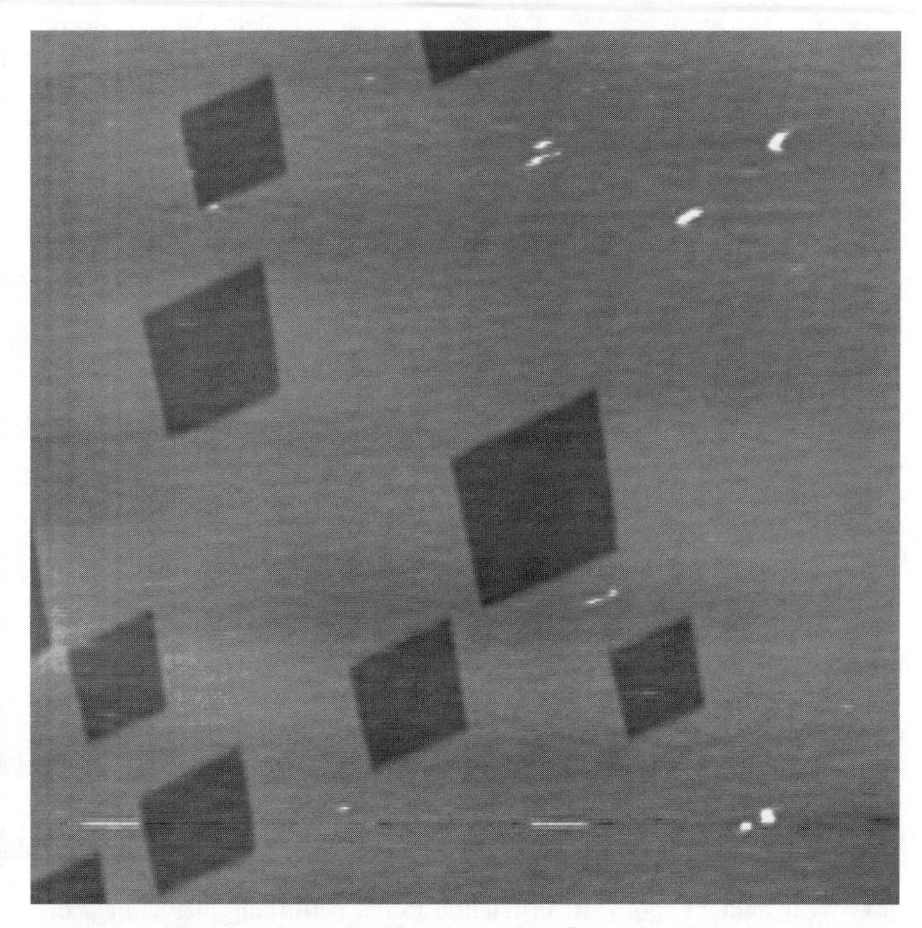

Figure 6. Rhombic etch pits formed on a calcite $(10\bar{1}4)$ surface in contact with pure water, image field 2.5 × 2.5 μm (D. Bosbach).

The ability of AFM to provide quantitative data on the kinematics of the dissolving surface is striking. AFM is therefore a powerful tool for investigating the mechanisms of chemical processes on surfaces, provided that it is combined with ancillary arrangements for controlling species transport across the surface being imaged. In the simplest liquid cells (see Ref. 30), the sample surface is in contact with a stagnant solution during imaging and the boundary condition for mineral-solution mass transfer (either for convection or diffusion) is not well defined. This was recognised by Hall and Cullen,[15] who introduced a simple modification to a standard cell that allowed steady laminar flow to be established (so that local flow velocities over the sample surface varied in direct proportion to the total flow rate). They showed that images could be obtained during flow and indeed that volume flow rates could be changed during imaging without any interference with image acquisition or impairment of image quality. A similar device was described by Jordan and Rammansee.[31] More recently, Compton and coworkers[32,33] have developed a true controlled-hydrodynamics cell with a known, if complicated, flow profile across the imaged surface and have used this to obtain absolute kinetic data for calcite dissolution under acid conditions. This work sets a new benchmark in quantitative AFM of the mineral-solution surface.

Recent work has also shown that AFM data on dissolution can support much more detailed crystallographic analysis, especially in relation to the energetics of surfaces and the role of defects such as dislocations. Thus Teng and Dove[34] used AFM to study the influence of low concentrations of sodium aspartate (an amino acid salt) on the dissolution of calcite. The etch pit geometry is completely altered, the normal rhombohedral forms being replaced by triangular pits having a different crystallographic alignment. This is evidence that the adsorption of aspartate strongly modifies the stability of particular lattice planes, probably by neutralizing the charge in dipolar surfaces. AFM studies on calcite dissolution in maleic acid[33] and phosphonic acids such as HEDP[35] also show a strong influence on etch pit morphology. Aspartate and other biomolecules are known to influence the morphology of calcite crystals and these microscopic observations provide further evidence for the close connection between dissolution and crystal growth phenomena, not least in relation to the perturbing effects of additives.

Crystal Growth

If calcite is exposed to a solution that is supersaturated rather than undersaturated with respect to calcium carbonate, we have the conditions for crystal growth. This experiment was carried out in the earliest AFM studies, also with striking results. Thus, at low supersaturations, Hillner and colleagues[22,36,37] observed the forward movement of step edges as ions from solution were integrated into the surface lattice, and described how these steps developed into spiral features (Fig. 7).

The observations were at least qualitatively consistent with the standard crystal growth model of Burton, Cabrera, and Frank[38] and its later elabora-

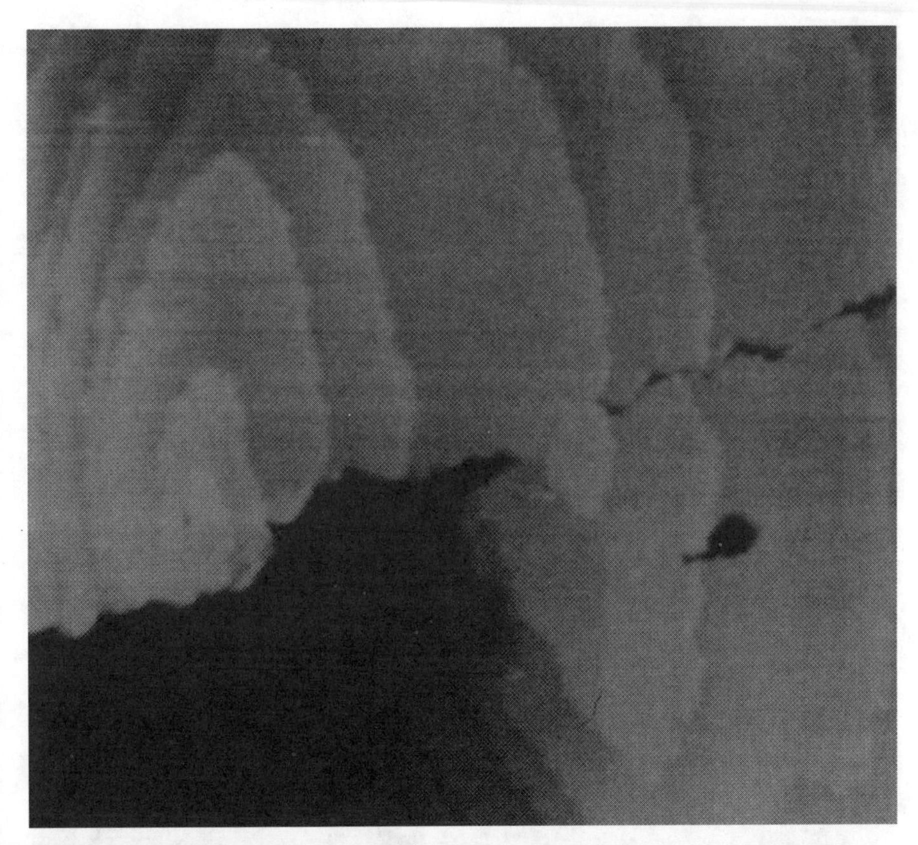

Figure 7. Growth spiral on calcite centered on a screw dislocation, image field 1 × 1 μm (Gratz93a).

tions. The origins of the growth spirals were presumed to mark the points of emergence on the surface of pre-existing screw dislocations. This is expected to be the dominant crystal growth mechanism at low supersaturations.

In the case of gypsum, spiral growth on dislocations has not been reported in AFM studies but growth by advance of existing steps occurs readily (Fig. 8). In addition, direct surface nucleation occurs rapidly above a certain threshold supersaturation (Fig. 9) and has been observed in situ.[39]

This is the "birth and spread" mechanism of the Ohara-Reid model,[40] a second mechanism by which new attachment sites are created on the surface. AFM observations of gypsum surfaces exposed ex situ to supersatu-

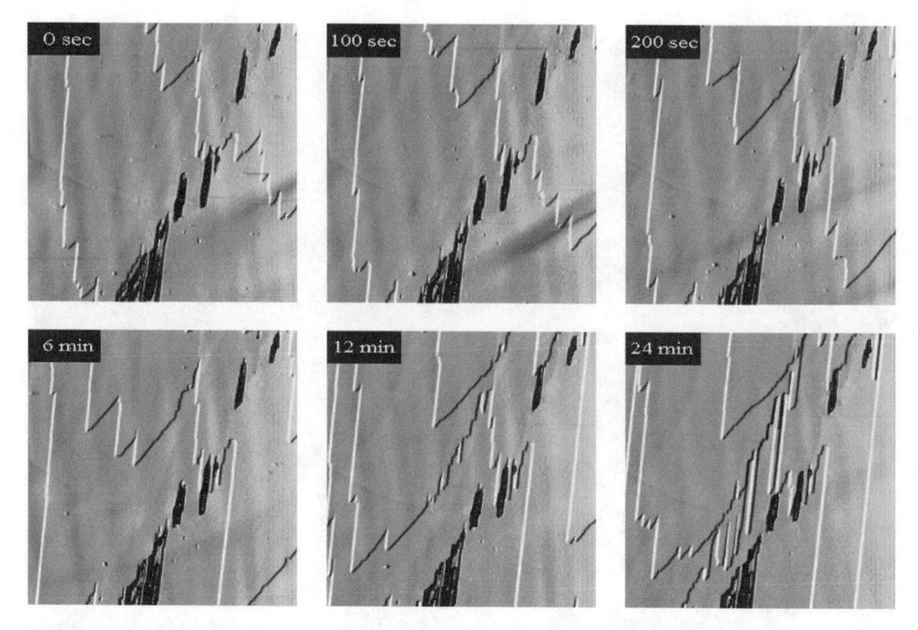

Figure 8. Crystal growth on gypsum (D. Bosbach).

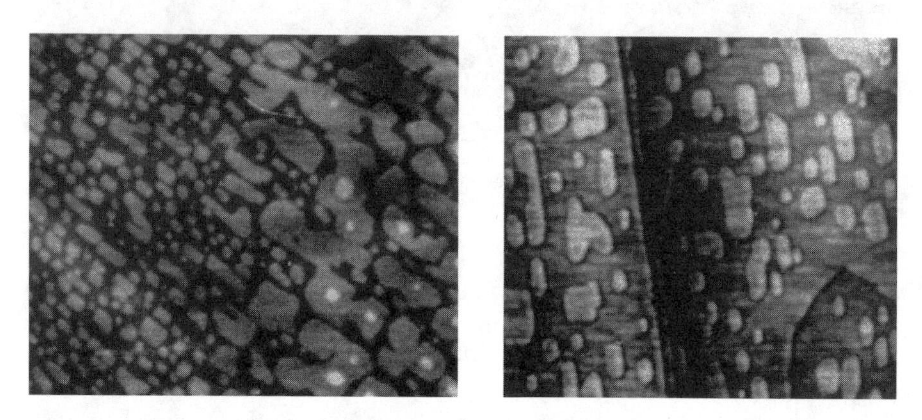

Figure 9. Surface nucleation on gypsum (010) surface, imaged in air after 1 min contact with supersaturated calcium sulfate solution (D. Bosbach).

rated solutions provided statistically reliable estimates of both nucleation and growth rates as a function of supersaturation. The smallest observed islands had a diameter of about 15 nm, interpreted as the critical size for nucleation.[41] Surface nucleation has not been observed on calcite. However, both surface nucleation and spiral growth have been observed on barium sulfate.[42]

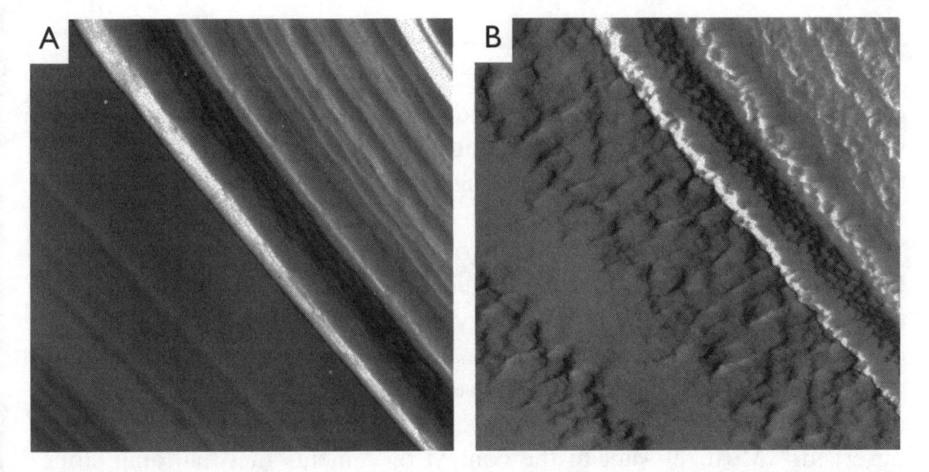

Figure 10. Gypsum crystal growth: (a) rapid growth of steps on the (010) face in supersaturated solution; (b) the effect of 50 ppm of phosphonate, killing step advance.[15]

Crystallization Inhibition

Gratz and Hillner[36,43] (see also Ref. 44) also made the first AFM observations on the important topic of crystal growth inhibition (undoubtedly important for cement hydration). The nucleation of solid particles from supersaturated solutions and the subsequent crystal growth rates are known to be sensitive to small amounts of foreign substances. Inhibitors are rather specific in their action: for calcite, both anions such as phosphates and phosphonates and cations such as Fe^{3+} are effective. Inhibition is observed at low (typically ppm in the aqueous phase) levels and clearly entails an adsorption reaction at critical surface sites. Several research groups have described the effect of inhibitors, mainly phosphates and phosphonates, on the topography of a growing surface. For calcite and gypsum,[4,15,41] the effect is dramatic. The step velocity changes abruptly and the generally straight and featureless steps become crenellated or scalloped in appearance (Fig. 10). This is consistent with a mechanism in which inhibitor molecules or ions are adsorbed at active kink sites, which are then pinned. When new kink sites are spontaneously created by thermal processes, these also are swiftly pinned by inhibitor. Kink nucleation is possible only on free edges above a critical length and eventually no more are produced. The qualitative — and to some extent semiquantative — AFM picture strongly supports this view. Estimates of the crenellation length scale must be approximately equal to the critical kink site density. Direct imaging of an inhibitor on a pinned kink has not yet been achieved.

Mineral Reactions

The work on inhibitors and on chemical dissolution points in an important direction: to the use of AFM as a unique tool for the direct observation of surface chemical reactions of many kinds. Of course, for the future we are thinking of the possibility of observing directly some or all the component reactions of portland cement hydration.

A few AFM observations of mineral reactions have already been published, although in none of these studies was AFM used in a real-time in-situ mode. We note three briefly. Nagy et al.[45] studied the epitaxial growth of periclase and brucite on the aluminosilicate muscovite (mica) as part of a study of clay diagenesis. The oxidative precipitation of manganese oxyhydroxides on to hematite and albite surfaces in Mn^{2+} solutions has been observed.[46] Most relevant in the context of cements (and building stone decay) is the work of Compton[47] on the epitaxial overgrowth of gypsum on calcite during acid dissolution in the presence of sulfate ion. This establishes a protective barrier layer that prevents further dissolution.

Extending the SPM methodology, we can imagine a number of experimental arrangements for setting up controlled in-situ reactions. We propose a few of these in Fig. 11. In Fig. 11(a) we observe a surface of anhydrous clinker mineral, say C_3A, in contact with a nonreactive fluid, say, isopropanol. Switching the flow to water initiates dissolution. Solution species generated by dissolution on one surface (e.g., gypsum) can be washed onto another surface, as in Fig. 11(b). Finally, in Fig. 11(c) we use a packed column to generate an aqueous phase essentially the same as that in a hydrating cement in direct flowing contact with the mineral surface of interest. A heated-cell AFM instrument has recently been built for hydrothermal minerals, able to work up to 150 and 6 bar.[48]

Cement Materials

The literature contains only a few publications describing AFM work undertaken directly for the purposes of cement research. The earliest report we have traced is that of Uchikawa[49] on cement clinkers. Hall[4] published images of a polished OPC clinker surface and of synthetic ettringite crystals. Mitchell et al.,[50] Papavassiliou et al.,[51] and Gauffinet et al.[52–54] have reported uses of AFM in support of hydration studies. Baumgarte et al.[56] have imaged the surface of a geological sample of ettringite, observing defects and dissolution processes [Fig 12(a)]. Hoffmann and Armbruster[57]

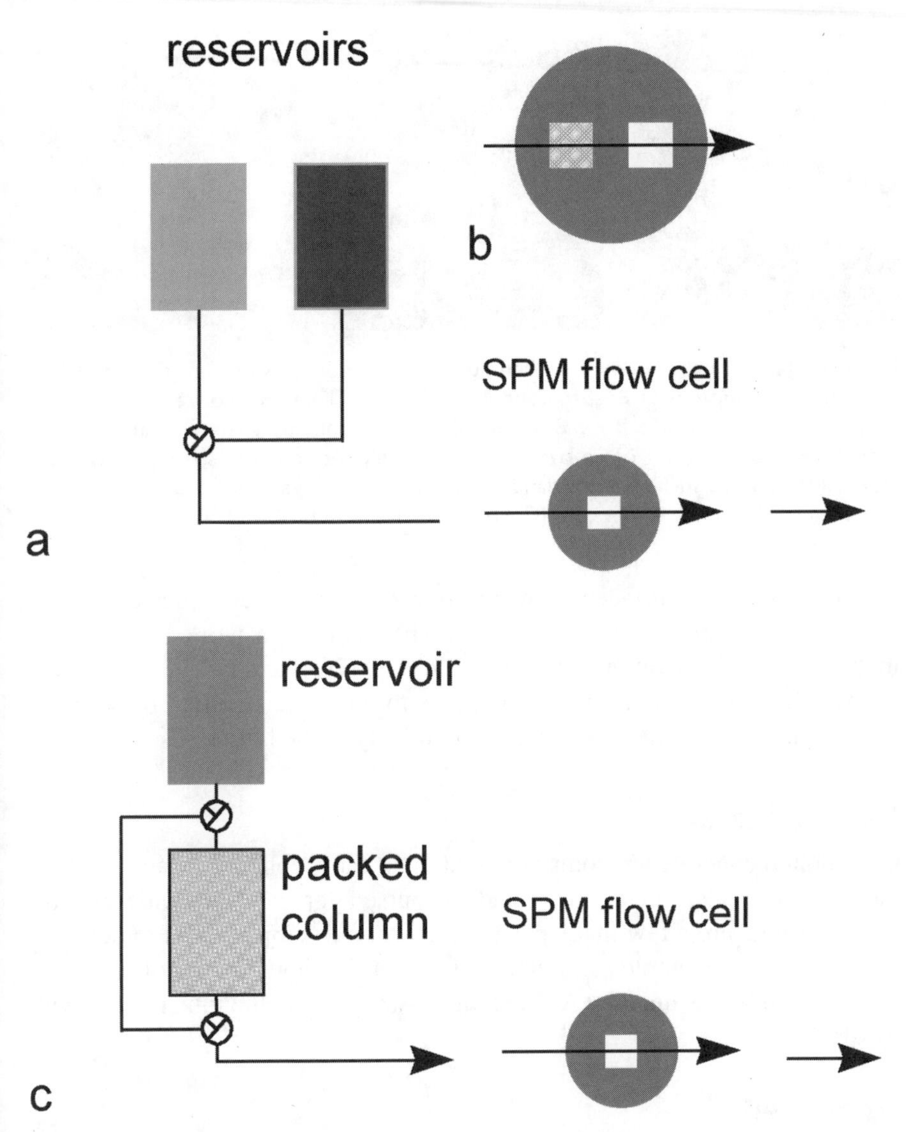

Figure 11. Concepts for in-situ reaction AFM: (a) flow switching; (b) grain-to-grain convection; (c) upstream packed column.

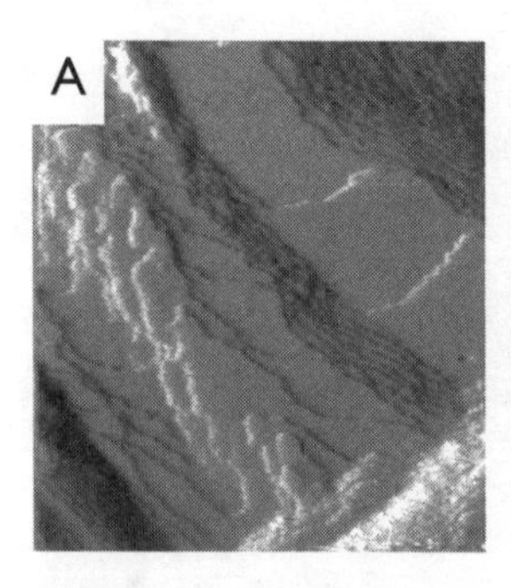

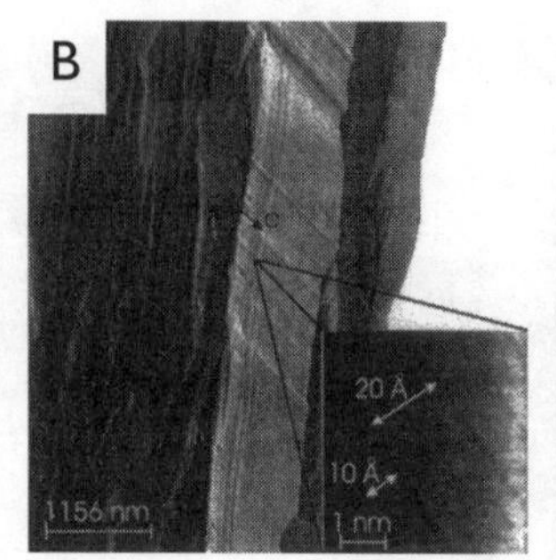

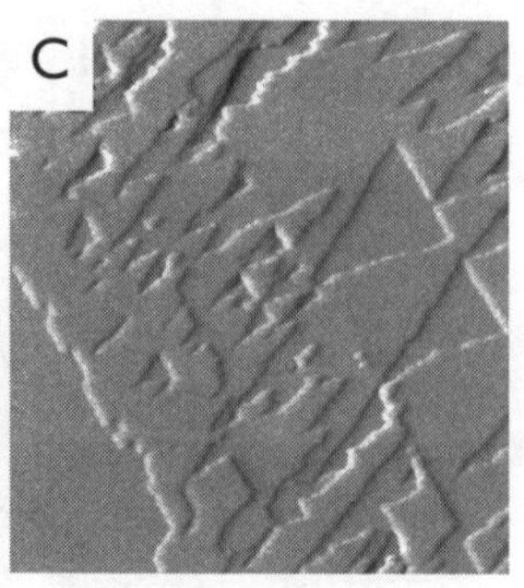

Figure 12. (a) Ettringite: view of surface during dissolution in water, image field 300 × 300 nm (C. Baumgarte, D. Bosbach, and A. Putnis). (b) Clinotobermorite imaged in air. Main picture: image field 5.8 × 5.8 μm. Inset: the *b-c* plane at higher resolution (5 × 5 nm), showing the stacking of Ca layers; a stacking fault with doubled spacing is visible.[57] (c) An 001 cleavage surface of portlandite imaged in air (image field 5 × 5 μm). The terrace step heights are 1–4 unit cells (C. Hall, D. Eger, and D. Bosbach).

have examined the surface of a naturally occurring calcium silicate hydrate closely related to tobermorite [Fig 12(b)]. Our own work in progress includes AFM observations of portlandite [Fig 12(c)] and calcium aluminoferrite. However, significant contributions to our understanding of the surface chemistry of cements from SPM lie mainly in the future.

Other Minerals

Cement and concrete are composite materials par excellence. Many supplementary materials, fillers, fibers, and of course aggregates are widely used. Of these there are a few that have already been the subject of AFM studies. We select three examples, the first of which shows how images of irregular surfaces can be obtained by AFM. Table II summarizes the minerals studied by SPM.

Fly-Ash Particles

Surface features on different types of fly ashes have been distinguished using AFM by Demanet.[58] By combining AFM and X-ray diffraction techniques, Bosbach and Enders[59] identified submicron crystallites of minerals such as anorthite, anhydrite, and gehlenite attached to glassy spheres in fly ashes. AFM phase imaging provided information about the heterogeneity of the glassy surfaces at exceptional resolution (Fig. 13).

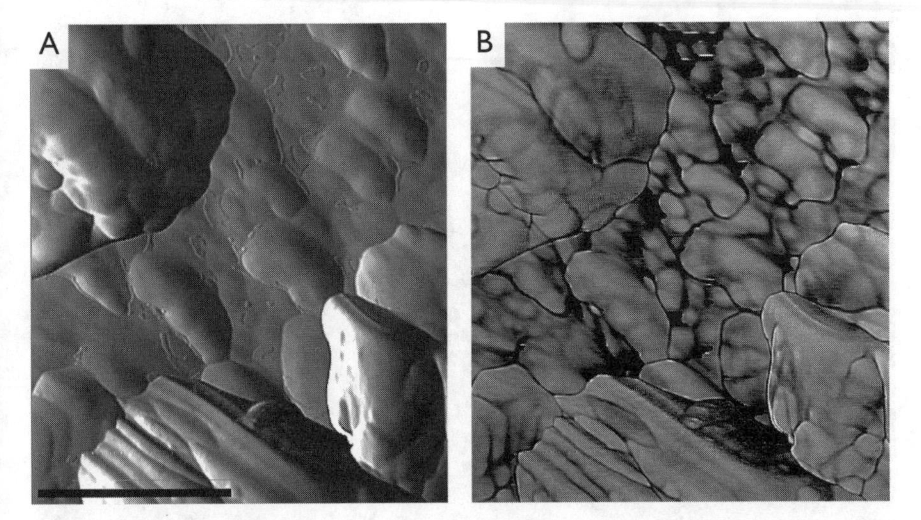

Figure 13. Fly ash. AFM tapping mode images of glassy matrix. (a) Deflection and (b) phase shift images of the same area , 1.2 × 1.2 μm.[59]

Clays

Clays present some problems in sample preparation. Clay platelets or aggregates have atomically flat basal surfaces but the particle size is small (typically around 1 μm). Procedures in which clays are deposited from dilute suspensions on to mica substrates have proved successful. Blum and Eberl[60] published images of kaolinite particles in which spiral growth was evident. Nagy et al.[61] and Blum[62] made direct measurements of particle thicknesses transverse to the basal planes.

Asbestos Fibers

Asbestos fibers have been imaged with AFM in order to characterize the microtopography and to understand the reactivity of their surfaces[63] in relation to carcinogenicity. The AFM images adequately resolve the individual fibrils (Fig. 14).

Microstructure

SPM finds its greatest use today in routine inspection tasks in the semiconductor and optics industries, rather than in the surface chemistry research laboratory. Therefore we might ask whether in cement and concrete

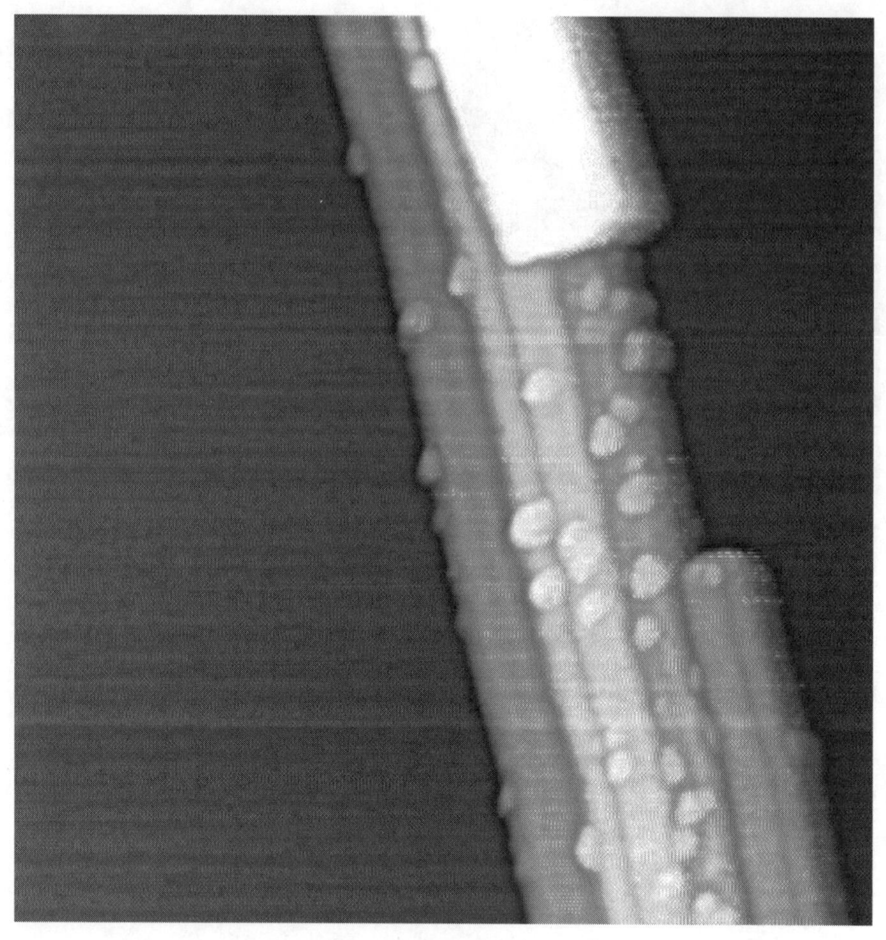

Figure 14. Crocidolite fibers in air on Si (111) substrate, 1 × 1 μm (D. Bosbach).

research AFM may have a role in the routine mapping of microstructure to delineate porosity, grain surface texture, and other physical features. Recently Colston[83,84] has shown that cement gels formed by the hydration of finely ground portland cement produce high-fidelity replicas when cast against a microfabricated template. Figure 15 shows an AFM view of the hydrated cement surface of such micromodel.

AFM provides a rapid means to obtain quantitative information on geometry, surface roughness, and texture. AFM is more sensitive to surface topography than electron microscopy is. The Colston microstructure was imaged in air, but could in fact have been imaged in water. This line of thinking also suggests the use of a casting process to obtain flat AFM surfaces from amorphous or microcrystalline hydration products for AFM study of surface chemical reactions, such as leaching and chemical degradation.

Table II. Minerals studied by SPM

Mineral	References
Gypsum	4, 15, 16, 29, 39, 41, 65, 66
Calcite	22, 24, 25, 26, 27, 28, 33, 34, 35, 36, 67, 68, 69
Brucite	31
Fluorite	16, 43, 70, 71
Barite	4, 42, 72, 92
Muscovite	73
Quartz	74
Albite	55, 75, 76
Kaolinite, illite	60, 62
Hydrotalcite	77
Zeolite	78, 79
Galena	90
Pyrite	91
Pyrrhotite	80
Hematite	81, 82
Clinotobermorite	57

Some Comments on Experimental Methods

AFM Imaging Modes

We should emphasize that there are several variants in AFM imaging, which differ in the way in which the tip-surface interaction is controlled and detected.[3,8,85]

Contact Mode Imaging

This is the primary method which we have already outlined. The cantilever carrying the tip deflects under the action of surface forces, the deflection is detected as the out-of-balance signal of a quadrant photometer and a feedback system maintains constant cantilever deflection (constant force imaging). The interaction between the tip and the surface may be either repulsive or attractive, depending on the nature of the surface or the desired tip-surface force.

Lateral Force Imaging

Like standard contact mode, except that we record the changes in cantilever twisting (torsion) by detecting the out-of-balance signal in the quadrant photometer. This method is sensitive to changes in the friction between tip and surface and may enhance contrast between regions of different chemical composition.

Noncontact or Tapping Mode Imaging

The tip-surface distance is modulated close to the cantilever natural resonance frequency; changes in tip-surface interaction are detected as shifts in frequency, amplitude, or phase. This mode of imaging greatly reduces the

Figure 15. AFM image of hydrated cement surface, showing shrinkage cracks and portlandite crystals, image field 14 × 14 μm. The surface relief is about 1.4 μm (cement sample S. Colston, Birkbeck College; image A. Murray, Topometrix Corporation).

time-average interaction force between tip and surface and minimizes tip-drag damage on soft surfaces (polymers, biological materials). In true non-contact mode, the z piezo is driven to maintain constant oscillation amplitude at a minimum distance of several nanometers and direct tip-surface contact is avoided. In tapping mode, the tip is in contact with the surface at its minimum distance of approach but is withdrawn for the rest of each oscillation cycle, again reducing the mean surface interaction and especially dragging during raster steps. In phase mode, image contrast is derived from the phase shift, which is related to the damping of the oscillation by

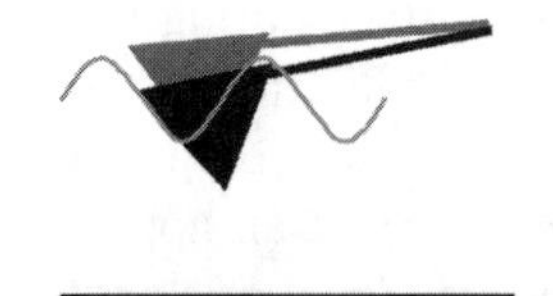
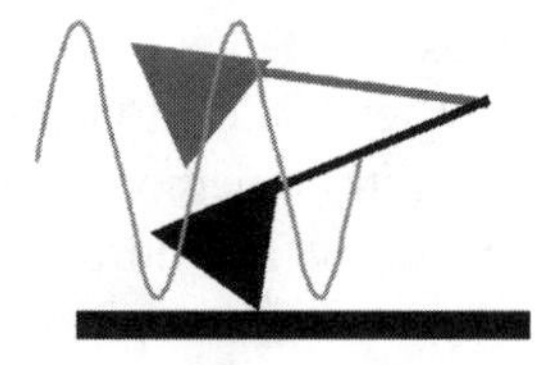

Figure 16. Schematic of contact, noncontact, and tapping mode imaging.

interaction with the surface. The AFM phase image is therefore sensitive to dynamic mechanical properties of the surface that control energy transfer through processes such as adhesion and viscoelastic deformation.

Figure 16 shows a schematic of contact, noncontact, and tapping modes.

Sample Preparation

Many minerals can be obtained as euhedral crystals and some show good or perfect cleavage parallel to at least one lattice plane. Cleavage is an easy way to form clean and often atomically flat surfaces, although the process of cleavage releases much mechanical energy and introduces defects that may in turn strongly influence surface processes such as dissolution. Powders have been imaged by pressing into pellets[86] or as in the case of clays by evaporation of a dilute suspension deposited on a flat substrate such as mica or graphite.[61,62] Clay particles on a sapphire Al_2O_3 surface attach sufficiently strongly via electrostatic forces to be imaged in aqueous media.[87] For complex polyphasic and porous materials such as cement clinker and hardened cement pastes, it appears that standard methods of petrographic polishing produce surfaces that can be imaged satisfactorily.

Beyond Imaging: Prospects and Problems

The best of recent SPM work on the chemistry of the mineral-solution interface shows clearly what could be done on cement materials. With an instrument equipped with controlled flow, there are now opportunities for systematic studies of the phenomena of dissolution on the surfaces of the major (and minor) cement minerals. By switching reactive species (say, sulfate or carbonate ions, or set-retarders such as gluconate ion) into the flow, the influence of adsorbed molecules and surface barrier layers becomes accessible. By injecting a pulse of water into a steady flow of propanol across a silicate or aluminate surface during imaging, it should be possible to observe the earliest stages of hydration. By the use of the grain-to-grain convection method we describe, more complex reactions can be engineered on the surface during imaging. Such an approach culminates in an AFM view at microscopic and unit-cell level of something close to the hydration of the whole cement.

This methodology can equally be applied to investigate the reactions that occur in the chemical alteration of hardened cement pastes and, for that matter, reactions occurring between mineral aggregates and environmental species.

Such experiments are not easy. SPM techniques have many virtues, but there are some challenges. Few cement minerals are obtainable as cleavable single crystals. AFM, and especially quantitative use of the kind we emphasize, works best with surfaces that are reasonably flat over an area at least a good fraction of a millimeter across. Much can no doubt be achieved by polishing and etching to prepare surfaces for examination, for example of unground clinker material. In any case, there is no fundamental reason why smaller particles should not be successfully imaged (and indeed there are plenty of examples of this, including some described above), but sample manipulation demands greater skill. Good crystalline samples of cement hydration products require individual synthesis, and some X-ray work to index specimens. Controlled-flow cells of the kind we have described are not yet commercially available.

Nevertheless, with SPM methods we take a large step toward the goal of observing the mineral surface at work, what Eggleston[88] calls the molecular and ionic "machinery." SPM is less a set of instruments than an approach to surface science that advocates the use of "local probes" of all ingenious kinds. Of special interest in cement and concrete is the prospect of making direct local measurements of mechanical properties at the microscopic level, as already demonstrated for gypsum.[66] This opens the way to controlled experiments in which the mechanical action of the tip is to create strain and strain-induced defects and to promote local transport in the fluid phase, thereby enhancing dissolution[89] or nucleation. These are new experiments that explore the coupling of mechanics and chemistry in processes such as durability, creep, and freeze-thaw damage in concrete.

The lesson from our research on simple minerals like calcite and gypsum is that these surfaces are more active than might have been imagined and that complicated processes at several length scales from the atomic to the microscopic work together. Similar processes must be active (and sometimes rate-determining) in cement hydration; the task is to track them down.

Acknowledgments

We thank T. Armbruster, S. Colston, D. Eger, M. Hochella, J. Leckenby, A. Murray, A. Nonat, and A. Putnis for contributions.

References

1. G. Binnig, H. Rohrer, C. Gerber, and E. Weibel, "Surface Studies by Scanning Tunneling Microscopy," *Physical Rev. Lett.,* **49**, 57–61 (1982).
2. G. Binnig and H. Rohrer, "Scanning Tunneling Microscopy — From Birth to Adolescence," *Rev. Modern Physics,* **59**, 615–625 (1987).
3. R. Wiesendanger, *Scanning Probe Microscopy and Spectroscopy: Methods and Applications.* Cambridge University Press, 1994.
4. C. Hall, "Scanning Probe Microscopy: Sulphate Minerals in Scales and Cements"; presented at the International Symposium on Oilfield Chemistry, San Antonio, February 1995, Society of Petroleum Engineers. Paper no. 28996
5. A.R. Lennie, N.G. Condon, F.M. Leibsle, P.W. Murray, G. Thornton, and D.J. Vaughan, "Structures of Fe_3O_4 (111) Surfaces Observed by Scanning Tunneling Microscopy," *Physical Rev. B,* **53**, 10 244–10 253 (1996).
6. G. Binnig, C.F. Quate, and C. Gerber, "Atomic Force Microscope," *Physical Rev. Lett.,* **56**, 930–933 (1986).
7. M.F. Hochella Jr., "Atomic Structure, Microtopography, Composition and Reactivity of Mineral Surfaces"; pp. 87–132 in *Mineral-Water Interface Geochemistry.* Reviews in Mineralogy vol. 23. Edited by M.F. Hochella Jr. and A.F. White. Mineralogical Society of America, 1990.
8. C.M. Eggleston. "High Resolution Scanning Probe Microscopy: Tip-Surface Interaction, Artefacts and Applications in Mineralogy and Geochemistry"; pp. 1–90 in *Scanning Probe Microscopy of Clay Minerals.* Clay Minerals Society Workshop Lectures, vol. 7. Edited by K.L. Nagy and A.E. Blum. Clay Minerals Society, Boulder, Colorado, 1994.
9. P.A. Maurice, "Applications of Atomic Force Microscopy in Environmental, Colloid and Surface Chemistry," *Colloids and Surfaces,* **107**, 57–75 (1996).
10. W. Gutmannsbauer, T. Huser, T. Lacoste, H. Heinzelmann, and H.-J. Guntherodt, "Scanning Near-Field Optical Microscopy (SNOM) and Its Application in Mineralogy," *Schweizerische Mineralogische und Petrographische Mitteilungen,* **75**, 259–264 (1995).
11. C.C. Williams and H.K. Wickramasinghe; p. 364 in *Photoacoustics and Photothermal Phenomena.* Edited by P. Hess and J. Peltzl. Springer, Berlin, 1988.
12. A.A. Gewirth and B.K. Niece, "Electrochemical Applications of In Situ Scanning Probe Microscopy," *Chemical Rev.,* **97**, 1129–1162 (1997).
13. J.V. Macpherson and P.R. Unwin, "Recent Advances in Kinetic Probes of the Dissolution of Ionic Crystals," *Prog. Reaction Kinetics,* **20**, 185–244 (1995).

14. G. Denuault, M.H. Frank, and S. Nugues, "A Description of the Scanning Electrochemical Microscope (SECM) and of Its Applications"; pp.69–82 in *Nanoscale Probes of the Solid/Liquid Interface.* Series E, vol. 288. Edited by A.A. Gewirth and H.S. Siegenthaler. Kluwer, Dordrecht, 1995.
15. C. Hall and D.C. Cullen, "Scanning Force Microscopy of Gypsum Dissolution and Crystal Growth," *Am. Inst. Chem. Eng. J.,* **42**, 232–238 (1996).
16. D. Bosbach and W. Rammensee, "In-Situ Investigation of Growth and Dissolution on the (010) Surface of Gypsum by Scanning Force Microscopy," *Geochim. Cosmochim. Acta,* **58**, 843–849 (1994).
17. F. Ohnesorge and G. Binnig, "True Atomic Resolution by Atomic Force Microscopy through Repulsive and Attractive Forces," *Science,* **260**, 1451–1456 (1993).
18. D.L. Patrick and R.M. Lynden-Bell, "Atomistic Simulations of Fluid Structure and Solvation Forces in Atomic Force Microscopy," *Surface Sci.,* **380**, 224–244 (1997).
19. Zhi Hui Liu and N.M.D. Brown. "The Influence of Imaging Conditions on the Appearance of Lattice-Resolved AFM Images of Mica Surfaces," *J. Physics D: Applied Physics,* **30**, 2503–2508 (1997).
20. M. Bammerlin, R. Lüthi, E. Meyer, A. Baratoff, J. Lü, M. Guggisberg, C. Loppacher, C. Gerber, and H.-J. Güntherodt, "Dynamic SFM with True Atomic Resolution on Alkali Halide Surfaces," *Appl. Physics A: Mater. Sci. Process.,* **66**, S293–S294 (1998).
21. R. Lüthi, E. Meyer, M. Bammerlin, and A. Baratoff, "Ultrahigh Vacuum Atomic Force Microscopy: True Atomic Resolution," *Surface Rev. Lett.,* **4**, 1025–1030 (1997).
22. P.E. Hillner, S. Manne, A.J. Gratz and P. Khansma, "AFM Images of Dissolution and Growth on a Calcite Crystal," *Ultramicroscopy,* **42-44**, 1387–1393 (1992).
23. P.E. Hillner, A.J. Gratz, S. Manne, and P.K. Hansma, "Atomic-Scale Imaging of Calcite Growth and Dissolution in Real-Time," *Geology,* **20**, 359–362 (1992).
24. Y. Liang, D.R. Baer, J.M. McCoy, J.E. Amonette, and J.P. Lafemina, "Dissolution Kinetics at the Calcite-Water Interface," *Geochim. Cosmochim. Acta,* **60**, 4883–4888 (1996).
25. Y. Liang, D.R. Baer, J .M. McCoy, and J.P. Lafemina, "Interplay Between Step Velocity and Morphology During the Dissolution of the $CaCO_3$ Surface," *J. Vacuum Sci. Tech.,* **60**, 4883–4887 (1996).
26. Y. Liang, A.S. Lea, D.R. Baer, and M.H. Engelhard, "Structure of the Cleaved $CaCo_3$ $(10\bar{1}4)$ Surface in an Aqueous Environment," *Surface Sci.,* **351**, 172–182 (1996).
27. Y. Liang and D.R. Baer, "Anistropic Dissolution at the $CaCO_3$ $(10\bar{1}4)$-Water Interface," *Surface Sci.,* **373**, 275–287 (1997).
28. G. Jordan and W. Rammensee, "Dissolution Rates of Calcite $(10\bar{1}4)$ Obtained by Scanning Force Microscopy: Microtopography-Based Dissolution Kinetics on Surfaces with Anisotropic Step Velocities," *Geochim. Cosmochim. Acta,* **60**, 5055–5062 (1996).
29. H. Shindo, M. Kais, H. Kondoh, C. Nisihara, H. Hayakawa, S. Ono, and H. Nozoye. "Structure of Cleaved Surfaces of Gypsum Studied with Atomic Force Microscopy," *Chem. Commun.,* **16**, 1097–1099 (1991).
30. P.M. Dove and J.A. Chermak, "Mineral-Water Interactions: Fluid Cell Applications of Scanning Force Microscopy"; pp. 140–169 in *Scanning Probe Microscopy of Clay Minerals.* Clay Minerals Society Workshop Lectures, Vol. 7. Edited by K. Nagy and A.E. Blum. Clay Minerals Society, Boulder Colorado, 1994.

31. G. Jordan and W. Rammensee, "Dissolution Rates and Activation Energy for Dissolution of Brucite (001): A New Method Based on the Microtopography of Crystal Surfaces," *Geochim. Cosmochim. Acta,* **60**, 5055–5062 (1996).
32. B.A. Coles, R.G. Compton, J. Booth, Q. Hong, and G.H.W. Sanders, "A Hydrodynamic AFM Flow Cell for the Quantitative Measurement of Interfacial Kinetics," *Chem. Commun.,* 619–620 (1997).
33. Q. Hong, M.F. Suarez, B.A. Coles, and R.G. Compton, "Mechanism of Solid/Liquid Interfacial Reactions. The Maleic Acid Driven Dissolution of Calcite: An Atomic Force Microscopy Study Under Defined Hydrodynamic Conditions," *J. Phys. Chem.,* **101**, 5557–5564 (1997).
34. H.H. Teng and P.M. Dove, "Surface Site-Specific Interactions of Aspartate with Calcite During Dissolution: Implications for Biomineralisation," *Am. Mineralogist,* **82**, 878–887 (1997).
35. D.W. Britt and V. Hlady, "In-Situ Atomic Force Microscope Imaging of Calcite: Etch Pit Morphology Changes in Undersaturated and 1-Hydroxyelthylidene, 1-Diphosphonic Acid Poisoned Solutions," *Langmuir,* **13**, 1873–1876 (1997).
36. A.J. Gratz and P.E. Hillner, "Poisoning of Calcite Growth Viewed in the Atomic Force Microscope (AFM)," *J. Cryst. Growth,* **129**, 789–793 (1993).
37. A.J. Gratz, P.E. Hillner, and P.K. Hansma, "Step Dynamics and Spiral Growth on Calcite," *Geochim. Cosmochim. Acta,* **57**, 491–495 (1993).
38. W.K. Burton, N. Cabrera, and F.C. Frank, "The Growth of Crystals and the Equilibrium Structure of Their Surfaces," *Philosophical Trans. Royal Soc.,* **A243**, 299–358 (1951).
39. D. Bosbach, J.L. Junta-Rosso, U. Becker, and M.F. Hochella Jr., "Gypsum Growth in the Presence of Background Electrolytes Studied by Scanning Force Microscopy," *Geochim. Cosmochim. Acta,* **60**, 3295–3304 (1996).
40. M. Ohara and R.C. Reid, *Modeling Crystal Growth Rates from Solution.* Prentice-Hall, Englewood Cliffs, New Jersey, 1973.
41. D. Bosbach and M.F. Hochella Jr., "Gypsum Growth in the Presence of Growth Inhibitors: A Scanning Force Microscopy Study," *Chem. Geol.,* **132**, 227–236 (1996).
42. D. Bosbach, C. Hall, and A. Putnis, "Mineral Precipitation and Dissolution in Aqueous Solution: In-Situ Microscopic Observations on Barite (001) with Atomic Force Microscopy," *Chem. Geol.* (in press).
43. P.E. Hillner, S. Manne, A.J. Gratz, and P.K. Hansma, "The AFM: A New Tool for Imaging Crystal Growth Processes," *Faraday Discussion,* **95**, 191–198 (1993).
44. P.M. Dove and M.F. Hochella Jr., "Calcite Precipitation and Inhibition by Orthophosphate: In-Situ Observation by Scanning Force Microscopy," *Geochim. Cosmochim. Acta,* **57**, 705–714 (1993).
45. K.L. Nagy, R.T. Cygan, N.C. Sturchio, and R.P. Chiarello, "Heterogeneous Nucleation and Growth of Clays: Gibbsite and Brucite on Muscovite," *J. Conf. Abs.,* **1**, 425 (1996).
46. J. Junta and M.F. Hochella, "Manganese(II) Oxidation at Mineral Surfaces: A Microscopic and Spectroscopic Study," *Geochim. Cosmochim. Acta,* **58**, 4985–4999 (1994).
47. J. Booth, Q. Hong, R.G. Compton, K. Prout, and R.M. Payne, "Gypsum Overgrowths Passivate Calcite to Acid Attack," *J. Colloid Interface Sci.,* **192**, 207–214 (1997).
48. S.R. Higgins, C.M. Eggleston, K.G. Knauss, and C.O. Boro, "A Hydrothermal Atomic Force Microscope for Imaging in Aqueous Solution up to 150°C," *Rev. Sci. Instruments,* **69**, 2994–2998 (1998).

49. H. Uchikawa, S. Hanehara, and D. Sawaki, "Observation of the Change of Composition and Surface Structure of Alite Crystals and Estimation of Hydration Reactivity by Etching"; pp. 202–207 in *Ninth International Congress on the Chemistry of Cements,* Vol. 4. National Council for Cement and Building Materials, New Delhi, India, 1992.
50. L.D. Mitchell, M. Prica, and J.D. Birchall, "Aspects of Portland Cement Hydration Studied Using Atomic Force Microscopy," *J. Mater. Sci.,* **31**, 4207–4212 (1996).
51. G. Papavassiliou, M. Fardis, E. Laganas, A. Leventis, A. Hassanien, F. Milia, A. Papageorgiou, and E. Chaniotakis, "Role of the Surface Morphology in Cement Gel Growth Dynamics: A Combined Nuclear Magnetic Resonance and Atomic Force Microscopy Study," *J. Appl. Physics,* **82**, 449–452 (1997).
52. S. Gauffinet, E. Finot, E. Lesniewska, S. Collin, and A. Nonat, "AFM and SEM Studies of C-S-H Growth on C_3S Surface During Its Early Hydration"; pp. 337–356 in *Proceedings of the 20th International Conference on Cement Microscopy, April 1998, Guadalajara, Mexico.* International Cement Microscopy Association, Duncanville, Texas, 1988.
53. S. Gauffinet, E. Finot, E. Lesniewska, and A. Nonat. "Observation Directe de la Croissance d'Hydrosilicate de Calcium sur des Surfaces d'Alite et de Silice par Microscopie à Force Atomique," *Comptes Rendues de l'Académie des Sciences, Paris, Série 2,* **327**, 231–236 (1998).
54. S. Gauffinet, E. Finot, and A. Nonat, "Experimental Study and Simulation of C-S-H Nucleation and Growth"; in *Proceedings of Second RILEM Workshop on Hydration and Setting, June 1997, Dijon, France.* Edited by A. Nonat and J.-C. Mutin. RILEM Editions, in press.
55. R. Hellmann, B. Drake, and K. Kjoller, "Using Atomic Force Microscopy to Study the Structure, Topography and Dissolution of Albite Surfaces"; pp. 149–152 in *Water-Rock Interaction.* Edited by J. Kharaka and A. Maest. A.A. Balkema, Rotterdam, 1992.
56. C. Baumgarte, D. Bosbach, and A. Putnis, "In-Situ Kristallisationsexperimente an Ettringiten mittels Rasterkraftmikroskopie," *Berichte der Deutschen Mineralogischen Gesellschaft* (Beihefte zum *European J. Mineralogy*), **8**, 16 (1996).
57. C. Hoffmann and T. Armbruster, "Clinotobermorite, $Ca_5[Si_3O_8(OH)]2.4H_2O$ – $Ca_5[Si_6O_{17}].5H_2O$, A Natural C-S-H(I) Type Cement Mineral: Determination of the Substructure," *Zeitschrift Kristallographie,* **212**, 864–873 (1997).
58. C.M. Demanet, "Atomic Force Microscopy Determination of the Topography of Fly-Ash Particles," *Appl. Surface Sci.,* **89**, 97–101 (1995).
59. D. Bosbach and M. Enders, "Microtopography of High-Calcium Fly Ash Particle Surfaces," *Adv. Cement Res.,* **10**, 17–23 (1998).
60. A.E. Blum and D.D. Eberl, "Determination of Clay Particle Thicknesses and Morphology Using Scanning Force Microscopy"; pp. 133–136 in *Water-Rock Interaction.* Edited by J. Kharaka and A. Maest. Balkema, Rotterdam, 1992.
61. K.L Nagy and A.E. Blum, eds., *Scanning Probe Microscopy of Clay Minerals.* Clay Minerals Society Workshop Lectures, Vol. 7. Clay Minerals Society, Boulder, Colorado, 1994.
62. A. Blum, "Determination of Illite/Smectite Particle Morphology Using Scanning Force Microscopy"; pp. 171–202 in *Scanning Probe Microscopy of Clay Minerals.* Clay Minerals Society Workshop Lectures, Vol. 7. Edited by K.L. Nagy and A.E. Blum. Clay Minerals Society, Boulder, Colorado, 1994.

63. Z.H. Shen, D. Bosbach, M.F. Hochella, D.L. Bish, M.G. Williams, R.F. Dodson, and A.E. Aust, "Using In-Vitro Iron Deposition on Asbestos to Model Asbestos Bodies Formed in Human Lung," *Chem. Res. Toxicol.,* **13**, 913–921 (2000).
64. P.M. Dove, M.F. Hochella Jr., and R.J. Reeder, "In Situ Investigation of Near-Equilibrium Calcite Precipitation by Atomic Force Microscopy"; pp. 141–144 in *Water-Rock Interaction.* Edited by J. Kharaka and A. Maest. A.A. Balkema, Rotterdam, 1992.
65. E. Finot, E. Lesniewska, J.-C. Mutin, and J.-P. Goudonnet, "Reactivity of Gypsum Faces According to the Relative Humidity by Scanning Force Microscopy," *Surface Sci.,* **384**, 201–217 (1997).
66. E. Finot, E. Lesniewska, J.-P. Goudonnet, and J.-C. Mutin, "Influence des Contraintes sur les Propriétés Élastiques de Surface du Gypse Sondées par Microscopie à Force Atomique," *Comptes Rendues de l'Académie des Sciences, Paris,* **IIb**, 577–586 (1997).
67. A.L. Rachlin, G.S. Henderson, and M.C. Goh, "An Atomic Force Microscope (AFM) Study of the Calcite Cleavage Plane: Image Averaging in Fourier Space," *Am. Mineralogist,* **77**, 904–910 (1992).
68. S.L.S. Stipp, C.M. Eggleston, and B.S. Nielsen, "Calcite Surface Structure Observed at Microtopographic and Molecular Scales with Atomic Force Microscopy (AFM)," *Geochim. Cosmochim. Acta,* **58**, 3023–3033 (1994).
69. S.L.S. Stipp, W. Gutmansbauer, and T. Lehmann, "The Dynamic Nature of Calcite Surfaces in Air," *Am. Mineralogist,* **81**, 1–8 (1996).
70. G. Jordan and W. Rammensee, "Growth and Dissolution on the CaF_2 (1110) Surface Observed by Scanning Force Microscopy," *Surface Sci.,* **371**, 371–380 (1997).
71. R. Bennewitz, M. Reichling, and E. Matthias, "Force Microscopy of Cleaved and Electron-Irradiated CaF_2 (111) Surfaces in Ultra-High Vacuum," *Surface Sci.,* **387**, 69–77 (1997).
72. D.D. Archibald, B.P. Gaber, J.D. Hopwood, S. Mann, and T. Boland, "AFM of Synthetic Barite Microcrystals," *J. Crystal Growth,* **172**, 231–248 (1997).
73. P.A. Johnsson, A.E. Blum, M.F. Hochella Jr., G.A. Parks, and G. Sposito, "Direct Observation of Muscovite Basal-Plane Dissolution and Secondary Phase Formation: An XPS, LEED and SFM Study"; pp. 159–162 in *Water-Rock Interaction.* Edited by J. Kharaka and A. Maest. A.A. Balkema, Rotterdam, 1992.
74. A.J. Gratz, S. Manne, and P.K. Hansma, "Atomic Force Microscopy of Atomic-Scale Ledges and Etch Pits During Dissolution of Quartz," *Science,* **251**, 1343–1345 (1991).
75. B. Drake and R. Hellmann, "Atomic Force Microscopy Imaging of the Albite (010) Surface," *Am. Mineralogist,* **76**, 1773–1776 (1991).
76. D. Nyfeler, R. Berger, and C. Gerber, "Scanning Force Microscopy on Albite Cleavage Surfaces," *Swiss Bull. Mineral. Petrol.,* **77**, 21–26 (1997).
77. H. Cai, A.C. Hillier, K.R. Franklin, C.C. Nunn, and M.D. Ward, "Nanoscale Imaging of Molecular Adsorption," *Science,* **266**, 1551–1555 (1994).
78. G. Binder, L. Scandella, A. Schumacher, N. Kruse, and R. Prins, "Microtopographic and Molecular Scale Observations of Zeolite Surface Structures: Atomic Force Microscopy on Natural Heulandite," *Zeolites,* **16**, 2–6 (1996).
79. M. Komiyama, "Adsorption Characteristics of Pyridine Bases on Zeolite (010) Examined by Atomic Force Microscopy (AFM)," *Studies Surface Sci. Catalysis,* **109**, 185–192 (1997).

80. U. Becker, A.W. Munz, A.R. Lennie, G. Thornton, and D.J. Vaughan, "The Atomic and Electronic Structure of the (001) Surface of Monoclinic Pyrrhotite (Fe_7S_8) as Studied Using STM, LEED and Quantum Mechanical Calculations," *Surface Sci.,* **380**, 66–87 (1997).
81. U. Becker, M.F. Hochella Jr., and E. Apra, "The Electronic Structure of Hematite {001} Surfaces: Applications to the Interpretation of STM Images and Heterogeneous Surface Reactions," *Am. Mineralogist,* **81**, 1301–1314 (1996).
82. C.M. Eggleston and M.F. Hochella, "The Structure of Hematite (001) Surfaces by Scanning Tunneling Microscopy: Image Interpretation, Surface Relaxation and Step Structure," *Am. Mineralogist,* **77**, 911–922 (1992).
83. S.L. Colston, P. Barnes, H. Freimuth, M. Lacher, and W. Ehrfeld, "Cements: A New Medium for Microengineering Structures?" *J. Mater. Sci. Lett.,* **15**, 1660–1663 (1996).
84. S.L. Colston, "Time-Resolved Studies on the Hydration Kinetics and Microstructure of Latex-Modified Cement," Ph.D. Dissertation. London University, 1997.
85. A. Noy, D.V. Vezenov, and C.M. Lieber, "Chemical Force Microscopy," *Ann. Rev. Mater. Sci.,* **27**, 381–422 (1997).
86. G. Friedbacher et al., "Imaging Powders with the AFM: From Biominerals to Commercial Materials," *Science,* **253**, 1261–1263 (1991).
87. B.R. Bickmore, personal communication, 1998.
88. C.M. Eggleston, "Mineral Surfaces in Geochemical and Environmental Processes: Direct Observation of the Microscopic Machinery," *J. Conf. Abs.,* **1**, 153 (1996).
89. N.-S. Park, M.-W. Kim, S.C. Langford, and J.T. Dickinson, "Atomic Layer Wear of Single-Crystal Calcite in Aqueous Solution Using Scanning Force Microscopy," *J. Appl. Physics,* **80**, 2680–2686 (1996).
90. C.M. Eggleston, "Initial Oxidation of Sulfide Sites on a Galena Surface: Experimental Confirmation of an ab-initio Calculation," *Geochim. Cosmochim. Acta,* **61**, 657–660 (1997).
91. C.M. Eggleston, J.J. Ehrhardt, and W. Strum, "Surface Structural Controls on Pyrite Oxidation Kinetics: An XPS-UPS, STM, and Modeling Study," *Am. Mineralogist,* **81**, 1036–1056 (1996).
92. A. Putnis, J.L. Junta-Rosso, and M.F. Hochella Jr., "Dissolution of Barite by a Chelating Ligand: An Atomic Force Microscopy Study," *Geochim. Cosmochim. Acta,* **59**, 4623–4632 (1995).

The Stereological and Statistical Properties of Entrained Air Voids in Concrete: A Mathematical Basis for Air Void System Characterization

Kenneth Snyder
National Institute of Standards and Technology, Gaithersburg, Maryland

Kumar Natesaiyer
USG Research Center, Libertyville, Illinois

Kenneth Hover
Cornell University, Ithaca, New York

> *"Yet I exist in the hope that these memoirs, in some manner, I know not how, may find their way to the minds of humanity in Some Dimension, an may stir up a rave of rebels who shall refuse to be confined to limited Dimensionality."*
>
> — Edwin Albott

To understand the freeze-thaw properties of hardened concrete, the air void system microstructure must be characterized. Studies of the stereological and statistical properties of entrained air voids in concrete have often involved a number of steps: sample preparation and air void identification, linear and planar analysis of a polished surface, uncertainty analysis of the recorded data, parametric and nonparametric estimates of the air void diameter distribution, and analysis of the air void system spatial statistics. Each of these steps has been discussed in detail in a number of engineering fields. For the civil engineering researcher, a comprehensive study of these properties requires consulting many varied texts, each addressing these steps individually. This chapter attempts to consolidate these topics, critically review them, and combine them into a single desk reference. The researcher can then realize the interdependencies of these topics, learn to accurately characterize the air void system microstructure, and develop an understanding of the basis for standardized test methods such as ASTM C 457.

Introduction

The effectiveness of air entrainment in providing frost resistance to concrete is well known to the industry. Though there are differing hypotheses on the mechanism of frost damage, all the mechanisms suggest that frost durability is a function of the material properties of the concrete, the freezing environment to which it is subjected, and the air void system in the concrete. Despite this knowledge, the frost resistance of concrete is commonly evaluated using air void parameters alone. The common air void parameters used are air content by volume, the specific surface area, and the Powers spacing factor. These values are calculated for hardened concrete using the standard test method ASTM C 457.[1]

When actual freeze/thaw testing is not feasible, it is common to predict the frost resistance of concrete on the basis of air void geometry alone. To make such a prediction, a small fraction of the air voids present in the hardened concrete are observed through a microscope, and statistical inferences are drawn about the population of the air voids as a whole.[2]

Given that a cubic meter of concrete may contain on the order of 10^{10} air voids, and that a conscientious microscopic evaluation might sample 10^3 voids, it is clear that the issues of statistical inference and uncertainty are central to geometric analysis. Accurately assessing the air void system is further complicated by the fact that information is needed about the three-dimensional air void characteristics, but observations are made on the two-dimensional intersection of the void with the plane. In a typical analysis performed in accordance with ASTM C 457, no attempt is made to measure the diameter of an actual air void, nor is any attempt made to measure the diameter of the intersection between the void and the two-dimensional plane. Even this most basic of geometric characteristics of a single void is inferred from a large number of observations, none which are intentionally of the actual air void diameter.

The descriptive geometry of the three-dimensional air void system is complicated still further by the need to know something about the spatial distribution of the voids. Once again, in a standard ASTM C 457 analysis, no data are collected about the actual location or spatial distribution of the voids. All of this is later inferred on the basis of a priori assumptions or structural models, irrespective of the actual concrete microstructure. The industry's ability to correlate actual frost resistance with air void system geometry attests to the fact that the mathematical approaches developed over the years have been able to deal with these complications to varying degrees.

There is considerable room for improvement of the current state of practice, however. First, there is not a broad understanding of the fundamental uncertainty that accompanies air void interpretation. The apparent complexity of the equations used in standard practice too often causes the practitioner to assume more accuracy and precision than can be justified. Second, the availability of computers means that we need not confine ourselves to oversimplified geometric models for reasons of computational expediency. More useful techniques are available than those currently used in practice. Perhaps this chapter will serve to clarify the mathematical background of the problem, emphasize the uncertainty inherent in various approaches, and suggest improved techniques.

Background

Since the 1940s, it has been known that the addition of a small volume of entrained air voids into concrete has a dramatic effect upon performance under freezing and thawing conditions. Since then, numerous reports have been published that discuss ways to characterize the air void system in concrete. One of the first attempts, and perhaps the most famous in the concrete research community, was the spacing equation proposed by Powers in 1949.[3] The equation that bears his name exists today as part of ASTM C 457, which is the standard test method for characterizing air voids in hardened concrete.

The Rosiwal technique[4] that is the basis for ASTM C 457 dates back to 1898. Brown and Pierson[5] were among the first researchers to apply this technique to the study of air voids. Today, the test is used to estimate the volume fraction of entrained air, the surface area of air voids per volume of air voids (specific surface area), and the Powers spacing factor. However, this is as far as the standardized test goes with respect to characterizing the air void system in concrete.

The methods for characterizing the uncertainty in measured volumetric quantities obtained from the ASTM C 457 method have been published in various sources. The techniques are rather straightforward, but have not been delineated for all three methods of ASTM C 457. Although there is an ASTM standard practice for calculating uncertainty (ASTM D 4356), this practice has not yet been incorporated into ASTM C 457. This is unfortunate, since the calculations for the point count, linear traverse, and modified point count methods have been published and are available. The individual results have been combined here for completeness.

More sophisticated techniques for characterizing spheres embedded within a matrix have been developed by concrete materials researchers and metallurgists, among others. Wicksell[6] was one of the first researchers to address the problem of reconstructing properties of the particle size distribution based upon data collected on a thin section. Others, such as Reid,[7] advanced these ideas to plane section analysis. Subsequent publications completed the remaining mathematical components required to allow one to easily transform among data collected in one-, two-, and three-dimensions.

The issue of "air void spacing" equations has been addressed a number of times since the 1949 equation by Powers.[3] Notably, Philleo,[8] Fagerlund,[9] Pleau and Pigeon,[10] and Attiogbe[11] have each contributed equations. Each equation addresses the topic differently. Also, the lack of consistent terminology has obfuscated discussions of the equations. The equations are discussed here with summary descriptions of each spacing equation. Results from previous numerical[12] and analytical[13] tests of the equations are also discussed.

Purpose

This chapter consolidates most of the information regarding the characterization of air voids in concrete. It is both interesting and unfortunate that the information contained in this chapter has not been summarized previously. The book by Underwood[14] is comprehensive, but it does not contain uncertainty equations sufficient for ASTM C 457, the inversion formulae in analytical form, sophisticated spacing equations, or the concept of protected paste volume. The hope is that this chapter will both serve as a clearinghouse for this information and also express all of this information within a single context. True understanding begins when one is able to look beyond a piece of information and see how this piece interrelates with the whole.

Once compiled, the information contained in this chapter should be of great value to anyone modeling the air void microstructure within concrete. To successfully model the air void system, one must be able to accurately characterize both the size distribution and the proximity of the voids to either one another or to points within the paste. This chapter discusses this in detail for both plane section and thin section analyses.

Finally, although this chapter serves as a useful starting point for future concrete research, it has also been developed with the intent of highlighting what information is still needed. To this end, each section within this chapter includes a discussion of existing research needs.

Quantitative Stereology

Quantitative stereology is the body of methods for quantifying properties of a three-dimensional composite using only the data present on either a plane surface or thin section that intersects that composite. Typically, the objects of interest are discrete entities within a host matrix. Historically, many of the advances in quantitative tools have been developed for solving problems in biology and metallurgy. The field traces most of the theoretical ground work back to Wicksell,[6] who was seeking to quantify the geometrical character of cells on a microscopic thin section of tissue. More recently, metallurgists have been making advances in quantifying irregularly shaped objects, as one might find in certain alloys.

The most useful single reference on this subject is the book *Quantitative Stereology* by Underwood.[14] The book covers much of the theoretical work on quantifying volumetric data. It is an excellent source for proofs of why linear and planar probes are unbiased estimators of volumetric quantities such as volume fraction.

Plane Section

In order to perform a quantitative analysis, a system is often penetrated by a plane section. The intersection of the plane with the spherical air voids creates circles. Since concrete is an opaque material, it may be difficult to visualize the connection between the circles and the spherical voids. Figure 1 shows a three-dimensional void system intersecting a plane. The intersection of the voids with the plane is depicted by a black circle; most of the spheres are below the plane and so do not intersect it. As one can readily see, the size of the intersected circle may not be representative of the intersected sphere radius. However, when considered on the whole, the distribution of circle diameters can yield useful information about the distribution of sphere radii.

The air void microstructure depicted in Fig. 1 shows a system of non-overlapping spheres. It will be assumed throughout this chapter that the air void system is composed of distinct spherical voids. This a priori assertion will receive further discussion in subsequent sections.

Probes

When performing a stereological analysis, the experimenter chooses the type of probe to use. Possible probes include points, lines, surfaces, or volumes. Concrete petrographers are typically most familiar with point probes (point count) and linear probes (linear traverse). ASTM C 457, based upon the work of Brown and Pierson,[5] includes both methods, and a third method (modified point count) that is a combination of the two. The use of a planar probe to analyze air voids in concrete has been discussed by Verbeck.[15] However, the standardized method has adopted the point and linear probe analyses due to their simplicity.

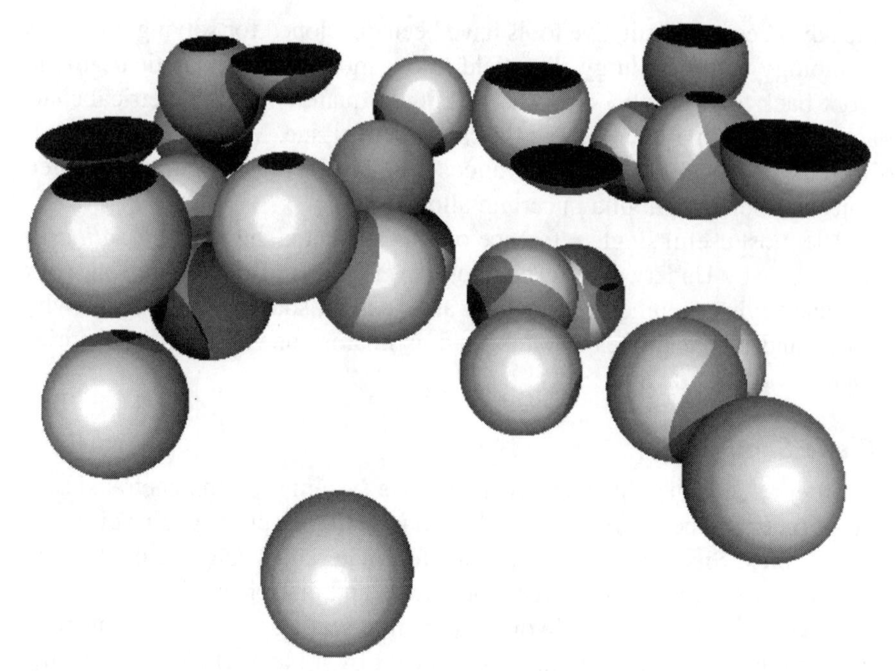

Figure 1. Three-dimensional view of a random plane intersecting a system of mono-sized spheres. The intersection of the plane with a sphere is denoted by a black circle.

The choice of a probe depends upon the quantity desired. Point probes are sufficient for determining the volume fractions. To characterize geometrical quantities, probes of higher dimensionality are required. Linear probes are sufficient for characterizing the size and spacing of the air voids. Estimates of the number of voids in a given volume of concrete can be calculated from planar probe data, but only approximated from linear probe data.

Since the point and linear probes are used in ASTM C 457, they are discussed in detail here. A schematic of a point probe is shown in Fig. 2(a), and a linear probe analysis is shown in Fig. 2(b). The figure represents the intersection of a plane with a matrix containing spheres of a constant radius. Since the location of each sphere is random (but not overlapping another sphere), the circles of intersection between the spheres and the plane do not all have the same radius. The image in Fig. 2(a) contains small filled circles that delineate the locations of the point probe analysis. Similarly, the image in Fig. 2(b) contains horizontal lines that delineate the path of a linear probe over the surface. One could also imagine performing a planar probe analysis by simply recording the radius of each circle intersecting the plane.

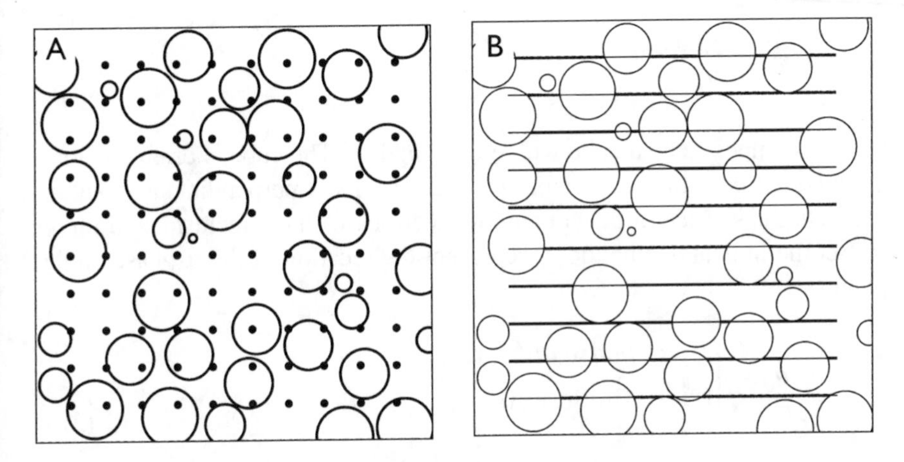

Figure 2. Schematic of a plane surface through a matrix containing mono-sized spheres. The points in (a) delineate a point probe analysis; the horizontal lines in (b) delineate a linear probe analysis.

Volumetric Statistical Properties

The methods of stereology are often used to characterize particles. The particles could either be separate entities or be part of a composite. The composite could be composed of either discrete particles in a matrix or space-filling particles like alloys in a metal. For the case of air voids in concrete, the following discussion will concentrate on discrete particles embedded in a matrix.

The methods discussed here are not used to characterize an individual particle. Since a composite may consist of millions of particles, this approach would be meaningless. Rather, a more useful goal is to characterize the collection of particles that are of interest by estimating useful statistical properties such as the volume fraction and the surface area.

Volume Fraction

One of the most informative statistical properties is the volume fraction of a particular phase of interest. According to Underwood,[14] the idea was outlined by Delesse,[16] and a number of mathematical derivations exist by Hilliard,[17] Chayes,[18] and Saltykov.[19] The following mathematical description follows those of Weibel[20] and McLean.[21] For some phase α within a composite of total volume V_T, the volume of the α-phase is V_α. The ratio of the volumes is the volume fraction ϕ_α of phase α:

$$\phi_\alpha = \frac{V_\alpha}{V_T} \tag{1}$$

Let the composite be a cube with edge length L. The cube is oriented along a Cartesian coordinate system, and a cut is made perpendicular to the z-axis. On the surface of the cut are intersections with the inclusions of phase α. Let the indicator function a_α represent the location of the α-phase on the plane:

$$a_\alpha = \begin{cases} 1 & \alpha \textit{ phase at } (x, y, z) \\ 0 & \textit{otherwise} \end{cases} \tag{2}$$

The surface area A_α of the α-phase on the plane located at z is the integral of a_α over the area A_T of the plane section:

$$A_\alpha = \int_{A_T} a_\alpha(x, y \mid z)\, dx\, dy \tag{3}$$

Replace the cut by a slice with differential thickness dz. The total volume of the α-phase is the integral of A_α:

$$V_\alpha = \int A_\alpha(z)\, dz \tag{4}$$

Let the total planar surface area be $A_T = L^2$. The ratio of volumes can be expressed as a function of $A_\alpha(z)$:

$$\frac{V_\alpha}{V_T} = \frac{\int A_\alpha(z)\, dz}{L A_T} \tag{5}$$

The integral is an average:

$$\frac{V_\alpha}{V_T} = \frac{\overline{A_\alpha}}{A_T} \tag{6}$$

Therefore, an estimate of the area fraction of the phase α is an unbiased estimator of the volume fraction of the phase α, in as much as $A_\alpha(z)$ is an unbiased estimator of $\overline{A_\alpha}$.

This approach can be extended to both linear and point probes. Consider the plane section A_T discussed above. Pass a line, perpendicular to the y-axis, across the surface. Let the indicator function l_α represent the location of the α-phase along the line:

$$l_\alpha = \begin{cases} 1 & \alpha \ phase \ at \ (x,y,z) \\ 0 & otherwise \end{cases} \tag{7}$$

The total length L_α of the α-phase along the line is the integral of l_α:

$$L_\alpha(y,z) = \int l_\alpha(x \mid y,z)\, dx \tag{8}$$

Replace the line with a ribbon with differential width dy. The total area of the α-phase on the surface is the integral of L_α:

$$A_\alpha(z) = \int L_\alpha(y,z)\, dy \tag{9}$$

The ratio of areas can be expressed as a function of $L_\alpha(y,z)$:

$$\frac{A_\alpha(z)}{A_T} = \frac{\int L_\alpha(y,z)\, dy}{L^2} \tag{10}$$

As in the case of the area average, this integral can also be expressed as an average:

$$\frac{A_\alpha(z)}{A_T} = \frac{\overline{L_\alpha}(z)}{L} \tag{11}$$

Therefore, the lineal fraction of phase α along a lineal probe is equal to the area fraction of that phase, which is in turn equal to the volume fraction. The mathematical framework for point probes proceeds in the same manner by constructing lines from points.

This entire development was performed without addressing randomness in either the α-phase or the locations of the probes. This issue arises when asserting that $A_\alpha(z) = \overline{A_\alpha}$. This statement assumes that the area fraction exposed at some value of z is representative of any other $A_\alpha(z)$. This kind of assumption is also made for the linear and point probes. The final result is that the linear fraction of phase α of a linear probe is an unbiased estimation of the volume fraction of phase α. Likewise, the number fraction of points landing on phase α is an unbiased estimation of the volume fraction of phase α.

Surface Area

Another useful characteristic of particles within a matrix is the particle surface area. Within a volume V_T there are particles with total surface area S_α and total volume V_α. The average surface area per particle is S_α / V_α. A linear probe analysis of total length L_T is composed of N_α chords over the α-phase. The total length of chords over the α-phase is L_α, and the average chord length is $\overline{L_\alpha} = L_\alpha / N_\alpha$. The total surface area of particles can be calculated using the equation of Saltykov[22]:

$$\frac{S_\alpha}{V_T} = \frac{4 N_\alpha}{L_T} \qquad (12)$$

This quantity is sometimes referred to as the specific surface. However, in the concrete research literature the quantity of interest has been $\overline{S_\alpha} \,/\, \overline{V_\alpha}$, and has been derived by Tomkeieff[23] and Chalkley et al.[24]:

$$\frac{\overline{S_\alpha}}{\overline{V_\alpha}} = \frac{4 N_\alpha}{L_\alpha} = \frac{4}{\bar{l}} \qquad (13)$$

The quantity $\bar{l}$ is the average chord length, as defined by ASTM C 457. This quantity is also referred to as the specific surface, and will be the default definition for subsequent discussions of air void specific surface.

ASTM C 457 Uncertainty Analysis

Interest in the expressed uncertainty in the results of a linear traverse analysis of air voids dates back to the paper of Verbeck.[15] Not long after, Brown and Pierson,[5] Willis,[25] and Mather[26] used uncertainty arguments for improving the test method. An interlaboratory comparison by Sommer[27] was one of the first attempts to relate theoretical and practical uncertainties.

While the uncertainty in point and linear analyses of volumetric data have been known for some time,[14] contribututions have been made to the specific tasks associated with ASTM C 457. Expressions for the uncertainty for the modified point count method have been given by Langan and Ward[28] and Pleau and Pigeon.[29] Expressions for the linear traverse method have been given by Snyder et al.[30] These results are presented here as a summary.

Details of the uncertainty calculations are given in Appendix A. In summary, let the calculated quantity y be a function of N different measured quantities x_n. The dispersion in each x_n value represents the standard uncertainty $u(x_n)$ in the estimate of x_n. The combined standard uncertainty $u_c(y)$ in y is a function of the individual uncertainties $u(x_n)$:

$$u_c^2(y) = \sum_{n=1}^{N} \left(\frac{\partial y}{\partial x_n} \right)^2 u^2(x_n) + 2 \sum_{m=1}^{N-1} \sum_{n=m+1}^{N} \frac{\partial y}{\partial x_m} \frac{\partial y}{\partial x_n} u(x_m, x_n) \tag{14}$$

To say something quantitative about the calculated value y, one must define a coverage factor k that is used to produce an expanded uncertainty $U_p = ku_c(y)$ that defines an interval $Y = y \pm U_p$ having an approximate level of confidence p. For a Gaussian error model, a coverage factor k equal to 2 will have nearly a 95 % level of confidence. This is the familiar "two-sigma" interval.

For brevity, the notation $u_c^2(y)$ will be shortened to s_y^2 and will serve as an estimated standard deviation. It will be assumed that when the subscript to s^2 is a calculated quantity, the uncertainty in question is a combined standard uncertainty. The coefficient of variation will be represented by the variable C.

It should be noted that these calculations characterize an idealization. The equations reflect the minimum uncertainty that a perfect operator, using equipment with infinite resolution, can state from the result of one planar analysis of a perfectly prepared specimen. Errors due to improper identification of voids, improper sample preparation, and insufficient magnification add to these minimum uncertainties. While each of these factors is an important component to the overall uncertainty, it is critical to at least be able to quantify the uncertainty in the ideal case. Once these have been outlined, one can then begin to quantify the impact of the other sources of error.

Random Processes

The measurement methods of ASTM C 457 include both point count and linear traverse. Although the results from both techniques are random, the statistics differ. The difference is small, but important.

The outcome from a point count experiment is characterized by the binomial distribution. Assume that an operator seeks to characterize the air content within a piece of concrete. The operator will perform a point count consisting of S_t total stops. The probability that an individual point lands over air is q. The probability of not landing over air is $1-q$. The outcome from an experiment is S_a points over air. The estimated variance in the number of points over air is calculated from the binomial distribution:

$$\begin{aligned} s_{S_a}^2 &= S_t \, q\,(1-q) \\ &= S_t \frac{S_a}{S_t}\left(1-\frac{S_a}{S_t}\right) \end{aligned} \tag{15}$$

The square of the coefficient of variation is a simple result:

$$\begin{aligned} C_{S_a}^2 &= \left(\frac{s_{S_a}}{S_a}\right)^2 \\ &= \frac{1}{S_a} - \frac{1}{S_t} \end{aligned} \tag{16}$$

The modified point count and the linear traverse require a similar quantity. During the test, the number of chords intersected N_a is recorded. The intersection of randomly placed air voids is a Poisson process.[31] Therefore, the variance in the number of chords recorded is equal to the number of chords:

$$s_{N_a}^2 = N_a \tag{17}$$

The coefficient of variation is similar to the point count value above:

$$C_{N_a}^2 = \frac{1}{N_a} \tag{18}$$

The difference between the result for C_{S_a} and for C_{N_a} is due to the difference between continuum and discrete sampling; this distinction seems to be missing in the mathematical development of Langan and Ward.[28] Note that as S_t approaches infinity, the value of $C_{S_a}^2$ approaches $1/S_a$, which is analogous to the result for N_a.

Point Count

The point count technique is probably the oldest, and best characterized, tool for petrographic characterization. Because it is a point probe, it is limited as to the volumetric information it can yield. Specifically, the point count method estimates the volume fraction of a particular phase of interest.

While performing an ASTM C 457 point count analysis, the operator records information regarding both the air and the paste:

S_a: number of stops over air

S_p: number of stops over paste

S_t: total number of stops

The desired quantities are the volume fraction of air A and paste p:

$$A = \frac{S_a}{S_t} \qquad p = \frac{S_p}{S_t} \tag{19}$$

Since the total number of stops S_t is fixed, the uncertainty in each volume fraction is proportional to the uncertainty in the number of stops over the phase of interest:

$$C_A^2 = C_{S_a}^2 = \frac{1}{S_a} - \frac{1}{S_t} \qquad C_p^2 = C_{S_p}^2 = \frac{1}{S_p} - \frac{1}{S_t} \tag{20}$$

Linear Traverse

The ASTM C 457 linear traverse technique is a linear probe analysis of concrete containing air entrainment. The operator records the "chords" where the line intersects the phase of interest. As currently written, the operator is required only to record the length of linear probe over the spherical air voids and the cumulative length of the chords. However, the analysis that follows assumes that the operator also records the lengths of the individual chords. This should not be an arduous task since modern laboratory equipment designed for this test is computerized. The uncertainty analysis that follows has been addressed previously.[28,30]

Upon completion of the linear traverse, the reported quantities are as follows:

N_a: number of air chords

T_t: length of entire traverse

T_a: length of traverse over air
$\bar{l}$: average chord length

For the purpose of this analysis, the uncertainty in the total traverse length T_t is assumed to be zero. The uncertainty in the reported quantities will depend upon the uncertainty in the measured quantities:

$$s_{N_a}^2 = N_a \qquad s_{\bar{l}}^2 = \frac{s_l^2}{N_a} \tag{21}$$

Air Content

The analysis of the uncertainty in the air volume fraction begins by expressing the air content A in terms of measured quantities:

$$A = \frac{N_a \bar{l}}{T_t} \tag{22}$$

The coefficient of variation in A is a function of these variables:

$$\left(\frac{s_A}{A}\right)^2 = \left(\frac{T_t}{N_a \bar{l}}\right)^2 \left[\left(\frac{\partial A}{\partial N_a}\right)^2 s_{N_a}^2 + \left(\frac{\partial A}{\partial \bar{l}}\right)^2 s_{\bar{l}}^2\right]$$
$$C_A^2 = \frac{1}{N_a}\left[1 + C_l^2\right] \tag{23}$$

This result has been reported elsewhere.[30,31]

Specific Surface

The specific surface is calculated from the average chord length $\bar{l}$:

$$\alpha = \frac{4}{\bar{l}} \tag{24}$$

Assuming a Gaussian error model for $\bar{l}$, the variance in α is infinite[32] because of the nonzero probability of a negative $\bar{l}$. Nonetheless, one proceeds under the assumption that the normal distribution that characterizes $\bar{l}$ is localized: $C_{\bar{l}} << 1$ This is reasonable since N_a is typically on the order of 1000.

On the basis of this assumption, the coefficient of variation in α can be calculated from the measured quantity $\bar{l}$:

$$\left(\frac{s_\alpha}{\alpha}\right)^2 = \left(\frac{\bar{l}}{4}\right)^2 \left(\frac{\partial \alpha}{\partial \bar{l}}\right)^2 s_{\bar{l}}^2$$

$$C_\alpha^2 = \frac{1}{N_a} C_l^2 \tag{25}$$

Spacing Factor

The analysis for the Powers spacing factor is divided into separate parts for the two equations that apply. When the ratio p/A is less than 4.342, for reasons to be discussed in a subsequent section, the spacing factor $\bar{L}$ has the following relationship:

$$\bar{L} = \frac{p\,T_t}{4\,N_a} \tag{26}$$

The coefficient of variation in $\bar{L}$ is

$$\left(\frac{s_{\bar{L}}}{\bar{L}}\right)^2 = \left(\frac{4\,N_a}{p\,T_t}\right)^2 \left[\left(\frac{\partial \bar{L}}{\partial p}\right)^2 s_p^2 + \left(\frac{\partial \bar{L}}{\partial N_a}\right)^2 s_{N_a}^2\right]$$

$$C_{\bar{L}}^2 = C_p^2 + \frac{1}{N_a} \tag{27}$$

When the ratio p/A is greater than 4.342, as it should be for most concretes containing entrained air,[3] the spacing factor has the following form:

$$\bar{L} = \frac{3}{\alpha}\left[1.4\left(1+\frac{p}{A}\right)^{1/3} - 1\right]$$

$$= \frac{3\bar{l}}{4}\left[1.4\left(1+\frac{p\,T_t}{N_a\,\bar{l}}\right)^{1/3} - 1\right] \tag{28}$$

The coefficient of variation can be expressed as

$$\left(\frac{s_{\overline{L}}}{\overline{L}}\right)^2 = \frac{1}{\overline{L}^2}\left[\left(\frac{\partial \overline{L}}{\partial \overline{l}}\right)^2 s_{\overline{l}}^2 + \left(\frac{\partial \overline{L}}{\partial p}\right)^2 s_p^2 + \left(\frac{\partial \overline{L}}{\partial N_a}\right)^2 s_{N_a}^2\right]$$

$$C_{\overline{L}}^2 = \frac{(1-\beta)^2}{N_a} C_l^2 + \beta^2\left(C_p^2 + \frac{1}{N_a}\right) \tag{29}$$

Here, the result has been made more compact through an intermediate quantity β:

$$\beta = \frac{\bar{l}}{\overline{L}}\left(\frac{1.4}{4}\right)\left(\frac{p}{A}\right)\left(1+\frac{p}{A}\right)^{-2/3} \tag{30}$$

These results represent the general approach of Snyder et al.,[30] Langan and Ward,[28] and Warris.[33] The final result of Langan and Ward is identical, but the answer was simplified by eliminating terms that have a negligible contribution[28]:

$$C_{\overline{L}}^2 = \beta^2\left[C_p^2 + C_A^2\right] \tag{31}$$

Modified Point Count

The modified point count method is a mixture of the point count and the linear traverse methods. The operator performs a point count on each line, then reanalyzes the line to record the number of air voids intersected. The uncertainty in this method has been addressed by Pleau and Pigeon.[29] The measured quantities are as follows:

- N_a: number of air chords
- S_t: total number of stops
- S_a: number of stops over air
- I: distance between stops

Air Content

The air content is simply the ratio of stops over air to the total number of stops, as in the point count method:

$$A = \frac{S_a}{S_t} \tag{32}$$

As discussed before, the coefficient of variation in the air content A follows from the binomial distribution:

$$C_A^2 = \frac{1}{S_a} - \frac{1}{S_t} \tag{33}$$

Specific Surface

The specific surface equation depends upon both point probe and lineal probe data:

$$\alpha = \frac{4\,N_a}{S_a\,I} \tag{34}$$

The coefficient of variation in α is calculated in the same manner as before:

$$\left(\frac{s_\alpha}{\alpha}\right)^2 = \left(\frac{N_a}{4\,S_a\,I}\right)^2 \left[\left(\frac{\partial\,\alpha}{\partial\,S_a}\right)^2 s_{S_a}^2 + \left(\frac{\partial\,\alpha}{\partial\,N_a}\right)^2 s_{N_a}^2\right]$$

$$C_\alpha^2 = \frac{1}{S_a} - \frac{1}{S_t} + \frac{1}{N_a} \tag{35}$$

Spacing Factor

When the ratio p/A is less than 4.342, the Powers spacing factor $\overline{L}$ has the following relationship:

$$\overline{L} = \frac{p}{\alpha\,A} = I\,\frac{p\,S_t}{4\,N_a} \tag{36}$$

The coefficient of variation will depend only upon the uncertainty in the paste content and the number of chords:

$$\left(\frac{s_{\overline{L}}}{\overline{L}}\right)^2 = C_p^2 + \frac{1}{N_a} \tag{37}$$

When the ratio p/A is greater than 4.342, the spacing factor $\overline{L}$ has the following relationship:

$$\overline{L} = \frac{3\,S_a\,I}{4\,N_a}\left[1.4\left(1+\frac{S_p}{S_a}\right)^{1/3} - 1\right] \tag{38}$$

The equation for the coefficient of variation resembles that for the linear traverse method:

$$\left(\frac{s_{\overline{L}}}{\overline{L}}\right)^2 = \left(\frac{\partial\,\overline{L}}{\partial\,N_a}\right)^2 s_{N_a}^2 + \left(\frac{\partial\,\overline{L}}{\partial\,S_a}\right)^2 s_{S_a}^2 + \left(\frac{\partial\,\overline{L}}{\partial\,p}\right)^2 s_p^2$$

$$C_{\overline{L}}^2 = C_{N_a}^2 + \beta^2\,C_p^2 + (1-\beta)^2\,C_{S_a}^2 \tag{39}$$

The value of β is the same as for the linear traverse case:

$$\beta = \frac{I\,S_a}{\overline{L}\,N_a}\left(\frac{1.4}{4}\right)\left(\frac{p\,S_t}{S_a}\right)\left(1+\frac{p\,S_t}{S_a}\right)^{-2/3} \tag{40}$$

This result for the Powers spacing factor coefficient of variation differs slightly from that of Pleau and Pigeon.[29]

Paste Content

In the preceding discussion, the relative uncertainty in the paste content C_p of the concrete has been left undefined. Currently, ASTM C 457 does not specify how the paste content shall be determined for either the linear traverse or the modified point count methods. Therefore, experimenters have been left to their own discretion. Two methods present themselves: record stops over paste during the modified point count method, or record individual paste chords during a linear traverse. By evaluating the paste volume fraction like the air volume fraction, the preceding sections that address uncertainty in air content A can be used to likewise determine uncertainty in the paste content p.

Example

The following example of an uncertainty calculation for ASTM C 457 is based upon data given in Ref. 30 that included results for individual paste chords. The average and standard deviation of the paste chords are represented by $\overline{\pi}$ and σ_π, respectively. The uncertainties reported below represent an estimate of two standard deviations. This corresponds to an approximate coverage factor of 95%.

The data in Ref. 30 are for a linear traverse that included measurements of individual paste chords. The data that appear below for the point count and modified point count are based upon these data. Also, the data are consistent among the methods.

Point Count

The following data are fictitious, but are based upon the data given in Ref. 30:

$$S_t = 1000 \qquad S_a = 57 \qquad S_p = 311$$

The coefficient of variation for the air content A and paste content p can be calculated directly:

$$C_A^2 = \frac{1}{S_a} - \frac{1}{S_t} = 0.0165 \qquad C_p^2 = \frac{1}{S_p} - \frac{1}{S_t} = 0.0022$$

The reported volume fractions are also calculated directly:

$$A = \frac{S_a}{S_t}\left[1 \pm 2\,C_A\right] = (5.7 \pm 1.4)\,\%$$

$$p = \frac{S_p}{S_t}\left[1 \pm 2\,C_p\right] = (31.1 \pm 2.9)\,\%$$

Linear Traverse

The following linear traverse data can be found in Ref. 30: $N_a = 1420$, $N_p = 3447$, $\bar{l} = 0.1126$ mm, $\overline{\pi} = 0.2569$ mm, $\sigma_l = 0.2536$ mm, and $\sigma_\pi = 0.8048$ mm. The total length of the traverse is L_T is 2656 mm.

The coefficient of variation for the volume fractions are as follows:

$$C_A^2 = \frac{1}{N_a}\left[1 + C_l^2\right] = 0.00428 \qquad C_p^2 = \frac{1}{N_p}\left[1 + C_\pi^2\right] = 0.00314$$

The reported volume fractions, to two standard deviations, are as follows:

$$A = \frac{N_a \bar{l}}{L_T}\left[1 \pm 2\, C_A\right] = (6.02 \pm 0.79)\,\%$$

$$p = \frac{N_p \bar{\pi}}{L_T}\left[1 \pm 2\, C_p\right] = (33.4 \pm 3.7)\,\%$$

The coefficient of variation for the specific surface is as follows:

$$C_\alpha^2 = \frac{1}{N_a} C_l^2 = 0.00357$$

The specific surface, reported to two standard deviations, is calculated from

$$\alpha = \frac{4}{\bar{l}}\left[1 \pm 2\, C_\alpha\right] = (35.5 \pm 4.2)\ \mathrm{mm}^{-1}$$

The calculation for the Powers spacing factor $\bar{L}$ is a little more involved. First, the spacing factor is calculated:

$$\bar{L} = \frac{3\bar{l}}{4}\left[1.4\left(1 + \frac{N_p \bar{\pi}}{N_a \bar{l}}\right)^{1/3} - 1\right] = 0.1366\ \mathrm{mm}$$

This is used to calculate β:

$$\beta = \frac{\bar{l}}{\bar{L}}\left(\frac{1.4}{4}\right)\left(\frac{p}{A}\right)\left(1 + \frac{p}{A}\right)^{-2/3} = 0.4574$$

The quantity β is then used to calculate the coefficient of variation in $\bar{L}$:

$$C_{\bar{L}}^2 = \frac{(1-\beta)^2}{N_a} C_l^2 + \beta^2\left(C_p^2 + \frac{1}{N_a}\right) = 0.00186$$

The final result, expressed with a relative uncertainty of $2C_{\bar{L}}$, is

$$\bar{L} = (0.137 \pm 0.012)\ \mathrm{mm}$$

These results are identical with those shown in Ref. 30.

Modified Point Count

The corresponding data for the modified point count method are as follows: $N_a = 1420$, $S_a = 57$, $S_t = 1000$, $S_p = 311$, and $I = 2.656$ mm

The calculation of the volume fractions is identical to that of the point count method: $A = (5.7 \pm 1.4)\%$ and $p = (31.1 \pm 2.9)\%$.

The coefficient of variation of the specific surface is

$$C_\alpha^2 = \frac{1}{S_a} - \frac{1}{S_t} + \frac{1}{N_a} = 0.0173$$

The specific surface to two standard deviations is

$$\alpha = \frac{4\,N_a}{S_a\,I}\left[1 \pm 2\,C_\alpha\right] = \left(37.5 \pm 9.9\right)\ \text{mm}^{-1}$$

As for the linear traverse method, the uncertainty in the Powers spacing factor begins with a calculation of $\overline{L}$:

$$\overline{L} = \frac{3}{4}\frac{S_a\,I}{N_a}\left[1.4\left(1 + \frac{S_p}{S_a}\right)^{1/3} - 1\right] = 0.129\ \text{mm}$$

Next, the quantity β is calculated:

$$\beta = \frac{I\,S_a}{\overline{L}\,N_a}\left(\frac{1.4}{4}\right)\left(\frac{S_p}{S_a}\right)\left(1 + \frac{S_p}{S_a}\right)^{-2/3} = 0.4570$$

The coefficient of variation in $\overline{L}$ can now be calculated:

$$C_{\overline{L}}^2 = \frac{1}{N_a} + \beta^2 C_p^2 + \left(1 - \beta\right)^2 C_{S_a}^2 = 0.00634$$

This result assumes the following:

$$C_{N_a}^2 = \frac{1}{N_a} \qquad C_{S_a}^2 = \frac{1}{S_a}$$

The final result, expressed with a relative uncertainty of $2C_{\overline{L}}$, is

$$\overline{L} = (0.129 \pm 0.021) \text{ mm}$$

Future Research

Although a thorough body of work exists on the topic of uncertainty for ASTM C 457, a number of important questions remain to be answered:

1. Monte Carlo methods could be used to demonstrate the validity of the uncertainty equations presented here. The analytical relationship between the air void radius distribution and the linear traverse chord distribution that will appear in subsequent sections can be used to choose chord lengths at random. These simulated data could then be used to estimate the various calculated quantities in ASTM C 457 and compared to the true value.
2. All of the equations presented in this section represent the smallest possible uncertainty. The equations do not account for measurement uncertainty due to finite optical resolution. The method in ASTM C 457 prescribes a minimum magnification of 50. A study is needed to determine the effects of equipment resolution.
3. A number of methods for determining the uncertainty in the paste content were proposed here. However, for someone performing a linear traverse, performing an additional test, such as a point count, to determine the paste content and its related uncertainty adds considerably to the overall effort. An alternative may be to determine information about the paste content from the air void data.
4. The assumption that the coefficient of variation in the specific surface is small should be tested using actual chord data. Also, for harmonic averages such as the specific surface, the confidence limits are not symmetric, which might have a bearing on compliance.

Inversion Formulae

There may arise situations in which the experimentalist requires information about the sphere radius distribution given information from either planar or linear probes. This is useful for the subsequent characterization of air voids in concrete containing air entrainment. This information is useful for computer simulations of an air void system[34] and studies of the effects of the void distribution on the measured air content.[35] It is also vital to the characterization of air void spacing to be discussed in a subsequent section.

The need for accurately characterizing the air void size distribution can also arise in situations where one wishes to compare two air void distributions. The different air void distributions could arise from differences in admixture type, admixture dosage, pumping and/or atmospheric pressure, vibration, and so on. It may be useful to know how a change in the mixture proportions or the construction process affects voids of a certain size.

Although the material in this section has been thoroughly investigated for use with spherical voids, there does not appear to be a single reference for this material. The book by Underwood[14] addresses numerical procedures, but the mathematics are implied and not discussed in detail.

This section relies heavily on the use of probability density functions (PDF) and cumulative distribution functions (CDF). Since notations vary among authors, Appendix B contains an explanation of the ideas used here, along with the corresponding notation.

Definitions

There does not appear to be a single nomenclature used for the development of the subsequent equations. For clarity, the following notation will be used:

- x: sphere diameter
- $f(x)$: sphere diameter probability density function
- $F(x)$: sphere diameter cumulative probability function
- y: circle diameter : circle diameter probability density function
- $g(y)$: circle diameter cumulative probability function
- z: chord length
- $h(z)$: chord length probability density function
- $H(z)$: chord length cumulative probability function

A number of the references given in this section use sphere and circle radii and half chord lengths in their theoretical developments. The choice of radius distributions seems natural given that the equations for expectations of surface areas and volumes are more easily recognizable. However, this naturally leads to the use of the half chord length, which seems awkward. Here the use of the diameter distributions and the chord length distributions are used because concrete petrographers have experience with chord length distributions and because the relationship among sphere diameters, circle diameters, and chord lengths is somewhat intuitive.

Spheres – Circles

For the case of spheres distributed throughout a volume, a random plane will intersect the spheres, creating circles on the plane. Given a known sphere diameter distribution $f(x)$, the circle diameter distribution $g(y)$ observed on the plane was originally derived by Wicksell,[36] and subsequently derived more rigorously by Nicholson,[37] Tallis,[32] and Watson,[38]

$$g(y) = \frac{y}{\langle X \rangle} \int_y^\infty \frac{f(x)}{\left(x^2 - y^2\right)^{1/2}} dx \tag{41}$$

and also appears elsewhere.[7,39] Note that only the spheres with diameters greater than y contribute to the integral. This equation is based on the model-based approach of assuming the centers of the spheres constituted a Poisson process, which is correct only for a highly dilute system. Jensen[40] used a design-based approach, which simply assumes the intersecting plane is randomly oriented, to derive the same equation.

Subsequent analyses of the Wicksell problem have shown that the Wicksell equation is valid under a wide variety of conditions. Mecke and Stoyan[41] used a marked Poisson process model to demonstrate that the Wicksell equation is valid under dense conditions. The only requirement is that the centers of the spheres are stationary under translations (randomly distributed with no preferred orientation). The derivation of Cruze-Orive[42] relaxed the condition of randomness on the sphere centers and showed that Wicksell's equation is valid for any arbitrary deterministic positions and particle sizes. Therefore, no matter how the non-overlapping spheres are placed, Wicksell's equation still holds. However, inhomogeneities in the system will require additional sampling to ensure precise estimates of the particle size distribution.

The more practical problem is to determine the sphere diameter distribution from the circle diameter distribution observed on the random plane. Using an intermediate function f_1 defined by $f(x) = xf_1(x^2)$, the Wicksell equation can be rewritten as

$$g(y) = \frac{y}{2\langle X \rangle} \int_{y^2}^\infty \frac{f_1(w)}{\left(w^2 - y^2\right)^{1/2}} dw \tag{42}$$

This is the form of the Abel integral equation.[43,44]

The inversion has been demonstrated by Watson.[38] Multiplying Eq. 41 by $(y^2 - w^2)^{-1/2}$ and integrating y from w to infinity gives

$$1 - F(w) = \frac{2\langle X\rangle}{\pi}\int_w^\infty \frac{g(y)}{\left(y^2 - w^2\right)^{1/2}}\,dy \tag{43}$$

which can be found in Wicksell[6] and Kendall and Moran.[39] In the limit where w approaches zero gives

$$1 = \frac{2\langle X\rangle}{\pi}\int_0^\infty \frac{g(y)}{y}\,dy \tag{44}$$

Therefore, the inversion equation for the sphere diameter CDF can be expressed as

$$1 - F(x) = \frac{\displaystyle\int_x^\infty \frac{g(y)}{\left(y^2 - x^2\right)^{1/2}}\,dy}{\displaystyle\int_0^\infty \frac{g(y)}{y}\,dy} \tag{45}$$

This equation is reported in Tallis[32] and Watson.[38]

The inversion equation for the sphere diameter PDF can be determined either from direct differentiation[45]:

$$f(x) = \frac{-2\langle X\rangle}{\pi}\frac{d}{dx}\int_x^\infty \frac{g(y)}{\left(y^2 - x^2\right)^{1/2}}\,dy \tag{46}$$

or from the solution to Abel's integral equation:

$$f(x) = \frac{-2\,x\langle X\rangle}{\pi}\int_x^\infty \left(y^2 - x^2\right)^{-1/2}\frac{d}{dy}\left[\frac{g(y)}{y}\right]dy \tag{47}$$

For completeness, Reid[7] reported yet another expression:

$$f(x) = \frac{1}{\pi\,x}\int_\infty^x \frac{y}{\left(y^2 - x^2\right)^{1/2}}\frac{d}{dy}\left[g(y)\right]dy \tag{48}$$

It is important to note that the circle diameter distribution on the plane cannot be monosized. The differential in the integrand would be zero except at the value of the single circle diameter. At this point the integrand is undefined. This is a mathematical reflection of the fact that it would be impossible for a random plane to intersect a collection of spheres and for all the circles on that plane have the same diameter.

Spheres – Chords

One of the earliest developments of an analytical relationship between air void diameters and chords measured along lines on a random plane is due to Reid.[7] Using similar arguments to the development of the equation relating circle diameters to sphere diameters, the chord length distribution can be calculated from the sphere distribution using the following equation[7,37,38,39]:

$$h(z) = \frac{2\,z}{\langle X^2 \rangle} \int_z^{\infty} f(x)\,dx \tag{49}$$

There are two important features of Eq. 49: the slope and the curvature of $h(z)$ near the origin. The slope can be determined through differentiation:

$$\frac{d}{dz} h(z) = \frac{2}{\langle X^2 \rangle}\left[1 - F(x)\right] - \frac{2\,z}{\langle X^2 \rangle} f(z) \tag{50}$$

Near the origin ($z \to 0$), the slope of the chord length distribution approaches a constant value of $2/\langle X^2 \rangle$. Since the chord PDF must be linear near the origin, the lognormal chord distribution used by Roberts and Scheiner[46] is unphysical because it has zero slope at the origin.

Differentiating Eq. 49 again gives the curvature:

$$\frac{d^2}{dz^2} h(z) = \frac{-4\,f(z)}{\langle X^2 \rangle} - \frac{2\,z}{\langle X^2 \rangle} \frac{d}{dz} f(z) \tag{51}$$

As a practical consideration, the stability of air voids assures that $f(x \to 0) = 0$. Given this fact, the chord distribution has no curvature at the origin. As the value of x increases, the curvature remains negative until after the derivative of $f(z)$ becomes negative at the modal sphere diameter. Therefore, the curvature of the chord length distribution must remain negative until z is greater than the model sphere diameter.

The derivation of the inversion equation can be seen by first rearranging Eq. 49:

$$\frac{h(z)}{z} = \frac{2}{\langle X^2 \rangle}[1 - F(z)] \tag{52}$$

The inverse of Eq. 49 can be determined from differentiating Eq. 52 with respect to z and using a change of variables to x:

$$f(x) = \frac{-\langle X^2 \rangle}{2} \frac{d}{dx}\left[\frac{h(x)}{x}\right] \tag{53}$$

As for the circles on the plane, it is impossible to have a monosized chord distribution along a random linear probe through a collection of spheres.

Circles – Chords

As a matter of completeness, the relationship between the circle diameter and the chord length distributions are given, even though this information is not currently used by ASTM C 457. The chord length distribution expressed as a function of the circle diameter distribution was reported by Reid[7]:

$$h(z) = \frac{z}{\langle Y \rangle} \int_z^{\infty} \frac{g(y)}{\left(y^2 - z^2\right)^{1/2}} dy \tag{54}$$

Since this equation is identical to the expression for $f(x)$ as a function of $g(y)$, the inversion formula can be derived in exactly the same manner:

$$g(y) = \frac{-2\langle Y \rangle}{\pi} \frac{d}{dy} \int_y^{\infty} \frac{h(z)}{\left(z^2 - y^2\right)^{1/2}} dz \tag{55}$$

Reconstruction

Reconstruction of the air void radius distribution can be accomplished using either parametric or nonparametric techniques.

Parametric

Using the parametric approach, the researcher starts with an air void radius distribution such as monosized, lognormal, Rayleigh, and so on. The probe data are then used to determine the parameters of the distribution chosen.

A parametric reconstruction reduces the problem to that of only trying to determine the moments of the distribution. Starting with some PDF for $f(x)$, one can solve for the analytical form for the probe data distribution and the associated moments. These moments will be a function of the PDF parameters, leading to a linear system of equations. The following section is devoted to this endeavor.

Nonparametric

The nonparametric approach makes no assumptions about the analytical form for $f(x)$. The probe data are integrated numerically to give a discrete version of $f(x)$. This approach gives the "true" $f(x)$, but the data also contain noise that must be controlled. This subject is complicated and warrants its own chapter. The reader will find useful discussions in Underwood[14] and in the proceedings publications of Elias[47] and DeHoff and Rhines.[48]

Future Research

Given the small body of research performed on the analysis of air void distributions in concrete, the field remains quite fertile.

1. A parametric approach to estimating the chord distribution seems like a profitable approach to air void characterization. However, until in situ measurements of air void radii can be performed, the experimentalist will have to choose a distribution for the air void radii. One could study the effects of guessing wrong by starting with a distribution for the air void radii, choosing a different distribution to approximate it, and then comparing the approximated distribution to the original distribution.
2. A simple parametric test of the lognormal distribution can be made by measuring the first three moments of a chord distribution and determine whether or not a lognormal sphere distribution exists that has a corresponding chord distribution with nearly the same three moments.

Estimating Sphere Diameter Moments

In some cases, knowledge of the specific sphere diameter distribution may be extraneous. Rather, many calculations require only knowledge of certain expectations, such as the average sphere radius or average sphere volume, from the sphere diameter distribution. The calculation of these expectations, referred to here as moments, is discussed in detail in Appendix B.

Useful information can be gained from only making a study of the moments of the sphere distribution. For example, the air void specific surface depends upon the ratio $\langle X^2\rangle/\langle X^3\rangle$. This technique has been employed in ASTM C 457. However, difficulties do exist. For planar probes, the moments $\langle X^n\rangle$ can be calculated from the averages $\overline{Y^n}$. But for linear probes, the moments can only be approximated from the averages $\overline{Z^n}$. Regardless, analysis of the moments of probe data facilitates parametric studies since one needs to solve only a linear system of equations.

Planar Probe

The straightforward means to determine the moments of the circle diameter distribution is by direct evaluation of the inversion equation[36,39]:

$$\langle Y^n \rangle = \frac{1}{\langle X \rangle} \int_0^\infty y^{n+1} \int_y^\infty \frac{f(x)}{\left(x^2 - y^2\right)^{1/2}} \, dx \, dy \tag{56}$$

The concise expression for the moments is given by Watson[38]:

$$\langle Y^n \rangle = \frac{\sqrt{\pi}}{2} \frac{\Gamma\left\{\frac{1}{2}(n+2)\right\}}{\Gamma\left\{\frac{1}{2}(n+3)\right\}} \frac{\langle X^{n+1} \rangle}{\langle X \rangle} \tag{57}$$

A more useful form for the equation is found in Wicksell[36] and in Kendall and Moran[39]:

$$\langle Y^n \rangle = \begin{cases} \dfrac{\langle X^{n+1} \rangle}{\langle X \rangle} \dfrac{1 \cdot 3 \cdot 5 \cdots n}{2 \cdot 4 \cdot 6 \cdots (n+1)} \dfrac{\pi}{2} & n \ \ is \ \ odd \\[2ex] \dfrac{\langle X^{n+1} \rangle}{\langle X \rangle} \dfrac{2 \cdot 4 \cdot 6 \cdots n}{1 \cdot 3 \cdot 5 \cdots (n+1)} & n \ \ is \ \ even \end{cases} \tag{58}$$

These equations are valid for $n \geq -1$. The equation for the harmonic mean ($n = -1$) gives a direct relationship between a single circle diameter moment and a sphere diameter moment:

$$\langle Y^{-1} \rangle = \frac{\pi}{2\langle X \rangle} \tag{59}$$

From this, the moments with positive exponents of the sphere diameter distribution can be calculated directly from the measured circle diameter moments. The first three moments of the sphere diameter distribution are as follows:

$$\langle X \rangle = \frac{\pi}{2} \frac{1}{\langle Y^{-1} \rangle}$$
$$\langle X^2 \rangle = 2 \frac{\langle Y \rangle}{\langle Y^{-1} \rangle}$$
$$\langle X^3 \rangle = \frac{3\pi}{4} \frac{\langle Y^2 \rangle}{\langle Y^{-1} \rangle} \tag{60}$$

Specific Surface Area

The specific surface area is the ratio of the expected sphere area to the expected sphere volume:

$$\alpha = \frac{\pi \langle X^2 \rangle}{\frac{\pi}{6} \langle X^3 \rangle} = \frac{16}{\pi} \frac{\langle Y \rangle}{\langle Y^2 \rangle} \tag{61}$$

Number Density

The number density of air voids (number of voids per unit volume) n is the ratio of the air volume fraction to the expected air void volume:

$$n = \frac{A}{\frac{\pi}{6} \langle X^3 \rangle} = \frac{8A}{\pi^2} \frac{\langle Y^{-1} \rangle}{\langle Y^2 \rangle} \tag{62}$$

This equation is referred to by Watson[38] as Fullman's formula.[49] The variance in the estimate of n is proportional to the variance in the quantity $\langle Y^{-1} \rangle$, which is infinite.[32] Therefore, estimates of n from planar data may be plagued by difficulties in controlling uncertainties.

Linear Probe

In a manner similar to the planar probe, the linear probe moments can be calculated from the inversion integral:

$$\langle Z^n \rangle = \frac{2}{\langle X^2 \rangle} \int_0^\infty z^{n+1} \int_z^\infty f(x)\, dx \tag{63}$$

The solution has been reported by Wicksell,[36] Tallis,[32] and Watson[38]:

$$\langle Z^n \rangle = \frac{2}{n+2} \frac{\langle X^{n+2} \rangle}{\langle X^2 \rangle} \tag{64}$$

This equation is valid for all $n \neq -2$. Therefore, a set of equations for m different chord moments will have $m + 1$ different sphere moments, and the system of equations is underdetermined. At best, only ratios of sphere moments can be determined from chord moments.

Specific Surface Area

Fortunately, the specific surface area α is a ratio of sphere diameter moments. Therefore, it can be estimated directly from chord moments:

$$\alpha = \frac{\pi \langle X^2 \rangle}{\frac{\pi}{6} \langle X^3 \rangle} = \frac{4}{\langle Z \rangle} \tag{65}$$

This is the equation for specific surface area that is used in the ASTM C 457 standard test method.

Number Density

Although the number density cannot be determined precisely from the chord moments, there are techniques for making estimates of the quantity. The most popular has been given by Lord and Willis,[50] Watson,[38] and Philleo.[8] The approach begins with Eq. 49:

$$\frac{h(z)}{z} = \frac{2}{\langle X^2 \rangle} \left[1 - F(z)\right] \tag{66}$$

As $z \to 0$, the left side of the equation approximates the derivative of the chord distribution at the origin:

$$\left. \frac{dh}{dz} \right|_{z \to 0} = \frac{2}{\langle X^2 \rangle} \tag{67}$$

For practical reasons, the value of z does not have to equal zero. Since the air voids are thermodynamically unstable below some minimum sphere diameter x_m, the above equation is valid for all z less than x_m. Philleo[8] noted that by tabulating chord lengths smaller than x_m into the first two or three chord length intervals, each of the intervals should have an equal value of $h(z)/z$.

Calculating the slope from the smallest chord length intervals is complicated by two factors: the smallest chord length interval typically has relatively few chords, and the measurement device has a limited resolution. The small number of chords in the interval can lead to excessive statistical uncertainty. The matter of equipment resolution has been addressed by Nicholson,[37] who concluded that an unbiased estimate of the sphere number density could not be achieved with a nonzero resolution limit apparatus.

Thin Section Analysis

A recent article by Aarre[51] suggests that there is interest in quantifying an air void distribution using transmitted light through a thin section of concrete. The process consists of placing a thin section over a light source and performing a linear traverse over the specimen, recording the locations where light passes through the specimen. This approach could simplify the task of automating a linear traverse. The drawback of the technique is that the analysis of the results is more complicated than the results from a plane polished section.

The thickness of the specimen determines the diameter of the smallest air void that is detectable. Since the modal diameter of entrained air voids may be as small as 30 μm, useful specimens must be thinner than this. As an analogy, consider trying to estimate the diameter distribution of voids in Swiss cheese from thick slices from the block of cheese.

A straightforward solution to the "Swiss cheese" problem has been presented by Coleman.[52] While others have developed nonparametric techniques for unfolding the distribution, the work of Coleman gives an analytical relationship between the observed circle diameter distribution and the sphere diameter distribution, given the specimen thickness. Further, his derivation included parameters to account for the effects of resolution that include the smallest measurable circle diameter and the accuracy with which one can determine the location of the circle's edge. However, these resolution considerations will not be discussed here.

The development of the thin section equation is quite similar to that of the corresponding inversion equation for a plane polished section. For a thin section of thickness τ, only the spheres with diameters greater than τ can penetrate the thin section. Also, for each diameter, there is a volume over which the sphere center can be located and still penetrate both sides of the thin section; larger spheres have a greater volume over which their centers can be located as compared to smaller spheres. This "weights" the original sphere diameter distribution $f(x)$, and gives an adjusted distribution $m(x)$ that characterizes the distribution of the penetrating spheres[52]:

$$m(x) = \frac{(x-\tau)\, f(x)}{\int_{\tau}^{\infty} (x-\tau)\, f(x)\, dx} \tag{68}$$

The denominator normalizes the distribution and can be expressed using a shorthand notation:

$$\langle X \rangle_{\tau} = \int_{\tau}^{\infty} (x-\tau)\, f(x)\, dx \tag{69}$$

The radius of the circle Y projected through the thin section depends upon the distance W the sphere center is located from the center of the thin section:

$$Y^2 = x^2 - (W+\tau)^2 \tag{70}$$

From this, one can calculate the probability of an exposed circle with diameter Y:

$$P(Y > y|x) = P\left(W < \left(x^2 - y^2\right)^{1/2} - \tau \,\middle|\, x\right) \tag{71}$$

Because the centers of the spheres are located at uniformly random distances from the center of the thin section, the distribution of W is uniform over the interval $[x - \tau]$. Using this, the previous probability can be expressed directly:

$$P(Y > y|x) = \frac{\left(x^2 - y^2\right)^{1/2} - \tau}{x - \tau} \qquad 0 \le y \le \left(x^2 - \tau^2\right)^{1/2} \tag{72}$$

The conditional dependence upon x can be eliminated by integration:

$$P(Y > y) = \int_{\tau}^{\infty} P(Y > y|x)\, m(x)\, dx \tag{73}$$

The resulting circle diameter distribution $g_\tau(y)$ projected through the thin section is the differential of this probability function[52]:

$$\begin{aligned} g_\tau(y) &= \frac{-d}{dy} P(Y > y) \\ &= \frac{y}{\langle X \rangle_\tau} \int_{(y^2+\tau^2)^{1/2}}^{\infty} \frac{f(x)}{(x^2 - y^2)^{1/2}}\, dx \end{aligned} \tag{74}$$

In the limit that τ approaches zero asymptotically, this equation becomes identical to Eq. 41 for the plane polished section. Similarly, the nth moment for the observed circle diameter distribution can be calculated by integrating the above inversion equation:

$$\langle Y^n \rangle_\tau = \frac{1}{\langle X \rangle_\tau} \int_0^{\infty} y^{n+1} \int_{(y^2+\tau^2)^{1/2}}^{\infty} \frac{f(x)}{(x^2 - y^2)^{1/2}}\, dx\, dy \tag{75}$$

Again, as $\tau \to 0$, this equation becomes identical to the corresponding plane section equation.

Monosized Sphere Diameters

Consider a monosized sphere diameter distribution: $f(x) = \delta(x - x_0)$. (A discussion of the Dirac delta function is given in Appendix C.) Assuming that the sphere diameter x is greater than the thin section thickness τ, the observed circle diameter distribution can be calculated from Eq. 74:

$$g_\tau(y) = \frac{y}{(x_0 - \tau)} \frac{1}{(x_0^2 - y^2)^{1/2}} \qquad 0 \le y \le (x_0^2 - \tau^2)^{1/2} \tag{76}$$

The moments of this distribution can be calculated from the following integral:

$$\langle Y^n \rangle_\tau = \frac{1}{(x_0 - \tau)} \int_0^{\sqrt{x_0^2 - \tau^2}} \frac{y^{n+1}}{(x_0^2 - y^2)^{1/2}}\, dy \tag{77}$$

This result is useful for exploring the results of the paper by Aarre[51] and the subsequent discussion by Snyder.[53] The purpose of the paper by Aarre was to calculate "correction" factors when measuring air content from a thin section. The calculations assumed a monosized air void diameter distribution. The original paper by Aarre contained errors that were corrected in the discussion by Snyder.

Although the equations by Snyder were correct, they contained a subtle oversight. Let the air void diameter distribution be expressed as a Delta function, as was done in the paragraphs above. Using planar probes on a plane polished section, the air content is a function of the number of voids per unit area n and the second moment of the circle diameters measured on the plane ($\tau = 0$):

$$A = n\pi \langle Y^2 \rangle \tag{78}$$

However, an analysis of a thin section with thickness $\tau \neq 0$ will contain fewer observed voids n_τ and a different second moment $\langle Y^2 \rangle_\tau$:

$$A_\tau = n\pi \langle Y^2 \rangle_\tau \tag{79}$$

The number density of circles n observed on the plane is proportional to the first moment $\langle X \rangle$ of the sphere diameters x:

$$\int_0^\infty x\, f(x)\, dx = x_0 \tag{80}$$

The number density n_τ of circles observable through a thin section is proportional to the corresponding first moment of observable spheres $\langle X \rangle_\tau = (x_0 - \tau)$, as was defined previously. The ratio of the number density of spheres observed through thin section to the number density circles on a plane section can be expressed as a function of the thin section thickness τ:

$$\begin{aligned} \frac{n_\tau}{n} &= \frac{x_0 - \tau}{x_0} \\ &= \left(1 - \frac{\tau}{x_0}\right) \end{aligned} \tag{81}$$

The ratio of the observed air content to the true air content can now be expressed as a function of thin section thickness τ:

$$\frac{A_\tau}{A} = \frac{n_\tau \langle Y^2 \rangle_\tau}{n \langle Y^2 \rangle}$$

$$= \left[1 - \frac{3}{2}\left(\frac{\tau}{x_0}\right) + \frac{1}{2}\left(\frac{\tau}{x_0}\right)^3\right] \tag{82}$$

This is identical to the correction factor reported by Snyder.[53] However, in the derivation of Snyder, the factor of $(1 - \tau / x_0)$ was implicit in the averaging used. This factor is absent from the correction factor for specific surface because the $\langle Y^2 \rangle_\tau / \langle Y^3 \rangle_\tau$ ratio cancels this. Therefore, the result of Snyder for the specific surface is also correct.

Lineal-Path Function

Lu and Torquato[54,55] have derived the lineal-path function that gives the probability that a line segment of length τ is entirely within a single phase of the microstructure. Equivalently, this is the area fraction of light passing through a section of concrete with thickness τ. Interestingly, the lineal-path function $L(\tau)$ can be calculated from the chord length distribution function $h(z)$ and the volume fraction of air A[55]:

$$L(\tau) = \frac{A}{\langle Z \rangle} \int_\tau^\infty (z - \tau)\, h(z)\, dz \tag{83}$$

The lineal-path function for a lognormal distribution of air void radii with a modal diameter of 30 μm and with $\sigma_0 = 0.7362$ is shown in Fig. 3.

For monosized spheres of diameter x_0, the chord distribution is $h(z) = 2z / x_0^2$. Substituting this into the previous equation yields the same correction factor as given above for monosized spheres. Conversely, the chord length distribution function can be derived from air volume fraction A and the lineal-path function $L(z)$ by differentiating this equation twice:

$$h(z) = \frac{\langle Z \rangle}{A} \frac{d^2}{dz^2} L(z) \tag{84}$$

This could be applied to a serial sectioning approach to automated air void analysis using transmitted light. If the only illumination was from below the specimen, the total transmission is equivalent to $L(z)$. Therefore, a single light intensity measurement could be made at each specimen thickness and the chord length distribution could calculated from the equation above.

Future Research

Relatively little research has been carried out on thin section analysis of air voids. Given that one could determine the chord distribution $h(z)$ using a single light intensity measurement after consecutive stages of a serial section, there are possibilities for automation. However, a number of theoretical studies could be performed to validate the approach and to test the applicability of the method in general:

1. Assuming a lognormal distribution of air void radii, determine the correction factor for air content, specific surface, and spacing factor from the parameters of the distribution and the thickness of the thin section.
2. Use the formulae in Coleman[52] to determine the effect of resolution limits on the results of an analysis. These formulae could also be used to analyze resolution limits on plane section analysis.

Air Void Spacing

One of the first attempts to characterize the "spacing" of air voids was by Powers.[3] This definition of spacing became a part of the standard test method for determining the air void parameters in hardened concrete (ASTM C 457),[1] and is quantified as the spacing factor $\overline{L}$. Since then, spacing equations have been proposed by Philleo,[8] Fagerlund,[9] Attiogbe,[11] and Pleau and Pigeon.[10] Each of these equations attempts to characterize the spacing of voids in air-entrained concrete, even though the Attiogbe equation estimates the spacing among air voids, and the other equations estimate the distance water must travel to reach the nearest air void.

At present, evaluation of an air void spacing equation consists of a comparison between the estimate of spacing and the results of laboratory freeze-thaw experiments.[56,57] The a priori assumption is that each equation is inherently correct in its estimate of spacing. Unfortunately, each of these spacing equations proposed for predicting freeze-thaw performance has inherent assumptions or simplifications built into its development. Until recently, no quantitative measure has been made of the effects due to these assumptions.

A numerical accuracy test of these equations was performed by Snyder.[12] The computer experiment measured various spacing quantities in a simulated paste-air system. Systems were composed of air voids with either monosized or lognormally distributed radii. Since the size and the location of each sphere were known exactly, the actual spacings could also be calculated numerically. To achieve acceptable statistics, the results from many system

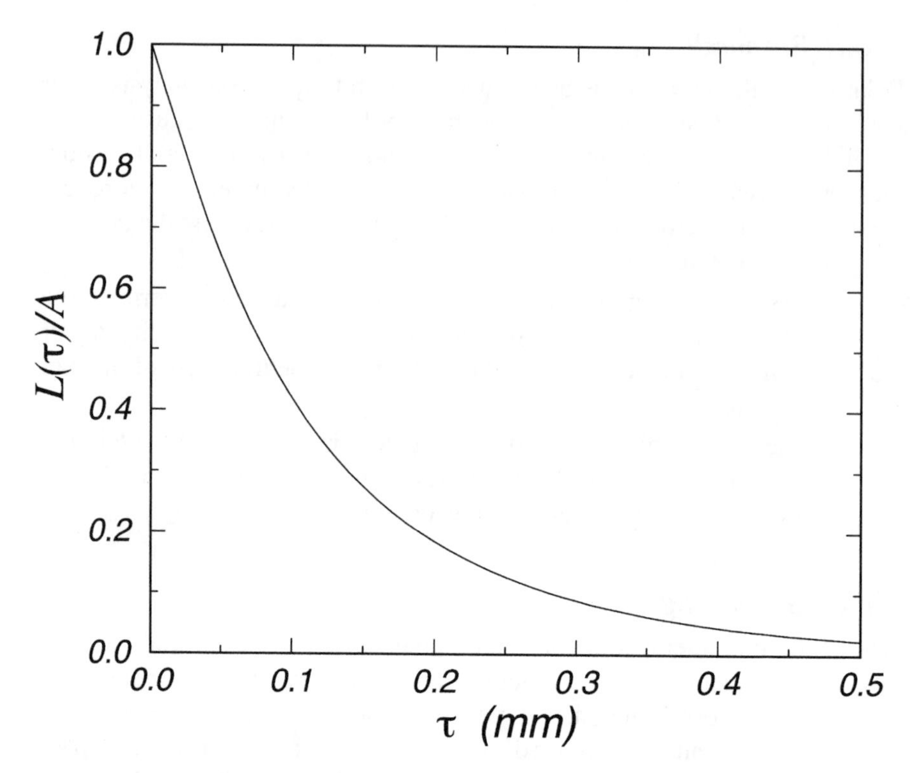

Figure 3. Lineal-path function $L(\tau)$, normalized by the air content A, for a lognormal radius distribution with a modal diameter of 30 μm and with σ_0 = 0.7362.

realizations were used to estimate averaged quantities. These results, along with the associated spacing equation predictions, were reported for comparison. In addition, an equation by Lu and Torquato[54] was included since it promised excellent performance for estimating various spacing quantities.

Spacing Distributions

This section also makes use of the ideas of probability density functions (PDF) and cumulative distribution functions (CDF), the details of which are discussed in Appendix B.

There are two classifications of spacing equations that will be discussed here. Some equations estimate the proximity of the paste to the voids, and others estimate the proximity of the voids to one another. Although this may seem a subtle distinction, it will be shown that the mathematical relationships that characterize these concepts have different behaviors.

Any reasonable concept of spacing should address the fact that there must exist a distribution of distances that characterize the spacing. Clearly, some regions of the paste are closer to an air void than other regions, and some voids have nearer neighbors than others. This characteristic can be represented by a distribution of distances, as depicted in Fig. 4 for a distance parameter s. This PDF represents the fraction of spacings found in the interval $[s,s + ds]$for some differential element ds. The CDF increases monotonically from zero to unity and represents the fraction of spacings less than s.

To illustrate the usefulness of the CDF, Fig. 4 shows two horizontal dashed lines intercept the ordinate axis at the 50th and 95th percentiles. These lines intercept the CDF at s values of 1.95 and 3.1, respectively; 50% of the spacings are less than 1.95, and 95% are less than 3.1. In theory, the CDF only asymptotes to unity; thus, to capture all of the spacings, s must increase to infinity. In practice, however, the quantity s can only increase to the size of the system. Therefore, the concept of a maximum spacing is an ill-defined quantity. Instead, a particular percentile must be chosen. In the report, the 50th and 95th percentiles of the spacing distributions were used to characterize both the measured and the estimated values, since these percentiles are intuitive to one's concept of spacing and protected paste.

Paste-Void Proximity

Paste-void proximity equations estimate the volume fraction of paste within some distance from the surface of the nearest air void. There are two simple ways to visualize this spacing, both shown schematically in Fig. 5:

1. Imagine surrounding each air void with a shell of thickness s. These shells may overlap one another, but may not overlap or penetrate air voids. The volume fraction of the paste that is within any shell is equivalent to the volume fraction of paste within a distance s of an air void surface.
2. Given an air void system, pick points at random throughout the paste that lie outside the air voids. For each point, find the distance to the nearest air void surface. The number fraction of the points that fall within a distance s of an air void surface is equal to the volume fraction of paste within a distance s of an air void surface. This second approach is the one used here to estimate the CDF of the spacing distribution.

This definition of the paste-void proximity distribution is the same as that used by proponents of the protected paste volume (PPV) concept.[58–60] The material parameters of the concrete determine the limiting spacing, and one wants to determine the fraction of paste within this distance to the nearest air void.

Void-Void Proximity

Void-void proximity spacing equations can be further classified into either nearest neighbor or mean free path calculations.

Nearest Neighbor

Nearest neighbor void-void proximity equations estimate the surface-surface distance between nearest neighbor air voids. This idea is shown schematically in Fig. 6(a). This distance is calculated by starting from a given air void and finding the shortest distance from the surface of that void to the surface of any other air void. This is repeated for a number of different air voids. This collection of random distances, when sorted and plotted versus its relative rank, form an estimated void-void proximity cumulative distribution function.

As will be demonstrated subsequently, void-void proximity spacings have a subtle complexity. For an air-void system composed of poly-dispersed sphere diameters, the average void-void spacing originating from large spheres is smaller than the average void-void spacing originating from small spheres. Therefore, the "mean void-void spacing" is an ill-defined quantity when stated without additional qualifiers, since it varies over the distribution of sphere diameters.

Mean Free Path

The mean free path is the average length of paste between adjacent air voids along a randomly chosen line passing through the air void system, and is shown schematically in Figure 6(b). If an ASTM C 457 linear traverse was performed on a paste specimen containing entrained air voids, the mean free path would be equal to the average paste chord length. It is important to note that this distance is neither the longest nor the shortest distance between air voids in an air void system.

Aggregate Effect

The effect aggregates have on the spacing distribution has been neglected for each of the spacing equations. The assumption is that the inter-aggre-

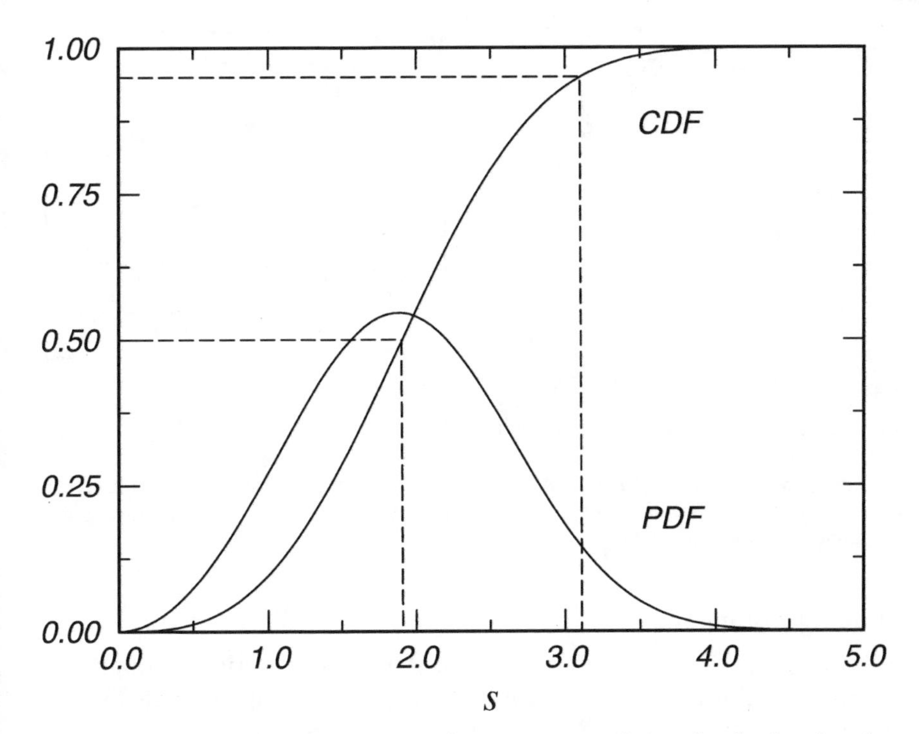

Figure 4. An idealized representation of a spacing cumulative distribution function (CDF) and the associated probability density function (PDF) for some distance s. The dashed lines demonstrate how to determine the 50th and 95th percentiles from CDF data.

gate paste regions are large enough to contain a statistically significant number of air voids. Based upon this assumption, the statistics calculated for the air voids in these inter-aggregate regions are unbiased estimates of the values calculated from the paste-air systems with the same number density of air voids. Measurements by Diamond et al.[61] indicate that the average inter-aggregate spacing on a plane section is on the order of 100 μm. Since this spacing in three dimensions may be less, the presence of aggregates may have a significant impact on spacing since there may only be a few air voids within many of the inter-aggregate regions. However, recent results by Bentz and Snyder[62] suggest that the effect upon relevant statistical properties due to aggregates is negligible if the aggregates are larger than the air voids of interest. Although further study is needed, this paper neglects the effect of aggregates, as do most air void spacing equations.

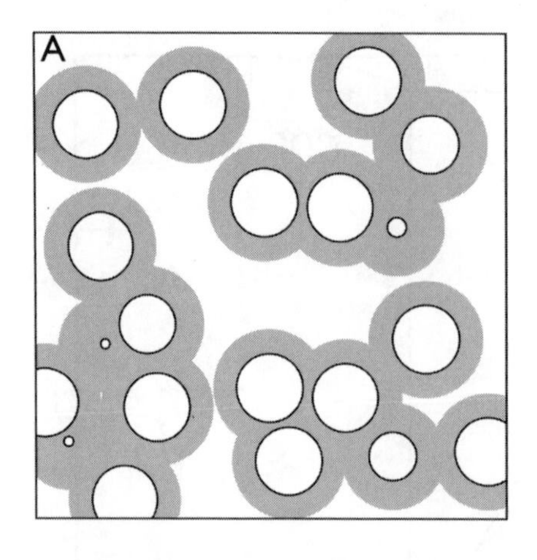

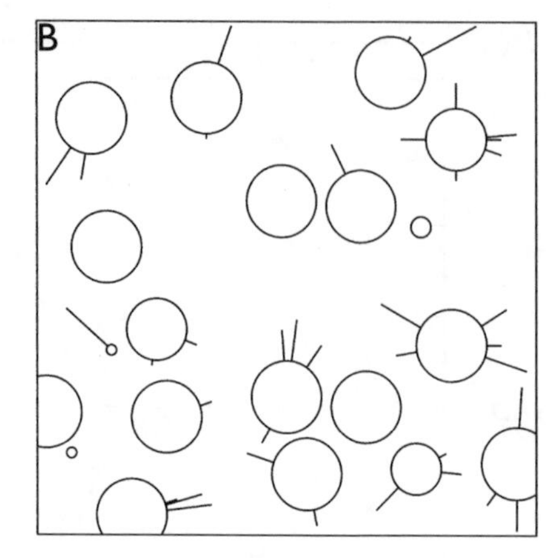

Figure 5. Two-dimensional schematic of the paste-void proximity. The quantity can determined from the volume of concentric shells (a) or from the nearest surface distance distribution (b).

Spacing Equations

Nomenclature for air void quantities differs among various authors. To express quantities with a common notation here, the following definitions are given:

- n: number of air voids per unit volume
- A: air void volume fraction
- p: paste volume fraction
- α: specific surface area of voids
- r: sphere radius
- $f(r)$: sphere radius probability density function
- $\langle R^k \rangle$: expectation of R^k for the radius distribution
- s: spacing distribution parameter

For the paste-air systems (no aggregate) considered here, these quantities can be defined analytically as follows:

$$A = \frac{4\pi}{3} n \langle R^3 \rangle$$
$$p = 1 - A$$
$$\alpha = 3 \frac{\langle R^2 \rangle}{\langle R^3 \rangle}$$
$$\langle R^k \rangle = \int_0^\infty r^k \; f(r) \, dr \tag{85}$$

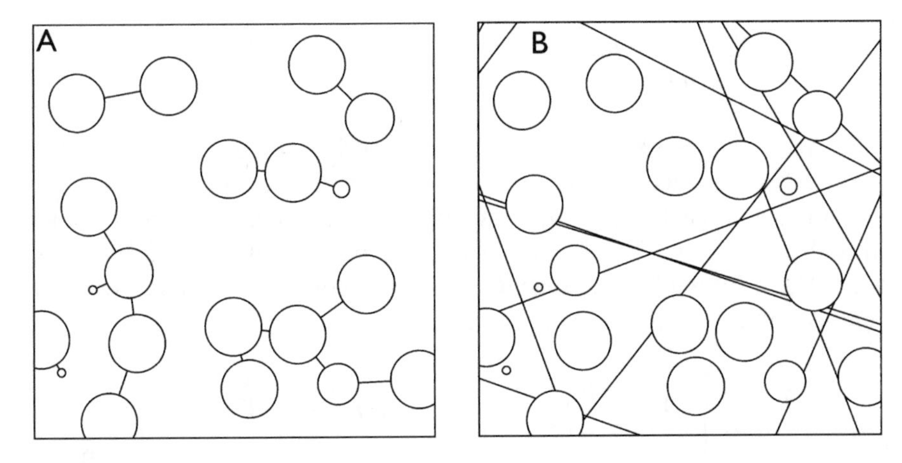

Figure 6. Two-dimensional schematic of the void-void proximity (a) and mean free path (b).

Powers Spacing Factor

The most widely used paste-void spacing equation is the Powers spacing factor.[3] Contrary to a popular misconception, it does not attempt to estimate the distance between air voids. Rather, it is an attempt to calculate the fraction of paste within some distance of an air void (paste-void proximity). The Powers equation approximates the constant distance from the surface of each air void surface, which would encompass some large fraction of the paste. However, the value of this fraction is not quantified.

The second misconception is that the Powers spacing factor represents the maximum distance water must travel to reach the nearest air void in a concrete specimen.[57,63] From the previous discussion of the distribution of paste-void and void-void spacings, it should be clear that there is no single theoretical maximum value for the paste-void spacings. One can only quantify percentiles of the distribution to characterize the fraction of paste within some distance to the nearest air void surface. In practice, the maximum paste-void spacing is the size of the sample.

The Powers spacing factor was developed using two idealized systems. For small values of the p/A ratio, there is very little paste for each air void. Powers used the "frosting" approach of spreading all of the paste in a uniformly thick layer over each air void. The thickness of this "frosting" is approximately equal to the ratio of the volume of paste to the total surface area of air voids:

$$\overline{L} = \frac{p}{4\pi \, n \, \langle R^2 \rangle} = \frac{p}{\alpha \, A} \qquad p/A < 4.342 \tag{86}$$

For large values of the *p*/*A* ratio, Powers used the cubic lattice model. The spheres are placed at the vertices of a simple cubic array. The air voids are monosized, each with a specific surface area equal to the bulk value. The cubic lattice spacing is chosen such that the air content equals the bulk value. The resulting Powers spacing factor is the distance from the center of a unit cell to the nearest air void surface:

$$\overline{L} = \frac{3}{\alpha}\left[1.4\left(1 + \frac{p}{A}\right)^{1/3} - 1\right] \qquad p/A \geq 4.342 \tag{87}$$

The *p*/*A* value of 4.342 is the point at which these two equations are equal.

The intent was that a large fraction of the paste should be within $\overline{L}$ of an air void surface. An acceptable value of $\overline{L}$ for good freeze-thaw performance is determined from the material properties of the concrete.

Philleo Spacing Equation

Philleo[8] extended the approach of Powers by attempting to quantify the volume fraction of paste within some distance of an air void system (paste-void proximity). Philleo started with an idealized air void system composed of n randomly distributed points per unit volume. The probability that the nearest randomly distributed point is a distance x from a random location was first given by Hertz[64]:

$$H(x) = 1 - \exp\left(\frac{-4\pi \, n}{3} x^3\right) \tag{88}$$

This is equivalent to the cumulative paste-void proximity distribution function for zero radius voids. Philleo then modified this distribution to account for finite-sized spheres by renormalizing the cumulative distribution to account for the air content. The rescaling is shown schematically in Fig. 7. For a void content of 0.20, the ordinate axis is rescaled from 0 to 1, and the abscissa axis is simply offset by $s = x - x_0$.

The result, although still only an approximation, characterizes the paste-void spacings for finite-sized air voids. For an air-paste system, the Philleo spacing factor for the volume fraction of paste within a distance s of an air void surface is

$$F(s) = 1 - \exp\left[-4.19x^3 - 7.80x^2\left(\ln p^{-1}\right)^{1/3} - 4.84x\left(\ln p^{-1}\right)^{2/3}\right] \quad (89)$$

Here, the substitution $x = sn^{1/3}$ has been made in order to simplify the appearance of the result.

Fagerlund Spacing Equation

The approach used by Fagerlund[9] is similar to that of Philleo. The air void system is approximated by non-overlapping voids. Each void is surrounded by a shell with thickness s. For small air contents and small values of s, the shells are essentially non-overlapping. As the shell thickness s increases, the number of shells that intersect increases. The value of s for which half of the void shells intersect another shell is denoted $\bar{s}$. Fagerlund defines the mean void spacing $\bar{d}$ as twice this value:

$$\bar{d} = 2\,\bar{s} \quad (90)$$

To estimate $\bar{s}$, Fagerlund uses the volume v_s of the shell system:

$$v_s = \int_0^\infty f(r)\,\frac{4\pi\,n}{3}\,(r+s)^3\,dr \quad (91)$$

The value $\bar{s}$ is approximated from that value of s required to make v_s equal to unity:

$$\int_0^\infty f(r)\,\frac{4\pi\,n}{3}\left(r+\bar{s}\right)^3\,dr = 1 \quad (92)$$

This equation is simplified by expanding the cubic term and substituting for $\bar{d}$ and the specific surface area α[9]:

$$\frac{4\pi\,n}{3}\langle R^3\rangle\left[1 + \left(\frac{\alpha}{2}\right)\bar{d} + \left(\frac{\alpha}{4}\right)\frac{\langle R\rangle}{\langle R^2\rangle}\bar{d}^2 + \left(\frac{\alpha}{8}\right)\frac{1}{3\,\langle R^2\rangle}\bar{d}^3\right] = 1 \quad (93)$$

For an air void radius distribution $f(r)$, one simply needs to solve this cubic equation for $\bar{d}$.

Attiogbe Spacing Equation

Attiogbe has proposed a spacing equation which estimates the "mean spacing of air voids" in concrete.[11] From the figures in the paper by Attiogbe, it appears as though the spacing equation attempts to estimate one-half of the average minimum surface-surface spacing among neighboring air voids. However, an accurate numerical test of the equation is complicated by ambiguity in the exact definition of what the Attiogbe spacing equation attempts to quantify. Figure 1 of Ref. 11 depicts the spacings considered. In that figure, the author has chosen the nearest three voids as neighbors. Attiogbe should have included the other six voids that are "visible" to the central void since, by Attiogbe's definition, "$\bar{d}$ is defined by considering only the distances, between adjacent air voids, which are entirely occupied by paste."[11]

A definitive numerical test of the Attiogbe spacing equation is complicated further by a choice between two spacing equations. The first equation published was, "valid for all values of p/A"[11]:

$$t = 2\frac{p^2}{\alpha A} \tag{94}$$

To avoid confusion with the other spacing equations presented here, the variable t has been substituted for $\bar{s}$ used in the original equation. Upon noting that this equation has peculiar properties for some values of p/A,[65] Attiogbe has proposed a more complete equation[57]:

$$t_G = 2G\frac{p^2}{\alpha A} \tag{95}$$

The variable G replaces the variable F to avoid confusion with the Philleo spacing factor. Attogbe states that, "[G] . . . is the fraction of the total paste volume within the distances of [t] from the edges of the air voids. . . . In this regard, [G] is equivalent to the probability factor defined in Philleo's 'protected paste volume concept.'"[11] The function is only a function of the paste-to-air ratio[11]:

$$G = \begin{cases} \dfrac{8}{p/A+1} & p/A \geq 7 \\ 1 & p/A < 7 \end{cases} \tag{96}$$

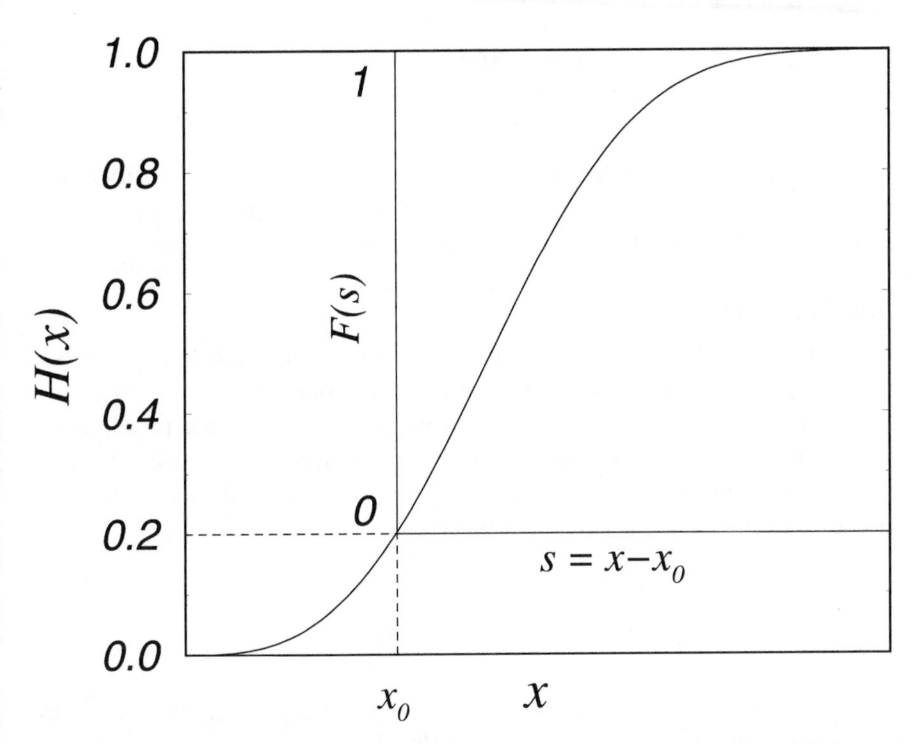

Figure 7. Schematic of Philleo method for rescaling the Hertz probability distribution function for a 20% volume fraction of voids.

However, the quantity G depends upon the air void radius distribution. Fortunately, Attiogbe[66] has recently given an explicit equation for G for an air void diameter probability density function $f(x)$ based upon the gamma function[67]:

$$f(x) = \frac{x^{a-1}\, e^{-x/b}}{b^{a}\, \Gamma(a)} \tag{97}$$

The parameters a and b can be related to the mean diameter $\overline{X}$ and the variance of the distribution σ^2:

$$a = \frac{\overline{X}^2}{\sigma^2} \qquad b = \frac{\sigma^2}{\overline{X}} \tag{98}$$

For any parameters (a,b), the equation for G has been given[66]:

$$G = \frac{(18/\pi)\left[1+(a+3)/4a\right]}{p/A+1} \tag{99}$$

This result can be applied to the lognormal air void radius distributions. Additionally, since G is an estimate of the fraction of paste within t of an air void, it will be compared to measured values.

Mean Free Path

The behavior of the Attiogbe function t for voids of zero radius suggested a relationship to the mean free path λ between air voids. As stated previously, the mean free path is numerically equivalent to the average paste chord lengths in a paste-air system. The probability density function of paste chords can be found in Lu and Torquato,[55] and is expressed as a function of moments of the air void radius distribution:

$$h_p(z) = \frac{n\pi\langle R^2\rangle}{1-A}\exp\left(\frac{-n\pi\langle R^2\rangle}{1-A}z\right) \tag{100}$$

This is the exponential distribution. For a paste-air system, the mean paste chord length λ can be written directly:

$$\lambda = \frac{p}{n\pi\langle R^2\rangle} \tag{101}$$

The same equation can be found in stereology books such as that of Underwood.[14] If the center of the air voids remain fixed, as the radii of the air voids decreases to zero, the mean free path diverges toward infinity, like the Attiogbe equation for t.

This similarity is more than coincidental. In fact, the Attiogbe equation for t is directly proportional to λ. Expressing the Attiogbe equation for t as

$$t = 2p\frac{p}{\alpha A} \tag{102}$$

the quantity αA can be simplified using Eqs. 85:

$$t = \frac{p}{2} \frac{p}{n\pi \langle R^2 \rangle}$$

$$= \frac{p}{2} \lambda \tag{103}$$

Therefore, at low air contents the value of t is approximately equal to one-half the mean free path between air voids in a paste-air system. It is interesting to note that the Powers equation for $p/A < 4.342$ is equal to $\lambda/4$. Therefore, at high air contents, there should be relatively little qualitative difference between the Attiogbe and the Powers equations. However, at low air contents, the two equations have different asymptotic behavior.

Pleau and Pigeon Spacing Equation

Pleau and Pigeon[10] have recently proposed a spacing equation for the paste-void spacing distribution. Their approach considers both the air void radius distribution and the distribution of distances between a random point in the paste and the nearest air void center. Let $h(x)$ represent the probability density function of the distance between a random point in the system and the center of the nearest air void. The joint probability β that this random point is a distance s from the surface of an air void with radius r is

$$\beta(s,r) = h(r+s)\, f(r) \tag{104}$$

As an approximation for $h(x)$, Pleau and Pigeon employ the PDF of the Hertz distribution[64] used by Philleo[8]:

$$h(x) = 4\pi\, n \exp\left(\frac{-4\pi\, n}{3} x^3 \right) \tag{105}$$

However, the centers of air voids are not entirely random since air voids do not overlap one another. The consequence of this choice for $h(x)$ is discussed subsequently.

The joint probability density function $\beta(s,r)$ depends upon $h(s + r)$. If a point chosen at random throughout the entire system lies at a distance x from the center of a sphere, the quantity s is defined as $x - r$. Therefore, if the random point lies within the sphere, the quantity s will be negative, but the argument x of $h(x)$ will be either zero or some positive number.

The parameter r may be eliminated from the joint probability $\beta(s,r)$ by integrating over the possible radii:

$$k(s) = \int_0^\infty h(r+s)\, f(r)\; \Theta(r+s)\, dr \tag{106}$$

The Heaviside function[68] insures that the argument of the function h remains positive. This equation is the fundamental equation of Pleau and Pigeon. The cumulative distribution function is

$$K'(s) = \int_{-\infty}^{s} k(s')\, ds' \tag{107}$$

and corresponds to the volume fraction of the entire system within s of an air void center. The volume fraction of the entire system that would lie within an air void is $K'(0)$, and corresponds to an estimate of the air void volume fraction. The volume fraction of paste within s of an air void surface would then be

$$K(s) = \frac{1}{Q}\int_0^{s} k(s')\, ds' \tag{108}$$

where Q normalizes the result by the volume fraction of paste.

The normalization factor Q should equal $1 - A$, or the paste volume fraction. Based upon the equations of Pleau and Pigeon, this is equivalent to

$$Q = 1 - K'(0) \tag{109}$$

which is used in their derivation. However, as demonstrated previously,[69] for monosized spheres the quantity $K'(0)$ corresponds to the air volume fraction for a system of overlapping spheres. This is a consequence of using the Hertz distribution for $h(x)$.

In the subsequent numerical experiment, two results will be reported for the Pleau and Pigeon equation corresponding to the normalization factors $1 - K'(0)$ and $1 - A$.

Lu and Torquato Equations

The paste-void and the void-void spacing distributions have application both inside the field of cementitious materials[70–75] and outside the field.[76–78]

Using various approximation techniques, the problems of the paste-void and the void-void spacing distributions have been solved for systems composed of monosized spheres.[79–85] These approximations have been compared to results of Monte Carlo method simulations[82,83] and they are in agreement. One method of approximation relies upon n-point correlation functions, and Torquato, Lu, and Rubenstein[83] have obtained exact expansions for monosized spheres. Lu and Torquato[86] developed a means to map these correlation functions to systems of polydispersed sphere radii, thereby making it possible to extend the approximations for monosized spheres. These approximations for poly-dispersed sphere radii are given in Lu and Torquato,[54] and are used here as estimates for both the paste-void and the void-void spacing distribution.

The results of Lu and Torquato[54] for both the paste-void and the void-void proximity calculations require the following defined quantities:

$$\begin{aligned}
\xi_k &= \frac{\pi}{3} n\, 2^{k-1} \langle R^k \rangle \\
c &= \frac{4 \langle R^2 \rangle}{1-A} \\
d &= \frac{4 \langle R \rangle}{1-A} + \frac{12\, \xi_2}{(1-\xi_3)^2} \langle R^2 \rangle \\
g &= \frac{4}{3(1-A)} + \frac{8\, \xi_2}{(1-A)^2} \langle R \rangle + \frac{16}{3} \frac{B\, \xi_2^2}{(1-A)^3} \langle R^3 \rangle
\end{aligned} \tag{110}$$

The value of B depends upon the exact way the system is constructed. For the calculations performed here, $B = 0$. Also, there was an error in the published value for g in Ref. 54, which has been corrected here.

Since Lu and Torquato were studying systems composed of a matrix containing solid spheres, they use the terms "void" and "particle" to represent the matrix and the spheres, respectively. Therefore, the Lu and Torquato "void exclusion probability" is used here to estimate the paste-void proximity distribution, and their "particle exclusion probability" is used here to estimate the void-void proximity distribution.

Paste-Void Proximity Distribution

The approach of Lu and Torquato[54] was to derive the probability that a point chosen at random throughout the entire system would have no part of an air void within a distance s from it. The region of thickness about the point constitutes a test sphere of radius s. This test sphere of radius s constitutes the Lu and Torquato "void." This void exclusion probability is given in Ref. 54:

$$e_V(s) = \begin{cases} 1 - \dfrac{4\pi}{3} n \left\langle (s+r)^3 \, \Theta(s+r) \right\rangle & s < 0 \\ (1-A) \exp\left[-\pi \, n\left(cs + ds^2 + gs^3\right)\right] & s > 0 \end{cases} \tag{111}$$

Having $s < 0$ corresponds to a sphere with radius $-(s)$ being entirely inside an air void. The averaged quantity in Eq. 111 has the same definition as before:

$$\left\langle (s+r)^3 \, \Theta(r+s) \right\rangle = \int_0^\infty (s+r)^3 \, \Theta(s+r) \, f(r) \, dr \tag{112}$$

This result can be recast into the air void problem. Since $e_V(s)$ represents the probability of a random point not being within a distance s of an air void surface, the probability of finding the nearest void surface within a distance s of a randomly chosen point is the complement of the void exclusion probability:

$$E'(s) = 1 - e_V(s) \tag{113}$$

The probability of finding the nearest air void surface a distance s from a random point in the paste portion requires only the air content A:

$$\begin{aligned} E_V(s) &= \frac{E'_V(s > 0) - A}{1 - A} \\ &= 1 - \exp\left[-\pi \, n\left(cs + ds^2 + gs^3\right)\right] \end{aligned} \tag{114}$$

The quantity $E_V(s)$ is the fraction of the paste volume within a distance s of an air void surface, which is equivalent to the definition of the paste-void proximity cumulative distribution function.

Void-Void Proximity Distribution

The approach used by Lu and Torquato[54] for the void-void proximity is similar to that for the paste-void proximity. For a point located at the center of an air void with radius R, the probability that the nearest air void surface is farther away than w is given in Ref. [54]:

$$e_P(w.R) = \begin{cases} 1 & w \le R \\ \exp\left\{-\pi\, n\left[c\,(w-R) + d\left(w^2 - R^2\right) + g\left(w^3 - R^3\right)\right]\right\} & w > R \end{cases} \tag{115}$$

Lu and Torquato refer to this as the particle exclusion probability.

The probability that the nearest air void surface is within a distance w from the center of an air void with radius R is

$$\begin{aligned} E_P'(w,R) &= 1 - e_P(w > R, R) \\ &= 1 - \exp\left\{-\pi\, n\left[c\,(w-R) + d\left(w^2 - R^2\right) + g\left(w^3 - R^3\right)\right]\right\} \end{aligned} \tag{116}$$

Let s represent the shortest surface-surface distance between two air voids. The probability $E_P(s,R)$ that the nearest air void surface is within s of the surface of the void with radius R requires only a substitution of variables:

$$E_P(s,R) = E_P'(s + R, R) \tag{117}$$

The function $E_P(s,R)$ is equivalent to the void-void spacing cumulative distribution function.

The most important feature of Eq. 117 is that $E_P(s,R)$ depends upon the size of the sphere one starts from. For monodispersed sphere diameters, R is a constant. However, for a system composed of polydispersed sphere diameters, $E_P(s,R)$ is a continuous function of R. Since a continuous distribution of sphere diameters would have an infinite number of possible diameters, there would exist an infinite number of possible $E_P(s,R)$ distributions. This complicates an evaluation of void-void spacing distributions for systems composed of polydispersed sphere radii.

One possible remedy is to calculate an ensemble average. Ensemble averages can be calculated based on either number density or volume density. This bulk value can then be compared to measured values. Here, the number density ensemble average was chosen:

$$\left\langle E_P(s)\right\rangle = \int_0^\infty E_P(s,r)\, f(r)\, dr \tag{118}$$

For a system of poly-dispersed sphere diameters one can also calculate the mean nearest surface-surface distance $l_P(R)$ as a function of sphere radius R[54]:

$$l_P(R) = \int_R^\infty e_P(w,R)\, dw \tag{119}$$

This gives the average surface-surface distance to the nearest air void surface when starting from spheres of radius R. The quantity $l_P(R)$ decreases as R increases. Therefore, on average, the larger the sphere one starts from, the shorter the distance one travels to reach the surface of the nearest air void.

Numerical Test

As a measure of spacing equation accuracy on a geometrical level, a numerical experiment was conducted using air void radii from a zeroth-order logarithmic distribution[87]:

$$f(r) = \frac{\exp\left[-\frac{1}{2}\left(\frac{\ln(r/r_0)}{\sigma_0}\right)^2\right]}{\sqrt{2\pi}\,\sigma_0\, r_0 \exp\left(\sigma_0^2/2\right)} \tag{120}$$

Since this distribution is quite useful for parametric studies of air voids, its details are discussed in Appendix D. Some of the results from Ref. 12 are shown here in the following tables. The systems were composed of lognormally distributed air void radii with a modal radius r_0 of 15 μm, and a value of 0.736 for the dispersion parameter σ_0 (standard deviation of the logarithms).

The measured values are the average and estimated standard deviation from 100 system realizations. Each realization consisted of a fresh collection of air void radii from the lognormal distribution, and new random locations for each void. For each new system, the voids were placed sequentially, largest to smallest, such that no two voids overlapped one another.

Paste-Void Proximity

The results reproduced in Table I suggest that the Powers spacing factor approximates some large percentile of the paste-void spacing distribution, as it was intended. For the logarithmic radius distribution used here, the Powers spacing factor is approximately 150% of the 95th percentile of the paste-void proximity distribution.

The Philleo equation consistently predicts the 95th percentile of the paste-void proximity distribution to within 10% error. This is somewhat remarkable given that it does not include information regarding the specific air void radius distribution.

The Fagerlund quantity $\bar{d}$ is meant to be an approximation, like the Powers factor $\bar{L}$. There is no a priori reason why $\bar{d}$ should predict the 95th percentile of the paste-void distribution. However, although the Fagerlund quantity $\bar{d}$ is not an accurate predictor of the 95th percentile, its ratio to the 95th percentile was also nearly constant for all the air contents tested.

The Pleau and Pigeon equation was also a reasonably accurate predictor of the 95th percentile of the paste-void proximity distribution. The value that was normalized by the air content for K_A was accurate to within 10%. The fact that the Pleau and Pigeon equation is accurate at the 95th percentile is due to the morphology of the system that contains 95% of the paste. This system would appear similar to a system of overlapping spheres, and would explain why their equation can work so well given that they assumed a system of overlapping spheres for the air void system.[13]

The Lu and Torquato estimate of the 95th percentile agreed with the measured value to within one standard deviation. It would appear as though the Lu and Torquato equation is sufficiently accurate for predicting the paste-void proximity distribution over the range of air contents expected in practice.

Void-Void Proximity

The results of the void-void proximity distribution estimates are shown in Table II. The Attiogbe equation for t is proportional to λ as expected. The Attiogbe quantity t_G is greater than the measured 50th percentile $vv50$ by at least a factor of five. Also, the value of t_G varied by a factor of three over the range of air contents, while the value of $vv50$ varied by a factor of 6. This would suggest that t_G is not proportional to $vv50$.

The Lu and Torquato estimate of the void-void distribution was based on Eq. 118. As for the paste-void proximity, the Lu and Torquato estimate of the 50th percentile of the void-void proximity distribution was within one standard deviation of the measured values, and appears to be sufficiently accurate over the range of air void volume fractions expected in concrete.

The performance of the function G in estimating the fraction of paste within t of an air void is shown in Table III. The value of G was calculated for the lognormal distribution by equating the parameters of the distribution in Eq. 97 to the statistical properties in Eq. 98. Because of the accuracy of the Lu and Torquato equations, the value given by E_V is used as a reference. The value of G varies by nearly an order of magnitude for the values of p/A

investigated. However, both $E_V(t)$ and $E_V(t_G)$ were equal to 1.000 over the same range.

The Lu and Torquato equation for E_P is an accurate estimate of the measured quantity *vv*50. The equation estimated *vv*50 to within one standard deviation over the range of air contents reported. The mean nearest surface-surface distance $l_P(R)$ for the two extreme air contents are shown in Fig. 8. The details of the numerical calculation from the simulated systems are given in Ref. 12. Once again, the Lu and Torquato equation is very accurate.

Lu and Torquato Equation

The performance of the Lu and Torquato equation is, by far, the most accurate estimate for every statistic considered. Not only does it predict these statistics well, it also predicts the average void-void spacing as a function of the radius for polydispersed sphere radii. It appears as though the Lu and Torquato equation is accurate to the level of precision required for investigations of air void spacing. These results also suggest that, at the air volume fractions investigated here, an air void distribution approximated by a collection of parked spheres has very similar spatial statistics to an equilibrium distribution of spheres, which has relevance to numerical tests of air void equations. It is also interesting to note that the Lu and Torquato equations do not require information about the entire air void radius distribution. Rather, only the values $\langle R \rangle$, $\langle R^2 \rangle$, and $\langle R^3 \rangle$ are needed.

Paste-Void Probability Density Function

A graphical performance comparison of the Philleo, the Pleau and Pigeon, and the Lu and Torquato estimates of the paste-void proximity probability density function is shown in Fig. 9. The sphere radii are lognormally distributed with a number density of 240 mm^{-3}. The Philleo estimate terminates at $s = 0$ because it is already normalized for the fraction of paste within s of an air void surface. The Lu and Torquato estimate is virtually exact at the resolution of this experiment. The Philleo estimate is fairly accurate for s greater than zero, while that of Pleau and Pigeon is noticeably in error. These qualitative differences are borne out in the previously reported results.

Analytical Test

Given the accuracy of the Lu and Torquato equations, subsequent tests of air void spacing equations can be performed without numerical simulation. Instead, the results can be compared directly to the Lu and Torquato equa-

Table 1. Estimates of the 95th percentile of the paste-void distribution for log-normally distributed sphere radii with number density *n* and air content *A*

n (mm)	A	$\bar{L}$ (mm)	F (mm)	$\bar{d}$ (mm)	K_A (mm)	$K_{K'}$ (mm)	E_V (mm)	$pv95$ (mm)
20	0.016	0.450	0.272	0.381	0.302	0.304	0.290	0.290±0.005
40	0.033	0.337	0.204	0.285	0.231	0.236	0.220	0.219±0.004
80	0.066	0.247	0.150	0.206	0.173	0.185	0.162	0.162±0.003
160	0.131	0.175	0.108	0.143	0.125	0.143	0.114	0.114±0.002
240	0.197	0.136	0.087	0.110	0.099	0.123	0.089	0.090±0.002

Table from Ref. 12. The estimates are from the equations of Powers ($\bar{L}$); Philleo (F); Fagerlund ($\bar{d}$); Pleau and Pigeon (K_A) and ($K_{K'}$); and Lu and Torquato (E_V). The measured values are labeled $pv95$ and have the one standard deviation uncertainties shown.

tion. Using this approach, a more thorough test of the spacing equations that uses a wider range of possible air void radius distributions can be investigated. The results presented here are from Ref. 13.

Air Void Radius Distributions

As was done for the numerical test, the analytical test used the zeroth-order logarithmic distribution because it was a reasonable representation of air void radii in concrete containing air entrainment. (Appendix D contains a detailed discussion of this distribution.) Figure 10 shows the distributions used in the analytical test. The distributions are plotted as a function of diameters to be more easily associated with chord lengths obtained from linear traverse.

An additional air void diameter distribution, not shown in Fig. 10, used in the study is a monosized distribution composed of 200 μm diameter spheres. The distribution corresponds to lognormal distribution parameters of $r_0 = 0.100$ mm, $\sigma_0 = 0$.

Results

The performance of the air void spacing equations was quantified by the accuracy with which they predicted the 95th percentile of the paste-void spacing distribution. The estimate of the distance at the 95th percentile s_{95} using the Lu and Torquato equation was used as a reference value, and is expressed mathematically as the inverse of the function $E_V(s)$:

$$s_{95} = E_V^{-1}(0.95) \tag{121}$$

Table II. Estimates of the 50th percentile of the void-void distribution for log-normally distributed sphere radii with number density *n*, air content *A*, and mean free path λ

n (mm^{-3})	A	λ (mm)	t (mm)	t_G (mm)	E_P (mm)	$vv50$ (mm)
20	0.016	7.969	3.919	0.642	0.134	0.134±0.003
40	0.033	3.918	1.895	0.621	0.092	0.093±0.002
80	0.066	1.892	0.884	0.580	0.060	0.060±0.002
160	0.131	0.880	0.382	0.382	0.035	0.035±0.001
240	0.197	0.542	0.218	0.218	0.024	0.024±0.001

Table from Ref. 12. The estimates are from the equations of Attiogbe (t and t_G) and Lu and Torquato (E_P). The measured values are labeled $vv50$ and have the one standard deviation uncertainties shown.

Table III. Estimates of the fraction of paste within either t or t_G of an air void surface for a lognormal air void distribution with number density *n* and air content *A*

n (mm^{-3})	A	p/A	t (mm)	t_G (mm)	G	$E_V(t)$	$E_V(t_G)$
20	0.016	59.84	3.919	0.642	0.164	1.000	1.000
40	0.033	29.42	1.895	0.621	0.328	1.000	1.000
80	0.066	14.21	0.884	0.580	0.656	1.000	1.000
160	0.131	6.605	0.382	0.382	1.000	1.000	1.000
240	0.197	4.070	0.218	0.218	1.000	1.000	1.000

Table from Ref. 12. The estimates are calculated from the Lu and Torquato equation for E_V.

The accuracy of the other spacing equations was quantified by the ratio of the spacing equation estimate to the Lu and Torquato value of s_{95}. For the Powers, Philleo, and Pleau and Pigeon equations, these ratios are, respectively:

$$Q_{\bar{L}} = \frac{\bar{L}}{s_{95}} \qquad Q_F = \frac{F^{-1}(0.95)}{s_{95}} \qquad Q_K = \frac{K^{-1}(0.95)}{s_{95}} \tag{122}$$

The results were reported as a function of the paste air content. The corresponding air contents for concrete would be approximately one-third of this value.

The values of $Q_{\bar{L}}$ for the Powers spacing equation are shown in Table IV.

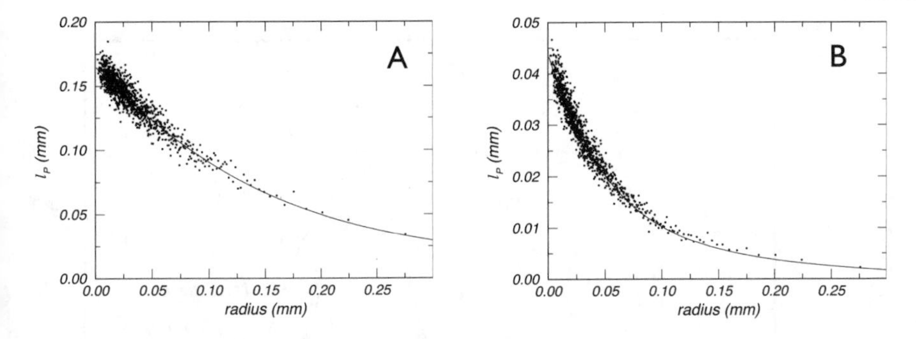

Figure 8. Mean void-void spacing l_p for lognormally distributed sphere radii with number density n: (a) 20 mm^{-3} and (b) 240 mm^{-3}. Measured values are shown as solid circles; the solid line is the estimate by Lu and Torquato. (From Ref. 12.)

The results indicate that $Q_{\bar{L}}$ for the Powers equation is a function of only the air content A and σ_0. The minimum value for $Q_{\bar{L}}$ is nearly unity for monosized spheres, and it increases to a value near two for the distributions used here. Note also that the non sequitur values for the 0.21 paste air content are due to the use of Eq. 86 as the ratio p/A becomes greater than 4.342.

Values of Q for the Philleo and the Pleau and Pigeon spacing equations are shown in Tables V and VI, respectively. These two equations have similar results. Over most of the parameter space investigated, the equations are typically within 10 % of s_{95}. Near the limits of the parameter space, the equations are still within approximately 20% of s_{95}. Also, as was true for the Powers spacing equation, the value of Q for these two spacing equations was only a function of σ_0 and A, and not a function of r_0.

The fact that all three spacing equations are only functions of σ_0 and A means that each equation can be corrected, given this information. One could perform a parametric study of linear traverse data, determine the corresponding parameters of the sphere radius distribution, and calculate σ_0. This, along with the air content A, could be used to establish an accurate estimate for s_{95}.

Upon reflection, it is not surprising that both the Philleo and the Pleau and Pigeon spacing equations can accurately predict the 95th percentile of the paste-void spacing distribution. As seen in Fig. 2(a), as the thickness of the shell surrounding the voids increases, the region within the shells begins to resemble a system of overlapping circles, which is the basis of their derivations. This is why the Philleo and the Pleau and Pigeon equations are more accurate at predicting larger percentiles of the paste-void spacing distribution.[12]

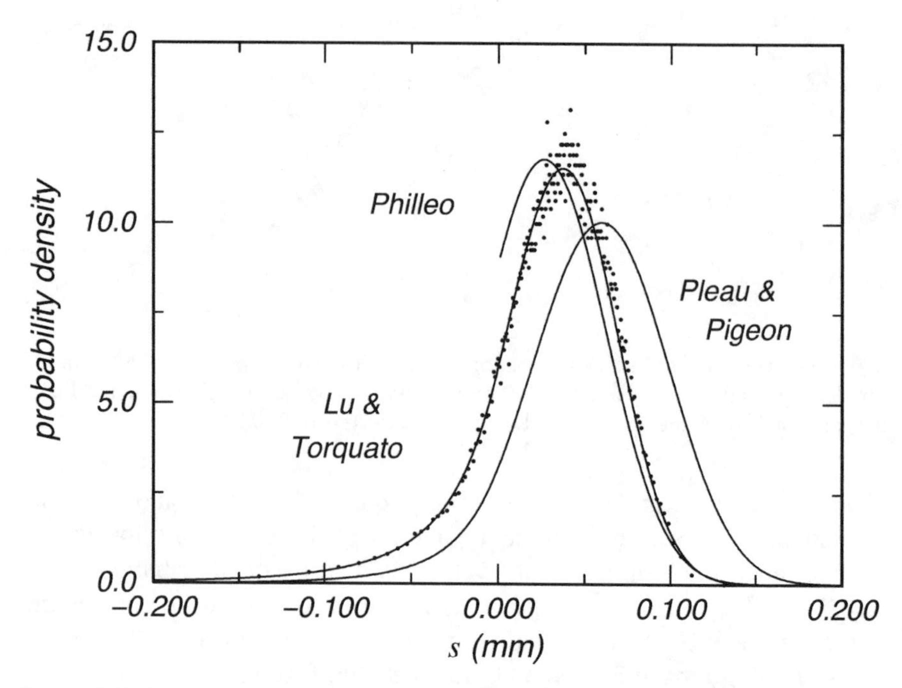

Figure 9. Estimates of the paste-void probability density function by Philleo, Pleau and Pigeon, and Lu and Torquato for lognormally distributed sphere radii with a number density of 240 mm^{-3}. Measured values are shown as filled circles. (From Ref. 12.)

Future Research

Given the recent interest in developing spacing equations, a number of issues should be resolved:

1. The results presented here assume that there exists sufficient inter-aggregate space to contain a statistical representation of the bulk air void system. This assumption should be tested for various aggregate volume fractions. For large aggregate fractions, one could expect that an insufficient number of voids exist within the inter-aggregate spaces, and that the critical void spacing should be smaller.
2. If the approximation of the air void radii by a lognormal distribution is valid, the parameters to the distribution could be used to convert the Powers spacing factor to an estimate of the 95th percentile of the paste-void spacing distribution.

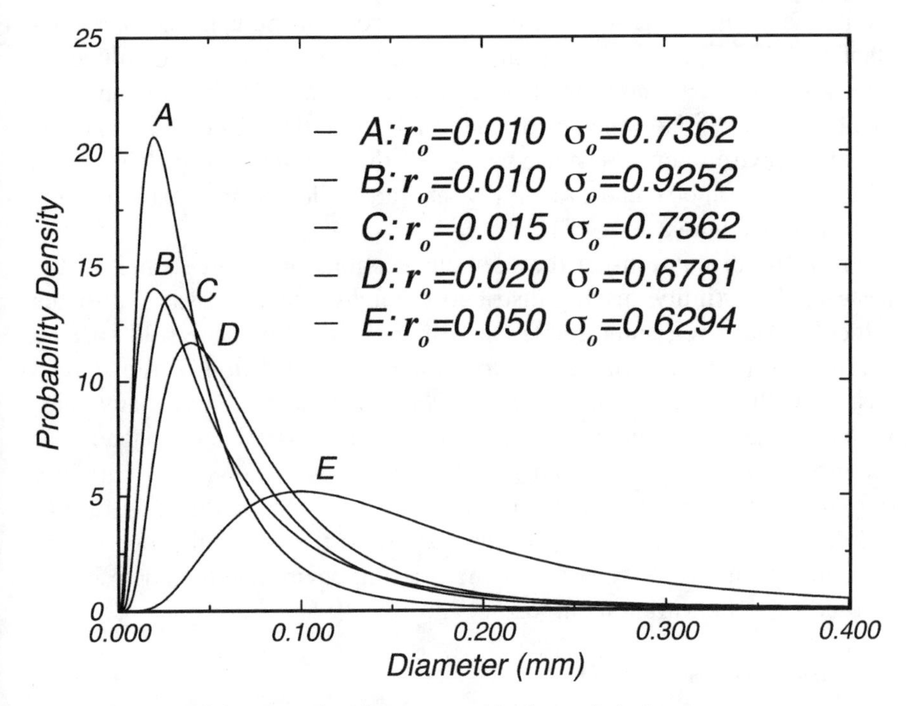

Figure 10. Air void diameter distributions used in the analytical test.

Protected Paste Volume

The concept of a "protected paste volume" (PPV) originated from Powers[3] and was extended by Larson et al.,[58] Fagerlund,[9] Philleo,[8] and Natesaiyer et al.[59,60] The concept assumes that each air void in concrete protects the spherical volume of paste that surrounds it from frost damage. Powers[3,88] called this zone the "sphere of influence" (SoI) of the air voids. The protected paste volume is then simply defined as the common volume of paste protected from frost damage by all the air voids present in the concrete. The ratio of the PPV to the total volume of air and paste is referred to as the protected paste ratio (PPR).[59]

Though the definition is simple, a researcher attempting to calculate the value for a given concrete is faced with an immediate decision: Is the zone of influence dependent on the material properties of the concrete, the freezing environment, and the size of the air void? Or is the zone of influence dependent only on the size of the air voids? The calculations are very simple if it is assumed that the thickness of the zone of influence is some constant value.[8,58] The calculations are more difficult if the zone of influence depends on the material properties of the concrete and the freezing environment as well.[59,60] However, if the material properties of concrete and the freezing environment are not considered, the calculations are simply an exercise in geometry and probability, and are valid for any material filled with spherical voids.

From the discussions in the previous sections, it might seem that the paste-void proximity spacing distribution might provide a way out of the difficulty. However, some reflection will show that the paste-void spacing distribution provides only a way to estimate what fraction of the cement paste is within some distance of an air void. They do not provide any guidance on what the critical distance should be for frost-resistant concrete. Of the proposed physical mechanisms of frost damage to concrete, currently only the hydraulic pressure theory has been developed sufficiently to provide estimates of the critical distance based on the material properties of the concrete and some measure of the freezing environment, and so it will be used for demonstration purposes.

Background

The first applications of the protected paste volume concept applied to air voids in concrete were performed by Warris[33,89] and by Larson, Cady, and Malloy.[58] They started with a parametric estimate of the chord distribution, and combined that with the Philleo spacing equation[8] to estimate the fraction of paste within some distance of an air void. The application was purely geometrical.

The concept was advanced by Natesaiyer et al.[59,60] by using the Powers hydraulic pressure theory.[3] The thickness of protected paste surrounding an air void is a function of the void radius. The volume of paste within any given shell would be the protected paste volume. However, this approach suffers from a number of drawbacks: Having a shell thickness that is a function of air void radius makes deriving analytic expressions for the protected paste volume difficult. The approach also depends upon a particular theory for freeze-thaw degradation.

The original protected paste volume concept of Larson, Cady, and Malloy has a number of advantages. For an arbitrary distribution of air void radii, numerically exact equations exist to calculate the fraction of paste within a given distance from each air void. The approach is independent of particular freeze-thaw degradation models. Either better materials science

Table IV. Values of $Q_{\bar{L}}$ for the Powers spacing equation*

r_0 (mm)	σ_0	α (mm^{-1})	Air content					
			0.06	0.09	0.12	0.15	0.18	0.21
0.100	0.0000	30	1.028	1.060	1.093	1.128	1.165	1.113
0.050	0.6294	15	1.360	1.370	1.386	1.406	1.430	1.349
0.020	0.6781	30	1.428	1.434	1.446	1.464	1.485	1.397
0.010	0.7362	45	1.522	1.522	1.530	1.544	1.562	1.465
0.015	0.7362	30	1.522	1.522	1.530	1.544	1.562	1.465
0.010	0.9252	15	1.965	1.941	1.928	1.925	1.927	1.791

*From Ref. 13.

Table V. Values of Q_F for the Philleo spacing equation*

r_0 (mm)	σ_0	α (mm^{-1})	Air content					
			0.06	0.09	0.12	0.15	0.18	0.21
0.100	0.0000	30	1.076	1.115	1.155	1.196	1.240	1.285
0.050	0.6294	15	0.958	0.970	0.985	1.004	1.024	1.048
0.020	0.6781	30	0.944	0.952	0.965	0.980	0.998	1.018
0.010	0.7362	45	0.926	0.931	0.940	0.952	0.967	0.984
0.015	0.7362	30	0.926	0.931	0.940	0.952	0.967	0.984
0.010	0.9252	15	0.874	0.867	0.865	0.867	0.872	0.878

*From Ref. 13.

Table VI. Values of Q_K for the Pleau and Pigeon spacing equation*

r_0 (mm)	σ_0	α (mm^{-1})	Air content					
			0.06	0.09	0.12	0.15	0.18	0.21
0.100	0.0000	30	1.074	1.108	1.140	1.169	1.196	1.221
0.050	0.6294	15	1.080	1.098	1.111	1.123	1.134	1.146
0.020	0.6781	30	1.072	1.085	1.096	1.106	1.115	1.125
0.010	0.7362	45	1.061	1.070	1.077	1.084	1.091	1.099
0.015	0.7362	30	1.061	1.070	1.077	1.084	1.091	1.099
0.010	0.9252	15	1.022	1.020	1.017	1.016	1.016	1.018

*From Ref. 13.

or simply laboratory tests could be used to establish both the fraction of paste that must be protected and the distance this fraction needs to be within an air void surface.

Constant Shell Thickness

As a first approximation, one could begin by assuming that, for some physical model of freeze-thaw, the SoI radius is a constant for all air void radii. Once this thickness is determined from the cement paste properties and the freezing rate, the calculation would then be identical to determining the paste-void spacing distribution as was done in the previous section.

While it would seem that a constant SoI radius is unlikely, this approach has a number of advantages. This approach can be developed independently of a freeze-thaw mechanism. In fact, the critical Powers spacing factor derived by Powers was based upon the hydraulic pressure theory, which has since been shown to have serious flaws. Yet it has been shown by Pigeon and coworkers that there exist critical values of $\overline{L}$ depend on the material properties of the concrete and the exposure conditions.

Variable Shell Thickness

A more rigorous approach is to calculate the protected paste thickness as a function of the void radius. Figure 11 depicts the idea schematically. Borrowing from the notation of Powers,[3] an air void with radius r_b protects a shell of paste out to a radius r_m from the center of the void. The shell thickness s is the difference $s = r_m - r_b$.

An estimate of the volume of paste protected was derived by Natesaiyer et al.[59] They approximated the system of r_m radius spheres by overlapping spheres. The protected paste ratio Ω follows from the Hertz[64] distribution:

$$\Omega = 1 - \exp\left[\frac{-4\pi\, n}{3}\left\langle R_m^3\right\rangle\right] \tag{123}$$

The performance of the Philleo and the Pleau and Pigeon spacing equations discussed in the previous section suggests that the approximation may be sufficiently accurate as the value of Ω approaches 1.

The air content of the paste A_p was used to simplify the equation[59]:

$$\Omega = 1 - \exp\left[-A_p\,\frac{\left\langle R_m^3\right\rangle}{\left\langle R_b^3\right\rangle}\right] \tag{124}$$

Certain limiting cases demonstrate similarities to spacing equations. In the limit $\langle R_b^3 \rangle \rightarrow 0$, the voids have zero radius and $r_m = s$:

$$\Omega\Big|_{\langle R_b^3 \rangle \rightarrow 0} = 1 - \exp\left[\frac{-4\pi\, n}{3} s^3\right] \tag{125}$$

This equation bears a resemblance to the Philleo approach. Conversely, in the limit that s is a constant, the quantity $\langle R_m^3 \rangle$ can be expanded in terms of $\langle R_b^3 \rangle$ and s:

$$\Omega\Big|_{r_m = s + r_b} = 1 - \exp\left[-A_P\left(1 + \alpha\, s + \alpha\frac{\langle R_b \rangle}{\langle R_b^2 \rangle} s^2 + \frac{\alpha}{3\langle R_b^2 \rangle} s^3\right)\right] \tag{126}$$

This result is directly related to that of Fagerlund.[9] The overall approach has the drawback that for $s \rightarrow 0$, the PPR Ω equals the volume of the corresponding system with overlapping spheres; this was the same deficiency as for the Pleau and Pigeon equation.

Freeze-Thaw Model

The development thus far for the variable SoI thickness has been done without regard to a specific model for freeze-thaw. Fundamentally, all one needs in order to proceed further is an analytical relationship between r_m and r_b. As an example, consider the Powers hydraulic pressure theory used in Natesaiyer et al.[59] The model requires a number of material and environmental quantities:

- k: cement paste permeability coefficient
- T: cement paste tensile strength
- μ: pore fluid viscosity
- S: capillary pore saturation
- u: volume of water frozen per volume paste, per degree
- θ: cooling rate

The radius r_m can be expressed as a function of the air void radius r_b and material properties of the concrete[3]:

$$\frac{r_m^3}{r_b} + \frac{r_b^2}{2} - \frac{3}{2} r_m^2 = \frac{3\,k\,T}{\mu\left(1.09 - S^{-1}\right) u\,\theta} \tag{127}$$

The left side of the equation is purely geometric, and the right side is purely material. For brevity, the materials term can be represented by Φ, and the material dependence is implied. Also, the geometric portion of the equation can be simplified slightly by expressing the result as a function of the shell thickness $s = r_m - r_b$ [3]:

$$\frac{s^3}{r_b} + \frac{3\,s^2}{2} = \Phi \tag{128}$$

The shell thickness s for a material parameter $\Phi = 0.1$ mm^2 is shown in Fig. 12(a). Assuming a capillary porosity $\varepsilon = 0.08$, the ratio of expelled volume of water to the volume of void is shown in Fig. 12(b). This ratio cannot be greater than unity. Using this physical limit, the corrected shell thickness in Fig. 12(a) is shown as a dashed line.

Once the functional relationship between the SoI radius r_m and the void radius r_b is established, the protected paste ratio Ω can be calculated using Eq. 124 above. Consider a concrete with a paste material parameter $\Phi = 0.1$ mm^2 and a porosity $\varepsilon = 0.08$. Using the five air void radius distributions shown in Fig. 10, the protected paste ratio Ω, calculated as a function of air content within the paste A_P, is shown in Fig. 13(a).

Similarly, the Powers spacing factor $\overline{L}$ can also be calculated for these systems as a function of paste air content A_P. Figure 13(b) shows a plot of Ω versus $\overline{L}$ for the five radius distributions. For these particular values of Φ and ε, the relationship between Ω and $\overline{L}$ seems to be relatively insensitive to the air void radius distribution. All the distributions but one appear to have a critical value of $\overline{L}$ near 0.300 μm for these values of Φ and ε. This result could help explain the success of the critical spacing factor approach to characterizing concrete performance to freezing and thawing.[90]

Future Research

New and more sophisticated freeze-thaw models are being developed. If air voids are the key to ideal performance under conditions of freezing and thawing, sufficiently sophisticated theories are needed to determine the quantity and "quality" of the air void distribution required for optimum performance.

1. It is reasonable to assume that other models for freeze-thaw will also show that the hydraulic pressures are relieved at the air voids. Therefore, the corresponding equations that relate geometry to material properties will resemble Eq. 128. It would be useful to

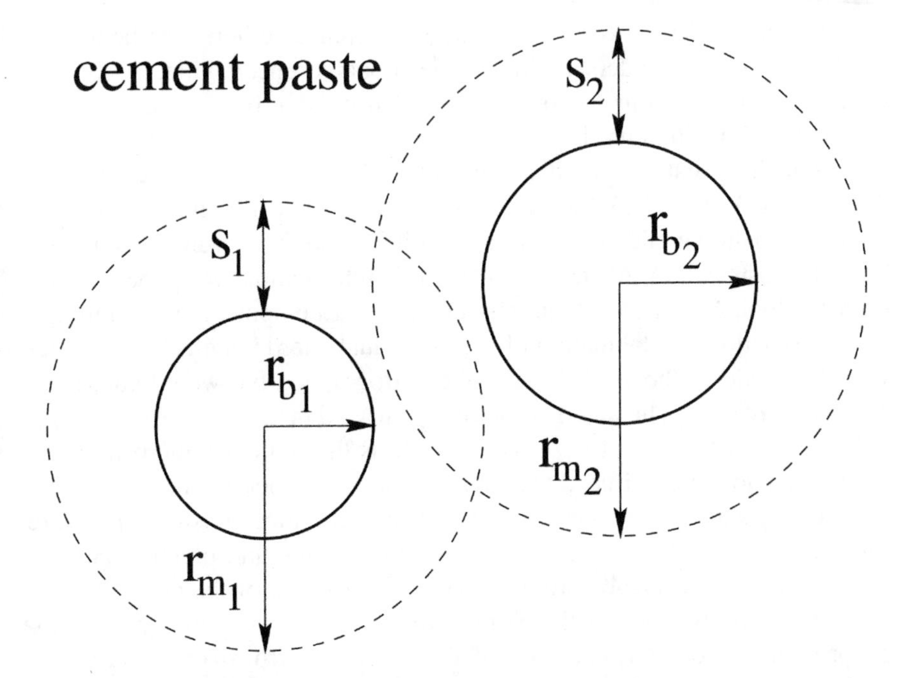

Figure 11. Sphere of influence for two air voids (r_{b1},r_{b2}) in proximity to one another. The radii of the protected paste zones are r_{m1} and r_{m2}, respectively.

generalize a pressure relief theory in order to obtain a generalized form for Eq. 128. Success in this endeavor would demonstrate why the Powers spacing factor has performed so well for so long.

2. The estimate of the protected paste ratio shown here is based upon a simple, overlapping sphere model. It would be useful to extend the Lu and Torquato equation to include varying shell thicknesses in order to develop a more accurate microstructural model.

Summary

This chapter is an introduction to the complete characterization of the air void system in hardened concrete. This characterization includes information regarding volumetric data and an ASTM C 457 analysis, the air void size distribution, and the paste-air system microstructure.

The present test methods in ASTM C 457 have been useful tools for the concrete industry. However, as use of performance requirements increases, the matter of performance acceptance will have to address uncertainty in test results. Further, rational performance acceptance criteria can be formulated only with an understanding of the probable uncertainties in the test method results. For this reason, a detailed discussion of uncertainty in the ASTM C 457 method has been addressed.

Although tabulating individual chord lengths is not currently required by the existing ASTM C 457 test method, it may become a future requirement. This information could be used to extract information regarding the air void diameter distribution. At the very least, this information would be a useful research tool. Given the ubiquitous presence of computers and computerized automation, the tabulation of the individual chord lengths by computer would not add to the work load of the petrographer, but would contribute substantially to the characterization of the air void distribution.

The last piece of the freeze-thaw puzzle is the paste-air microstructure. There are two parts to freeze-thaw performance in concrete: the mechanism of freezing and the pressures generated, and the mechanism of pressure release by the air voids. The mechanism of freezing and pressure generation is a materials science problem. The effectiveness of the air voids, due to their proximity to either the paste or to one another, is purely a stereology problem. Accurate prediction of freeze-thaw performance requires an understanding of both parts, which are brought together in the formulation of the protected paste ratio.

The compilation and distillation of available knowledge within a particular field of study serves two purposes. Expressing what is known delineates the problems that are currently solvable. At the same time, it delineates where additional work is needed to either verify existing assumptions or to advance understanding. Some of the remaining problems in characterizing air voids have been enumerated in the previous sections, so that this chapter serves not only as a primer on air void characterization, but also as a starting point for future research.

Appendix A: Uncertainty

The definitions for expressing uncertainty in measurements have been standardized by ISO.[91] These definitions have been summarized in a shorter document published by the National Institute of Standards and Technology.[92] There are a number of good books that address uncertainty analysis. Classic texts include Mandel[93] and Wilson.[94] The books by Taylor[95] and Dieck[96] are ISO compliant.

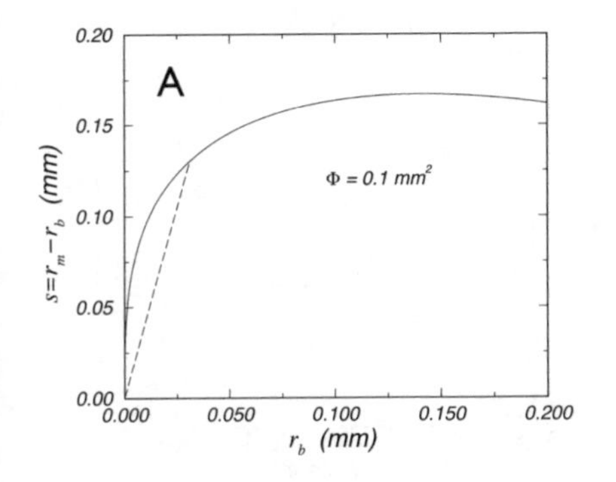

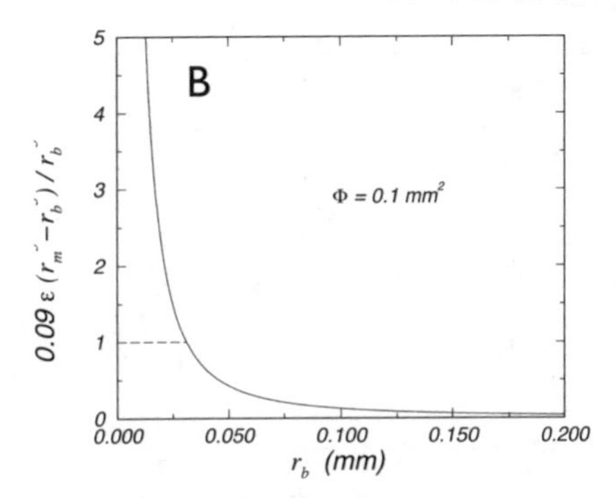

Figure 12. The shell thickness s (a) and the ratio of expelled ice volume to the volume of the air void (b). The solid line is the original equation by Powers, the dashed line accounts for the finite volume of the air void. (From Ref. 59. Used with permission.)

Let there be a physical process from which values of X are measured. The individual values X_i differ from the true x by the random error ε_i:

$$X_i = x + \varepsilon_i \tag{129}$$

Fundamentally, the error ε_i is unknowable. Repeated measurements will yield a distribution of X_i values. The dispersion in these values is represented by the standard uncertainty $u(x)$. The particular functional form for $u(x)$ depends upon the distribution of the errors ε_i. In principle, the experimenter chooses an appropriate distribution function. Here, all errors will be characterized by a normal distribution with mean zero and variance σ_X^2. Therefore, σ_X represents the standard uncertainty in determining x.

For most uncertainty analyses, the final quantity of interest is calculated from a number of measured quantities. The uncertainty in the final quantity is the combined standard uncertainty u_c. One of the most lucid derivations for u_c, based upon a Taylor expansion, is given in the ISO document.[91] What follows is taken nearly verbatim from this document:

Let the response y depend upon N variables, each labeled x_i. Assuming that $\langle Y\rangle = f(\langle X_1\rangle,\langle X_2\rangle,\ldots,\langle X_N\rangle)$, the response y can be expanded about the expected value $\langle Y\rangle$ for a single experiment:

$$y - \langle Y\rangle = \sum_{n=1}^{N} \frac{\partial f}{\partial x_n}\left(x_n - \langle X_n\rangle\right) \tag{130}$$

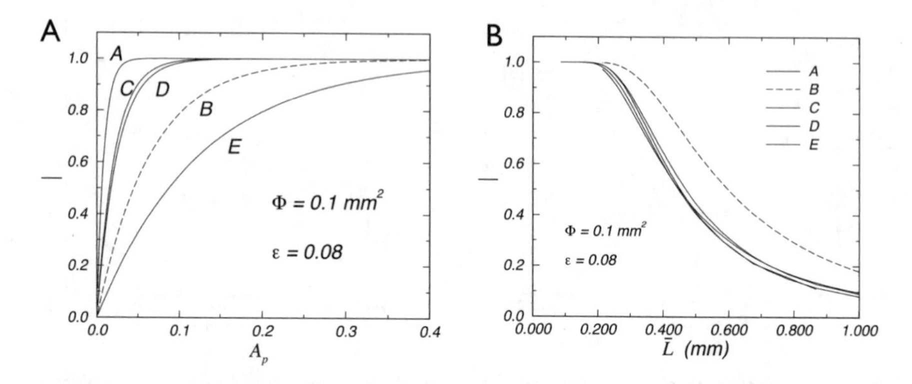

Figure 13. The protected paste ratio Ω as a function of paste air content A_P (a) and Powers spacing factor $\bar{L}$ (b). The air void radius distributions are the same as those shown in Fig. 10.

The square of this deviation can be expressed succinctly:

$$(y - \langle Y \rangle)^2 = \sum_{n=1}^{N} \left(\frac{\partial f}{\partial x_n} \right)^2 (x_n - \langle X_n \rangle)^2$$

$$+2 \sum \sum \frac{\partial f}{\partial x_n} \frac{\partial f}{\partial x_m} (x_n - \langle X_n \rangle)(x_m - \langle X_m \rangle) \tag{131}$$

One next takes the expectation of both sides. The result is simplified because of the Gaussian error model. The expectation of the left side is simply the variance in y:

$$\sigma_y^2 = \sum_{n=1}^{N} \left(\frac{\partial f}{\partial x_n} \right)^2 \sigma_n^2 + 2 \sum_{n=1}^{N-1} \sum_{m=n+1}^{N} \frac{\partial f}{\partial x_n} \frac{\partial f}{\partial x_m} \sigma_{nm} \tag{132}$$

The quantity σ_n^2 replaces $\langle (x_n - \langle X_n \rangle)^2 \rangle$. The quantity σ_{nm} is the covariance between x_n and x_m and is shorthand for $\langle (x_n - \langle X_n \rangle)(x_m - \langle X_m \rangle) \rangle$. In summary, the combined standard uncertainty is the standard deviation in y. Note that σ_y does not use the random variable notation to distinguish it from a standard uncertainty. The corresponding estimated standard deviation in y is expressed as s_y.

Coefficient of Variation

The coefficient of variation η is the ratio of the standard deviation to the expected value:

$$\eta_X = \frac{\sigma_X}{\langle X \rangle} \tag{133}$$

The corresponding quantity based upon measured values is denoted C_X, and is the ratio of the estimated standard deviation s_X to the average value $\overline{X}$:

$$C_X = \frac{s_X}{\overline{X}} \tag{134}$$

Many of the equations considered in the subsequent sections are multiplicative. For example, consider a calculated quantity z that is a simple function of the random variables X and Y: $z = XY$. The estimated combined standard uncertainty in z, assuming no covariance in X and Y, is calculated from Eq. 132:

$$\begin{aligned} s_z^2 &= \left(\frac{\partial z}{\partial X}\right)^2 s_X^2 + \left(\frac{\partial z}{\partial Y}\right)^2 s_Y^2 \\ &= Y^2 s_X^2 + X^2 s_Y^2 \end{aligned} \tag{135}$$

Since $\bar{z} = \overline{XY}$, the above expression be expressed compactly using the coefficient of variation C_z:

$$C_z^2 = C_X^2 + C_Y^2 \tag{136}$$

Appendix B: Probability and Statistics

The topics discussed in this chapter rely upon the concepts of random variables, probability theory, and statistics. It will be useful to discuss these topics in detail to ensure uniformity throughout the sections. A number of excellent books exist for additional information and discussion.[97,98]

Random Variables

A useful way to introduce the idea of random variables is through an example. Consider an experiment that has inherent randomness. The result of the experiment is a single value, out of a number of possible values. The result of a single experiment may be represented by the random variable X. The possible values that X may take can be represented by the variable x. For an experiment consisting of a coin toss, the possible outcomes are either heads H or tails T. The variable x can represent the set of outcomes, $x = \{H,T\}$. The random variable X may take on the value of either H or T. From experience, one knows that the probability P of either event is one-half:

$$P(X{=}H) = P(X{=}T) = \frac{1}{2} \tag{137}$$

Probability Density Function

In contrast to the coin toss experiment, consider an experiment where X can be any real number on the interval [0,1]. In this case, there are an infinite number of possible values X can take, and X is a continuous random variable. Since the number of possible outcomes is infinite, the probability that $X = 0.5$ is zero. For a continuous random variable, one can only define a probability density function $f(x)$ such that

$$P(a \le X \le b) = \int_a^b f(x)\,dx \tag{138}$$

under the constraints

$$\int_{-\infty}^{+\infty} f(x)\,dx = 1 \qquad f(x) \ge 0 \tag{139}$$

This definition is consistent with the fact that the probability of a particular outcome such as $X = a$ is zero:

$$P(a \le X \le a) = \int_a^a f(x)\,dx = 0 \tag{140}$$

Cumulative Distribution Function

When analyzing measured data, it is often more convenient to work with the cumulative distribution function (CDF), also referred to as the probability function. For a probability density function $f(x)$, the cumulative distribution function $F(x)$ is the probability that a particular outcome can be less than x:

$$F(x) = P(x \le X) \tag{141}$$

For a continuous probability density $f(x)$, the quantity has the following mathematical definition:

$$F(x) = \int_{-\infty}^{x} f(x')\, dx' \tag{142}$$

Point Probability Function

When the possible outcomes are either finite or countably infinite (e.g., all positive prime numbers), the discrete random variable X is characterized by a point probability function $p(x)$:

$$P(X = a) = p(a) \tag{143}$$

Here, the point probability function will be treated as a special case of the continuous probability density function. While investigating air void radius distributions, it may be useful at times to consider the monosized sphere radius distribution having radius r_0:

$$P(R = r_0) = p(r_0) = 1 \tag{144}$$

However, the subsequent sections will summarize equations for continuous random variables. As a solution, one can represent the monosized probability density function $f(x)$ by a Dirac delta function $\delta(x)$ (see Appendix C). For a system of monosized sphere radii, the PDF $f(r)$ can be expressed as a delta function:

$$f(r) = \delta(r - r_0) \tag{145}$$

Mixtures of monosized sphere radii, each denoted by radius r_i, are composed of weighted delta functions

$$f(r) = \sum_i \gamma_i \, \delta(r - r_i) \tag{146}$$

such that $\gamma_i = 1$. The coefficients γ_i represent the number fraction of radius r_i.

Statistics

Under many circumstances one can sufficiently characterize the air void radius distribution in concrete from certain statistical properties, and the functional form for the radius probability density function is not required.

Expected Values

Let the random variable X take values from the continuous probability density function $f(x)$. The expectation of X has the following definition:

$$\langle X \rangle = \int_{-\infty}^{+\infty} x \, f(x) \, dx \tag{147}$$

with the following constraint:

$$\int_{-\infty}^{+\infty} |x| \, f(x) \, dx < \infty \tag{148}$$

Similarly for higher powers of X:

$$\langle X^n \rangle = \int_{-\infty}^{+\infty} x^n \, f(x) \, dx \tag{149}$$

Let $f(r)$ represent the air void radius probability density function. Quantities like the expected average surface area S and the expected volume V can be expressed as expectations of the radius distribution:

$$S = 4\pi \langle R^2 \rangle$$
$$V = \frac{4\pi}{3} \langle R^3 \rangle \tag{150}$$

Averages

The preceding section is distinguished from averages. Expectations are calculated from equations. Averages are calculated from measured data. Consider an experiment with random variables X_i from N trials. The average value $\overline{X}$ is

$$\overline{X} = \frac{1}{N}\sum_{i=1}^{N} X_i \tag{151}$$

Similarly for higher powers of X_i:

$$\overline{X^n} = \frac{1}{N}\sum_{i=1}^{N} X_i^n \tag{152}$$

An experiment is conducted with the assumption that $\overline{X^n}$ is an unbiased estimator of $\langle X^n \rangle$.

Variance

The variation in a population of random variables X_i are typically expressed as a standard deviation σ_X. However, it is the variance $\sigma_X{}^2$ that more easily expressed mathematically:

$$\begin{aligned} \sigma_X^2 &= \left\langle \left(x - \langle X \rangle\right)^2 \right\rangle \\ &= \langle X^2 \rangle - \langle X \rangle^2 \end{aligned} \tag{153}$$

The kth moment is defined as $\langle (x - \langle X \rangle)^k \rangle$.[97] Therefore, the variance is the second moment about the mean $\langle X \rangle$. However, it is common for publications to use the term "moment" to mean expectation. In that context, the term is correct if one is calculating the moment about zero mean.

Appendix C: Delta Function Distribution

The Dirac delta function $\delta(x)$ is a useful tool for representing discrete probability distributions as continuous distributions. A useful discussion of the Dirac delta function can be found either in Lighthill[99] or in any suitable mathematical physics text such as Arfken.[68] For purposes of this chapter, the Dirac delta function can be thought of as a Gaussian in the limit that the variance approaches zero.

The Dirac delta function can be described rigorously using a few definitions. The function itself has the following nondifferentiable characteristic:

$$\delta(x - x_0) = \begin{cases} \infty & x = x_0 \\ 0 & otherwise \end{cases} \tag{154}$$

However, the area under the Dirac delta function is unity:

$$\int_{-\infty}^{+\infty} \delta(x)\, dx = 1 \tag{155}$$

The function also has the following convolution property for an arbitrary function $f(x)$:

$$\int_a^b \delta(x - x_0)\, f(x)\, dx = \begin{cases} f(x_0) & a \leq x_0 \leq b \\ 0 & otherwise \end{cases} \tag{156}$$

A point probability distribution can be replaced by a Dirac delta function. Given a point probability function $p(x)$ that describes the outcomes of a random trial, let $p(x_0)$ represent the probability that the random variable X equals x_0. The corresponding continuous function representation would be $p(x_0)\delta(x - x_0)$.

Monosized Spheres

The same applies to monosized sphere radius distributions. A distribution of monosized radius r_0 can be represented by $f(r) = \delta(r - r_0)$. The nth moment can be calculated directly:

$$\begin{aligned} \langle R^n \rangle &= \int_0^{\infty} r^n\, \delta(r - r_0)\, dr \\ &= r_0^n \end{aligned} \tag{157}$$

This demonstrates the greatest usefulness of the Dirac delta function as a representation of a monosized void system. Using the delta function, one can then use the tools developed for continuous functions.

Point Probability

The Dirac delta function is also useful for describing a distribution composed of a finite number of sphere radii. This would be useful for approximating data from a sieve analysis. Let $f(r)$ represent the combined sphere radius distribution. Let γ_i represent the probability that there exists a sphere with radius R_i. Let there be N such possible radii. The PDF $f(r)$ can be expressed as a linear sum of each radius:

$$f(r) = \sum_{i=1}^{N} \gamma_i \, \delta(r - R_i) \tag{158}$$

The values of γ_i are constrained to sum to one:

$$\sum_{i=1}^{N} \gamma_i = 1 \tag{159}$$

Circle Distribution

Consider a monosized sphere diameter distribution $f(x)$, with every sphere of diameter x_0. The corresponding circle diameter distribution $g(y)$ is calculated from Eq. 41:

$$\begin{aligned} g(y) &= \frac{y}{\langle X \rangle} \int_y^{\infty} \frac{\delta(x - x_0)}{\left(x^2 - y^2\right)^{1/2}} \, dx \\ &= \frac{y}{x_0} \frac{1}{\left(x_0^2 - y^2\right)^{1/2}} \qquad y \le x_0 \end{aligned} \tag{160}$$

Chord Distribution

The corresponding chord distribution $h(z)$ can be calculated from Eq. 49:

$$\begin{aligned} h(z) &= \frac{2z}{\langle X^2 \rangle} \int_z^{\infty} \delta\,(x - x_0) \, dx \\ &= \frac{2z}{x_0^2} \qquad z \le x_0 \end{aligned} \tag{161}$$

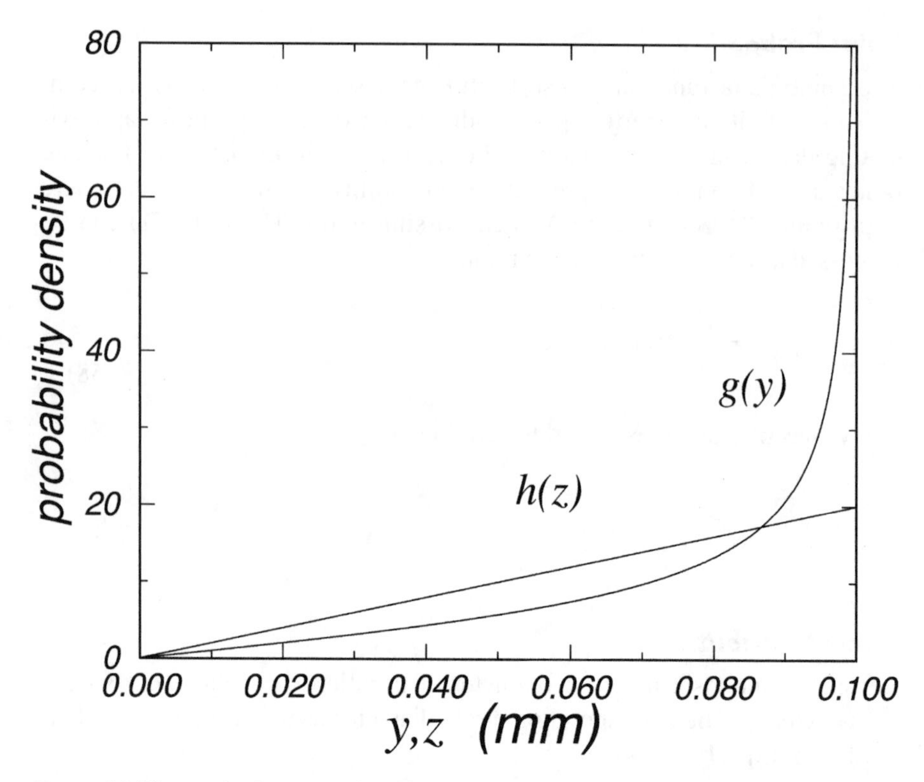

Figure 14. The circle diameter distribution $g(y)$ and the chord distribution $h(z)$ for a monosized 0.100 mm sphere diameter distribution.

Example

The circle diameter distribution $g(y)$ and the chord distribution $h(z)$ for a monosized 100 μm diameter sphere are shown in Fig. 14. The corresponding sphere diameter distribution would appear as a vertical line at 0.100 mm. The shape of $g(y)$ suggests that there is a propensity for intersecting spheres near their equators. The function $h(z)$ in the figure shows the contribution to the chord distribution from a particular sphere size. The sum total of a number of a finite number of different monosized spheres would be the sum total of the corresponding $h(z)$. It is this property that was exploited by Lord and Willis.[50]

Appendix D: Lognormal Distribution

A thorough discussion of the logarithmic family of distributions can be found in Ref. 87. The general nth order logarithmic distribution has the following form[87]:

$$f_n(x) = \frac{x^n \exp\left[-\frac{1}{2}\left(\frac{\ln x - \ln x_n}{\sigma_n}\right)^2\right]}{\sqrt{2\pi}\,\sigma_n\, x_n^{n+1} \exp\left[(n+1)^2\, \sigma_n^2 / 2\right]} \tag{162}$$

The expectations can also be expressed in a generalized form:

$$\left\langle X^k \right\rangle_n = x_n^k \exp\left[\left(k^2 + 2k + 2kn\right) \sigma_n^2 / 2\right] \tag{163}$$

The function $f_n(x)$ is the same for all values of n. Only the meaning of x_n varies. There are three values of n that are typically of interest to experimenters:

$x_{-3/2}$: mean

x_{-1}: median

x_0: mode

The parameters can be mapped onto one another. Compare the ratio of the first moment to the second moment for an nth order and an mth order distribution:

$$\frac{x_n\, e^{(2n+3)\sigma_n^2/2}}{x_m\, e^{(2m+3)\sigma_n^2/2}} = \frac{x_n^2\, e^{(4n+8)\sigma_n^2/2}}{x_m^2\, e^{(4m+8)\sigma_n^2/2}} \tag{164}$$

Solving this equation for σ demonstrates that the meaning of σ is identical for all n and m:

$$\sigma_m = \sigma_n = \sigma \tag{165}$$

Given this, the parameters x_n and x_m can be mapped to one another:

$$\frac{x_n}{x_m} = e^{(m-n)\sigma^2} \tag{166}$$

Specific Surface

From the material presented so far, the specific surface can be calculated directly. Let $f_n(x)$ represent a sphere diameter distribution. The specific surface is the ratio of two moments:

$$\begin{aligned}\alpha &= 6\frac{\langle X^2\rangle}{\langle X^3\rangle}\\ &= \frac{6}{x_n}\exp\left[-(2n+7)\,\sigma_n^2/2\right]\end{aligned} \tag{167}$$

Chord Distribution

The corresponding chord length distribution can be calculated from Eq. 49:

$$\begin{aligned}h(z) &= \frac{2z}{\langle X^2\rangle}\int_z^{\infty} f_n(x)\,dx\\ &= \frac{z}{\langle X^2\rangle}\,erfc\left(\frac{\ln z - \ln x_n}{\sqrt{2}\,\sigma_n} - \frac{(n+1)}{\sqrt{2}}\,\sigma_n\right)\end{aligned} \tag{168}$$

The function $erfc(x)$ is the complementary error function.[67] For parametric studies using lognormally distributed sphere diameters, the moments of the chord distribution are quite useful:

$$\begin{aligned}\langle Z^k\rangle_n &= \frac{2}{k+2}\frac{\langle X^{k+2}\rangle}{\langle X^2\rangle}\\ &= \frac{2\,x_n^k}{k+2}\exp\left[\left(k^2+6k+2nk\right)\sigma_n^2/2\right]\end{aligned} \tag{169}$$

Example

The circle diameter distribution $g(y)$ and the chord length distribution $h(z)$ for a zeroth-order logarithmic sphere diameter distribution $f(x)$ are shown in Fig. 15. As the dimensionality of the probe decreases (spheres–circles–chords), the corresponding distribution widens due to the cumulative contribution of larger sphere diameters. This is true for the logarithmic distribution, but can be far different for other distributions. In fact, some distributions are invariant: $f(x) = g(y) = h(z)$.

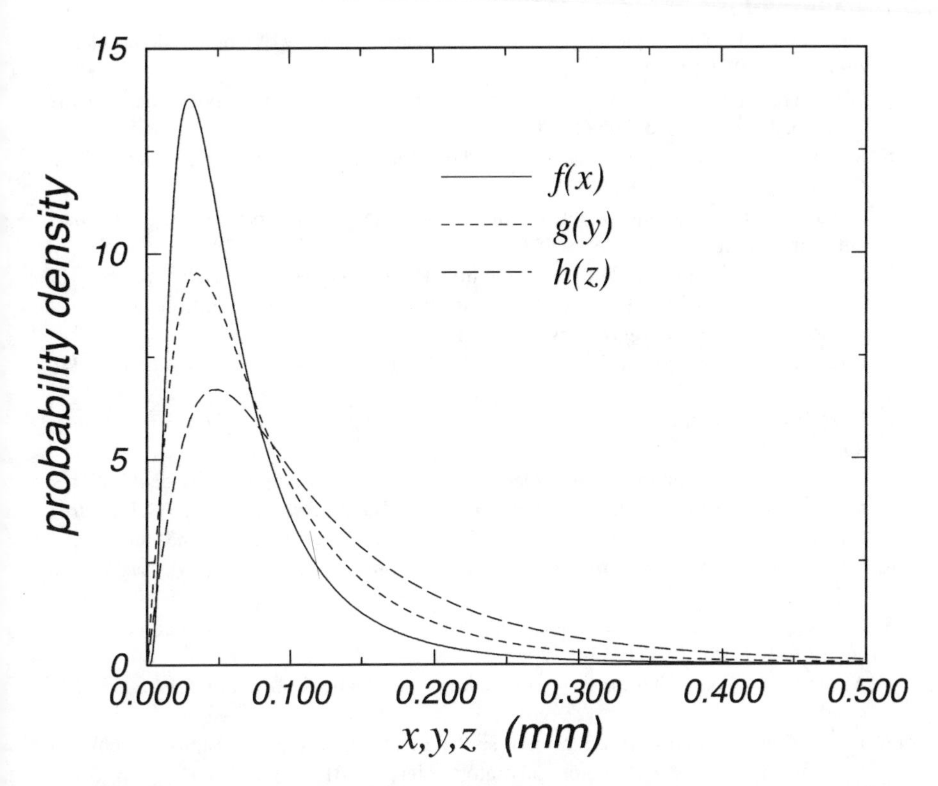

Figure 15. The circle diameter distribution g(y) and the chord length distribution h(z) for a zeroth-order logarithmic sphere diameter distribution f(x) with modal diameter 0.030 mm and σ_0 = 0.7362.

References

1. ASTM C457: Standard test method for microscopical determination of parameters of the air-void system in hardened concrete. American Society for Testing and Materials, West Conshohocken, Pennsylvania, 1990.
2. K.C. Hover, "Air content and unit weight of hardened concrete"; p. 296 in *Significance of Tests and Properties of Concrete and Concrete-Making Materials,* STP 169C. Edited by Lamond and Kleiger. American Society for Testing and Materials, Philadelphia, 1994.
3. T.C. Powers, "The air requirement of frost-resistant concrete," *Proc. Highway Res. Board,* **29**, 184, (1949).
4. A. Rosiwal, "Über geometrische gesteinsanalysen. ein einfacher weg zur ziffermässigen feststellung des quantitätsverhältnisses der mineralbestandtheile gemengter gesteine," *Verhandl. der K.-K. geologischen Reichsanstalt,* 5/6, 143 (1898).
5. L.S. Brown and C.U. Pierson, "Linear traverse technique for measurement of air in hardened concrete," *J. ACI,* **22**, 117 (1950).

6. S.D. Wicksell, "The corpuscle problem: A mathematical study of a biometric problem," *Biometrika,* **17**, 84 (1925).
7. W.P. Reid, "Distribution of sizes of spheres in a solid from a study of slices of the solid," *J. Math. Phys.,* **34**, 95 (1955).
8. R.E. Philleo, "A method for analyzing void distribution in air-entrained concrete," *Cem. Concr. Aggregates,* **5**, 128 (1983).
9. G. Fagerlund, "Equations for calculating the mean free distance between aggregate particles or air-pores in concrete," *CBI Forskning/Research,* **8**, 77 (1977).
10. R. Pleau and M. Pigeon, "The use of the flow length concept to assess the efficiency of air entrainment with regards to frost durability: Part I — Description of the test method," *Cem. Concr. Aggregates,* **18**, 19 (1996).
11. E.K. Attiogbe, "Mean spacing of air voids in hardened concrete," *ACI Mater. J.,* **90**, 174 (1993).
12. K.A. Snyder, "A numerical test of air void spacing equations," *Adv. Cem. Based Mater.,* **8**, 28 (1998).
13. K.A. Snyder, "Estimating the 95th percentile of the paste-void proximity distribution"; to be published in *RILEM Workshop on Freeze-Thaw Performance of High Performance Concrete.* Edited by D. Janssen and M. Snyder. Minneapolis, Minnesota, 1999.
14. E.E. Underwood, *Quantitative Stereology.* Addison-Wesley, Reading, Massachusetts, 1970.
15. G.J. Verbeck, "The camera lucida method for measuring air voids in hardened concrete," *J. ACI,* **18**, 1025 (1947).
16. A. Delesse, "Pour déterminer la composition des roches," *Ann. des Mines,* **13**, 379 (1848).
17. J.E. Hilliard, "Volume fraction analysis by quantitative metallography," Technical report, General Electric Research Laboratory, March 1961.
18. F. Chayes, *Petrographic Modal Analysis.* John Wiley, New York, 1956.
19. S.A. Saltykov, *Stereometric Metallography,* 2nd ed. Metallurgizdat, Moscow, 1958.
20. E.R. Weibel, *Morphometry of the Human Lung.* Springer-Verlag, Berlin, 1963.
21. D. McLean, *Grain Boundaries in Metals.* Oxford University Press, London, 1957.
22. S.A. Saltykov, "A stereological method for measuring the specific surface area of metallic powders"; p. 63 in *Stereology.* Edited by H. Elias. Springer-Verlag, 1967.
23. S.I. Tomkeieff, "Linear intercepts, areas and volumes," *Nature,* **155**, 24 (1945).
24. H.W. Chalkley, J. Cornfield, and H. Park, "A method for estimating volume-surface ratios," *Science,* **110**, 295 (1949).
25. T.F. Willis, "Measuring air in hardened concrete," *Proc. Am. Concr. Inst.,* **48**, 901 (1952).
26. K. Mather, "Measuring air in hardened concrete," *Proc. Am. Concr. Inst.,* **49**, 61 (1953).
27. H. Sommer, "The precision of the microscopical determination of the air-void system in hardened concrete," *Cem. Concr. Aggregates,* **1**, 49 (1979).
28. B.W. Langan and M.A. Ward, "Determination of the air-void system parameters in hardened concrete - an error analysis," *ACI J.,* **83**, 943 (1986).
29. R. Pleau and M. Pigeon, "Precision statement for ASTM C 457 practice for microscopical determination of air-void content and other parameters of the air-void system in hardened concrete," *Cem. Concr. Aggregates,* **14**, 118 (1992).

30. K. Snyder, K. Hover, and K. Natesaiyer, "An investigation of the minimum expected uncertainty in the linear traverse technique," *Cem. Concr. Aggregates,* **13**, 3 (1991).
31. J.E. Hilliard and J.W. Cahn, "An evaluation of procedures in quantitative metallography for volume fraction analysis," *Trans. Metal. Soc. AIME,* **221**, 344 (1961).
32. G.M. Tallis, "Estimating the distribution of spherical and elliptical bodies in conglomerates from plane sections," *Biometrics,* **26**, 87 (1970).
33. B. Warris, "The Influence of Air-Entrainment on the Frost Resistance of Concrete — Part A. Void Distribution"; in *Swedish Cement and Concrete Research Institute Proceedings NR 35.* Stockholm, 1963.
34. K.A. Snyder, K.C. Hover, K. Natesaiyer, and M. Simon, "Modeling air-void systems in hydrated cement paste," *Microcomp. Civ. Eng.,* **6**, 35 (1991).
35. K.C. Hover, "Analytical investigation of the influence of air bubble size on the determination of the air content of freshly mixed concrete," *Cem. Concr. Aggregates,* **10**, 29 (1988).
36. S.D. Wicksell, "The corpuscle problem: A case of ellipsoidal corpuscles," *Biometrika,* **18**, 152 (1926).
37. W.L. Nicholson, "Estimation of linear properties of particle size distributions," *Biometrika,* **57**, 273 (1970).
38. G.S. Watson, "Estimating functionals of particle size distributions," *Biometrika,* **58**, 483 (1971).
39. M.G. Kendall and P.A.P. Moran, *Geometrical Probability.* Hafner Publishing Co., New York, 1963.
40. E.B. Jensen, "A design-based proof of Wicksell's integral equation," *J. Microscopy,* **136**, 345 (1984).
41. J. Mecke and D. Stoyan, "Stereological problems for spherical particles," *Mathematische Nachrichten,* **96**, 311 (1980).
42. L.M. Cruze-Orive, "Distribution-free estimation of sphere size distribution from slabs showing overprojection and truncation, with a review of previous methods," *J. Microscopy,* **131**, 265 (1983).
43. E.T. Whittaker and G.N. Watson, *A Course in Modern Analysis,* 4th ed. Cambridge University Press, Cambridge, 1952.
44. R. Courant and D. Hilbert, *Methods of Mathematical Physics.* Wiley, 1989.
45. R.S. Anderssen and A.J. Jakeman, "Abel type integral equations in stereology — II. Computational methods of solution and the random spheres approximation," *J. Microscopy,* **105**, 135 (1975).
46. L.R. Roberts and P. Scheiner, "Microprocessor-based linear traverse apparatus for air-void distribution analysis"; p. 135 in *Proceedings of the Third International Conference on Cement Microscopy.* Edited by G.R. Gouda. International Cement Microscopy Association, 1981.
47. H. Elias, ed., *Stereology.* Springer-Verlag, 1967.
48. R.T. DeHoff and F.N. Rhines, eds., *Quantitative Microscopy.* McGraw-Hill, 1968.
49. R.L. Fullman, "Measurement of particle sizes in opaque bodies," *J. Metals*, **5**, 447 (1953).
50. G.W. Lord and T.F. Willis, "Calculation of air bubble size distribution from results of a Rosiwal traverse of aerated concrete," *ASTM Bull.,* **177**, 56 (1951).

51. T. Aarre, "Influence of measurement technique on the air-void structure of hardened concrete," *ACI Mater. J.*, **92**, 599 (1995).
52. R. Coleman, "The sizes of spheres from profiles in a thin slice, II. Transparent spheres," *Biometrical J.*, **25**, 745 (1983).
53. K. Snyder, "Discussion of 'Influence of measurement technique on the air-void structure of hardened concrete,' by T. Aare," *ACI Mater. J.*, **93**, 512 (1996).
54. B. Lu and S. Torquato, "Nearest-surface distribution functions for polydispersed particle systems," *Phys. Rev. A*, **45**, 5530 (1992).
55. B. Lu and S. Torquato, "Chord-length and free-path distribution functions for many-body systems," *J. Chem. Physics*, **98**, 6472 (1993).
56. R. Pleau, M. Pigeon, J.L. Laurencot, and R. Gagné, "The use of the flow length concept to assess the efficiency of air entrainment with regards to frost durability: Part II — Experimental results," *Cem. Concr. Aggregates*, **18**, 30 (1996).
57. E.K. Attiogbe, "Predicting freeze-thaw durability of concrete — A new approach," *ACI Mater. J.*, **93**, 457 (1996).
58. T.D. Larson, P.D. Cady, and J.J. Malloy, "The protected paste volume concept using new air-void measurement and distribution techniques," *J. Mater.*, **2**, 202 (1967).
59. K. Natesaiyer, K.C. Hover, and K.A. Snyder, "Protected-paste volume of air-entrained cement paste. Part I," *J. Mater. Civ. Eng.*, **4**, 166 (1992).
60. K. Natesaiyer, K.C. Hover, and K.A. Snyder, "Protected-paste volume of air-entrained cement paste. Part II," *J. Mater. Civ. Eng.*, **5**, 170 (1993).
61. S. Diamond, S. Mindess, and J. Lovell, "On the spacing between aggregate grains in concrete and the dimension of the aureole de transition"; in *Liaisons Pâtes de Ciment/Matériaux Associés.* International RILEM Colloquium (Toulouse), C42. 1982.
62. D.P. Bentz and K.A. Snyder, "Protected paste volume in concrete: Extension to internal curing using saturated lightweight fine aggregate," to be published in *Cem. Concr. Res.*
63. J. Marchand, R. Pleau, and R. Gagné, "Deterioration of concrete due to freezing and thawing"; p. 283 in *Materials Science of Concrete IV.* Edited by J. Skalny and S. Mindess. American Ceramic Society, Westerville, Ohio, 1995.
64. P. Hertz, "Über den gegenseitigen durchschnittlichen Abstand von Punkten, die mit bekannter mittlerer Dichte im Raume angeordnet sind," *Mathematische Annalen,* **67**, 387 (1909). See also Section VII of S. Chandrasekhar, "Stochastic problems in physics and astronomy," *Rev. Modern Phys.*, **15**, 1 (1943).
65. K. Natesaiyer, M. Simon, and K. Snyder, "Discussion of 'Mean spacing of air voids in hardened concrete' by E. Attiogbe," *ACI Mater. J.*, **91**, 123 (1994).
66. E.K. Attiogbe, "Volume fraction of protected paste and mean spacing of air voids," *ACI Mater. J.*, **94**, 588, 1997.
67. M. Abramowitz and I.A. Stegun, eds., *Handbook of Mathematical Functions.* Dover Publications, Inc., New York, 1972.
68. G. Arfken, *Mathematical Methods for Physicists.* Academic Press, New York, 1970.
69. K.A. Snyder, "Discussion of 'The use of the flow length concept to assess the efficiency of air entrainment with regards to frost durability: Part I — Description of the test method,'" *Cem. Concr. Aggregates,* **19**, 116, 1997.
70. K.A. Snyder, D.N. Winslow, D.P. Bentz, and E.J. Garboczi, "Effects of interfacial zone percolation in cement-aggregate composites"; p. 265 in *Advanced Cementitious Systems: Mechanisms and Properties.* Materials Research Society, 1992.

71. K.A. Snyder, D.N. Winslow, D.P. Bentz, and E.J. Garboczi, "Interfacial zone percolation in cement-aggregate composites"; p. 259 in *Interfaces in Cementitious Composites.* Edited by J.C. Maso. RILEM, 1992.
72. D.N. Winslow, M.D. Cohen, D.P. Bentz, K.A. Snyder, and E.J. Garboczi, "Percolation and pore structure in mortars and concrete," *Cem. Concr. Res.,* **24**, 25 (1994).
73. D.P. Bentz, J.T.G. Hwang, C. Hagwood, E.J. Garboczi, K.A. Snyder, N. Buenfeld, and K.L. Scrivener, "Interfacial zone percolation in concrete: Effects of interfacial zone thickness and aggregate shape"; p. 437 in *Microstructure of Cement-Based Systems: Bonding and Interfaces in Cementitious Materials.* Edited by S. Diamond, S. Mindess, F.P. Glasser, L.W. Roberts, J.P. Skalny, and L.D. Wakeley. Materials Research Society, Pittsburgh, 1995.
74. E.J. Garboczi and D.P. Bentz, "Analytical formulas for interfacial transition zone properties," *Adv. Cem. Based Mater.,* **6**, 99 (1997).
75. E.J. Garboczi and D.P. Bentz, "Multi-scale analytical/numerical theory of the diffusivity of concrete," *Adv. Cem. Based Mater.,* **8**, 77, 1998.
76. S.B. Lee and S. Torquato, "Porosity for the penetrable-concentric-shell model of two-phase disordered media: computer simulation results," *J. Chem. Phys.,* **89**, 3258 (1988).
77. J. Vieillard-Baron, "Phase transitions of the classical hard-ellipse system," *J. Chem. Phys.,* **56**, 4729 (1972).
78. A.L.R. Bug, S.A. Safran, G.S. Grest, and I. Webman, "Do interactions raise or lower a percolation threshold?" *Phys. Rev. Lett.,* **55**, 1896 (1985).
79. H. Reiss, H.L. Frisch, and J.L. Lebowitz, "Statistical mechanics of rigid spheres," *J. Chem. Phys.,* **31**, 369 (1959).
80. H. Reiss and R.V. Casberg, "Radial distribution function for hard spheres from scaled particle theory, and an improved equation of state," *J. Chem. Phys.,* **61**, 1107 (1974).
81. J.R. MacDonald, "On the mean separation of particles of finite size in one to three dimensions," *Molec. Phys.,* **44**, 1043, 1981.
82. S. Torquato and S.B. Lee, "Computer simulations of nearest-neighbor distribution functions and related quantities for hard-sphere systems," *Physica A,* **167**, 361 (1990).
83. S. Torquato, B. Lu, and J. Rubinstein, "Nearest-neighbor distribution functions in many-body systems," *Phys. Rev. A,* **41**, 2059 (1990).
83. P.A. Rikvold and G. Stell, "D-dimension interpenetrable-sphere models of random two-phase media: Microstructure and an application to chromatography," *J. Colloid Interface Sci.,* **108**, 158 (1985).
85. P.A. Rikvold and G. Stell, "Porosity and specific surface for interpenetrable-sphere models of two-phase random media," *J. Chem. Phys.,* **82**, 1014 (1985).
86. B. Lu and S. Torquato, "General formalism to characterize the microstructure of polydispersed random media," *Phys. Rev. A,* **43**, 2078 (1991).
87. W.F. Espenscheid, M. Kerker, and E. Matijevic, "Logarithmic distribution functions for colloidal particles," *J. Phys. Chem.,* **68**, 3093 (1964).
88. T.C. Powers, "Void spacing as a basis for producing air-entrained concrete," *Proc. ACI,* **50**, 741 (1954).
89. B. Warris, "The Influence of Air-Entrainment on the Frost Resistance of Concrete — Part B. Hypothesis and Freezing Experiments"; in *Swedish Cement and Concrete Research Institute Proceedings NR 36.* Stockholm, 1964.

90. M. Pigeon, R. Gagné, and C. Foy, "Critical air-void spacing factors for low water-cement ratio concretes with and without condensed silica fume," *Cem. Concr. Res.,* **17**, 896 (1987).
91. "Guide to the expression of uncertainty in measurement," Technical report, International Organization for Standardization (ISO), Genève, Switzerland, 1993.
92. B.N. Taylor and C.E. Kuyatt, "Guidelines for evaluating and expressing the uncertainty of NIST measurement results," Technical report, National Institute of Standards and Technology, Gaithersburg, Maryland, 1993.
93. J. Mandel, *The Statistical Analysis of Experimental Data.* Interscience Publishers, New York, 1964.
94. E.B. Wilson, *An Introduction to Scientific Research.* McGraw-Hill, New York, 1952.
95. J.R. Taylor, *An Introduction to Error Analysis,* 2nd ed. University Science Books, 1997.
96. R.H. Dieck, *Measurement Uncertainty: Methods and Applications,* 2nd ed. Instrument Society of America, 1997.
97. P.L. Meyer, *Introductory Probability and Statistical Applications,* 2nd ed. Addison-Wesley, Reading, 1970.
98. W. Feller, *An Introduction to Probability Theory and its Applications,* vol. I. John Wiley & Sons, New York, 1950.
99. M.J. Lighthill, *Introduction To Fourier Analysis and Generalised Functions.* Cambridge University Press, Cambridge, 1958.

Fresh Concrete Rheology: Recent Developments

C. Ferraris
National Institute of Standards and Technology, Gaithersburg, Maryland

F. de Larrard
Laboratoire Central des Ponts et Chaussées, Nantes, France

N. Martys
National Institute of Standards and Technology, Gaithersburg, Maryland

"In science it is not enough to think of an important problem on which to work. It is also necessary to know the means which could be used to investigate this problem."

— Leo Szilard

The design of concrete with specified properties for an application is not a new science, but it has taken on a new meaning with the wide use of high-performance concretes. The following properties are related to fresh concrete: ease of placement and compaction without segregation. "Ease of placement" covers various other properties of fresh concrete, such as workability, flowability, compactibility, stability, finishability, pumpability, and/or consistency. These words are often used interchangeably without definition based on fundamental measurements of properties. Several attempts were made to better relate fresh concrete properties with measurable entities. Some researchers treated fresh concrete as a fluid and used the fluid rheology methods to describe concrete flow. This approach, the most fundamental one, is reviewed in this paper. The fundamental definitions of entities used to uniquely describe the flow of concrete are reviewed. An overview of tests that are commonly used to measure the rheology of fresh concrete is given. Methods to predict the flow of concrete from either composition or laboratory tests and the main parameters that affect the flow of fresh concrete, such as composition, placement, and mixing methods, are discussed. Two special applications of rheology are also discussed: pumpable concrete and self-compacting concrete.

Introduction

The design of concrete with specified properties for an application is not a new science, but it has taken on a new meaning with the wide use of high performance concretes (HPCs). Recently, an ACI task group (a subcommittee of the Technical Activities Committee on HPC, THPC) published a new definition of HPC.[1] This definition states: "HPC is a concrete meeting special combinations of performance and uniformity requirements that cannot always be achieved routinely using conventional constituents and normal mixing." They continue by citing some of the properties that are critical for an application. The following properties are related to fresh concrete: ease of placement and compaction without segregation. The term "ease of placement" covers various other properties of fresh concrete, such as workability, flowability, compactibility, stability, finishability, pumpability, and/or consistency. These words are often used interchangeably without a definition based on fundamental measurements of properties. Typical qualitative definitions are "the ease with which concrete can be mixed, placed, compacted, and finished"[2] and "consistency is the ability of freshly mixed concrete to flow."[3]

Several attempts were made to better relate fresh concrete properties with measurable quantities. Ritchie[4] attempted to define the flow of concrete by linking it to various effects such as bleeding, sedimentation, and density. He distinguished three properties: stability, compactibility, and mobility. Stability is linked to bleeding and segregation, compactibility is equivalent to density, and mobility is linked to internal friction angle, bonding force, and viscosity. These descriptions, although subjective, at least link commonly used words with physical factors that can be measured. Other researchers[5–7] treat fresh concrete as a fluid and use fluid rheology methods to describe concrete flow. This approach, the most fundamental one, is reviewed in this paper. The fundamental definitions of entities used to uniquely describe the flow of concrete are reviewed. An overview of tests that are commonly used to measure the rheology of fresh concrete is given. Methods to predict the flow of concrete from either composition or laboratory tests and the main parameters that affect the flow of fresh concrete, such as composition, placement, and mixing methods, are discussed. Two special applications of rheology are also discussed: pumpable concrete and self-compacting concrete.

Concrete Flow Using Rheological Parameters

Concrete in its fresh state can be thought of as a fluid, provided that a certain degree of flow can be achieved and that concrete is homogeneous. This constraint could be defined by a slump of at least 100 mm and no segregation. This requirement would exclude, for example, roller-compacted concretes. The description of flow of a fluid uses concepts such as shear stress and shear rate as described in Refs. 5 and 8. Concrete, as a fluid, is most often assumed to behave like a Bingham fluid. In this case, its flow is defined by two parameters: yield stress and plastic viscosity. The Bingham equation is:

$$\tau = \tau_0 + \mu\dot{\gamma}$$

where τ is the shear stress applied to the material, $\dot{\gamma}$ is the shear strain rate (also called the strain gradient), τ_0 is the yield stress, and μ is the plastic viscosity. To determine the Bingham parameters, there are two possibilities:

1. The stress applied to the material is increased slowly and the shear rate is measured. When the stress is high enough the concrete will start flowing. The point at which the materials flow is the yield stress and the slope of the curve above this stress is the plastic viscosity.
2. The fresh concrete is sheared at high rate before the rheological test. Then, the shear rate is decreased gradually and the stress is measured. The relationship between the stress and shear rate is plotted and the intercept at zero shear rate is the yield stress, while the slope is the plastic viscosity.

In this review, we will assume that the fresh concrete has been sheared at a high rate before the rheological test, because this is the most commonly used procedure. The main reason that this procedure is the most widely used is that it is easier to develop a rheometer that is shear-rate-controlled (Procedure 2) than stress-controlled (Procedure 1).

In addition, some concretes, such as self-compacting concrete (SCC), do not follow the linear function described by Bingham.[9] In fact, the calculation of a yield stress using a Bingham equation in the case of SCC will result in a negative yield stress, as shown in Fig. 1. De Larrard et al.[10] use another equation that describes the flow of suspensions, the Herschel-Bulkley (HB) equation. This provides a relationship of shear stress to shear strain rate based on a power function:

$$\tau = \tau'_0 + a\dot{\gamma}^b$$

where τ is the shear stress, $\dot{\gamma}$ is the shear strain rate imposed on the sample, τ'_0 is the yield stress, and a and b are the new characteristic parameters describing the rheological behavior of the concrete. In this case, plastic viscosity cannot be calculated directly.

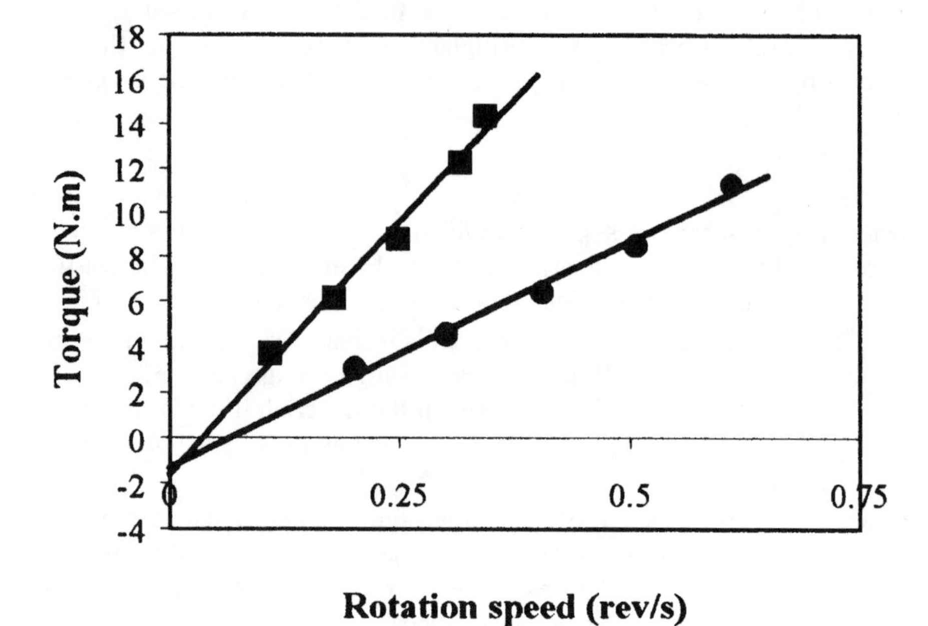

Figure 1. Self-compacting concrete flow measured with a parallel plate concrete rheometer, BTRHEOM.*[,9] The torque is a measure of the shear stresses and the rotation speed is related to the shear rates.

De Larrard et al.[10] also investigated the possibility of reducing the number of parameters to two while still using the HB equation. The HB equation could be considered as a linear relationship for a "short" range of shear strain rate. The yield stress is calculated by the HB equation, while the viscosity is calculated using the following equation:

$$\mu' = \frac{3a}{b+2}\dot{\gamma}_{max}^{\,b-1}$$

where μ' is the slope of the straight dotted line in Fig. 2, $\dot{\gamma}_{max}$ is the maximum shear strain rate achieved in the test, and a and b are the parameters as calculated by the HB equation. This Bingham-modified equation is deter-

*Certain commercial equipment, instruments, or materials are identified in this review to foster understanding. Such identification does not imply recommendation or endorsement by the National Institute of Standards and Technology, nor does it implies that the materials or equipment identified are necessarily the best available for the purpose.

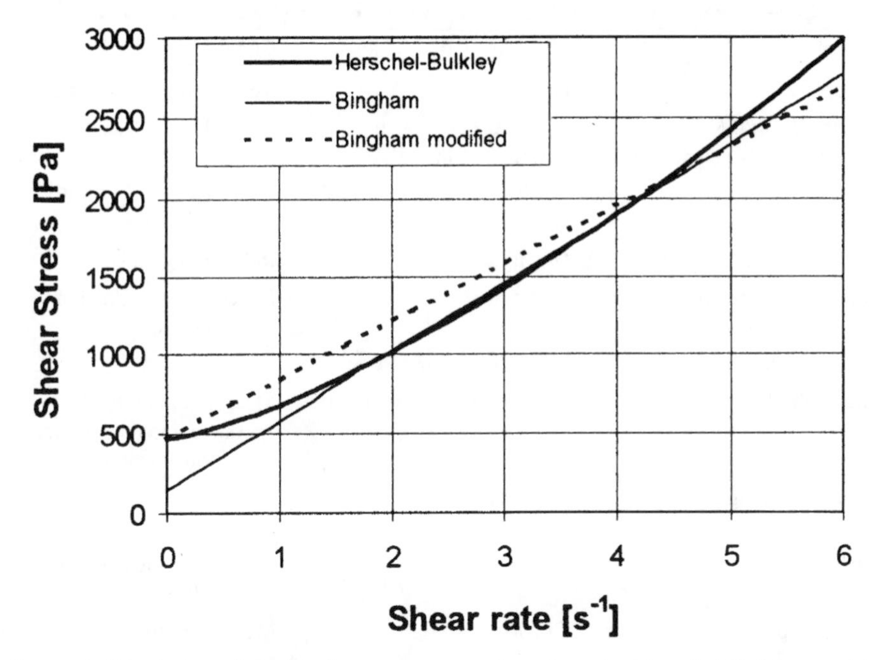

Figure 2. Calculation of the Bingham parameters based on the Herschel-Bulkley model. The dotted straight line departs from the same point as the HB model $(0, \tau'_0)$.

mined by approximation of the HB equation with a straight line, minimizing (using a least-squares method) the deviation between the two models, that is, HB and straight line (modified Bingham equation).

In summary, it should be noted that concrete must be defined by at least two parameters because it shows an initial resistance to flow, yield stress, and a plastic viscosity that governs the flow after it is initiated. Nevertheless, most commonly used tests to describe concrete flow are limited to the measurement of only one parameter, often not directly related to either of the Bingham parameters. Only recently were some instruments designed to better describe concrete flow.[11] A description of the available tests is given below.

Measurement Techniques for Fresh Concrete

As discussed earlier, a test characterizing the flow of concrete should be able to determine at least two parameters, such as yield stress and plastic viscosity. The design of a rheometer for concrete must take into account the dimensions of the coarse aggregate. The smallest gap in the instrument

Table I. Tests that measure only one parameter, either yield stress or viscosity

Test	Stress applied	Comments
Slump[13]	Gravity	Related to yield stress
Penetrating rod: Kelly ball,[14] Vicat,[15] DIN penetration test[24]	Applied stress, i.e. the weight of the ball or other device	Related to yield stress
K-slump test[16,17]	Gravity	Related to segregation
Turning tube viscometer[18]	Gravity	Related to viscosity
Ve-Be time or remolding test (Powers apparatus)[19]	Vibration	For concretes with high yield stress
LCL apparatus[20]	Vibration and gravity	
Filling ability[20,21]	Applied pressure or gravity	Measure of ability of concrete to flow between reinforcement bars
Vibration testing apparatus or settling curve[22]	Vibration	
Flow cone[23]	Gravity	Measure of the ability to flow through an opening
Orimet apparatus[19]	Gravity	Measure of the ability to flow through an opening
Slump drop test[24]	External pressure/gravity	

should be at least three times the largest diameter of the coarse aggregate to obtain a representative sample and to avoid interlocking of the aggregates, which will prevent flow. The difficulty in meeting this requirement led to the design of empirical tests that do not allow for the calculation of the yield stress and plastic viscosity in fundamental units. The design of such tests was to imitate the method of placement in the field. These tests very often measure only one value, which is not necessarily related to the fundamental parameter defined by Bingham. It is only recently that some instruments were designed to obtain two values that are related to the fundamental parameters.

There are numerous standard and nonstandard empirical tests to measure the flow of concrete. Because the results of such tests are not expressed in fundamental units, it is difficult to relate results from different tests. They can be used only for a direct comparison between concretes when using the same test.

As a full description of all the tests is beyond the scope of this review, we will limit ourselves to a list of the tests with some comments. There are two broad categories of tests: those that provide one parameter and those that provide two.

Table I gives a list of most common tests with some comments on the type of result that can be obtained. A discussion of the merits and results obtained can be found elsewhere.[11,12] To measure the viscosity, the yield stress must be exceeded. This can be achieved by various methods, but the two most common are gravity or vibration. In the gravity method, the stress applied is the weight of the materials, as opposed to an external applied stress. In the vibration method, the yield stress and flow behavior of the concrete are completely different from those observed without vibration. These tests are intended to simulate field performance in the laboratory.

The design of a rheometer for concrete allowing measurements of a flow curve describing the relationship between shear stress and shear rate can be taken from the science of fluid rheology. The most common rheometers are coaxial or parallel plate.

A coaxial rheometer is composed of two concentric cylinders. The outer cylinder is usually stationary and the inner cylinder rotates at a controlled speed. The shear stresses generated by the fluid are measured on the inner cylinder. To be able to compute the shear stress and shear rates as well as calculate the yield stress and plastic viscosity according to the Bingham equation, the gap between the cylinders needs to be relatively small as compared to their diameters. It is generally accepted that the ratio of the radii of the two cylinders should be between 1 and 1.1. For concrete, the gap needs to be at least three to five times the size of the coarse aggregate to avoid interaction between the aggregates and the walls of the rheometer. Therefore, for an aggregate maximum size of 10 mm, the minimum radius is 0.5 m, which will require the diameter of the outer cylinder to be between 0.53 and 0.55 m. These dimensions would have to be increased with the maximum size of the aggregate, rendering this type of instrument unsuitable for field use because it would not be easily transportable outside the laboratory. Such a rheometer was built by Coussot[25] and used for fresh concrete by Hu et al.[26] to validate the results obtained with the BTRHEOM rheometer developed at the Laboratoire Central des Ponts et Chaussées (LCPC).

To overcome the dimension limitations of the coaxial cylinder rheometer while maintaining the possibility of estimating the two Bingham parameters, Tattersall[5] designed a rheometer that consisted of a shaft with blades that rotated in a bucket of concrete at a controlled speed. The torque generated by the concrete is measured on the shaft. This method does not allow for the calculation of viscosity and yield stress in fundamental units, but it enables the sudy of concrete flow under various shear rates. This rheometer, referred to as the "two-point test," was modified and computerized by Wallevik and Gjørv.[27] The commercially available BML rheometer by Wallevik† has another modification involving the shape of the blades. The blades are fins attached radially on the shaft. This rheometer can be used to estimate the Bingham parameters in fundamental units if no plug flow occurs. Hu et al.[28] showed that plug flow occurs in concretes with a slump (measured according to the standards[13]) less than 200 mm. Beaupré[7] also developed a rheometer, referred to as IBB, with a different blade/shaft assembly and the shape of the letter H. The IBB rheometer is also a modification of the original two-point test. Further descriptions of these tests are found elsewhere.[8,12]

Another geometry that is commonly used for rheological measurements is a parallel plate. Here, an upper plate rotates at a preselected speed and the torque generated by the shear resistance of the material is recorded on the same plate. The bottom plate is stationary. The shear rate in such an instrument is not constant and depends on the radial position, that is, the shear rate is zero at the center of the plate and maximum at the edge. In most cases the shear rate and the shear stress at the edge are the measurements considered for the calculation of viscosity. This is not a serious problem if the fluid is Newtonian, because the viscosity does not depend on the shear rate, but it is for non-Newtonian fluids. For non-Newtonian fluids, an analytical calculation needs to be carried out. As before, the distance between the two plates must reflect the size of the aggregates. This distance should be at least three to five times the diameter of the largest aggregate.

There is only one rheometer that uses the parallel plate geometry: the BTRHEOM,[29] which was developed by de Larrard et al. at LCPC. It consists of a bucket that has a capacity of about 7 L, with a fixed wheel at the bottom and a wheel at the top rotating at any selected speed. The bottom wheel records the torque generated by the material reaction to shearing. The results of this test can be computed to obtain viscosity and yield stress in fundamental units.

Whereas the concentric cylinder rheometers described above are too large for field use, the BTRHEOM is relatively small and can be carried by one person. Data acquisition is made with a portable computer.

Nevertheless, there is a need for a simple, inexpensive test to be used in the field for quality control of the concrete. In a survey conducted by the National Ready-Mixed Concrete Association in 1997,[30] more than half of

†BML viscometer, the Icelandic Building Research Institute, O. Wallevik.

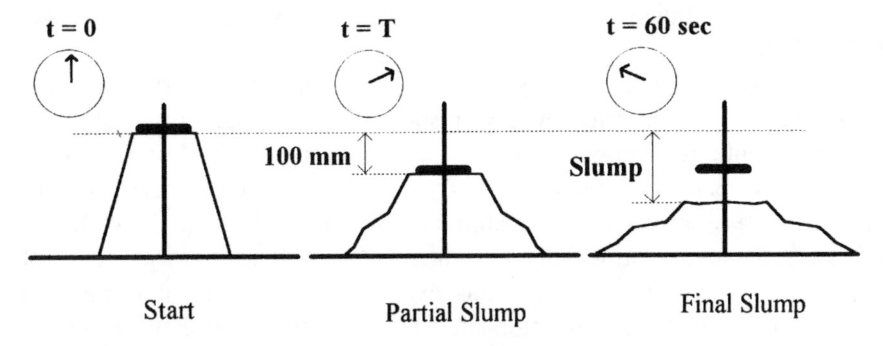

Figure 3. Schematic of the modified slump test. *T* is the slump time.

the participants indicated that although they considered the slump test adequate to describe workability, they felt that a better test was needed. They indicated that the slump test[13] did not give them a full description of the flow of concrete. For this reason, Ferraris and de Larrard developed at NIST a modified slump cone test.[31,32] Figure 3 shows the schematic of this test. The modification consists of measuring not only the final slump height but also the time it takes for the concrete to slump the first 100 mm, that is, the speed of slumping. There are two methods to measure the speed:

1. The original method consists of measuring the time for a plate to slide down with the concrete[32] a distance of 100 mm.
2. Researchers at Sherbrooke University[33] eliminated the plate and shortened the central rod so that its top was 100 mm below the full slump cone height. The test consists of measuring the time it takes for the concrete to slump to the height where the rod first becomes visible.

The second method has the advantage that there is no risk of the plate getting stuck, but has the disadvantage that it may be difficult to see the appearance of the rod. Painting the end of the rod in a bright color does not solve the problem,[33] because it is covered with cement paste.

From the final slump and slump time, the yield stress and plastic viscosity (in fundamental units) can be calculated using an empirical equation that was developed by comparing the modified test measurement with the values obtained with the BTRHEOM.[9,31,32]

This test is being evaluated in various laboratories in United States and France to determine the reproducibility of the results and the correlation of the slumping time and the final slump with the yield stress and plastic viscosity. When sufficient data have been collected, this test will be proposed to ASTM for consideration as a standard test.

In summary, while it can be seen that there are numerous tests to characterize the flow of concrete, few give results in fundamental units and therefore the rheological properties of concretes measured using different tests cannot be directly compared. Recently, new tests for characterizing concrete using a more fundamental approach have been developed. While not all researchers agree on which test is the most suitable for the wide range of concretes in use today, tests that can give results in fundamental units and that can be used on a construction site should be favored, because comparison between test results can be achieved.

Models to Predict Rheological Properties

For the engineer who needs to design concrete for a specific application or for a specific placement method, the challenge lies in the prediction of the fresh concrete's properties from its composition. Generally, there are procedures or codes to estimate the slump values depending on factors such as w/c ratio and chemical and mineral admixture dosage, but most of the time several trial batches are needed.

A model that could predict the rheological parameters, yield stress and viscosity, from the composition or from minimal laboratory tests would be beneficial. Three promising models will be reviewed here: the compressible packing model (CPM) developed by de Larrard, simulation of flow of suspensions developed by NIST researchers, and a semi-empirical model based on the Krieger-Dougherty equation developed by Struble.

Compressible Packing Model

LCPC developed this model for predicting concrete properties from its composition. Concrete is defined as a granular mixture (from cement to the coarse aggregates) in a water suspension. A concrete with no workability, that is, no flow, is defined as a concrete where the porosity is filled with water. This statement implies that there is no excess water between the solid components. Therefore, the yield stress can be correlated with the stress needed to initiate flow by overcoming the friction forces between the particles. These forces depend on the number and type of contacts between the particles.

Each component i of the mixture is defined by its close packing density, ϕ^*_i, and the volumetric fraction of solid material (with respect to a total volume of one), ϕ_i. A close packing density is defined as the maximum possible value of ϕ_i, with all the other ϕ_j ($j \neq i$) being constant. Also, the whole mixture is characterized by a close packing maximum, ϕ^*, and the volumetric fraction of the solid materials, ϕ.

The yield stress, τ_0, can then be defined as:

$$\tau_0 = f\left(\frac{\phi_1}{\phi_1^*}, \frac{\phi_2}{\phi_2^*}, \ldots \frac{\phi_n}{\phi_n^*}\right)$$

where f is an increasing function because the yield stress will increase with increasing value of ϕ_i / ϕ^*_i.

To determine the viscosity dependence on the volumetric concentration, we can assume that the speed of each particle under shear is the same and equal to the macroscopic speed. Therefore, it is assumed that the flow of the fluid between the particles is laminar and that the shear resistance will remain proportional to the overall gradient. Thus, if the Bingham equation is assumed to be valid, the plastic viscosity can be deduced to be:

$$\mu = \mu_0 g \frac{\phi}{\phi^*}$$

where μ_0 is the plastic viscosity of the suspending fluid and g is an increasing function, because the viscosity will increase with increasing concentration of particles.

These equations were tested by comparison with a series of 78 concrete batches in which rheological parameters were measured using the BTRHEOM. The close packing and the volumetric fraction of each component were calculated using the CPM.[9] The plastic viscosity was determined by a best-fit equation, given by the equation below, from the data shown in Fig. 4.

$$\mu' = \exp\left\{26.75\left[\left(\frac{\phi}{\phi^*}\right) - 0.7448\right]\right\}$$

The yield stress can be calculated by a linear combination of all the components' volume fraction/close-packing ratios. It appears that different coefficients need to be calculated for concrete with and without high-range water reducing admixtures (HRWRA). The data used were from the same set as used for the viscosity.

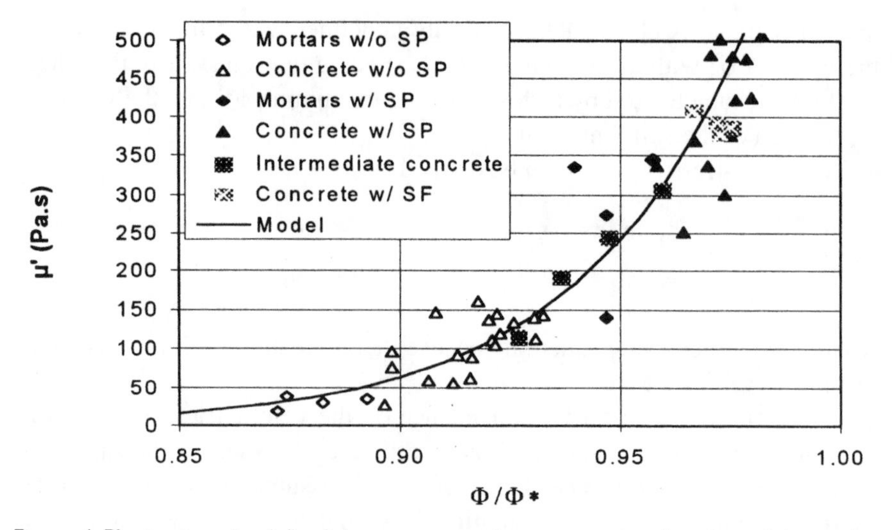

Figure 4. Plastic viscosity (μ') of the mortars and concretes as a function of their relative solid concentrations. SP = superplasticizers or HRWRA; SF = silica fume.

For mixtures without HRWRA the yield stress was:

$$\tau'_0 = \exp\,(2.537 + 0.540\,K'_g + 0.854\,K'_s + 1.134\,K'_c)$$

And, for the mixtures with 1% HRWRA (without silica fume), it was:

$$\tau'_0 = \exp\,(2.537 + 0.540\,K'_g + 0.854\,K'_s + 0.224\,K'_c)$$

In these equations, τ'_0 is the yield stress obtained by fitting the rheometer results in accordance with the Herschel-Bulkley model. The indices g, s, and c relate to gravel, sand, and cement, respectively. K_x is equal to $(1 - \phi_x / \phi^*_x)$.

These results were confirmed with other data sets resulting from variation of the coefficients used in fitting the data (Fig. 5).[34]

This model is part of a larger set of models that can take into account other properties of both fresh and hardened concrete. This model links the composition of the concrete with its performance.[34]

Simulation of Flow of Suspensions

Ferraris and Martys are developing a new procedure that includes simulation of the concrete flow. The procedure is based on the chart illustrated in Fig. 6. The cement paste rheological parameters, yield stress and viscosity, are measured using a laboratory fluid rheometer. The cement paste in this

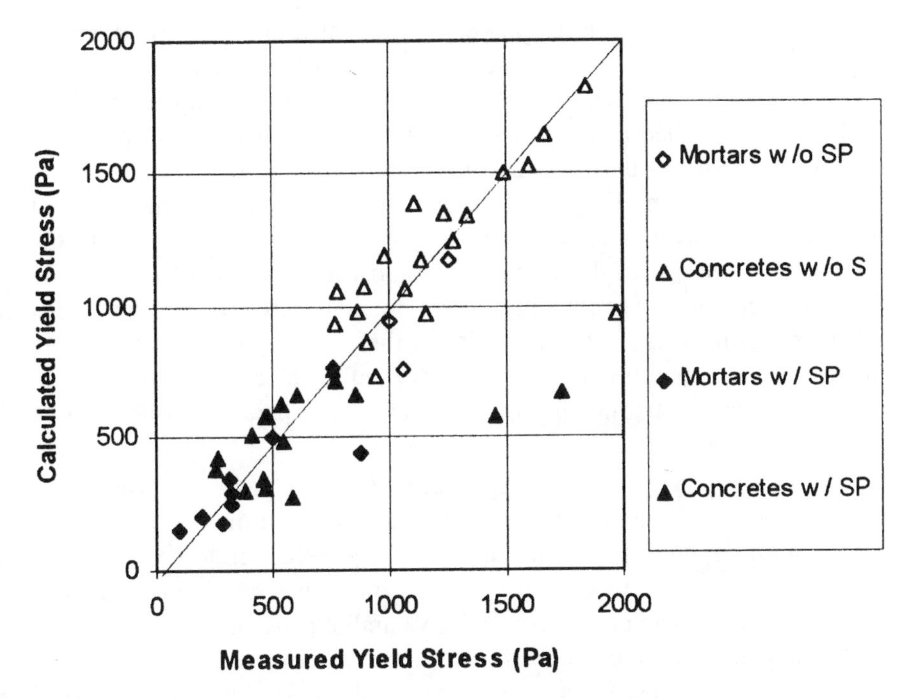

Figure 5. Comparison between experimental values and model values of the yield stress.[9]

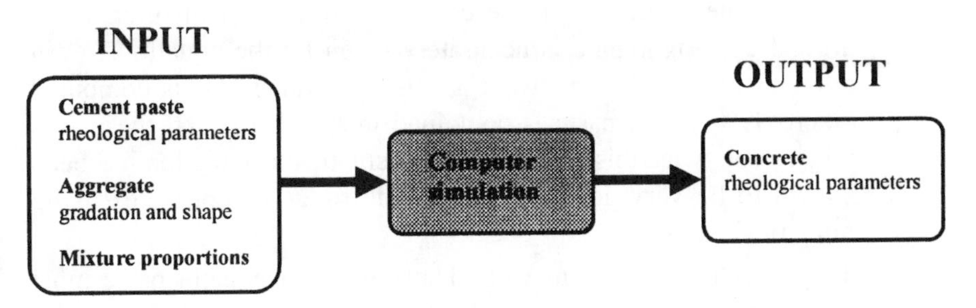

Figure 6. Principle of the procedure.

case includes any chemical and mineral admixtures that are selected. Then these values along with the mixture design of the mortar or concrete (i.e., gradation, shape, and total content of aggregates) are used as input to a computer simulation. The results are the rheological parameters of concrete. The simulation method is in the process of being developed and is presented below.

To be able to succeed in the procedure illustrated in Fig. 6, the rheology of the cement paste must be measured in the same conditions experienced in concrete, that is, at the same shear rate and the same temperature. Barrioulet et al.[35] studied a set of concrete mixtures having cement pastes of various compositions but with the same viscosity, and with various aggregates having the same shape and gradation. They found that the flows of these concretes were not the same. They attributed this difference to the fact that the rheology of the whole is not equal to the rheology of the parts if the interactions between the parts were not considered. It is believed that the error was to measure the viscosity of the cement paste without taking into account the condition of shear experienced by the cement paste in concrete. Therefore, the following three experimental parameters need to be monitored:

1. The gap between the aggregates. Cement paste is "squeezed" between the aggregates. The distance between the aggregates (the gap) depends on the paste content of the concrete considered.[36] Therefore, the rheometer geometry must be a variable geometry in which the gap can be changed. A parallel plate rheometer is a suitable device. To estimate the average gap between aggregates, NIST researchers used a mathematical method developed by Garboczi and Bentz[37] based on equations developed by Lu and Torquato.[38] The aggregates are treated as spheres suspended in a cement paste matrix. The volume of paste contained in a shell of thickness r around each aggregate is accurately given by the equations, even allowing for overlaps between shells. The value of r is computed where 99% of the paste is contained in the shells, and the gap is taken to be twice this value. The mathematical calculation has been shown to be very accurate for a wide range of concretes using numerical analysis.[39]
2. The shear rate during mixing: The mixing of cement paste must replicate the shear stresses experienced in concrete. Helmuth et al.[40] developed a methodology and identified the hardware required.
3. The temperature of the cement paste. That of the cement paste mixed alone is usually higher than the temperature of the cement paste in concrete. This discrepancy is due to the heat sink that the aggregates represent in concrete. Therefore, the cement paste needs to be correctly cooled during mixing. Helmuth et al.[40] identified the hardware needed to achieve such control.

Figure 7. A configuration of densely packed monosized spheres at a solid fraction (volume of spheres/system volume) of 0.5. The system is composed of 52 spheres.

In Fig. 6, the link between the mixture design/cement paste rheology and the concrete rheology is a simulation model. The simulation of concrete flow is based on a mesoscopic model of complex fluids called dissipative particle dynamics (DPD),[41] which blend together cellular automata ideas with molecular dynamics methods. The original DPD algorithm used symmetry properties such as conservation of mass, momentum, and Galilean invariance to construct a set of equations for updating the position of particles, which can be thought of as representing clusters of molecules or "lumps" of fluid. Later modifications to the DPD algorithm resulted in a more rigorous formulation and improved numerical accuracy. An algorithm for modeling the motion of arbitrary shaped objects subject to hydrody-

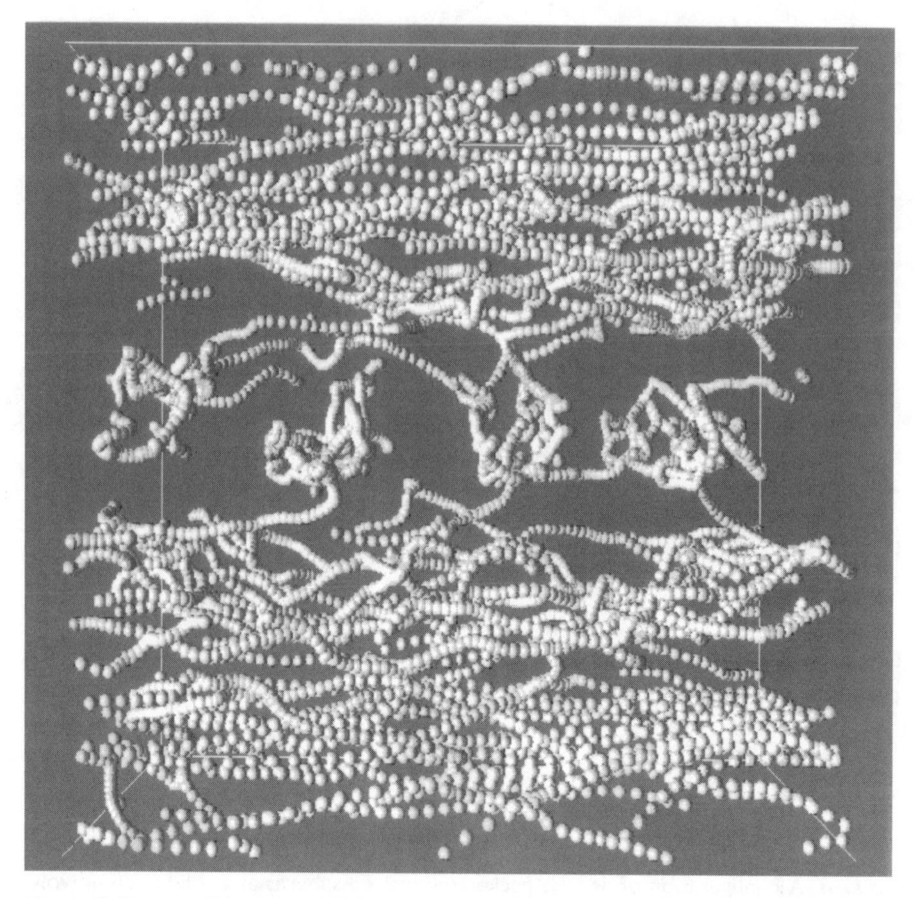

Figure 8. Traces of the centers of spheres moving under an applied shear strain. Near the top and bottom layers, the spheres move along relatively smooth paths. Near the center, the spheres move more slowly as the mean velocity is close to zero in this region.

namic interactions by DPD was suggested by Koelman and Hoogerbrugge.[42,43] The rigid body is approximated by "freezing" a set of randomly placed particles where the solid inclusion is located and updating their position according to the Euler equations.

Martys is currently extending the DPD method to model the flow of concrete as a function of mixture design. The Koelman-Hoogerbrugge procedure for representing rigid bodies is ideally suited for modeling the flow of concrete because this technique easily allows for the representation of wide

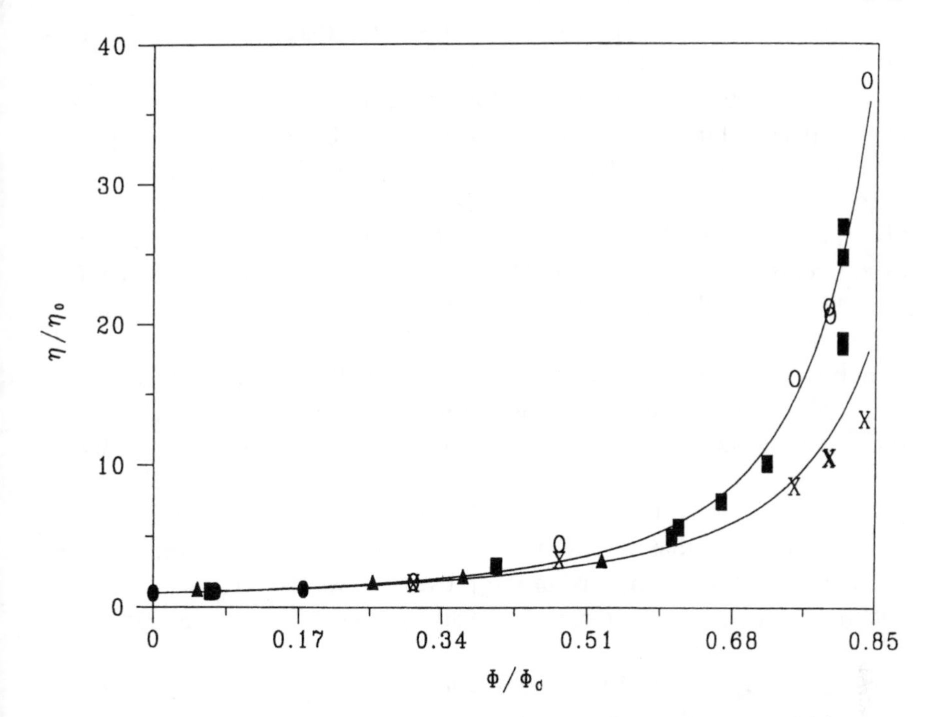

Figure 9. Viscosity, η, of suspension normalized to solvent viscosity, η_0 vs. solid fraction, ϕ, normalized to maximum random packing fraction, ϕ_0. The solid data points (squares and triangles) are computer simulation data, while the Xs and 0s are derived from experiments[44] on sheared hard sphere colloids. The Xs correspond to an infinite shear rate limit whereas the 0s correspond to the zero shear rate limit. The three simulation data points at the highest solid fraction correspond to different shear rates (approximately a factor of ten greater with decreasing viscosity on the figure).

distributions of particle size and shape such as are common in concrete. As a test of the DPD method, Martys studied the flow of a suspension of spheres of equal radii under shear as a function of solid fraction and shear rate. Figure 7 shows a configuration of densely packed spheres at a solid fraction (volume of spheres/system volume) of 0.5. A constant rate of strain is applied in opposite directions at the top and bottom of the system.

Figure 8 shows a trace of the sphere positions over several time steps. Near the top and bottom, the spheres flow more smoothly as they respond to the applied strain. In the middle horizontal region the spheres appear to diffuse more as the velocity is zero on average here. Figure 9 shows the effective viscosity as a function of the solid fraction divided by the maximum packing fraction of random spheres. Note that at higher solid fractions the suspension exhibits shear thinning.

We have begun to examine the effect of varying the distribution of sizes of the spheres. For instance, at a solid fraction of 0.4, 10% of the spheres were replaced with smaller spheres (about one-sixth of the radius) while fixing the total solid fraction. For this simple change in composition the viscosity decreased by about 8%. Similar results were found in physical experiments examining the flow of cement paste where some of the cement particles were replaced by fly ash, which are about a factor of 10 smaller in diameter.

Future studies will include evaluation of the effects of particle shape, as ellipsoids with varying aspect ratios replace spheres in the DPD simulations. We will also investigate the effects of the roughness of the walls on the plastic viscosity and yield stress.

To validate the DPD method, rheological properties should be measured using a rheometer that allows the calculation of the rheological parameters in fundamental units. We are in the process of validating the model with data from concrete and cement paste made using the same raw materials.

Krieger-Dougherty Modified Model

Another approach in predicting the viscosity of concrete from the measurements of cement paste was developed by Struble.[45] She based her model on the Krieger-Dougherty equation. This equation shows that there is an increase in the viscosity of the medium when particles are added. This increase depends on the concentration of the particles:

$$\frac{\eta}{\eta_0} = \left[1 - \left(\frac{\phi}{\phi_m}\right)\right]^{-[n]\phi_m}$$

where $[\eta]$, the intrinsic viscosity, is equal to 2.5 for spheres, ϕ is the volume concentration of particles, ϕ_m is the maximum packing, η is the viscosity of the suspension, and η_0 is the viscosity of the medium. Therefore, if the viscosity of the cement paste and the concentration of the aggregates are known and the maximum packing of the particles is determined, then the viscosity of the concrete can be calculated. Struble used a coaxial rheometer with a gap between the cylinders of 0.7 mm. This gap, although small, is not quite as small as the mean gap between the aggregates in concrete (0.12–0.26 mm).[11] The concrete rheological properties are measured using a rheometer similar to the BTRHEOM. This model has not been validated for concrete. Struble is currently designing a program to validate the model.

Factors Affecting the Rheology of Concrete

As stated earlier, workability and other flow properties are related to the rheology of concrete, which requires at least two parameters, such as the Bingham parameters, for adequate description. What are the principal factors that influence the rheological parameters of concrete? The first factors are the composition of the concrete, including the chemical and mineral admixture dosage and type; the gradation, shape, and type of the aggregates; the water content; and the cement characteristics. The same mixture design can result in different flow properties if secondary factors are not taken into account. These are:

- Mixer type: pan, truck, and so on. These may induce various levels of deflocculation and air entrainment.
- Mixing sequence, that is, the sequence of introduction of the materials into the mixer.
- Mixing duration.
- Temperature.

To determine the rheological characteristics needed for a specific application, the following items need to be considered:

- Method of delivering the concrete to the forms, for example, pumping, bucket.
- Method of consolidation, for example, vibration, tamping, none.
- Type of finishing method.

In considering the application, some of these items will be automatically selected. For instance, if a structure with a very high amount of reinforcement is built, the concrete needs to be self-consolidating because it will be impossible for a vibrator to reach all the concrete.

Another variable that should be addressed is the time dependence of the rheological parameters. This phenomenon is often called slump loss or excessive retardation. The placement of the concrete becomes either difficult (slump loss) or the demolding is retarded and strength development is delayed.[46]

A detailed review of how the various factors mentioned above will influence the flow, and specifically the Bingham parameters, will not be attempted here, but a good review is given by Khayat et al.[47] We will examine in more depth some special cases that are of interest for high-performance concrete usage: pumpable concrete and self-consolidating concrete.

Pumpable Concrete

Pumping is one of the most popular techniques worldwide to transport fresh concrete. Until recently, pumping engineering has been an empirical process. Now, a systematic scientific research is being carried out, which should lead eventually to optimization of concrete-pumping systems.

Practical Problems Dealing with Concrete Pumping

For placing large quantities of fresh concrete, piston pumps are generally used.[48] Concrete is pushed alternately by two pistons acting in cylinders. The first problem to be solved is filling the total length of the pumping network without creating a blockage. Blockage may occur due to leakage (generally at the joint between several pipes, where cement paste may flow out, provoking an accumulation of aggregate) or segregation. Also, after a stop in the pumping process, concrete setting may begin. To avoid blockage, a number of precautions must be taken:

- The network must be concrete tight.
- Fresh concrete should have a minimal susceptibility to segregation.
- A sufficient volume of cement grout has to be pumped before fresh concrete, so that the coarse aggregate particles that are expelled out of concrete by the pumping strokes remain in a suspending fluid.
- The time of practical use of the concrete, during which the consistency is soft enough, should be higher than the time necessary to pump and to place it.

However, even when these conditions are met, the pump may be unable to convey the concrete up to the end of the pumping network at the required flow rate. This emphasizes the need for a thorough understanding and application of rheological principles.

Tribology of the Steel/Concrete Interface

When concrete is pushed through a steel pipe of constant diameter, coarse aggregate particles tend to concentrate in the axis of the pipe, so that a slip motion happens in a thin layer of cement paste located at the steel/concrete interface (see Fig. 10). In general, all shear deformations are located in this zone with the rest of the concrete being transported as a plug. Thus, the study of the interface mechanical behavior, which is the purpose of tribology,[49] must be carried out. It is possible to reproduce the slip phenomenon in the laboratory, using a large-gap coaxial viscometer with a smooth inner cylinder. At low rotation speed, a linear relation is found between shear stress and slip rate, corresponding to the following equation:

$$\tau = \tau_{0,i} + k v$$

where τ is the shear stress, $\tau_{0,i}$ is the interface yield stress, v is the slip rate, and k is a viscous interface constant. At higher rotation speeds, the shear deformation tends to propagate into the bulk of concrete. Therefore, a rheometer test must be performed first in order to know the Bingham constants of the concrete considered. Then, a numerical analysis of the experimental relationship between torque and rotation speed in the coaxial viscometer leads to the determination of the two interface parameters. It is especially important to have a gap large enough in order to be able to keep the external concrete layer at rest in the viscometer.

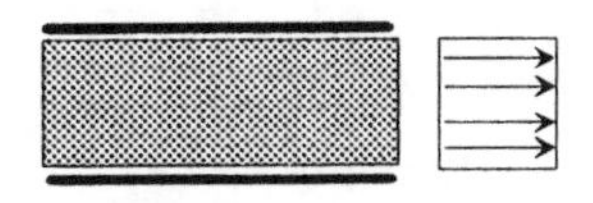

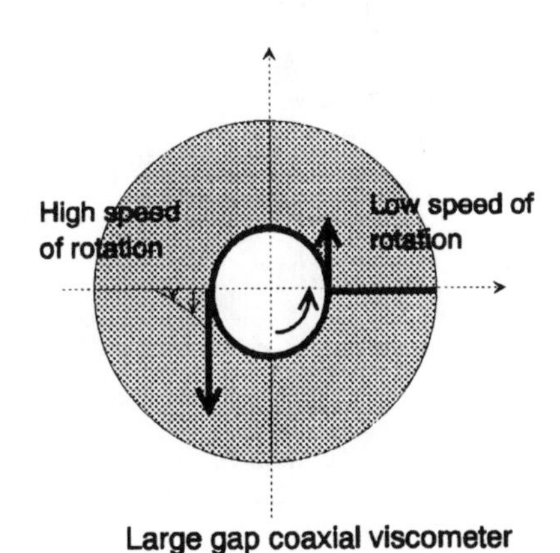

Figure 10. Flow of fresh concrete in a pump and in a large-gap coaxial viscometer. The arrows represent for the speed of particles.

The CALIBE Project

In the CALIBE national project carried out in France from 1995 to 1999, one of the goals was to develop special devices in order to better control concrete pumping. An experimental 150-m (500-ft) pumping network with 12 elbows was installed, in which more than 50 concrete compositions were pumped. The flow rate and the pressure in the concrete versus length were measured during each test. It was found that the pressure versus length patterns were always linear, which is experimental evidence of the validity of the equation in the previous section. Moreover, the rheological

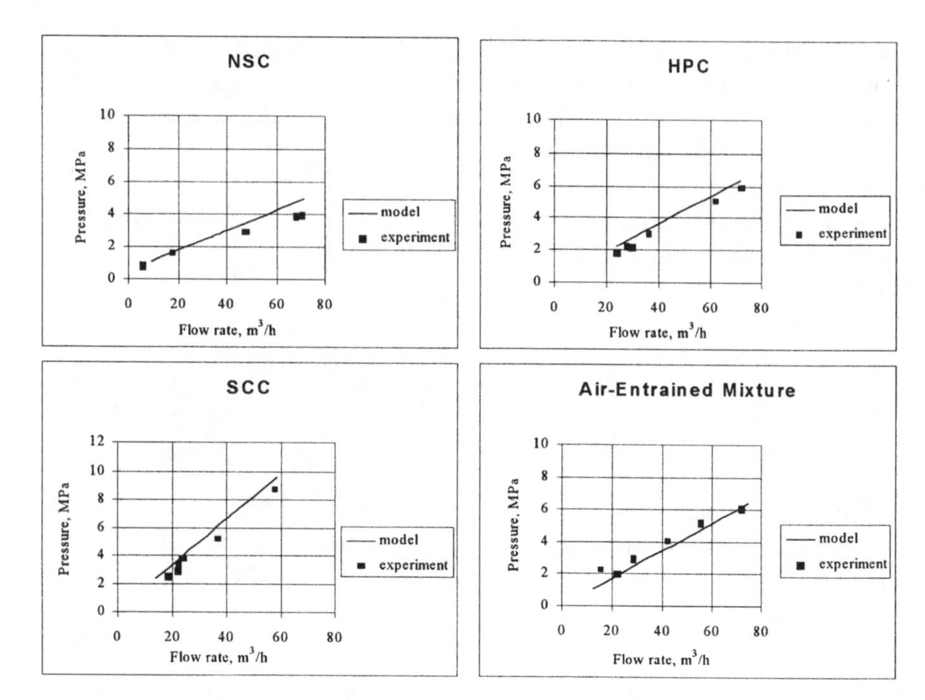

Figure 11. Comparison between experiments and theory from the CALIBE pumping tests.[49] NSC: normal-strength concrete (w/c = 0.53, slump = 190 mm); HPC: high-performance concrete (w/c = 0.33, slump = 240 mm); SCC: self-compacting concrete (w/c = 0.25, slump = 270 mm); air-entrained mixture: 10% air (w/c = 0.40, slump = 110 mm).

parameters measured with the BTRHEOM rheometer and the interface parameters assessed with a large-gap coaxial viscometer were used to predict the relationship between pressure and flow rate, accounting for the total length and diameter of the pipes. The agreement between a calculation and experiment was found to be very satisfactory (see Fig. 11). In addition, it was found that elbows play a negligible role in the pressure losses, but are important only to the extent to which they may facilitate blockage.

Self-Consolidating Concrete

Self-consolidating concrete (SCC) can be placed without vibration and flows easily in very narrow gaps. It is widely used in Japan and has recently been used in Europe, but is used very little in the United States. Nevertheless, it must be expected that SCC will be used more in the future because of its good performance, such as ease of placement. Due to its rheological properties, the expense of vibration can be eliminated while still obtaining good consolidation. Heavily reinforced structures can be designed and built using this material. According to Kim et al.[50] the major factors defining SCC are rheological properties of the cement paste, the volume ratio of paste to coarse aggregates, and the unit volume of the coarse aggregates. If we examine the rheological properties that characterize SCC, the yield stress must be zero or very low and the viscosity must be controlled. The range of viscosities needed to obtain good consolidation without vibration and without segregation has been the topic of various papers.[50–53] Most of them used semi-empirical tests such as the filling ability test to characterize concrete flow behavior. The properties of the cement paste or the mortar of SCC were found to be very important to avoid segregation. If the viscosity of the mortar is high enough, the coarse aggregates will be supported by the mortar, thus avoiding segregation. Often, viscosifiers such as welan gum or mineral admixtures are added to increase the viscosity of the paste, without significantly increasing the yield stress.[53,54] NIST research should provide a tool for use in linking cement paste rheology to the performance of SCC and lead to a full specification for SCC.

Conclusion

This review has presented recent developments in the rheology of concrete, including:

- Tests that are commonly used.
- Models and simulation of concrete flow.
- Factors that affect the rheological properties of concrete.
- Examination of two special applications: pumpable concrete and self-compacting concrete.

Concrete workability can be characterized in terms of the rheological parameters in either the Bingham or Herschel-Buckley equation. The theory exists but the measurements are not easy to obtain. The flow of a granular material such as concrete needs to be defined by at least two parameters, for instance, yield stress and plastic viscosity, as defined by the Bingham equation. While there are numerous tests to characterize the flow of concrete, few give results in fundamental units and therefore the measured rheological properties of concretes, using different tests, cannot be directly compared. Recently, new tests have attempted to characterize concrete using a more fundamental approach. While not all researchers agree on which test is the most suitable for the wide range of concretes in use today, all agree that tests that give results in fundamental units and that can be used on a construction site are needed.

Three different models allow the prediction of the flow of concrete from the mixture composition. The approaches of these models are based on:

- Packing density of all solids included in concrete, that is, cement, sand, and aggregates.
- Simulation of granular flow using dissipative particle dynamic methods.
- Application of the Krieger-Dougherty equation.

Factors affecting the flow of concrete extend beyond mixture composition, with processing and the environmental factors also affecting rheological behavior.

The application of rheological principles to pumpable concrete and SCC seems to be a particularly promising field of investigation. The prediction of the pressure/flow rate relationship in pumping networks is critical for today's construction industry. Such a calculation appears possible, based upon scientific measurements of the concrete flow, but of interaction at the steel/concrete interface, which can be readily performed with a large-gap coaxial viscometer. In the case of SCC, the properties are heavily based on the viscosity of the cement paste and of the concrete. These properties must be properly characterized to be able to design concrete mixtures with improved performance.

Progress in the areas discussed in this review will facilitate specification of concrete based on performance instead of prescriptive criteria. Research is being performed to validate the models and simulations that will provide new tools for use in optimization of day-to-day processing of concrete in ready-mixed concrete plants and other concreting operations.

References

1. H. G. Russell, "ACI Defines High-Performance Concrete," *Concr Int.,* **21** [2] (1999).
2. N. Iwasaki, "Estimation of Workability — Why Has the Sump Remained Being Used So Long?" *Concr. J.,* **21** [10] (1983).
3. S.H. Kosmatka and W.C. Panarese, *Design and Control of Concrete Mixtures.* PCA, 1994.
4. A.G.B. Richtie, "The Triaxial Testing of Fresh Concrete," *Mag. Concr. Res.,* **14** [40] 37–41 (1962).

5. G.H. Tattersall, *The Workability of Concrete, A Viewpoint Publication.* PCA, 1976.
6. F. de Larrard, J.-C. Szitkar, C. Hu, and M. Joly, "Design of a Rheometer for Fluid Concretes"; pp. 201–208 in *Special Concretes — Workability and Mixing.* RILEM, 1993.
7. D. Beaupré, "Rheology of High-Performance Shotcrete," Ph.D. Thesis, University of British Columbia, 1994.
8. D. Beaupré and S. Mindess, "Rheology of Fresh Concrete: Principles, Measurements, and Applications"; pp. 149–180 in Materials Science of Concrete V. Edited by J. Skalny and S. Mindess. American Ceramic Society, Westerville, Ohio, 1998.
9. C.F. Ferraris and F. de Larrard, "Testing and Modelling of Fresh Concrete Rheology," NISTIR 6094, February 1998.
10. F. de Larrard, C.F. Ferraris, and T. Sedran. "Fresh Concrete: A Herschel-Bukley Material," *Mater. Struct.,* **31** [211] 494–498 (1998).
11. C.F. Ferraris, "Measurements of Rheological Properties of High-Performance Concrete: State of the Art Report," NIST-IR 5869, 1996.
12. C.F. Ferraris, "Test Methods to Measure the Rheological Properties of High-Performance Concrete: State of the Art Report," *J. Res. NIST,* **104** [5] 461–478 (1999).
13. "Standard Test Method for Slump of Hydraulic Cement Concrete," ASTM C143-97, Vol. 04.02.
14. "Standard Test Method for Ball Penetration in Freshly Mixed Hydraulic Cement Concrete," ASTM C360-92, Vol. 04.02.
15. "Standard Test Method for Time Setting of Hydraulic Cement by Vicat Needle," ASTM C191–92, Vol. 04.01.
16. K.W. Nasser, "New and Simple Tester for Slump of Concrete," *ACI J. Proc.,* **73** [10] 561–565 (1976).
17. "Standard Test Method for Flow of Freshly Mixed Hydraulic Cement Concrete," ASTM C1362-97, Vol. 04.02.
18. C.J. Hopkins and J.G. Cabrera, "The Turning-Tube Viscometer: An Instrument to Measure the Flow Behavior of Cement-pfa Pastes," *Mag. Concr. Res.,* **37**, 101–109 (1985).
19. P. Bartos, *Fresh Concrete: Properties and Tests.* Elsevier, 1992.
20. F. De Larrard, "Optimization of HPC"; in *Micromechanics of Concrete and Cementitious Composites.* Ed. Huet, Switzerland, 1993.
21. S. Kuroiwa, Y. Matsuoka, M. Hayakawa, and T. Shindoh, "Application of Super Workable Concrete to Construction of a 20-Story Building"; pp. 147–161 in *High-Performance Concrete in Severe Environments,* ACI SP-140. Edited by Paul Zia. American Concrete Institute, 1993.
22. S. Popovics, *Fundamentals of Portland Cement Concrete: A Quantitative Approach.* John Wiley, 1982.
23. N. Miura, N. Takeda, R. Chikamatsu, and S. Sogo, "Application of Super Workable Concrete to Reinforced Concrete Structures with Difficult Construction Conditions"; pp. 163–186 in *High-Performance Concrete in Severe Environments,* ACI SP-140. Edited by Paul Zia. American Concrete Institute, 1993.
24. "Concrete and Reinforced Concrete: Design and Construction," DIN 1045. Deitches Institut fur Normung E. V., Berlin, 1988.
25. P. Coussot, "Rhéologie des boues et laves torrentielles — Etudes de dispersions et suspensions concentrées," Thèse de doctorat de l'Institut national Polytechnique de Grenoble, et Etudes du CEMAGREF, Série Montagne #5, 1993.

26. C. Hu, "Rhéologie des bétons fluides"; in *Etudes et Recherches des Laboratoires des Ponts et Chaussées,* OA 16. Paris, France, 1995.
27. O.H. Wallevik and O.E. Gjørv, "Rheology of Fresh Concrete"; pp. 133–134 in *Advances in Cement Manufacture and Use.* Eng. Found. Conf., Potosi, Michigan, 1988.
28. C. Hu, F. de Larrard, and O. Gjørv, "Rheological Testing and Modelling of Fresh High Performance Concrete," *Mater. Struct.,* **28**, 1–7 (1995).
29. F. De Larrard, C. Hu, J.C. Szitkar, M. Joly, F. Claux, and T. Sedran, "A New Rheometer for Soft-to-Fluid Fresh Concrete," LCPC Internal Report, 1995.
30. C.F. Ferraris and C. Lobo, "Processing of HPC," *Concr. Int.,* **20** [4] 61–64 (1998).
31. F. de Larrard and C.F. Ferraris,"Rhéologie du béton frais remanié. III: L'essai au cône d'Abrams modifié," *Bulletin des Laboratoire des Ponts et Chaussées* (France), no. 215, 53–60 (1998).
32. C.F. Ferraris and F. de Larrard, "Modified Slump Test to Measure Rheological Parameters of Fresh Concrete," *ASTM Cement, Concr. Aggregates,* **20** [2] 241–247 (1998).
33. L. Brower, private communication.
34. F. de Larrard, *Concrete Mixtures Proportioning: A Scientific Approach.* Modern Concrete Technology Series. E&FN Spon, 1999.
35. M. Barrioulet and C. Legrand, "Les interactions mécaniques entre pâte et granulats dans l'écoulement du béton frais,"; pp. 263–270 in *Proceedings of RILEM Coll. Properties of Fresh Concrete.* Edited by H.-J. Wierig. RILEM, 1990.
36. F. Ferraris and J. Gaidis, "Connection Between the Rheology of Concrete and Rheology of Cement Paste," *ACI Mater. J.,* **89** [4] 388–393 (1992).
37. E.J. Garboczi and D.P. Bentz, "Analytical Formulas for Interfacial Transition Zone Properties," *Adv. Cem. Based. Mater.,* **6**, 99–108 (1997)
38. B. Lu and S. Torquato, *S. Phys. Rev. A,* **45**, 5530–5544 (1992).
39. E. Garboczi and D.P. Bentz, "Multi-Scale Analytical/Numerical Theory of the Diffusivity of Concrete," *Adv. Cem. Based Mater.,* **8**, 77–88 (1998).
40. R. Helmuth, L. Hills, D. Whitting, and S. Bhattacharja, "Abnormal Concrete Performance in the Presence of Admixtures," PCA # 2006, 1995.
41. R.D. Groot and P.B. Warren, "Dissipative Particle Dynamics: Bridging the Gap Between Atomistic and Mesoscopic Simulation," *J. Chem Phys.,* **107**, 4423–4435, (1997).
42. P.J. Hoogerbrugge and J.M.V.A. Koelman, "Simulating Microscopic Hydrodynamic Phenomena with Dissipative Particle Dynamics," *Europhys. Lett.,* **19**, 155–160 (1992).
43. J.M.V.A. Koelman and P.J. Hoogerbrugge, "Dynamic Simulations of Hard-Sphere Suspensions Under Steady Shear," *Europhys. Lett.,* **21**, 363–368 (1993).
44. C.G. de Kruif, E.M.F. van Iersel, A. Vrij , and W.B. Russel., *J. Chem. Phys.,* **83** [9] (1985).
45. L.J. Struble and G.K. Sun, "Cement Viscosity as a Function of Concentration"; pp. 173–178 in *Flow and Microstructure of Dense Suspensions.* Edited by Struble, Zukoski, and Maitland. Materials Research Society, Pittsburgh, 1993.
46. P.-C. Aïtcin, *High-Performance Concrete.* E&FN Spon, London, 1998.
47. K. Khayat and J.-P. Ollivier, "Viser une consistance adaptee aux moyens de mise en œuvre"; pp. 187–221 in *Les bétons: Base et données pour leur formulation.* Edited by J. Baron and J.-P. Ollivier. Eyrolles, Paris, 1997.

48. ACI Committee 304, "Proposed Report: Placing Concrete by Pumping Methods," *ACI Mater. J.,* July–August 1995, pp. 441–464.
49. D. Kaplan, T. Sedran, F. de Larrard, J.P. Busson, G.R. Sarraco, F. Cussigh, J.L. Duchene, and A. Thomas, "Contrôler le pompage du béton avec les outils de la rhéologie," proposed to *Bulletin des Laboratoires des Ponts et Chaussées,* May 1999.
50. H. Kim, Y.-D. Park, J. Noh, Y. Song, C. Han, and S. Kang, "Rheological Properties of Self-Compacting, High-Performance Concrete"; pp. 653–668 in *Third Int. ACI Conf. Proc. on High-Performance Concrete: Design and Materials, and Recent Advances in Concrete Technology.* Edited by Malhotra. 1997.
51. T. Shindoh, K. Yokota, and K. Yokoi, "Effect of Mix Constituents on Rheological Properties of Super Workable Concrete"; pp. 263–270 in *Proc. of RILEM Int. Conf. Production Methods and Workability of Concrete.* Edited by P.J.M. Bartos, D.L. Marrs, and D.J. Cleland. Scotland, 1996.
52. P.L. Domone and H.-W. Chai, "Design and Testing of Self-Compacting Concrete"; pp. 223–236 in *Proc. of RILEM Int. Conf. Production Methods and Workability of Concrete.* Edited by P.J.M. Bartos, D.L. Marrs, and D.J. Cleland. Scotland, 1996.
53. S. Nagataki, "Present State of Superplasticizers in Japan"; in *International Symposium on Mineral and Chemical Admixtures in Concrete.* Toronto, 1998.
54. N. Sakata, K. Maruyama, and M. Minami, "Basic Properties and Effects of Welan Gum on Self-Consolidating Concrete"; pp. 237–253 in *Proc. of RILEM Int. Conf. Production Methods and Workability of Concrete.* Edited by P.J.M. Bartos, D.L. Marrs, and D.J. Cleland. Scotland, 1996.

Early Age Behavior of Cement-Based Materials

B. Bissonnette, J. Marchand, and J.P. Charron
Université Laval, Sainte-Foy, Québec, Canada

A. Delagrave
Lafarge Canada Inc., Montréal, Québec, Canada

L. Barcelo
Laboratoire Central de Recherche – Lafarge, St-Quentin Fallavier Cedex, France

"Keep your early enthusiasm . . . Worship the spirit of Criticism."
— Louis Pasteur

The improvement of mixture design methods, the widespread use of supplementary cementing materials, and the development of new chemical admixtures have largely contributed to increase the overall performance of hydrated cement systems. However, despite their superior mechanical properties and their improved durability, high-performance cement-based materials are apparently more sensitive to early age cracking. An overview of the early age behavior of hydrated cement systems is presented in this paper. The mechanisms of chemical shrinkage are first reviewed. A second section is devoted to the phenomenon of autogeneous shrinkage of neat paste and mortar mixtures. Emphasis is placed on the volumetric changes happening during the first 24 h of hydration. The final section is specifically devoted to early age stress buildup in concrete. Special attention is paid to the mechanisms that may induce or limit cracking. Laboratory test procedures designed to study the early age behavior of cement systems are also critically reviewed.

Introduction

Concrete, like any other cement system, may undergo significant volume changes during the first hours that follow the initial contact of water with cement particles. In some practical cases, these volume changes may even lead to the premature cracking of concrete structures. Although early age cracking is often related to improper construction practices, the problem is increasingly observed in structures for which concrete has been produced, placed, and cured according to state-of-the art procedures.[1,2]

Given the importance of the problem, significant effort has been made to understand the basic mechanisms that control the sensitivity of concrete to early age cracking. Over the past decade, the number of scientific and technical reports published on the subject has increased dramatically. This paper is intended to be a modest attempt to review the most recent developments in this field. Of necessity, coverage here has been selective. Although many valuable reports had to be set aside, it is hoped that the most important contributions have been reviewed.

It should also be emphasized that the scope of the present review is limited to autogeneous phenomena. Although temperature (or, more precisely, temperature gradients) can be, in certain cases, the main cause of the premature failure of some concrete structures, the present survey is solely restricted to the behavior of cement systems kept under isothermal conditions. Furthermore, the review does not cover problems related to plastic and drying shrinkage. Comprehensive, state-of-the-art reports on these topics can be found elsewhere.[1,3,4]

The review is divided into four distinct sections. The mechanisms of chemical shrinkage and self-desiccation are reviewed in the first section. The second section is devoted to the autogeneous (or external) shrinkage of hydrated cement systems. Emphasis is placed on the phenomena occurring during the first 24 h of hydration. The third section is dedicated to mechanisms of stress development in concrete at early age. Special attention is paid to the mechanisms that may induce or limit cracking. Laboratory test procedures designed to study the early age behavior of cement systems are also critically reviewed. Finally, the issue of modeling the early age behavior of concrete is briefly addressed.

Chemical Shrinkage

The contact of water with portland cement triggers a series of complex dissolution/precipitation reactions that eventually lead to the progressive formation of a rigid structure.[5–7] As can be seen in Fig. 1, the hydration of cement systems can be schematically divided in four stages,[8] the first three being of particular interest in the context of the early age behavior of concrete.

The initial stage (Period 1) is characterized by a sudden exothermic reaction immediately followed by a marked drop of the heat dissipation rate. During this initial period, the Ca^{2+} ion concentration in the surrounding solution significantly increases.

The solvation proceeds during the second stage (Period 2) of the hydration process (also called the "dormant period"). During this second period,

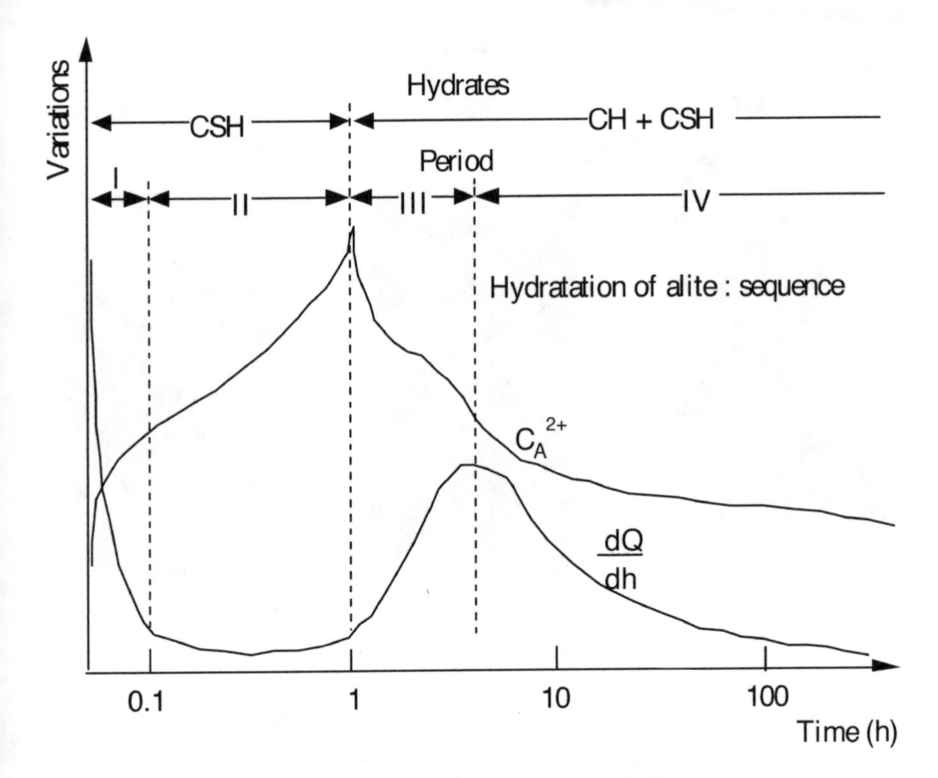

Figure 1. Isothermal calorimetry and conductivity test results.[8]

the Ca^{2+} ion concentration slowly increases, but very little heat is liberated.

Period 3 (the "acceleration period") is characterized by a sudden drop in the Ca^{2+} ion concentration. As emphasized by Paulini,[9] the starting point of this third stage is defined by a state of supersaturation in the aqueous solution. Once nucleation is initiated, a simultaneous process of solvation, transport, and crystallization determines the reaction kinetic. During this period, the exothermic heat rate reaches a peak.

The fourth and last period corresponds to the progressive "densification" of the material. As previously mentioned, the phenomena occurring upon this last stage are less of concern here.

If the hydration process results in the formation of new chemical compounds, it also significantly modifies the physical properties of the system. After a few hours of reaction, the initially liquid cement paste gradually sets and hardens to eventually develop significant mechanical strength. The

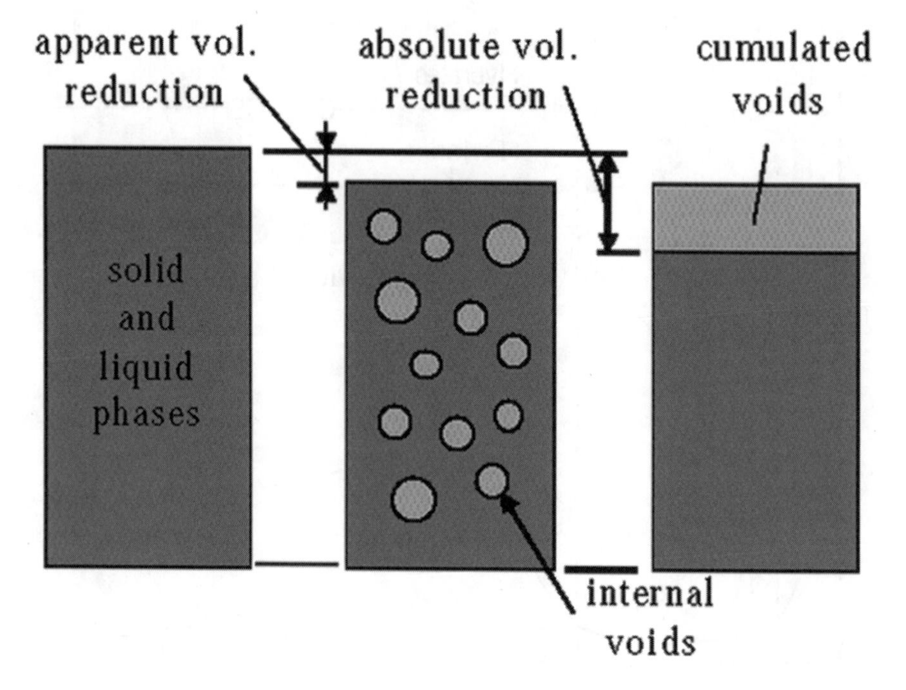

Figure 2. Schematic distinction between absolute volume and apparent volume.

hydration reaction also contributes to reducing the total absolute volume of the material.

The latter phenomenon, called chemical shrinkage,* was first described by Le Chatelier in 1900. One of the first chemists to study the basic properties of hydraulic binders, Le Chatelier is well known for his pioneering work on the production and hydration of plaster and cement.[10] He was also among the first to note that the hydration of cement comes with an important decrease of absolute volume,[11] and further demonstrated that the rate of volume change was a reliable indicator of the kinetics of hydration.

In his study of the mechanisms of chemical shrinkage, Le Chatelier[11] emphasized the basic distinction between the apparent and the absolute volume of a solid. Figure 2 presents schematically the difference between the two concepts. On the one hand, the apparent volume is essentially the external volume of the specimen, which includes the space occupied by the solid, the liquid, and the gaseous phases. On the other hand, the absolute

*In many textbooks, chemical shrinkage is also referred to as Le Chatelier contraction.

volume corresponds to the sum of the volumes occupied by the liquid and the solid phases.

As previously mentioned, the chemical shrinkage refers to a reduction in the absolute volume of the material. It results from a densification of the solid phases during the hydration process. As will be emphasized in the following section, the autogeneous shrinkage rather corresponds to a reduction of the apparent (external) volume of the material.

Thermodynamics of Hydration

Paulini[9,12,13] has analyzed the mechanisms of early age shrinkage from a thermodynamic point of view. This original approach is particularly interesting because it helps understanding the basic relationship between the exothermic nature of the hydration reaction and the simultaneous volume change of the hydrating materials.

As emphasized by the author, hydraulic reactions are irreversible processes evolving from an energetic state of nonequilibrium toward a (metastable) equilibrium. During the reaction, the system tends to go from a high free energy level toward a lower one. This implies that Helmholtz free energy *(F)* and Gibbs free enthalpy *(G)* must continuously decline during the chemical reaction:

$$dF = -pdV - SdT \leq 0 \quad (1)$$

$$dG = Vdp - SdT \leq 0 \quad (2)$$

where p stands for pressure, V for volume, S for entropy and T for temperature.

Upon hydration, the internal energy *(U)* of the system can be dissipated through heat production (Q_r) and volume work (W_B):

$$U = Q_r + W_B \quad (3)$$

From the law of energy conservation, it is required that the internal energy of a closed system remain unchanged:

$$dU = dQ_r + dW_B = 0 \quad (4)$$

The energy dissipated during an irreversible chemical reaction through heat production (dQ_r) should be equal to the amount of volume work (dW_B) done on the system:

$$dQ_r = -dW_B \quad (5)$$

According to the second law of thermodynamics, the change in the entropy of the system must be positive:

$$dS = dS_i + dS_e \geq 0 \tag{6}$$

As indicated by Eq. 6, the entropy production can be divided into an internal term dS_i and a flux term dS_e. It can be shown that the internal entropy term should always be positive[9,12,13]:

$$dS_i \geq 0 \tag{7}$$

Therefore, the dissipative flux energies and the volume work done on the system should respect the following relationship:

$$TdS_e < 0 < TdS_i \tag{8}$$

This implies that, for an exothermic reaction such as the hydration of cement, the energy loss term (dQ_r) is always negative and the volume work transferred to the system (dW_B) remains positive:

$$dQ_r < 0 < dW_B \tag{9}$$

Equation 9 clearly emphasizes the two counteracting processes occurring during the hydration of cement. The exothermic heat of reaction dQ_r can be considered as an excess energy that is released from the system. This dissipative energy flux results in a net energy loss, which is counterbalanced by a positive bonding energy dW_B. The competing nature of these two phenomena is known as the Le Chatelier principle.[9,12,13]

According to Eq. 9, the kinetics of hydration can therefore be followed either by measuring the heat of reaction or by measuring the absolute volume change of the material. The good correlation between the two series of measurements is illustrated in Fig. 3.[13] This simple relationship between the two phenomena is the basis of most of the work done on the chemical shrinkage of hydrating cement systems.

This description of the hydration mechanisms based on thermodynamic considerations has led Paulini to develop a model for the prediction of the strength development of hydrating cement systems.[13,14] This interesting approach is based on the assumption that the hardening process of cement paste can be estimated by the amount of irreversible volume work done on the system. More information on this original approach can be found in Refs. 13 and 14.

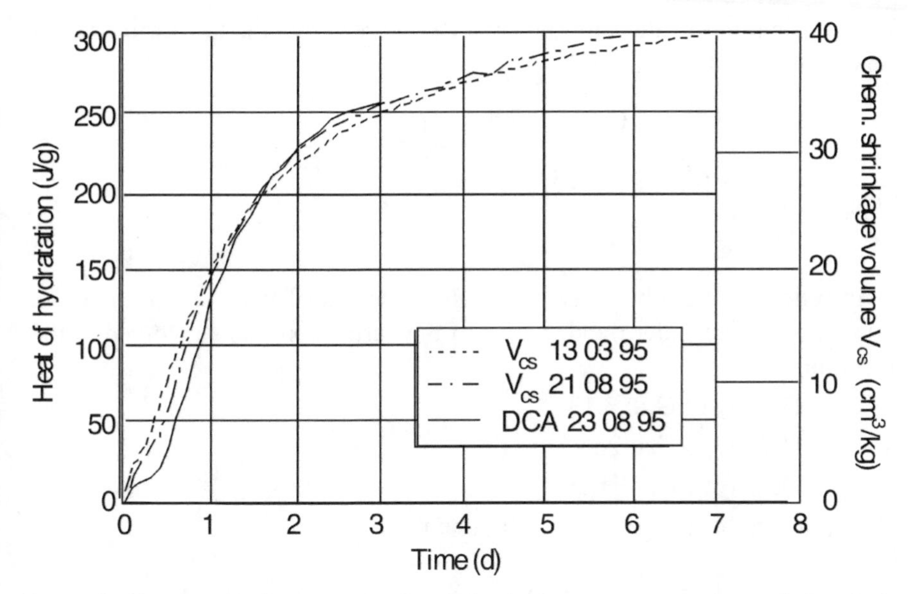

Figure 3. Comparison between isothermal calorimetry test results and chemical shrinkage data.[12]

Chemical Shrinkage: A Problem of Volume Conservation

The densification process at the origin of the chemical shrinkage of cement systems can be more easily understood by considering the hydration of tricalcium silicate (C_3S)[15]:

$$C_3S + 5.2H \rightarrow C_{1.7}SH_{3.9} + 1.3CH \quad (10)$$

where, according to the notation traditionally used in cement chemistry, H stands for water, C stands for CaO, and S stands for SiO_2.

The molar masses of the various products involved in this reaction can be calculated as follows:

$$M_{C_3S} = 3M_{Ca} + M_{Si} + 5M_O = 120 + 28 + 80 = 228 \text{ g·mol}^{-1}$$

$$M_H = 2M_H + M_O = 2 + 16 = 18 \text{ g·mol}^{-1}$$

$$M_{C_{1.7}SH_{3.9}} = 1.7M_{Ca} + M_{Si} + 7.8M_H + 7.6\,M_O = 68 + 28 + 7.8 + 121.6 = 225.4 \text{ g·mol}^{-1}$$

$$M_{CH} = M_{Ca} + 2M_H + 2M_O = 40 + 2 + 32 = 74 \text{ g·mol}^{-1}$$

By inserting these values in Eq. 10, it can be seen that the law of mass conservation is verified during the hydration of C_3S:

$$C_3S + 5.2H \rightarrow C_{1.7}SH_{3.9} + 1.3CH$$

$$228\ g + 93.6\ g \rightarrow 225.4\ g + 96\ g$$

$$321.6\ g \rightarrow 321.6\ g$$

In order to determine if Eq. 10 verifies the law of volume conservation, one must first consider the densities of the various products involved in the reaction:

- Density of C_3S: 3.13
- Density of CH: 2.24
- Density of $C_{1.7}SH_{3.9}$: 2.35

By replacing these values in Eq. 10, one finds:

$$C_3S + 5.2H \rightarrow C_{1.7}SH_{3.9} + 1.3CH$$

$$72.8\ cm^3 + 93.6\ cm^3 \rightarrow 95.9\ cm^3 + 42.9\ cm^3$$

$$166.4\ cm^3 \rightarrow 138.9\ cm^3$$

As can be seen, the hydration of C_3S results in a reduction of the absolute volume of the system. If the reactants are found in their stoichiometric proportions, the decrease in volume is equal to:

$$(166.4 - 138.9) / 166.4 = 16.5\%$$

A similar calculation can be made for a paste of C_3S prepared at a water/solid ratio of 0.4. According to these proportions, 123 cm^3 of water is mixed with 100 cm^3 of C_3S. According to Eq. 10, 64.3 cm^3 of water is needed to hydrate 50 cm^3 of C_3S (i.e., to reach a degree of hydration of 0.5). The hydration produces 65.9 cm^3 of C-S-H and 29.5 cm^3 of CH.

As illustrated in Fig. 4, the hydration of 50% of the C_3S leads to a net increase of 45% in the volume of solids (which goes from 100 to 145 cm^3). However, this phenomenon is not sufficient to totally counterbalance the volume of water consumed by the reaction. At the end, the system undergoes an absolute volume reduction of approximately 8.4%. This is the curious paradox of the chemical shrinkage phenomenon.

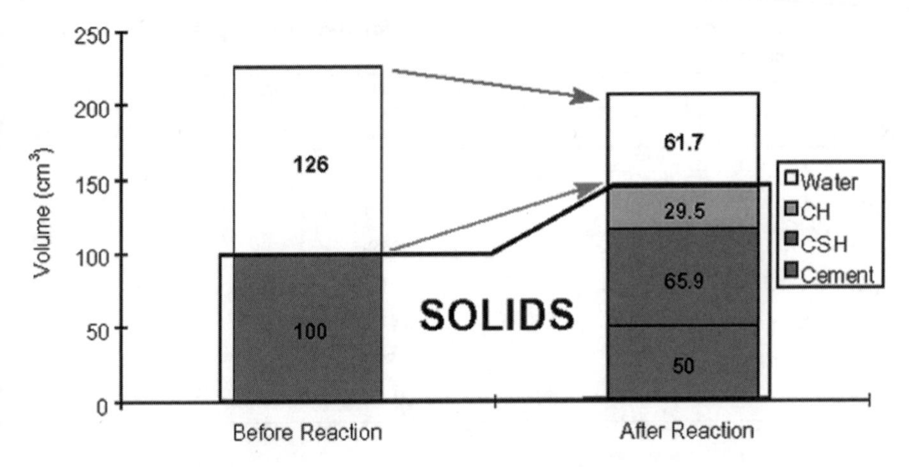

Figure 4. Absolute volume reduction during the hydration of a 0.4 water/solid ratio C_3S paste. Degree of hydration was 0.5.

Theoretical Evaluation of the Chemical Shrinkage of Silicate and Aluminate Phases

In theory, the same approach can be applied to all the phases found in an ordinary portland cement. For instance, according to Justnes et al.,[16] the reaction of gypsum with water leads to an absolute volume reduction of 0.04 cm^3 per kilogram of gypsum. Justnes et al.,[16] Paulini,[9,12,13] and Tazawa and Miyazawa[17] performed similar calculations for the various phases found in ordinary portland cement systems. Their results are summarized in Table I. Additional information on their calculations is given in Appendix A.

As can be seen, there is good agreement between the results obtained by the different investigators for the chemical shrinkage of C_3S. The data presented in Table I also emphasize the importance of the addition of gypsum on the behavior of C_3A. When compared to the formation of C_3AH_6, the precipitation of ettringite should theoretically increase the absolute volume deformation of the system. As will be discussed in the following sections, the influence of ettringite formation on the early age behavior of hydrating cement systems is imperfectly understood.

Globally, the data presented in Table I indicate that the chemical shrinkage of hydrating cement systems should be (at least in theory) significantly influenced by the mineralogical composition of the binder. This point will be extensively discussed in the following sections of this chapter.

Table I. Theoretical chemical shrinkage values for the various phases found in cement systems

Reactant	Absolute volume change (cm^3/kg of reactant)		
	Justnes[16]	Paulini[9,12,13]	Tazawa and Miyazawa[17]
C_3S	–49	–53.2	–60
C_2S		–40	
$C_3A \rightarrow C_3AH_6$		–178.5	
$C_4AF \rightarrow C_3AH_6$		–111.3	
$C_3A \rightarrow$ ettringite	–273	–72	
$C_3A \rightarrow$ monosulfate		–72.2	
Gypsum	–40	–37.9 ~ –55.5	

Chemical Shrinkage Measurements

Over the past decades, the chemical shrinkage of hydrating cement systems also has been the subject of numerous experimental studies. Researchers have relied on two types of measurement: the dilatometry method and the weighing method. The dilatometry method is very similar to the simple procedure originally developed by Le Chatelier.[10] It essentially consists of introducing a small volume of fresh mixture (paste or mortar) into an Erlenmeyer flask extending into a narrow neck.[18–21] The material is then covered with water (or with a lime-saturated solution).[21] A silicon rubber stopper with a calibrated capillary tube is fixed under water (to avoid the entrapment of air bubbles) to the flask, which is then secured in a temperature-controlled water bath (Fig. 5). A drop of paraffin oil is usually added on top of the water meniscus within the capillary tube to prevent any risk of evaporation. The absolute volume change of the hydrating material is determined by following the vertical displacement of water in the Erlenmeyer neck (see Fig. 5).

The main advantage of the dilatometry method lies in its simplicity. Although frequent readings can be made, the behavior of the material can hardly be followed continuously. In order to get continuous information, many authors have relied on the weighing method,[9,12,13,20,22–24] which is essentially a modified version of the dilatometry procedure. As can be seen from Fig. 6, the operating principle of the two methods is similar. In the weighing method, the Erlenmeyer flask filled with water is immersed in a thermo-regulated bath. The flask is suspended under a high-precision balance, and the chemical shrinkage of the material is determined by measuring the change in the apparent weight of the system. The apparent weight of

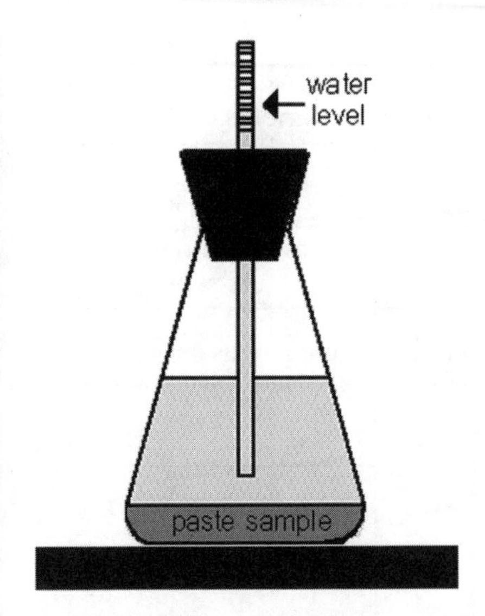

Figure 5. Schematic representation of the equipment used to measure chemical shrinkage according to the dilatometric method.[20]

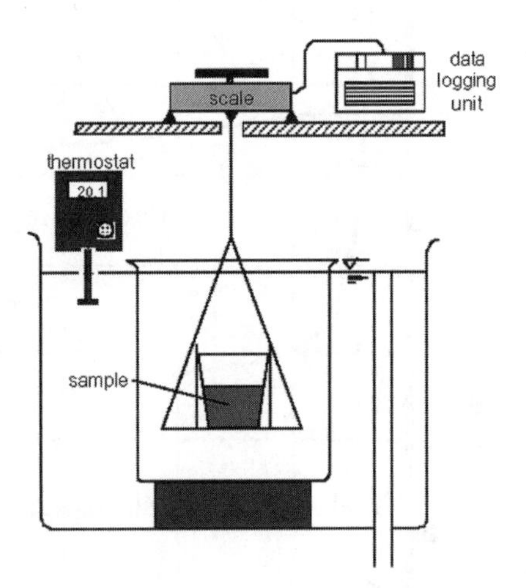

Figure 6. Schematic representation of the equipment used to measure chemical shrinkage according to the weighing method.[12]

the flask (which increases during the experiment) is equal to the weight of the hydrating mixture minus the buoyancy force. The increment in the apparent weight thus corresponds to the amount of water penetrating the sample during the experiment. According to Boivin et al.,[20] both procedures tend to yield the same results.

Recently, a new procedure based on the dual-compartment cell was proposed by Gagné et al.[25] The main originality of this interesting technique lies in its ability to measure the chemical shrinkage and the autogeneous (external) deformation of a neat cement paste sample simultaneously. A relatively important volume of fresh paste (approximately 800 cm^3) is introduced in the internal compartment of the cell, which consists of a PVC cylindrical base and a PVC cover joined by a flexible cylindrical latex membrane. The autogeneous shrinkage is determined by measuring the external deformation of the internal compartment during the experiment. The chemical shrinkage is assumed to be equal to the sum of the autogeneous shrinkage and the volume of air sucked up (through the five sintered microporous rods) by the hydrating paste during the experiment. A set of typical results obtained with this technique is given in Fig. 7.

In appearance relatively simple, chemical shrinkage experiments must be

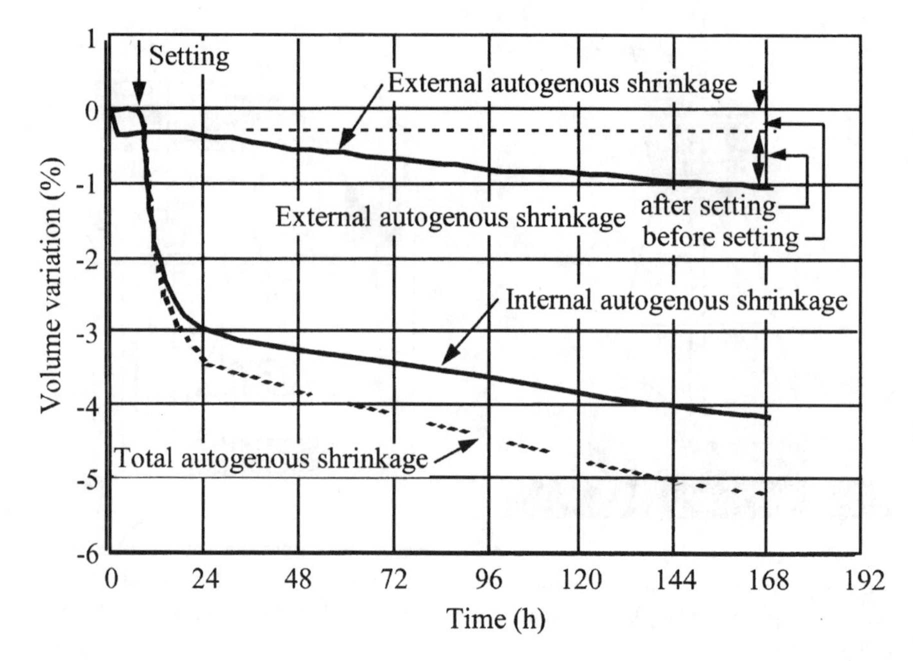

Figure 7. Typical chemical shrinkage yielded by the method proposed by Gagné et al.[25]

performed with great care. For instance, the behavior of hydrating cement systems is very sensitive to temperature variations. Therefore, special care should be taken to maintain the system in isothermal conditions throughout the entire duration of the experiment (usually a few days).

As emphasized by many authors, the validity of these measurements rests solely on the assumption that all the volume created by hydration of the cement and the simultaneous contraction of the solid is filled up with the external water penetrating the material.[18–21] In that respect, the thickness of the layer of fresh mixture introduced in the flask has a critical influence on the final result. The layer should be thin enough to allow water to penetrate the entire thickness of the material and therefore avoid any self-desiccation of the porous solid. This might be particularly difficult to achieve at the later stages of the experiment, when the permeability of the material tends to decrease markedly.

The importance of this parameter can be clearly seen in Fig. 8, where the results of a study carried out by Knudsen[26,27] are illustrated. As can be seen, the amount of chemical shrinkage tends to be reduced as the thickness of

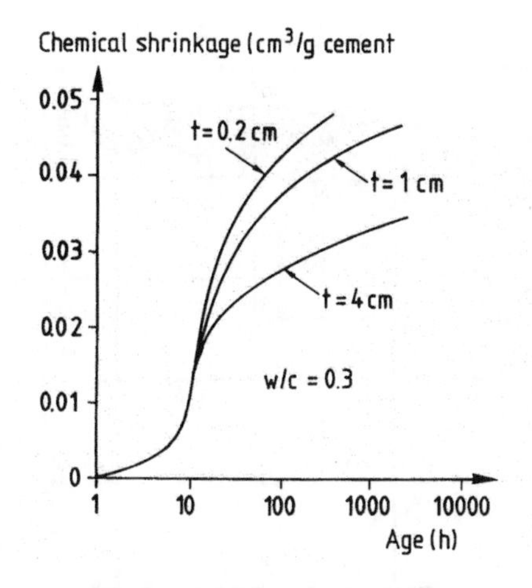

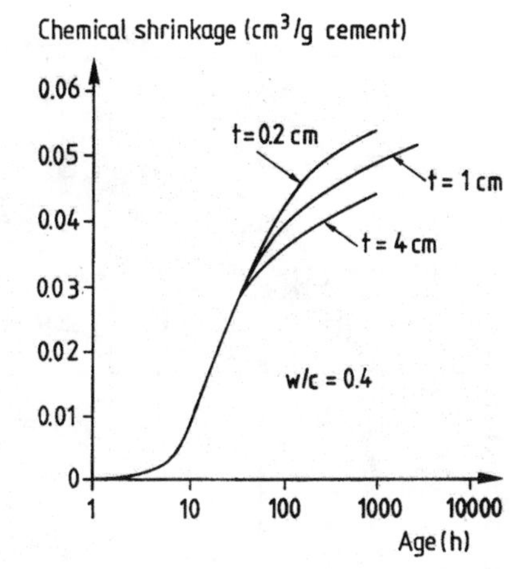

Figure 8. Effect of the thickness of the paste layer on chemical shrinkage measurements.[26,27]

the cement paste layer is increased. This reduction in chemical shrinkage is a clear indication that water could not penetrate the entire thickness of the material. Similar results were obtained by Boivin et al.[20] (see Fig. 9). According to both series of data, self-desiccation can occur in a layer of cement paste as thin as 10 mm. This phenomenon appears to be particularly significant for lower water/cement ratio mixtures (i.e., w/c $\leq$ 0.30) that are characterized by a low permeability after only a few days of hydration.

Factors Affecting the Chemical Shrinkage of Hydrating Cement Systems

Cement mineralogical composition

Powers[18] was probably the first researcher to systematically study the influence of cement mineralogical composition on chemical shrinkage. In 1935, Powers performed a series of dilatometry experiments on neat paste mixtures of C_2S, C_3S, C_3A, and C_4AF prepared with and without gypsum (the gypsum content of these mixtures varied from 0 to 30% of the total mass of solid materials).

Powers' results are summarized in Table II and Fig. 10. As can be seen, chemical shrinkage tends to be significantly more important for the C_3A and C_4AF mixtures than for pastes made of C_2S and C_3S. The amount of water absorbed during the experiment varies from 1% for C_2S to 16% for

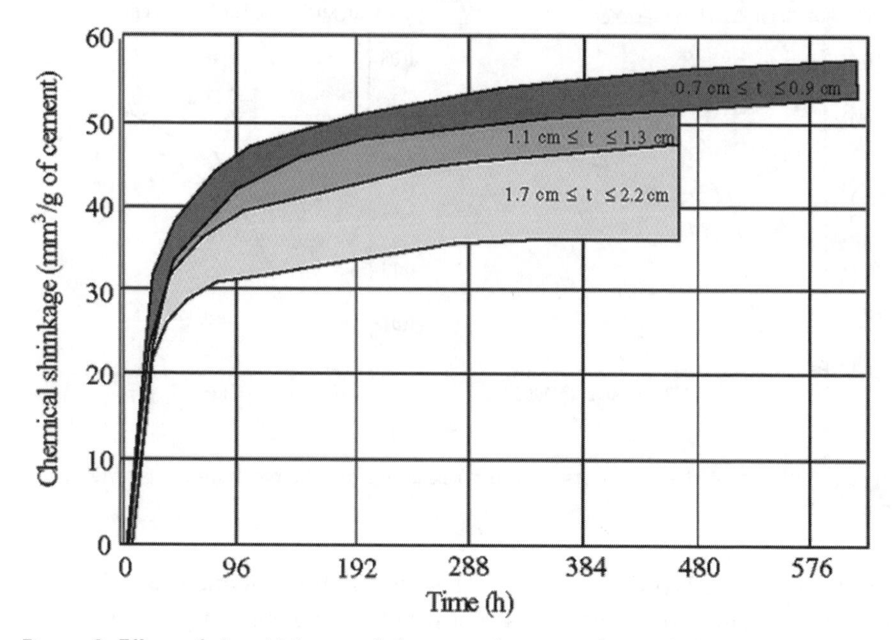

Figure 9. Effect of the thickness of the paste layer on chemical shrinkage measurements.[20]

C_3A. For the silicates, the absorption of water is more important for C_3S than for C_2S.

The addition of gypsum appears to significantly modify the absorption of water for the aluminates. For C_3A, an increase in gypsum reduces the amount of absorbed water, while for the C_4AF there seems to be an optimum gypsum content for which the amount of absorbed water is maximized. This optimum corresponds to a gypsum content of approximately 5% (by of the total mass of reactant). For the silicates, there is a slight increase in the amount of absorbed water when gypsum is added to the mixture. However, this effect remains limited.

In his investigation of chemical shrinkage, Powers[18] also tested 22 different laboratory cements. Based on a statistical analysis of the data obtained in this test series and assuming that the synergetic effects between the hydration of the various compounds were minimal, Powers proposed the following empirical equation to estimate the chemical shrinkage of cement:

$$A = a\mathrm{C_3S}\ (\%) + b\mathrm{C_2S}\ (\%) + c\mathrm{C_3A}\ (\%) + d\mathrm{C_4AF}\ (\%) \qquad (11)$$

Table II. Experimental chemical shrinkage values measured by Powers[18]

Reactant	Absolute volume change (cm^3/kg of reactant) 1 day	3 days	7 days	14 days	28 days
C_3S	–18.8	–30.0	–33.6	–40.9	–48.1
C_2S	–11.0	–12.6	–10.6	–14.0	–20.2
C_3A	–63.2	–75.9	–113.3	–120.1	–109.1
C_4AF	–19.0	–20.2	–41.5	–35.2	–24.7

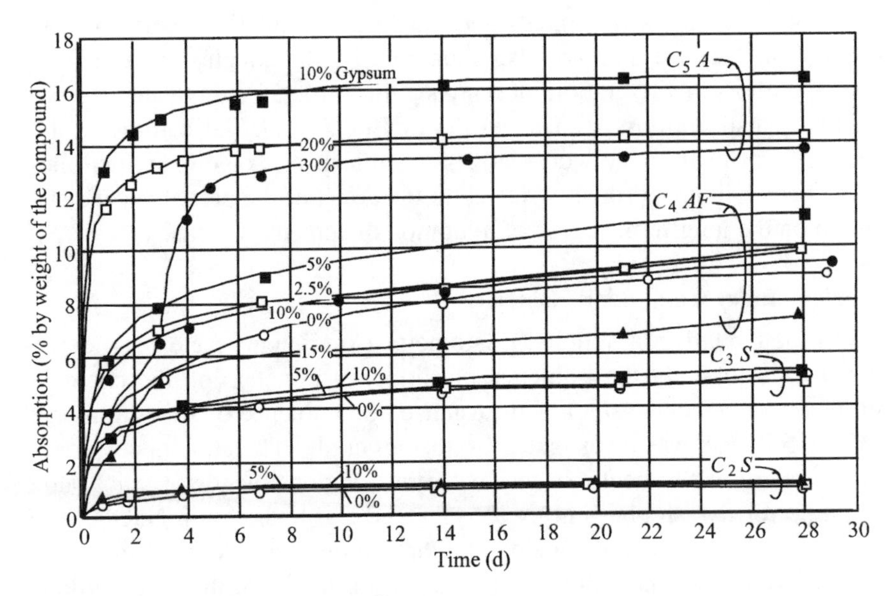

Figure 10. Chemical shrinkage results obtained by Powers.[18]

where *A* stands for the mass of water absorbed during the experiment (expressed as a percentage of the initial mass of solid). The parameters *a, b, c,* and *d* are empirical coefficients.

The experimental results obtained by Powers for the hydration of C_3S are in relatively good agreement with the theoretical values summarized previously in Table I. However, the shrinkage measured by Powers during the hydration of C_2S and C_3A are not that close to the corresponding values calculated by Paulini,[9,12,13] Justnes et al.,[16] and Tazawa and Miyazawa.[17] This discrepancy is difficult to explain. One possible explanation might be linked to the fact that the pure aluminate compounds tested by Powers were not fully hydrated (even after 28 days) as assumed by Paulini[9,12,13] and Justnes et al.[16] in their calculations.

Cement Particle Size Distribution

The influence of the fineness of cement on chemical shrinkage was recently investigated by Justnes et al.[28] Their results, obtained using the dilatometry technique, indicate that the kinetics of shrinkage, and particularly the initial rate of volume change, is markedly accelerated by an increase in cement fineness. These observations are in good agreement with the results of a series of laboratory experiments and numerical simulations performed by Bentz et al.[29] These data are also supported by the conclusions drawn by several authors that hydration rates are significantly affected by cement particle size distributions.[30,31] The results of Bentz et al.[29] also indicate that coarser cements ultimately develop less chemical shrinkage than finer ones. This phenomenon is probably related to the effect of the particle size distribution on the long-term degree of hydration of cement systems.

Supplementary Cementing Materials

Given their relative simplicity, dilatometry experiments have been extensively used to study the influence of various parameters on the chemical shrinkage of cement systems. For instance, the influence of silica fume and a Class F fly ash was investigated by Justnes et al.[32] The authors performed two series of experiments. In the first series, various blends of silica fume and lime (or fly ash and lime) were prepared and hydrated in NaOH/KOH solutions. The pozzolanic reaction and the chemical shrinkage of these systems were found to be quite sensitive to the initial pH of the test solutions. On one hand, the rate of chemical shrinkage of the silica fume/lime systems was found to decrease with an increase of the initial pH. This phenomenon was attributed to the effect of pH on the solubility of calcium hydroxide. On the other hand, the rate of chemical shrinkage of the fly ash/lime systems was found to increase with an increase of the test solution pH. This effect was explained by the influence of pH on the dissolution of the glassy aluminosilicate phase of the fly ash particles.

The total chemical shrinkage was estimated to be 8.8 cm^3/100 g of reacted silica fume and 10.0 cm^3/100 g of reacted fly ash, as compared to 6.25 cm^3/100 g of hydrated cement. It should be emphasized that these are most probably lower bound values. According to the authors, fly ash particles

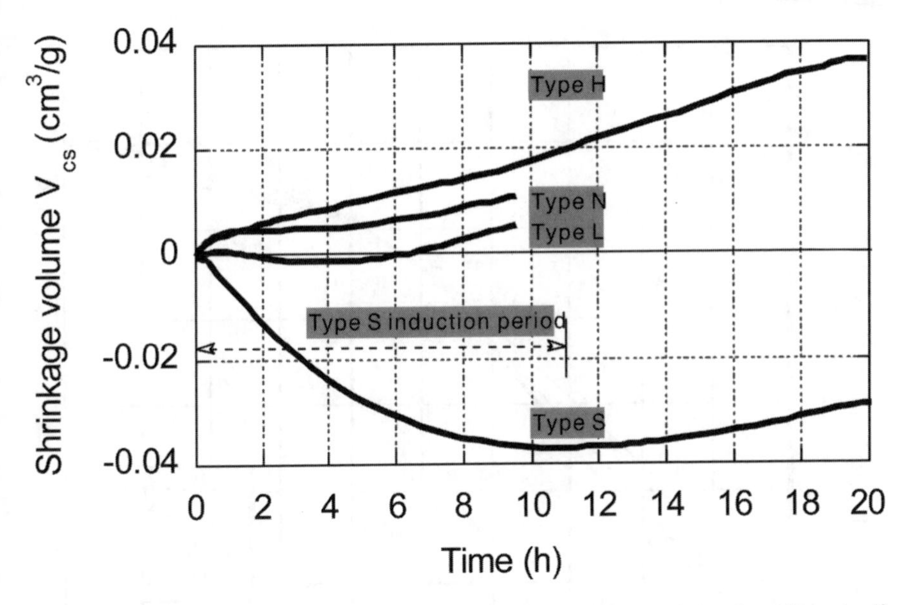

Figure 11. Schematic representation of the various types of chemical shrinkage behavior.[12]

were still hydrating after 56 days of immersion in the test solutions. Furthermore, the reaction of minor amounts of silicon metal in the silica fume particles was found to produce hydrogen gas and induce expansion.

Justnes et al.[32] also found that the addition of supplementary cementing materials, such as silica fume and Class F fly ash, generally contributes to increase the initial and long-term chemical shrinkage of blended cement systems. The authors attributed this phenomenon to the pozzolanic reaction and to the fact that silica fume and fly ash particles tend to increase the number of nucleation sites for hydration products.

According to Paulini,[9,12,13] hydrating cement systems can be divided into four different categories according to their kinetics of chemical shrinkage. As can be seen in Fig. 11, Type H systems are characterized by their high initial rate of shrinkage. Finely ground cements usually fall into that category. Portland cements with a more conventional particle size distribution tend to follow either curve L or curve N. As can be seen in the figure, type N cements develop very little chemical shrinkage during the first hours of hydration. Type S systems tend to swell during the same period. According to the author, sulfate-resistant (ASTM Type V) and slag modified cements usually exhibit type S behaviors.

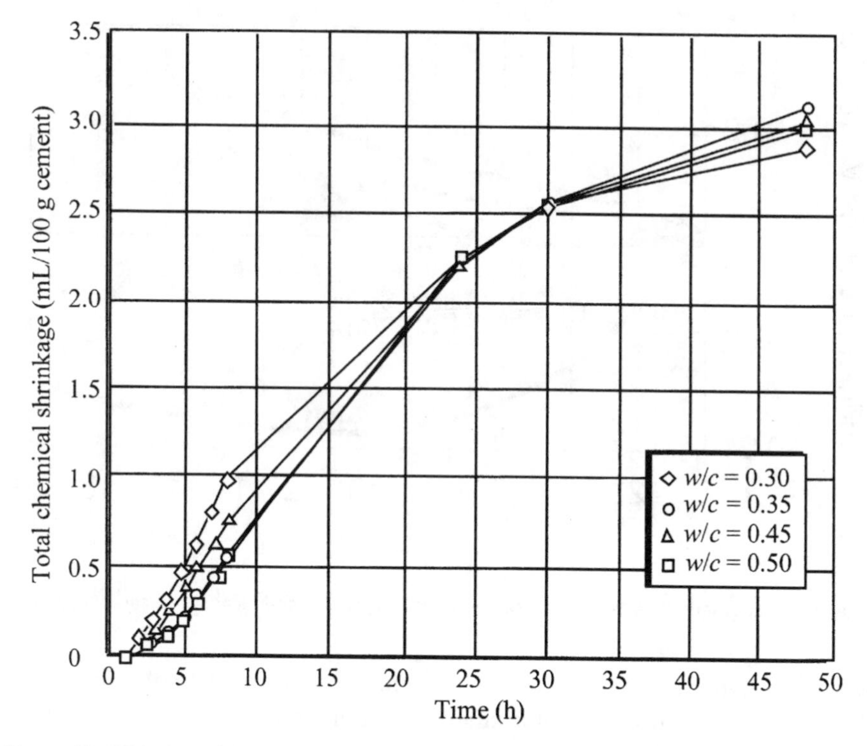

Figure 12. Influence of water/cement ratio on chemical shrinkage.[19]

Chemical Admixtures

Paulini[9] also studied the influence of various chemical admixtures on the chemical shrinkage of neat cement paste mixtures. According to the author, accelerators (such as calcium chloride) tend to markedly accelerate the initial rate of shrinkage of the material. Mixtures prepared with a water-reducing admixture, retarding agent, or superplasticizer usually tend to have S-type behaviors.

Water/Cement Ratio

The influence of water/cement ratio was investigated by Powers in his 1935 study.[18] The author tested four different systems made of the same ordinary portland cement. For the pastes prepared at a water/cement ratio ranging from 0.42 to 0.65, the initial rate and the total chemical shrinkage were found to be slightly affected by the water/cement ratio of the mixture. A significant increase in the amount of total chemical shrinkage was observed for the paste prepared at a water/cement ratio of 0.93. This phenomenon was attributed to the reduced amount of water-filled space of the relatively low water/cement ratio pastes, which limits the volume reaction that can take place.

More recently, Justnes et al.[19] studied the behavior of neat paste mixtures prepared at water/binder ratios ranging from 0.30 to 0.50. Their results are illustrated in Fig. 12. As can be seen, the important amount of chemical shrinkage measured during the first hours after casting clearly emphasizes the fact that the dormant period, prior to setting, is quite active from the standpoint of volume change. According to these results, chemical shrinkage appears to be largely insensitive to water/binder ratio. Similar results were obtained by Boivin.[33] Both series of results are in fair agreement with the experimental data reported by Taplin[34] and Bentz,[35] which indicate that water/cement ratio has little influence on the kinetics of cement reaction during the first few days of hydration, at least for systems prepared at a water/cement ratio ranging from 0.30 to 0.50. Taplin's results also demonstrate that water/cement ratio has a significant influence on the kinetics of hydration after 4 days of curing. This phenomenon might explain the apparent discrepancy between the results of Justnes et al.[19] and the early data of Powers.[18]

Concluding Remarks

Globally, the various investigations devoted to the subject tend to indicate that there is a good correlation between the kinetics of hydration and the rate of chemical shrinkage. The abundant data recently published on the topic also emphasize the intricate nature of the problem. In that respect, more research is certainly required to clearly understand the influence of cement mineralogical composition on the mechanisms of chemical shrinkage. More attention should also be paid to the effect of secondary constituents, such as alkalis and sulfates, which can have a significant influence on the kinetics of cement hydration.[36] Additional information is also needed on the effects of chemical and mineral admixtures that are regularly used in the production of high-performance concrete mixtures.

Although it was not within the scope of this work, it became clear during the course of this literature review that very little research has been devoted to the influence of temperature on the kinetics of chemical shrinkage. The sparse information available on the subject indicates that temperature can significantly influence the mechanisms of chemical shrinkage. This aspect of the problem could be particularly critical to understanding the behavior of concrete materials in the field, which are often subjected to significant temperature rises during the first hours of hydration.

Autogeneous Shrinkage and Self-Desiccation

As previously mentioned, the hydration of cement not only leads to a reduction of the absolute volume of the system but also results in a decrease in the bulk volume of the material. The latter phenomenon, called autogeneous or external shrinkage, is accompanied by a marked reduction of the internal relative humidity of the porous solid. In some practical cases, the autogeneous deformation of concrete can contribute to the formation of cracks in restrained structures. The mechanisms of autogeneous shrinkage are discussed in the following paragraphs.

Autogeneous Shrinkage

Figure 13 describes, from a qualitative point of view, the evolution of the autogeneous deformations of a neat paste mixture with an increase of the degree of hydration of the cement. The autogeneous shrinkage of a hydrating cement system can be simply divided in two phases: before and after the setting of the material. The setting period is here defined as the time required to reach the rigidity percolation threshold, that is, to form the first continuous mineral pathway in the material. For cement pastes prepared at a water/cement ratio in the 0.30–0.50 range, this percolation threshold is assumed to correspond approximately to a degree of hydration of 5%.[37–39] During these two periods (before and after setting), the chemical shrinkage is the driving force for the autogeneous deformations. This aspect will be discussed further in the text.

As can be seen in Fig. 13, in the first few hours (i.e., from 5 to 10 h after the initial contact of water with the cement particles), chemical shrinkage measurements essentially coincide with the autogeneous (external) deformations, thus indicating that the material still behaves as a suspension. The moment at which the autogeneous shrinkage curve begins to depart from the chemical shrinkage deformation corresponds to the formation of a self-supporting skeleton. From that moment on, the hydrating paste cannot deform freely to accommodate the volume reduction, and significant internal stresses are generated in the material. As illustrated by Fig. 13, the dif-

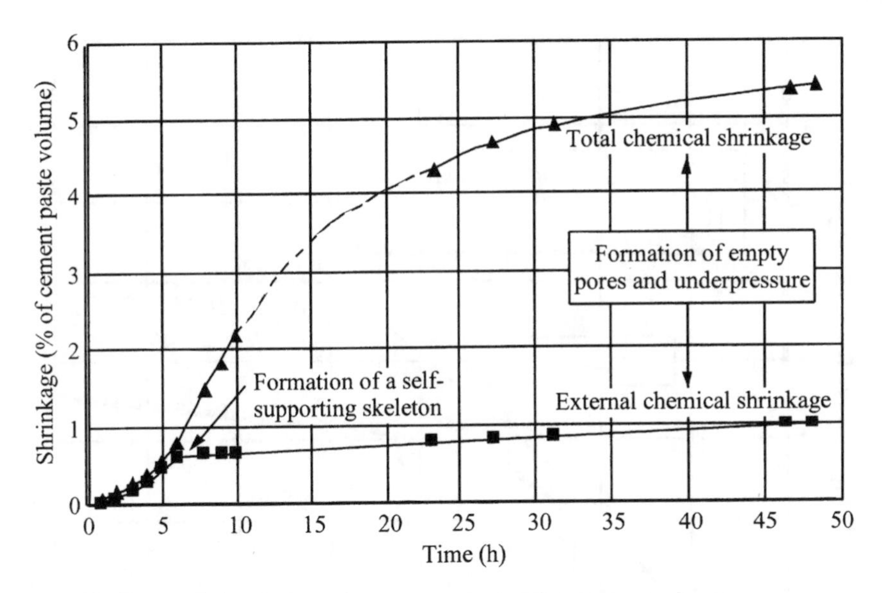

Figure 13. Chemical shrinkage and autogeneous shrinkage test results.

ference between the two types of shrinkage will become more important as hydration continues and the rigidity of the material increases.

From the standpoint of crack sensitivity, it is the beginning of a crucial period. As pointed out by Sellevold et al.,[40] the transition from a fluid paste to a solid skeleton occurs while the strain capacity of the material is still very low and the solidifying system is thus vulnerable to any additional internal (thermal effects) or external influence.

In a comprehensive investigation of the chemical shrinkage and self-desiccation of cement-based materials, Boivin[33] has shown that the point where the autogeneous shrinkage curve deviates from the chemical shrinkage deformation occurs between the initial and the final setting times of the cement paste as measured by the Vicat apparatus (Fig. 14). Moreover, using a pore water pressure measurement technique, Radocea[41] established that the transition between the semi-liquid and solid states corresponds more or less to the time when the rate of autogeneous shrinkage starts to decelerate as compared to the kinetics of chemical shrinkage. As can be seen in Fig. 15, the lower portion of the graph, where the pore pressure becomes less than the hydrostatic pressure, indicates that the solid skeleton is sufficiently rigid and strong to allow unsaturated pores to form.

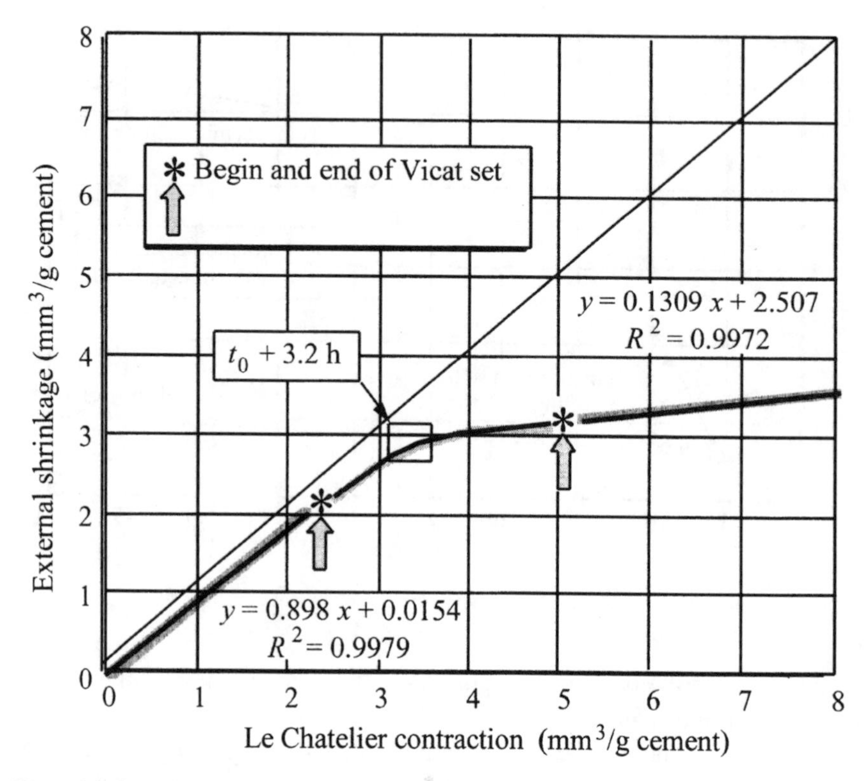

Figure 14. Correlation between chemical shrinkage and setting.[20]

The marked difference between the behavior of concrete before and after setting is a direct consequence of the progressive modification of the microstructure of the material.[37–41] A thorough discussion of the various mechanisms taking place during the setting of cement paste is beyond the scope of this review. However, detailed discussions of this subject can be found in Refs. 37–41.

In certain cases, the initial shrinkage is followed by a swelling period (see Fig. 16).[42] This swelling phase generally begins a few hours after mixing (approximately 4–10 h after the initial contact of water with the cement particles) and can last for 5–20 h. According to Baron[42] and Takahashi et al.,[43] this swelling period is related to the formation of ettringite. This conclusion is supported by the results of Beaudoin et al.[44,45] who studied the early volume change of C_3A and C_4AF mineral pastes using an overhanging-beam method. The hydration of these pure compounds in the presence of

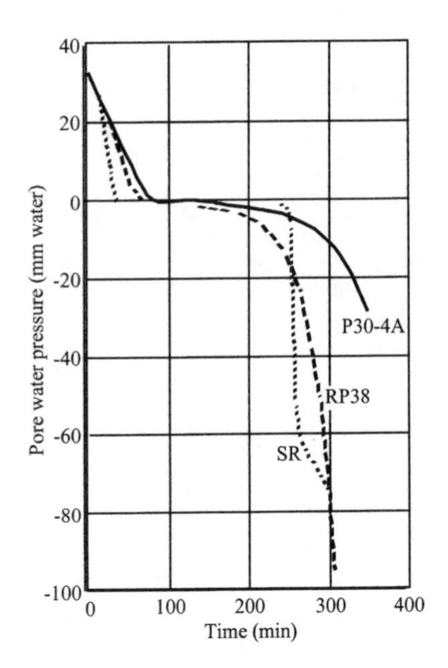

Figure 15. Evolution of internal pore pressure during setting.[41]

gypsum was found to lead to expansion. The rate of expansion was generally lower for the C_4AF systems than for the C_3A systems. Test results also indicate an increase in the rate of expansion for systems containing calcium chloride.

The fact that the initial shrinkage is not systematically followed by a swelling period emphasizes the complexity of the problem. Various competing mechanisms are probably going on during the first hours that follow the setting of the material. This view is supported by the results of Bentz et al.,[29] who studied the early age behavior of neat paste systems made of cements of different particle size distributions. Although the various cements tested had the same mineralogical compositions, the authors could measure expansion only for the systems prepared with the coarser cements.

The detection of the swelling phase also appears to be quite sensitive to the experimental technique used to follow the volume change of the material. Curiously, swelling is usually observed by linear length change measurements, and is rarely recorded by volume change measurements. The various experimental techniques used to measure the autogeneous shrinkage of hydrating cement systems will be discussed in detail in a later section.

Self-Desiccation

Another representation of the autogeneous deformations of a cement system upon hydration is given in Fig. 17.[46] Before setting, the mixture is in a liquidlike state and the apparent volume change (autogeneous shrinkage) is equal to the decrease in absolute volume (chemical shrinkage). Once the rigidity percolation threshold is reached, two phenomena take place simultaneously: a certain fraction of the bleeding water is absorbed by the material, and the volume of air in the capillaries increases.

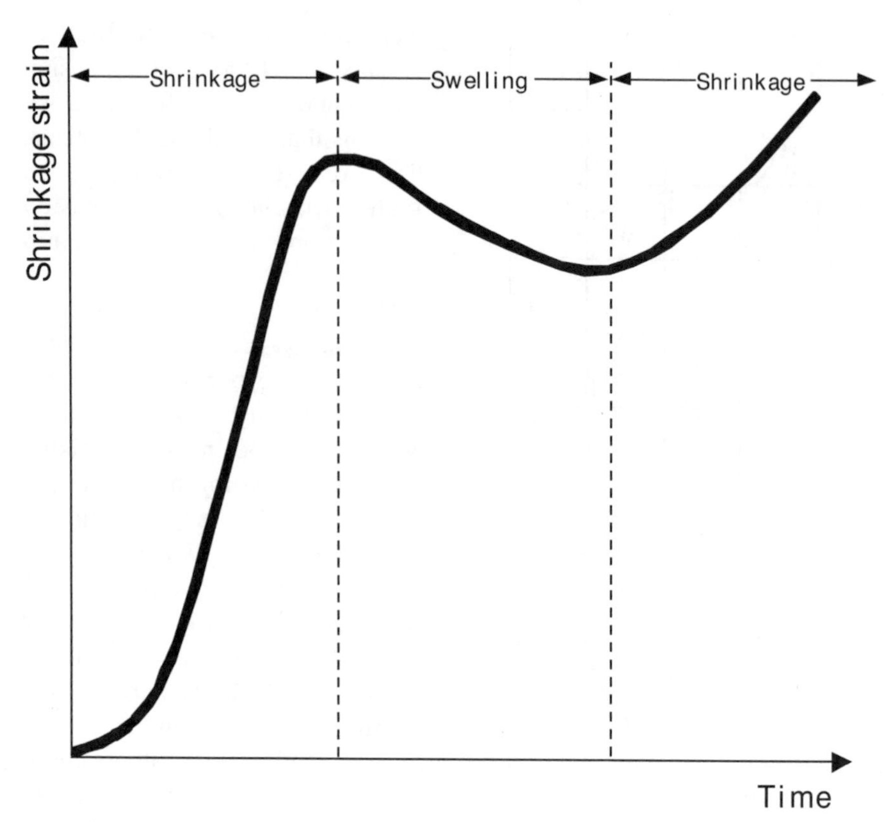

Figure 16. Autogeneous deformation (shrinkage and swelling) of cement paste.[42]

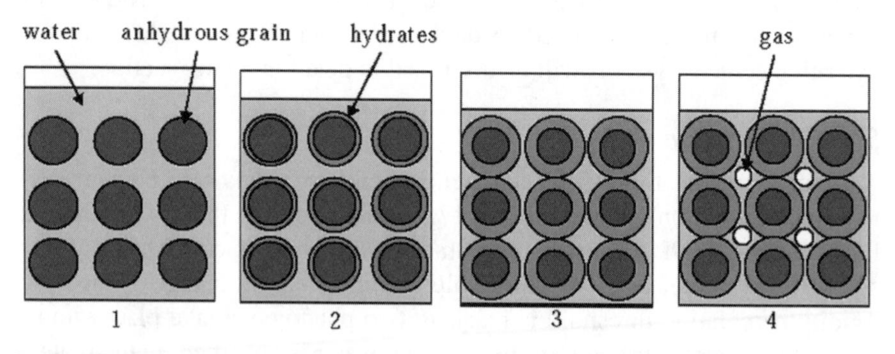

Figure 17. Autogeneous deformation and self-desiccation of cement paste.[46]

The absorption of bleeding water (if there is any) is caused by the consumption of internal water resulting from the hydration process (Le Chatelier phenomenon). However, it can be reasonably assumed that the penetration of the bleeding water remains limited to the surface layers of the material. As will be seen in the following section, the bleeding effects readily complicate the measurement of the autogeneous shrinkage of neat cement pastes.

The volume of air created in the capillaries (initially saturated with water) is equal to the difference between the chemical shrinkage and the autogeneous deformation. If the material is isolated from any external source of water, the volume of air will increase as the hydration continues, the net result being a gradual reduction of the internal relative humidity of the porous solid. This phenomenon is called self-desiccation.[47,48]

The reduction in the relative humidity in hydrating cement pastes was first measured by Gause and Tucker in 1940.[49] More systematic analyses of this phenomenon were later carried out by Powers and Brownyard[47] and Copeland and Bragg.[50] Both studies clearly indicated that self-desiccation effects were more pronounced for low water/cement ratio systems. For instance, Copeland and Bragg[50] demonstrated that the reduction in the internal relative humidity of sealed systems was minimal for mixtures prepared at water/cement ratios exceedings 0.53, while significant drops in relative humidity were measured for mixtures prepared at a water/cement ratio of 0.44.

The development of high-performance concrete mixtures has prompted many researchers to study self-desiccation effects in low water/binder ratio systems. Typical results recently reported by Baroghel-Bouny et al.[51] are presented in Fig. 18. As can be seen, the relative humidity in the high-performance materials decreases sharply during the first 60 days of hydration. It reaches approximately 82% for a neat cement paste and 75% for a concrete mixture. For ordinary concrete mixtures (water/binder ratio near 0.50), the relative humidity is still very high (approximately 95%) after six months of hydration in sealed conditions. Similar results have also been reported by Sellevold et al.[40]

As pointed out by Copeland and Bragg,[50] self-desiccation has direct consequences on the hydration kinetics of cement systems. The reduction in the internal relative humidity of the system contributes to reduce the rate of hydration.[29,50] It is generally considered that useful hydration ceases when the relative humidity of a sealed paste drops to about 75–80%.[47–50] The

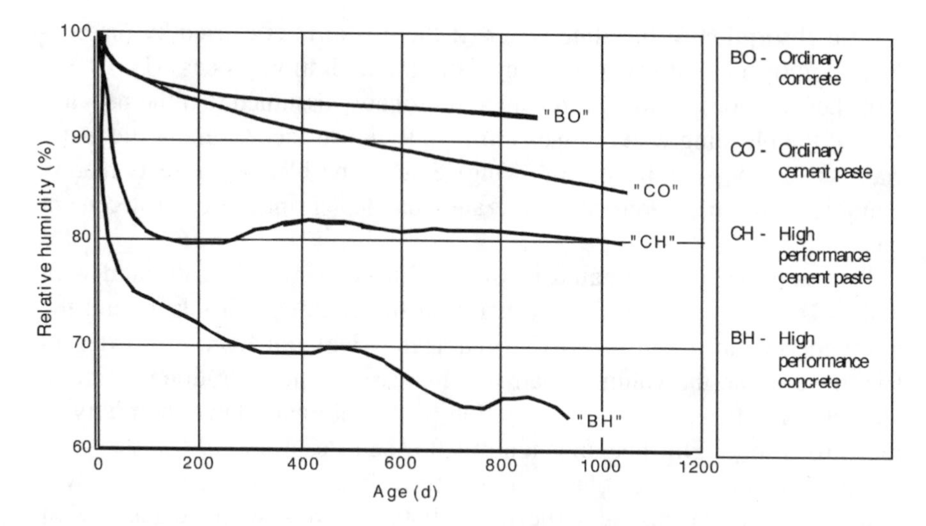

Figure 18. Evolution of the internal relative humidity of various cement systems.[51]

hydration of cements rich in C_2S and binders containing fly ash appears to be particularly sensitive to self-desiccation effects.[52,53]

Basic Mechanisms

Over the years, the basic mechanisms at the origin of the autogeneous shrinkage of hydrating cement systems have curiously not been the subject of much investigation. However, this aspect of the question is particularly important since a better understanding of the problem could eventually lead to the development of practical solutions to reduce the early age cracking of concrete structures.

Three principal basic mechanisms are most generally considered to be responsible for the shrinkage of portland cement paste induced by the removal of internal moisture, either by drying or self-desiccation, over the entire relative humidity range.[46,54–57] These mechanisms are associated with:

- The variation of the surface free energy (Gibbs-Bangham effects).
- The hindered adsorption in restricted areas (disjoining pressure effects).
- Capillary condensation effects.

A brief discussion of each of these three mechanisms is given in the following paragraphs.

Variation of the Surface Free Energy (Gibbs-Bangham Effects)

The free surface of any solid particle is under stress because of the non-symmetry of the attraction forces between atoms or molecules at this location. As emphasized by many authors,[6,54–59] the volume stability of highly divided solids, such as hydrated cement paste, is directly influenced by the water molecules adsorbed on the solid surface and present in the interlayer spaces. The presence of adsorbed water on the surface modifies the surface energy of the solid.

As explained by Ramachandran et al.,[6] the reversible change in free energy (ΔG) of a pure adsorbent from its initial state (and under its own vapor pressure, p) to its combining state is given by Gibbs's adsorption equation:

$$\Delta G = RT\int_0^p n\frac{dp}{p} \qquad (12)$$

where n is the number of moles of adsorbate on a fixed mass of adsorbent, R the ideal gas constant, and T the temperature. If it is assumed that the constant solid surface area (δ expressed in m^2/g) has undergone a change in surface tension ($\Delta\gamma$), one can write:

$$\Delta G = \delta\Delta\gamma \qquad (13)$$

and thus,

$$\Delta\gamma = \frac{-RT}{\delta}\int_0^p n\frac{dp}{p} \qquad (14)$$

As emphasized by Ramachandran et al.,[6] $\Delta\gamma$ in fact indicates changes in the state of stress of the solid induced by the interaction of adsorbed water molecules with forces at the liquid/solid interface, placing the solid in a state of compressive stress. The stress exerted on the solid surface is maximum in vacuum. Adsorption of water on cement gel particles reduces the free energy and the surface tension, thus causing a net expansion. On the contrary, the removal of adsorbed water causes the particles to contract.

The deformation ($\Delta L/L$) induced to the solid during the reversible adsorption process can be approximated by the Bangham equation[58,59]:

$$\frac{\Delta L}{L} = k_1 \Delta\gamma \tag{15}$$

where k_1 is a material parameter. According to Bangham and Maggs,[59] the Young's modulus (E) of a solid may be calculated by:

$$E = \rho \frac{\delta}{k_1} \tag{16}$$

with ρ being the density of the solid expressed in g/cm^3. Equation 15 was derived for plates, or infinitively long cylinders.

The variation of surface free energy is considered to be a significant shrinkage mechanism only at low relative humidities (i.e., the lower region of the adsorption/desorption isotherm). Beyond the second or third layer of adsorbed water, the effect flattens and variations in relative humidity do not much affect the surface tension. As pointed out by Hua et al.,[46,54] Gibbs-Bangham effects are unlikely to have any significant influence on the early age behavior of concrete. As previously mentioned, the internal relative humidity of a hydrating cement paste kept in sealed conditions cannot drop below approximately 75–80%. Below this critical value, the hydration process stops, and no additional self-desiccation can take place.

Hindered Adsorption in Restricted Areas (Disjoining Pressure Effects)

The second mechanism is also related to the tendency of (water) molecules to be adsorbed on a solid surface. This second mechanism is based on the assumption that the hydrated cement paste is made of a collection of discrete particles separated by narrow spaces.[6] As can be seen in Fig. 19, at relative humidities above 0% the surfaces bounding the narrow spaces can be covered with an adsorbed film of water molecules. In locations where the distance between two surfaces is restricted, the adsorption of water molecules may induce a pressure and cause an expansion. Because of the physical restriction of the film thickness, Powers[60] called these molecules "hindered adsorbed water."

According to the theory, the pressure in the hindered adsorbed water is a function of the thickness of the layer. This pressure is different in the hindered and adjacent free layers, and an additional pressure (P_D), called dis-

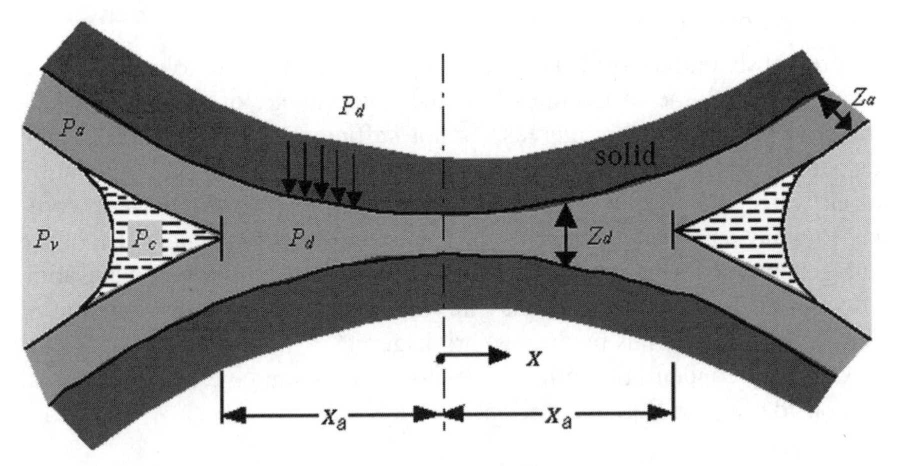

Figure 19. Disjoining pressures in restricted spaces.[63]

joining pressure, develops on the solid surfaces in order to make the total pressure (P_d) equal to P_a (see Fig. 19). The disjoining pressure (P_d) will increase with an increase of the relative humidity of the system and lead to expansion. Conversely, a diminution in relative humidity will result in shrinkage.

According to Ferraris and Wittmann[61] and Derjaguin and Churaev,[62] the disjoining pressure is essentially the sum of a series of attractive forces (Van der Waals forces) and repulsive forces (the double layer repulsion and the structural force) acting on the surfaces of the particles. Ferraris and Wittmann[61] established that, for a highly divided solid such as the hydrated cement paste, the volume change due to the disjoining pressure develops steadily in the range of relative humidity between 100% and approximately 40%. Hindered adsorption effects have also been associated with the viscoelastic behavior of cement systems.[60,63,64]

Despite the fact that Ferraris and Wittmann[61] clearly demonstrated that disjoining pressures are still active above 80% RH, many authors[46,54,56] tend to believe that they do not have any significant influence on the autogeneous shrinkage of cement systems. According to Hua et al.,[46,54] at relative humidities above 60% the shrinkage deformation of cement-based materials is essentially dominated by capillary condensation effects. More research is needed to clarify this point.

Capillary Condensation Effects

As previously emphasized, as hydration proceeds, the solid volume increases (i.e., the volume of hydrates is higher that volume of anhydrous products). Nevertheless, this increase is not sufficient to compensate for the amount of water that has reacted. This lack of volume leads to the creation of air/water interfaces within the material capillary pore structure. According to many authors, the presence of these menisci is the main cause of the autogeneous deformation of cement systems.[46,54,56] Capillary condensation effects are also considered to play an important role in the deformation of cement-based materials by drying shrinkage.[65,66]

Capillary condensation effects in porous solids can be easily understood by considering the Kelvin equation:

$$p_c - p_v = \frac{RT}{Mv}\ln(RH) \tag{17}$$

and the Laplace equation:

$$p_v - p_c = \frac{2\sigma}{r}\cos\theta \tag{18}$$

where p_c is capillary pressure in the pore water, p_v is vapor pressure, R is the ideal gas constant, T is temperature (°K), M is the mass of a mole of water, RH is relative humidity, σ is surface tension of water in contact with air, r is the radius of the pore, and θ is the liquid-solid contact angle.

According to the Kelvin equation (Eq. 17), for any given unsaturated state, there is a threshold pore radius r_0 such that all pores with a smaller radius are filled with water and those having a larger radius are empty. From the Laplace equation (Eq. 18), it follows that the existence of a meniscus at the liquid/vapor interface induces a tensile stress in the liquid phase and, in turn, the solid phase is subjected to a compressive stress. As a result, the material undergoes shrinkage.

Hua et al.[46,54] made some interesting calculations to assess the validity of the capillary tension theory as the main driving mechanism of self-desiccation shrinkage in isothermal conditions and without moisture exchange. Assuming (1) that the liquid phase is continuous, so that the capillary pressure induced in it is uniform, and (2) that macroscopically, the cement paste structure is homogeneous and isotropic, the following relationship between the macroscopic stress (Σ^s) acting on the solid phase and the capillary pressure (p_c) was derived on the basis of the virtual work principle:

$$\left(\Sigma^S\right) = p_c\phi \tag{19}$$

where ϕ is the total porosity of the paste. Then, using test data obtained from an in-depth characterization[67] of a 0.42 w/c ratio cement paste that included modulus of elasticity, creep, shrinkage, porosimetry (MIP), and degree of hydration measurements at different times starting at 24 h, simulations were performed and compared to the experimental self-desiccation shrinkage results. The main results of those simulations are summarized in Table III together with the experimental shrinkage results recorded from 24 h to 25 days of hydration.

The calculated self-desiccation shrinkage shows good agreement with the experimental values. While some other physicochemical mechanisms might be active, it appears that the change in capillary pressure could account for most of the observed autogeneous shrinkage. Besides, the effect of creep could be quite significant with the calculated contribution of the viscous strain to the overall shrinkage strain reaching more than 60% at 25 days. It should also be emphasized that the results of Hua et al.[46,54] do not totally exclude the existence of another mechanism, such as the disjoining pressure.

It can further be observed in Table III that the internal compressive stress that develops with self-desiccation (and that is responsible for the observed shrinkage contraction) is of the order of magnitude of the material tensile strength. The occurrence of such a uniform compressive prestress is consistent with the increase in strength observed upon drying in compression and also in tension, in the last case, provided that there is no important moisture gradient in the element.[68] Nevertheless, the increase in the measured tensile strength is not equivalent to the calculated prestress. The capillary-induced stress is assumed uniform at a macroscopic scale, but at a microscopic scale, it is non-uniform. Hence, some areas within the porous structure are subjected to much lower compressive stresses and, perhaps, even to tensile stresses.

In view of a macro-scale stress analysis of concrete at early age, the previous considerations on the capillary-induced stress are interesting in so far as it affects the material strength. While the change in pore water pressure during hydration can be measured with a very sensitive method,[41] it is not

Table III. Calculated and experimental autogeneous shrinkage results obtained by Hua et al.[54]

Age (days)	α (%)	ϕ (%)	p_c (MPa)	Σ^s (MPa)	RH (%)	$\varepsilon_{elastic}$ (× 10^6)	ε_{creep} (× 10^6)	$\varepsilon_{shr.\ cal.}$ (× 10^6)	$\varepsilon'_{shr.\ cal.}$ (× 10^6)	$\varepsilon_{shr.\ exp.}$ (× 10^6)
1	46.7	56.6	1.18	–0.54	99.1	72	24	96		
2	54.7	45.4	2.34	–1.02	98.3	107	47	154	58	60
4	60.6	43.4	4.55	–1.91	96.7	143	88	231	135	160
7	63.5	42.0	6.83	–2.82	95.1	171	141	322	226	260
15	67.9	41.4	8.64	–3.50	93.8	202	248	450	354	400
25	68.0	40.3	9.71	–3.91	93.1	212	332	544	449	480

KEY: α = degree of hydration; ϕ = total porosity; $\varepsilon'_{shr.\ cal.}$ = calculated shrinkage from 24 h; $\varepsilon_{shr.\ exp.}$ = measured shrinkage from 24 h.

possible to directly measure the stress induced in the solid skeleton. Furthermore, the calculations rest on different assumptions and hypothesis. As previously emphasized, the calculated volumetric compressive prestress does not linearly increase with the apparent tensile strength of the material. Measuring the development of strength in time on specimens cured in appropriate conditions is the simplest and likely the most effective way to take into account this effect that in turn appears to be the main cause of autogeneous shrinkage.

The previous discussion on the disjoining pressure theory and capillary condensation effects emphasizes the potential influence of the pore solution chemistry on the kinetics of autogeneous shrinkage. The double-layer repulsion, one of the three components of the disjoining pressure, is most likely to be affected by the chemical composition of the pore solution.[69] Similarly, evidence of the significant effect of the pore solution chemistry on the surface tension at the air/liquid interface has been reported.[70] This point has been systematically overlooked in the past. However, the introduction on the market of shrinkage-reducing admixtures (which modify the interfacial tension and the pore solution chemistry) will hopefully prompt more researchers to study this important aspect of the problem.

Experimental Measurements

Linear Shrinkage Measurements

Measurement of autogeneous deformations can be done using linear length change or volume change measurements. Linear measurements are usually carried out by recording the length change between two plugs attached to the specimen. Experimental methods to measure the autogeneous length change of cement systems have recently been critically reviewed.[71,72] The authors mention several parameters that may influence the experimental results. For example, the geometry of the sample, the position (horizontal or vertical) of the specimen, and the friction between the sample and the mold during the first hours of hydration are among the points that need to be taken into account in the analysis of the results.

Volume Change Measurements

The measurement of autogeneous volume deformations is usually done using a weighing method. The procedure consists of filling a latex membrane with fresh cement paste or mortar and placing it in a rigid tube. This tube is then immersed in a controlled-temperature water bath and continuously rotated to avoid bleeding.[19,73] The creation of empty spaces leads to a modification of the buoyancy, making it possible to follow the variations in volume of the sample by simple weight measurements.

One of the most important parameters influencing the autogeneous shrinkage measured using volumetric techniques is the reabsorption of the bleeding water. This point was clearly emphasized in the early work by Setter and Roy.[74] Figure 20 shows typical results of volumetric deformations measured for rotated and static samples. As can be seen, after approximately 10 h of hydration, both samples exhibit similar autogeneous shrinkage. When a rigid skeleton is formed (corresponding to 10 h in this case), the two curves separate. Beyond that point, the autogeneous shrinkage of the static sample is always lower than for the rotated sample. This is caused by the reabsorption of the bleeding water, which reduces significantly the self-desiccation in the static sample. This effect increases, as expected, with the water/binder ratio of the material.[73] Low water/binder ratio mixtures are less susceptible to bleeding.

It should be emphasized that very little information exists on the correlation between linear and volume change measurements. The only systematic comparison of the two series of measurements reported by Barcelo et al.[72] tends to indicate that linear shrinkage measurements systematically underestimate the deformation. Furthermore, the initial deformation induced to the solid prior to setting is not recorded by these measurements. More studies are needed to clarify these points.

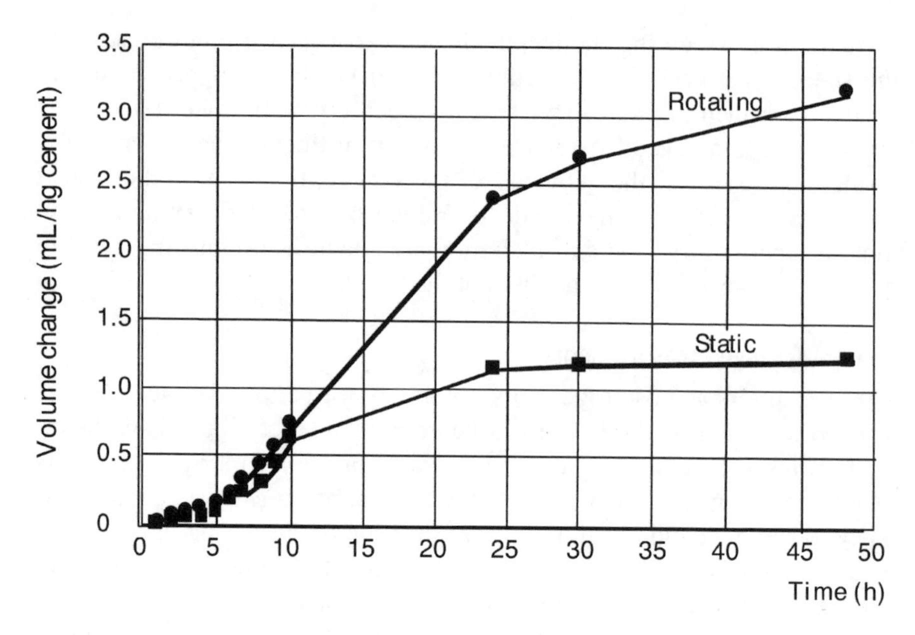

Figure 20. Influence of test procedure on autogeneous shrinkage.[73]

Relative Humidity Measurements

Internal relative humidity measurements are usually done according to the procedure described by Jensen and Hansen.[75] A sample of fresh cement paste is simply placed on a sample holder that fits tightly into a sealed measurement cell equipped with a humidity probe. The probe must be regularly calibrated using salt solutions with relative humidity between 75 and 100%. The cell is then placed in a thermostatically controlled box. The temperature of the box is usually controlled to within ±0.1°C. The drop in relative humidity can be monitored continuously or by regular measurements.

Factors Influencing Autogeneous Shrinkage

Given the nature and the complexity of the mechanisms of autogeneous shrinkage, it is obvious that the behavior of the material is likely to be affected by a wide range of parameters. A brief discussion of the main factors that influence the kinetics and the magnitude of the autogeneous deformations of hydrating cement systems is presented in the following paragraphs.

Mineralogical Composition of Cement

The influence of the type of cement was investigated by Tazawa and Miyazawa.[76] The autogeneous shrinkage results of neat paste mixtures prepared with different cements at a fixed water/binder ratio of 0.30 are presented in Fig. 21. As can be seen, for some mixtures, autogeneous shrinkage is still active after 700 days of hydration. Compared to normal portland cement (N), larger autogeneous shrinkage was observed for the high–early strength cements and less for moderate heat cement (M) and low heat cement (L). Some of these cements showed expansion in the first days after casting. Similar results were also reported by Takashi et al.[43]

Later, Tazawa and Miyazawa[77] reported additional findings from a comprehensive investigation of the influence of the cement mineralogical composition on autogeneous shrinkage. The authors tested seven different cements of known compositions. Mixtures were prepared at three different water/cement ratios: 0.23, 0.30, and 0.40. The autogeneous shrinkage was assessed by linear length change measurements. In their analysis, Tazawa and Miyazawa assumed, like Powers had previously,[18] that the autogeneous deformation of a paste was a linear function of the shrinkage of each individual constituent (C_2S, C_3S, C_3A, C_4AF) of the cement. They further assumed that the autogeneous deformation of each compound was directly dependent on its degree of hydration. They proposed the following linear equation to estimate the autogeneous shrinkage of neat cement pastes:

$$\begin{aligned}\varepsilon_p(t) &= a\alpha_{C_3S}(t)\cdot(C_3S\%) + b\alpha_{C_2S}(t)\cdot(C_2S\%)\\ &\quad + c\alpha_{C_3A}(t)\cdot(C_3A\%) + d\alpha_{C_4AF}(t)\cdot(C_4AF\%)\end{aligned} \tag{20}$$

where $\varepsilon_p(t)$ is the autogeneous shrinkage of the cement paste (expressed in microstrains), $\alpha_i(t)$ is the degree of hydration of compound i (%), t is the age of the material, (i%) is the percentage of compound i. Tazawa and Miyazawa[77] estimated the kinetics of hydration of each compound on the basis of the data reported by Copeland and Bragg.[50] The empirical coefficients *a, b, c,* and *d,* which depend on the water/cement ratio of the paste, were determined by a least-square analysis.

In all cases, the coefficients calculated for the aluminate compounds were at least an order of magnitude larger than those derived for the silicates, with the biggest one being that of C_3A , and the coefficients for C_2S and C_3S had a negative sign. According to the authors, this suggests that autogeneous shrinkage depends to a large degree on the hydration of C_3A and C_4AF.

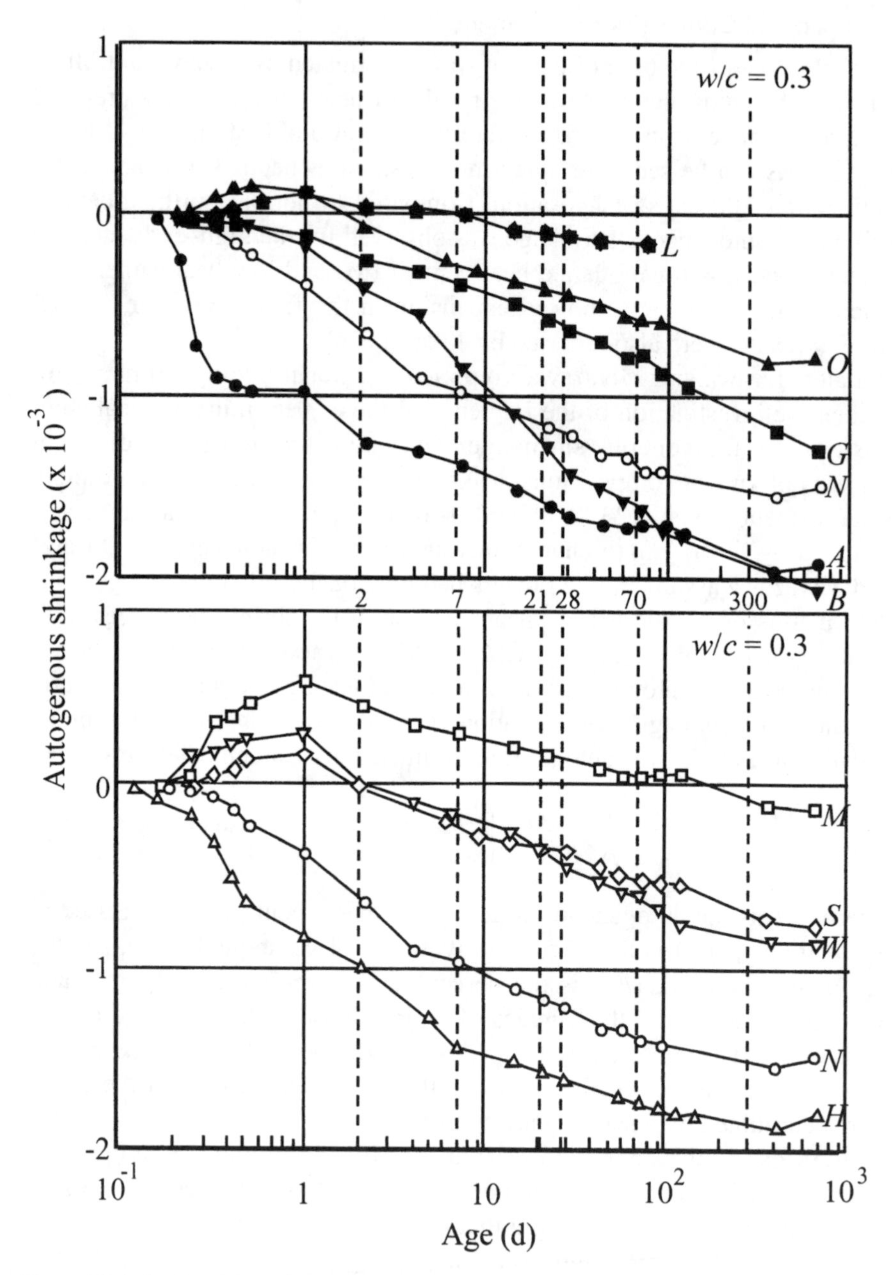

Figure 21. Influence of cement type on autogeneous shrinkage.[76]

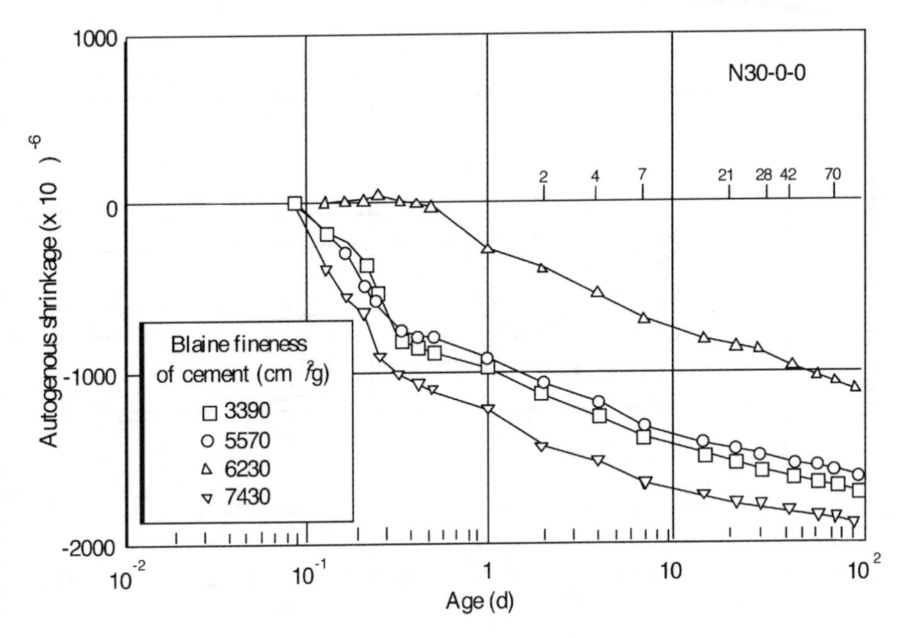

Figure 22. Influence of cement fineness on autogeneous shrinkage.[76]

Particle Size Distribution

As emphasized by Bentz et al.,[29,66] the cement particle size distribution has a strong influence on the kinetics and intensity of autogeneous shrinkage. This phenomenon was first studied by Tazawa and Miyazawa.[76] As shown in Fig. 22, an increase in the cement fineness clearly increases the autogeneous shrinkage measured during the first 24 h of hydration. Later on, the rate of autogeneous shrinkage appears to be independent of the cement.

The significant influence of the cement particle size distribution on the early kinetics of shrinkage has also been observed by Justnes et al.[28] These authors systematically measured the chemical shrinkage and the autogeneous (external) volume deformation of various cements. Similarly to Tazawa and Miyazawa,[76] they noted that the early rate of deformation prior to setting was markedly affected by the cement fineness. However, the autogeneous deformation measured after setting was found to be little affected by the cement particle size distribution.

Recent results by Bentz et al.[29,66] clearly confirm the direct influence of the particle size distribution on the early age deformations of hydrating

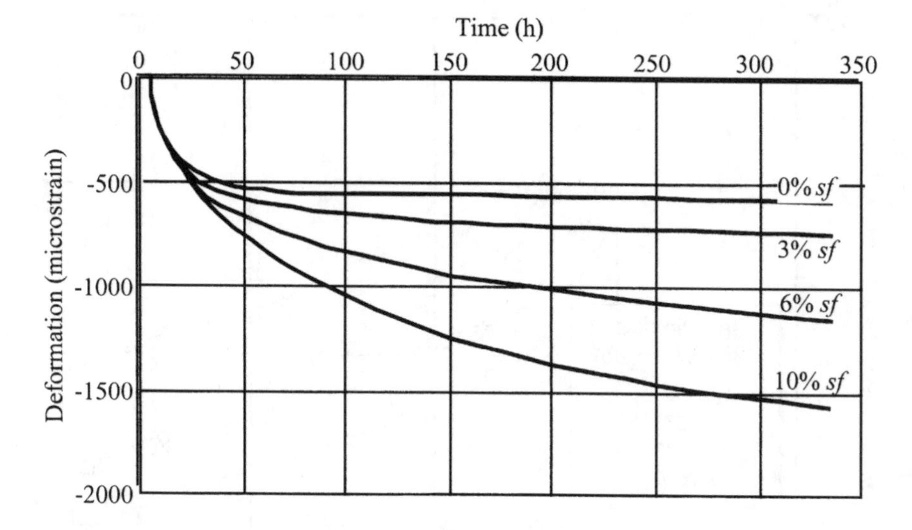

Figure 23. Influence of silica fume addition on autogeneous shrinkage of neat cement pastes.[79]

cement systems. The authors tested neat paste mixtures made of cements with similar mineralogical compositions but ground to different particle size distributions. The chemical shrinkage and the autogeneous linear deformation were both found to be sensitive to the cement particle size distribution. The mixtures prepared with the coarser cements were even found to swell after a few hours of hydration.

From an engineering point of view, cements with a coarser particle size distributions appear to be well suited for the production of high-performance concrete mixtures. As stressed by Bentz et al.,[66,78] the selection of a coarser cement does not have any detrimental influence of the mechanical and transport properties of low water/cement ratio mixtures, and it allows the reduction of the risk of early age cracking. This aspect of the problem should certainly be further investigated.

Supplementary Cementing Materials

Supplementary cementing materials, such as silica fume and fly ash, are increasingly used for the production of low water/binder ratio mixtures. Their influence on autogeneous shrinkage has therefore been the subject of a lot of attention. Jensen and Hansen[79] measured the linear autogeneous

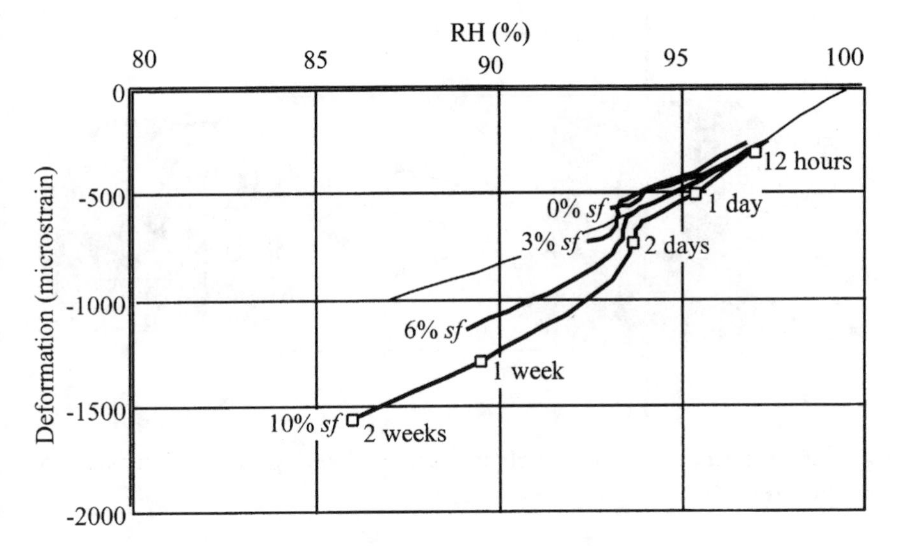

Figure 24. Influence of silica fume addition on autogeneous shrinkage and relative humidity of neat cement pastes.[79]

shrinkage of neat cement pastes containing 0–10% silica fume. All mixtures were prepared at a fixed water/cement ratio of 0.35. Figure 23 presents the main results of their investigation. As can be clearly seen, the addition of silica fume markedly increases the autogeneous shrinkage. For instance, the addition of 10% silica fume was found to increase the autogeneous shrinkage by almost 1000 μm/m.

Jensen and Hansen[79] also measured the evolution of the internal relative humidity of these various systems. These measurements were performed on identical samples. Figure 24 presents the evolution of the autogeneous shrinkage as a function of the relative humidity. A nick point can be clearly distinguished in all the curves. Nuclear magnetic resonance experiments revealed that this nick point corresponds to the disappearance of silica fume (pozzolanic reaction). This break is also associated with a vertical displacement of the curves that appears to be related to the amount of silica fume added to the mixtures.

Jensen and Hansen[79] proposed three mechanisms to explain the increase in autogeneous shrinkage associated with the addition of silica fume:

1. The disappearance of restraining components (silica fume or calcium hydroxide, see Fig. 25).

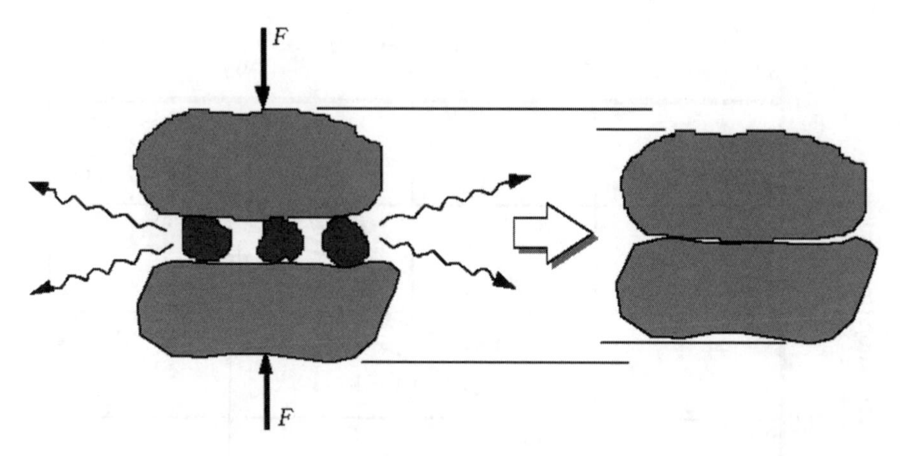

Figure 25. Schematic representation of the effect of calcium hydroxide consumption on autogeneous shrinkage.[79]

2. The chemical shrinkage of the pozzolanic reaction, which is similar to that observed for pure cement systems.
3. The pore structure refinement that increases the self-desiccation of silica fume systems.

The influence of silica fume on autogeneous shrinkage has also been investigated by Tazawa and Miyazawa.[76] These authors measured the linear autogeneous shrinkage of neat cement pastes with a water/binder ratio of 0.40 containing up to 20% silica fume. The silica fume was used as cement replacement. According to their results, an increase in the silica fume content clearly contributes to increase the autogeneous shrinkage of hydrating cement pastes.

The influence of silica fume was further confirmed by a series of results reported by Justnes et al.[28,32] These data also indicate that the effect of silica fume can be quite different from one cement to another. More recently, silica fume was also found to increase the autogeneous shrinkage of mortar and microconcrete mixtures.[80,81]

Few investigations have been specifically devoted to the influence of other mineral additives on the autogeneous shrinkage of cement systems. Tazawa and Miyazawa[17] measured the linear length change of neat paste mixtures made of different amounts of blast furnace slags of different fineness. Their data indicate that slags tend to increase autogeneous shrinkage, particularly when they are used at a level of cement replacement of 50–

75% of the total mass of binder. The influence of the particle size distribution of slag has also been underlined in this series of experiments. As for cements, increasing the fineness of slag leads to an increase in autogeneous shrinkage.

The influence of fly ash was investigated in an early study by Setter and Roy.[74] Fly ash was not found to have any significant effect. In a more recent study, Justnes et al.[32] clearly established that the addition of a Class F fly ash could significantly increase the autogeneous shrinkage of neat cement paste mixtures.

Chemical Admixtures

The addition of chemical admixtures has been the subject of only a limited number of studies. In an early investigation of the problem, Edmeades et al.[82] found the autogeneous shrinkage of cement pastes containing 0.1% calcium lignosulfonate to be more important than that of reference cement pastes, particularly after 10 h of hydration. The admixture was also found to delay setting.

Tazawa and Miyazawa[17] investigated the influence of the type and dosage of superplasticizing agents and the effect of shrinkage-reducing admixtures on the behavior of neat cement pastes prepared at a fixed water/binder ratio of 0.30. According to their results, superplasticizing agents do not have any marked effect on autogeneous shrinkage. However, shrinkage-reducing admixtures tend to reduce the autogeneous shrinkage of neat cement pastes. The authors found that both alcohol alkylene oxide and glycol ether types of admixtures could contribute to reduce the linear autogeneous shrinkage.

Water/Binder Ratio

Justnes et al.[19] have conducted a series of experiments on neat paste samples with water/binder ratios ranging from 0.30 to 0.50. They measured the total chemical shrinkage and the external shrinkage. As can be seen in Fig. 26, a reduction of the water/cement ratio tends to increase the rate and the intensity of autogeneous shrinkage. Furthermore, the pastes with lower water/binder ratios show a more gradual transition from the liquid to the solid state. For high water/binder ratio mixtures, this transition appears later and is clearly more abrupt. This tends to indicate that the earlier formation of a rigid skeleton for low water/binder ratio pastes. However, this solid is probably insufficiently rigid to totally impede the deformation. According to the authors, the more gradual transition of the lower water/binder pastes

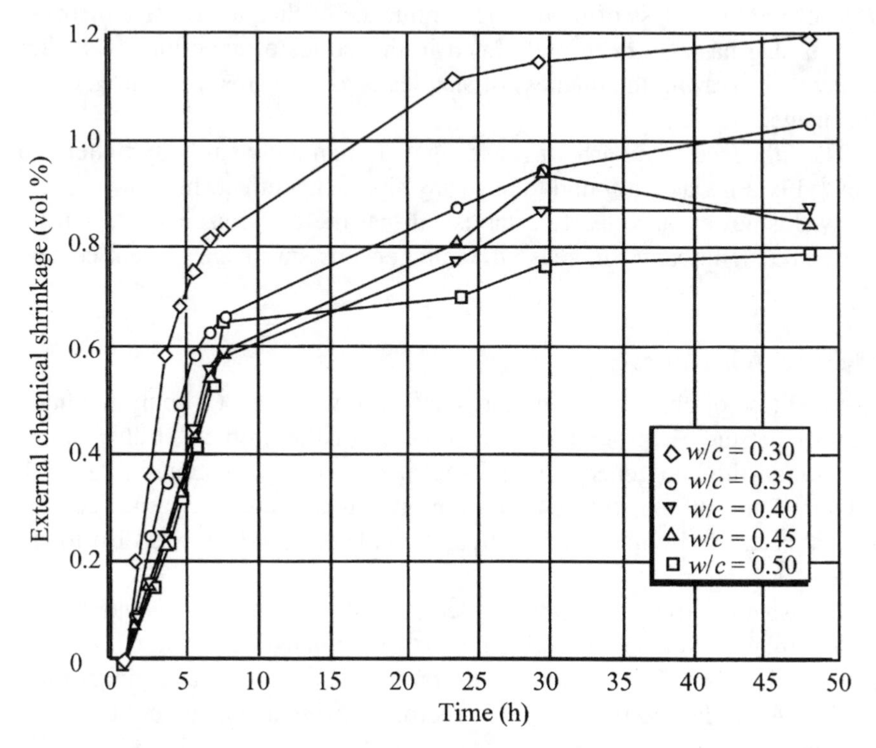

Figure 26. Influence of water/cement ratio on autogeneous shrinkage.[19]

may contribute to increase the risk of early age cracking in high-performance concrete.

Toma et al.[83] and Igarashi et al.[84] have also investigated the influence of the water/binder ratio on autogeneous shrinkage of microconcrete specimens using linear length change measurements. They both observed an increase in the kinetics and the intensity of autogeneous shrinkage with a reduction of the water/binder ratio. Figure 27 presents the results obtained by Toma et al.[80] of free autogeneous shrinkage measurements for three concrete mixtures prepared at different water/cement ratios.

Aggregates

Another parameter of concrete composition that should influence the autogeneous shrinkage is the volume fraction of aggregates. Knowing that the

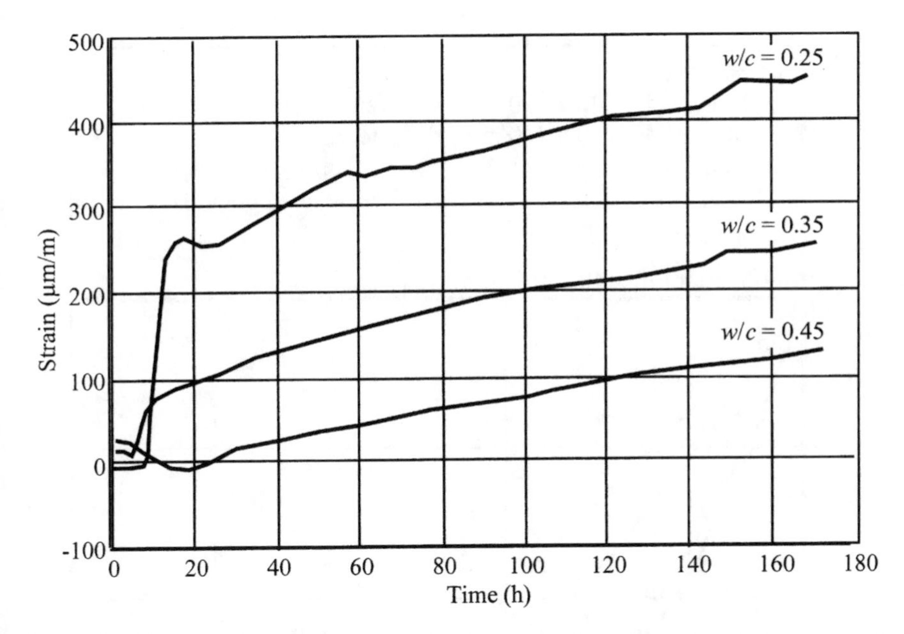

Figure 27. Influence of water/cement ratio on autogeneous shrinkage of microconcrete mixtures.[80]

shrinkage takes place only in the paste fraction, an increase of the aggregate concentration of a concrete mixture should contribute to the reduction of the autogeneous shrinkage. In a comprehensive study of the early age behavior of paste and mortar mixtures, de Haas et al.[85] found that the addition of sand does, in fact, contribute to reduce the intensity of autogeneous shrinkage. The presence of sand particles was also found to slightly increase the initial rate of shrinkage of very low water/cement ratio (0.20) mixtures.

Similar results were later reported by Tazawa and Miyazawa,[17] who tested concrete mixtures with different volume fractions of aggregates. Their results clearly show that an increase in the aggregate content leads to a decrease of autogeneous shrinkage. The authors also evaluated the ability of various models to predict the autogeneous shrinkage of concrete. Series models, parallel models, and the model developed by Hobbs for predicting drying shrinkage were considered. According to their analysis, the model proposed by Hobbs tends to yield the best results (see Fig. 28). The following equations describe the various models that were tested:

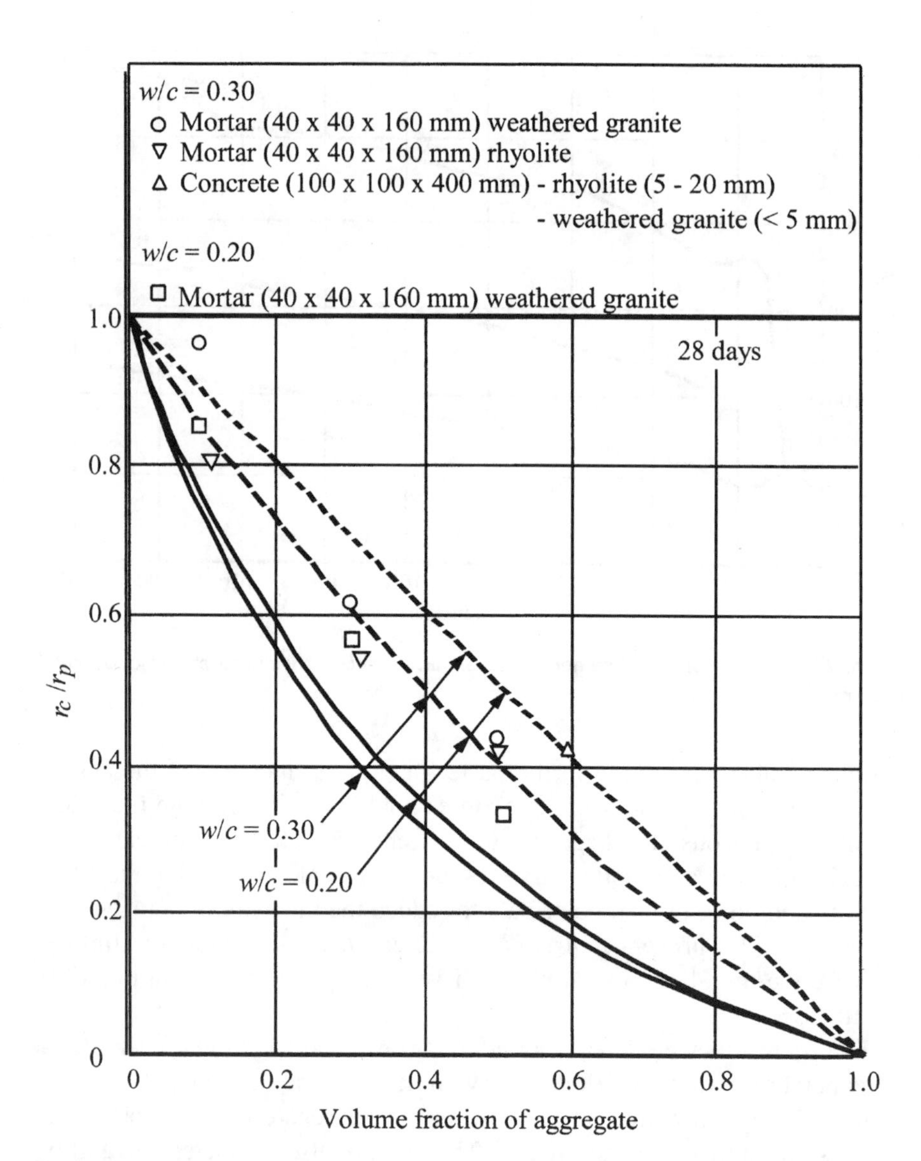

Figure 28. Influence of aggregate volume fraction on autogeneous shrinkage of concrete.[17]

$$\frac{\varepsilon_c}{\varepsilon_p} = 1 - V_a \qquad \text{(series model) (21)}$$

$$\frac{\varepsilon_c}{\varepsilon_p} = \frac{1 - V_a}{\left(E_a / E_p\right)V_a + 1} \qquad \text{(parallel model) (22)}$$

$$\frac{\varepsilon_c}{\varepsilon_p} = \frac{\left(1 - V_a\right)\left(K_a / K_p + 1\right)}{1 + K_a / K_p + V_a\left(K_a / K_p - 1\right)} \qquad \text{(Hobbs' model) (23)}$$

where a, p, and c stand for aggregate, paste, and concrete, respectively; $K = E/3(1-2\nu)$; and E is the modulus of elasticity.

Another parameter that seems to influence the autogeneous shrinkage of concrete is the water content of the aggregates. Recently, Bentur et al.[86] measured the autogeneous shrinkage of concrete mixtures prepared at a fixed water/binder ratio of 0.33. The mixtures were produced with and without saturated lightweight aggregates. The main results of their experiments are shown in Fig. 29. The normal aggregate mixture shows a little early expansion and then undergoes shrinkage, which increases linearly with time. In contrast, the mixture containing saturated lightweight aggregates shows a rapid swelling soon after the water/cement contact followed by a little shrinkage that remains small. Even after 7 days of autogeneous curing, the lightweight aggregate mixture shows no significant shrinkage deformation. The beneficial influence of using saturated lightweight aggregates as partial replacement has also been reported by Hammer et al.,[87] Van Breugel and de Vries,[88] Bentz and Snyder,[89] and Weber and Reinhardt.[90]

The beneficial use of saturated aggregates (lightweight or normal aggregates) can probably be explained by the fact that these porous particles act as internal water reservoirs that counteract the self-desiccation. As previously explained, the hydration of cement leads to a decrease of the internal relative humidity of the cement paste. A humidity gradient exists between the hydrating cement paste and the internal relative humidity of the aggregates that are saturated. This gradient will act as a driving force to suck up

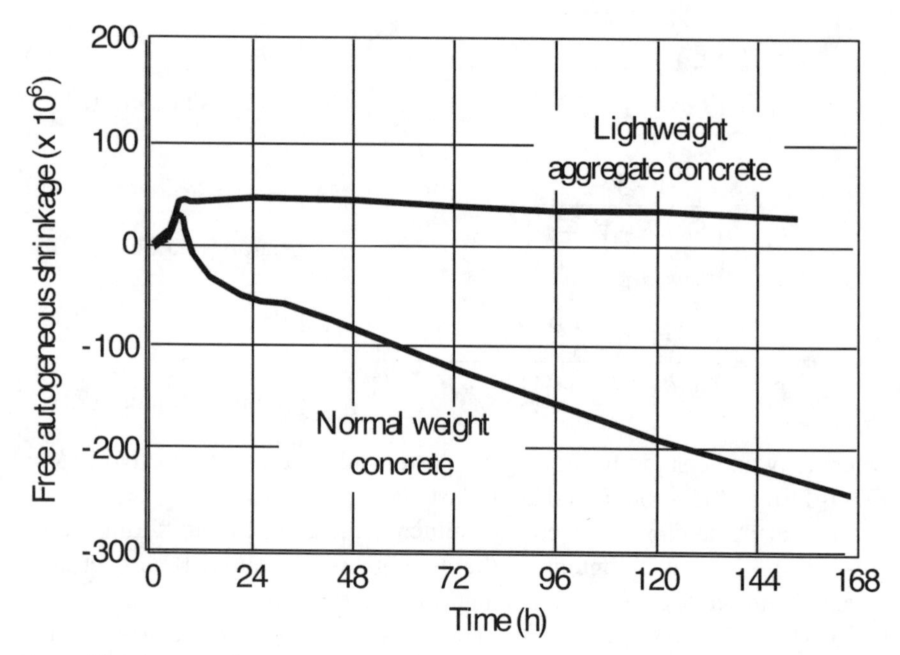

Figure 29. Influence of aggregate saturation on autogeneous shrinkage of concrete.[86]

the water contained in the aggregates. The incoming water in the cement paste will help pursue the hydration of cement and decrease the autogeneous shrinkage.

Concluding Remarks

As can be seen, the autogeneous shrinkage of hydrating cement systems has been the subject of a great deal of research over the past decade. Most of the recent investigations have been focused on the influence of various material parameters (cement mineralogical composition, particle size distribution, etc.) on the kinetics and the intensity of the autogeneous deformations. Unfortunately, researchers have relied on different techniques to study the mechanisms of autogeneous shrinkage. Recent results reported by Barcelo et al.[72] clearly indicate that there is no direct correlation between linear and volume change measurements. As previously emphasized, this point needs to be further investigated.

It should also be pointed out that very little work has been specifically focused on the basic mechanisms of autogeneous shrinkage. For instance, the relative importance of disjoining pressure effects still needs to be clearly understood. The lack of information on the fundamental aspects of the problem is most unfortunate. It impedes the development of reliable, practical solutions to minimize the risk of early age cracking.

Early Age Stress Buildup

While the advantages of low water/cement ratio mixtures are numerous, either in terms of durability or mechanical behavior, recent experience clearly indicates that the early age volume behavior of high-performance concrete calls for special attention. In fact, it also true for usual good quality concrete, but obviously to a lesser extent. The apparently more frequent early cracking problems associated with high-performance concrete mixtures have drawn attention from the cement and concrete community.

A certain number of years ago, thermal-related problems in mass concrete dams were first addressed.[91] The autogeneous shrinkage is an additional issue that must now be considered in situations where low water/cement concrete mixtures are used. In what follows, we will deal more specifically with autogeneous shrinkage-induced stress. As previously emphasized, early age thermal effects have been treated in a previous volume of this series.[4,92]

General Considerations

Although the nature of the autogeneous deformation of concrete is similar to that of drying shrinkage, that is, a contraction of the material resulting from a reduction of its internal relative humidity, the two phenomena are different and the actual consequences on stress and cracking are not the same. First, the two phenomena tend to occur at different periods of the service life of the material. As pointed out in the previous section, autogeneous deformations develop during the first hours and days of hydration. When adequate curing is performed, drying shrinkage will not begin before concrete has reached a certain level of maturity. Furthermore, autogeneous shrinkage does not necessarily generate strain gradients within the material. If hydration takes place in isothermal conditions and without local restraint, self-desiccation and autogeneous shrinkage will be relatively uniform within the element. On the contrary, drying shrinkage most generally occurs non-uniformly, as drying proceeds from the concrete surface inward.

The fact that autogeneous volume changes begin almost instantly upon casting brings us to the heart of this early age matter. When does it begin and what is the extent of that so-called period? It is a significant issue as far as stress development and cracking are concerned. No definite answers to those questions exist, but recent experimental results have shed new light on the problem. As mentioned earlier, the comparison of total chemical shrinkage with external shrinkage of cement pastes shows that, in the first few hours (until 5–10 h), the two types of measurements coincide (see Fig. 13), thus indicating that concrete still behaves as a suspension. The moment at which the external shrinkage curve begins to depart from the total chemical shrinkage curve corresponds to the formation of a self-supporting skeleton. From that moment on, it appears that significant internal stresses can be generated provided that the hydrating paste cannot deform freely to accommodate the volume reduction. It is the beginning of a crucial period regarding the cracking sensitivity of the material.

As emphasized in the previous section, it appears quite obvious that self-desiccation can lead, or at least contribute significantly, to early age cracking. Beyond the liquid-to-solid transition, hydration continues, further increasing the autogeneous deformation. The occurrence of cracking is then the result of a competition between autogeneous shrinkage, strength, and creep. The shrinkage-induced stresses are dependent on the intensity of shrinkage, but also on creep and on the degree of restraint provided by the reinforcement and the supports of the element.

It was said earlier that autogeneous shrinkage occurs uniformly within the element, whereas drying shrinkage is systematically nonuniform. It should be emphasized that the absence of such an internal restraint is true as long as hydration progresses in isothermal conditions and no significant moisture exchange with the surroundings takes place. Hydration being dependent on both the temperature and the moisture content, a gradient of either one of these parameters will lead to a nonuniform degree of hydration and, therefore, to a nonuniform autogeneous strain distribution.

Shrinkage-Induced Stress Development

When considering early age stress development in concrete, it is necessary to separate and identify the different strain components that need to be considered. The internal stresses induced in concrete structures at that time of their life are almost exclusively due to restricted strains. At a given moment and location in concrete, the resulting total uniaxial strain (ε_{total}), which could be interpreted as the measurable strain, is given by the following equation:

$$\varepsilon_{total} = (\varepsilon_{elastic} + \varepsilon_{creep}) + \varepsilon_{shrinkage} + \varepsilon_{thermal} \tag{24}$$

The sum of the elastic and creep strain is the mechanical response of the material to the fact that shrinkage and thermal strains cannot occur freely. Two different types of restraint exist, provided that concrete is regarded as homogeneous. The first and most obvious type is external to concrete and is associated with the direct obstruction to its free movement by a rigid body. The concrete volume changes are first of all locally restricted by the reinforcement. At the structural level, the dead weight and the connection with adjoining rigid elements hinder the free movement of the aging material.

The other type of restraint is internal and occurs as a result of a nonuniform strain distribution within the concrete element. This nonuniform strain distribution may originate from temperature or moisture effects. This second type of restraint is observed in many concrete structures.

In the strain balance equation (Eq. 24), the resulting total strain depends on the effective degree of restraint. In a perfectly restrained system, $\varepsilon_{total} = 0$ and no apparent deformation can be observed. All imposed strains are then accommodated by the elastic and viscoelastic strains (mechanical strains), which correspond to the first and the second term respectively on the right side of Eq. 24. Conversely, if there is no external restraint and the shrinkage and thermal strains are both uniformly distributed, no mechanical stresses and strains are induced and, in each point of the system, the total strain is equal to the sum of the local shrinkage and thermal strains. In real structures, the actual degree of restraint is usually difficult to evaluate. It strongly depends on many factors and tends to vary with aging. However, in most practical cases, the degree of restraint is usually quite significant.

When free shrinkage is prevented by any form of restraint, the level of stress (σ) that develops can be estimated using the following equation[93]:

$$d\varepsilon_{total}(t,t') = d\sigma(t')\left[\frac{1}{E(t')} + \frac{\phi(t,t')}{E_{28d}}\right] + d\varepsilon_{shrinkage}(t') \qquad (25)$$

where t and t' refer respectively to the age of concrete and the time at which the strain increment occurred, $E(t')$ is the modulus of elasticity at time t', E_{28d} is the modulus of elasticity at 28 days, $\phi(t,t')$ is the creep function, and $\varepsilon_{shrinkage}$ here represents the sum of autogeneous, drying, and thermal shrinkage. As previously mentioned, the stress development and, accordingly, the resulting total strain $[\varepsilon_{total}(t,t')]$ are dependent upon the effective degree of restraint (λ). Rearranging Eq. 25, the stress development can be expressed as a function of the parameter λ:

$$d\sigma(t')\left[\frac{1}{E(t')}+\frac{\phi(t,t')}{E_{28d}}\right]=\lambda d\varepsilon_{\text{shrinkage}}(t') \tag{26}$$

with

$$\lambda=\frac{\left[d\varepsilon_{\text{total}}(t,t')-d\varepsilon_{\text{shrinkage}}(t')\right]}{d\varepsilon_{\text{shrinkage}}(t')} \tag{27a}$$

and

$$0\leq\lambda\leq 1 \quad \text{(a contraction is negative)} \tag{27b}$$

The above expressions clearly emphasize the intricate nature of the problem. The analysis must take into account the viscoelastic behavior of concrete. It should also take into consideration the fact that all the material properties are significantly changing as hydration continues. The level of restraint, which depends on the ratio between the local stiffness of concrete and that of the rest of the structure, is thus a time-dependent variable.

Nonetheless, simulations can be made on the basis of a constant level of restraint in order to evaluate the sensitivity of cracking to this parameter. Such calculations have been performed by Weiss et al.[93] using a viscoelastic cracking model for three concrete mixtures assumed to shrink at different rates. Their results are presented in Fig. 30. As can be seen, a reduction of the level of restraint delays the time of cracking, which eventually reaches an asymptotic value. Obviously, the effective degree of restraint appears to be a key parameter in the cracking response of a concrete element. Thus, to predict whether concrete will crack or not, an accurate knowledge of the restraint provided by adjoining structural elements in the field is needed.

In addition to the paramount importance of the restraint, the results summarized in Fig. 30 illustrate the influence of the shrinkage rate on the stress development and the predicted age of cracking. The simulations indicate that for a high degree of restraint, a faster rate of shrinkage will cause earlier cracking. Conversely, it shows that for low degrees of restraint, the threshold being approximately 60% in that case, the time to cracking increases with the rate of shrinkage. It is quite significant in view of early age cracking problems. As previously mentioned, a reduction of the water/cement ratio tends to increase the rate and the intensity of autoge-

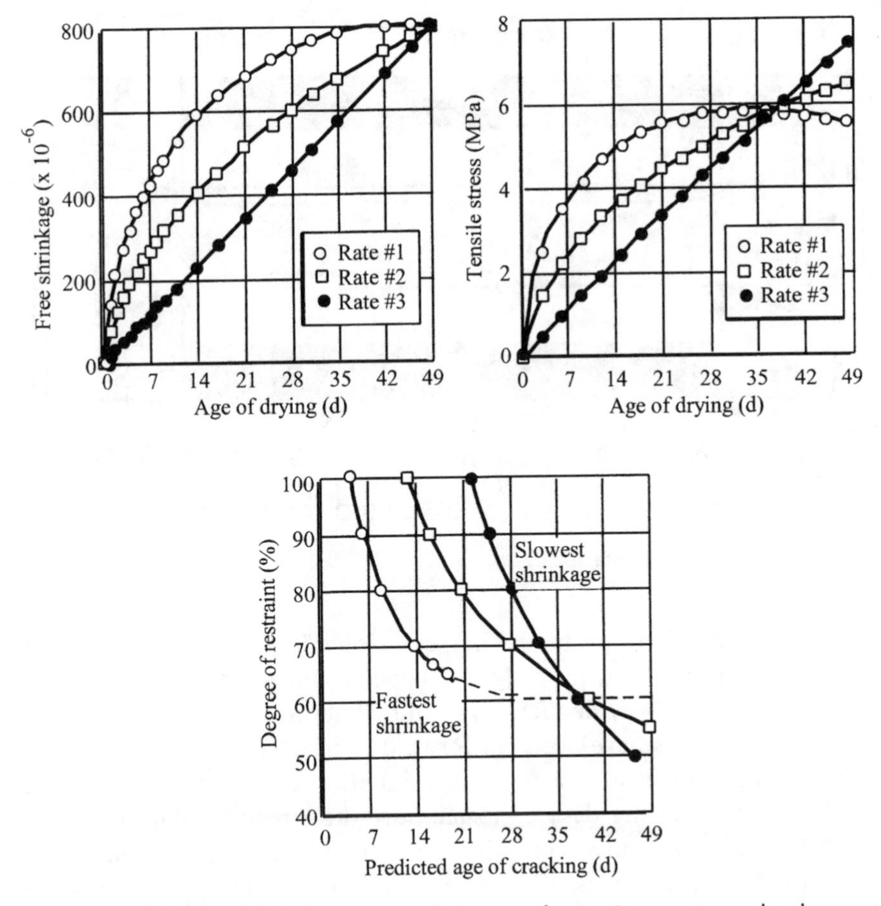

Figure 30. Influence of shrinkage rate and degree of restraint on stress development and time to cracking.[93]

neous shrinkage. Low water/binder concrete mixtures are thus more susceptible to the development of harmful internal stresses while the structure is still extremely weak.

Measurement of Early Age Stress Development

Engineers and practitioners are used to deal with stresses in concrete only once it has reached the final set or, most generally, once the curing operations are completed. Accordingly, mechanical and volume change tests are rarely performed during the first 24 h of hydration. The increasing number

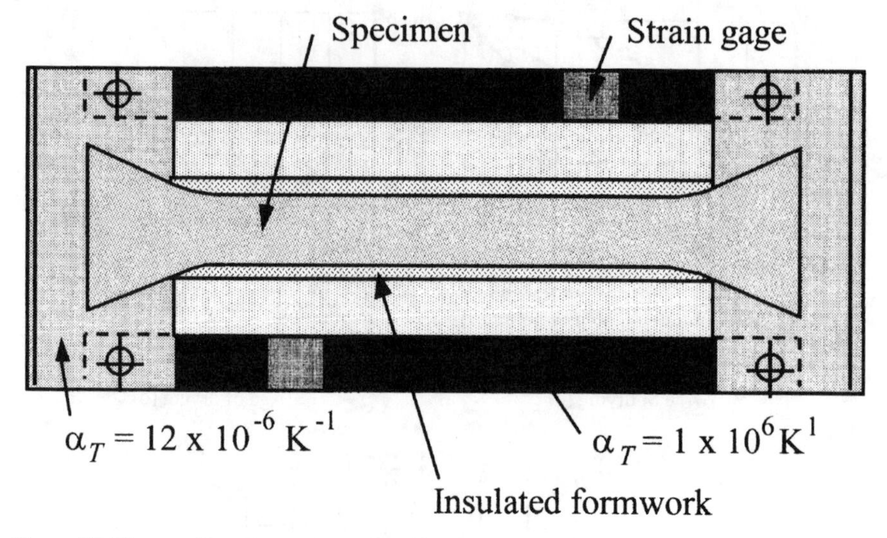

Figure 31. The cracking frame apparatus.[94]

of early age cracking problems has contributed to the development of experimental procedures to study and better characterize the behavior of concrete during the very first few hours of hydration.

While techniques have been put forward to measure accurately the free early volume changes of concrete, a special effort has been focused on the development of testing devices to measure the stress buildup occurring under field representative conditions (i.e., under restrained conditions). Basically, two main types of uniaxial restrained shrinkage devices have been set up for that matter, allowing measurements almost right after casting. The first type is based on the so-called cracking frame, which was first developed at the Institute for Building Materials (IBM) in Germany.[94] It consists of a concrete specimen (100×100 mm^2 or 150×150 mm^2 cross section) cast horizontally and having a straight prismatic central part with flared ends fitting into rigid grips. As shown in Fig. 31, the restraint is provided by stiff lateral steel members intended to keep the distance between the grips as constant as possible. The load induced by the restrained thermal and shrinkage strains is measured through strain gages mounted on the lateral members. The cracking frame was built mainly to study thermal cracking in concrete, so the apparatus formwork was insulated in order to reproduce the temperature development in a thick cross section. The output

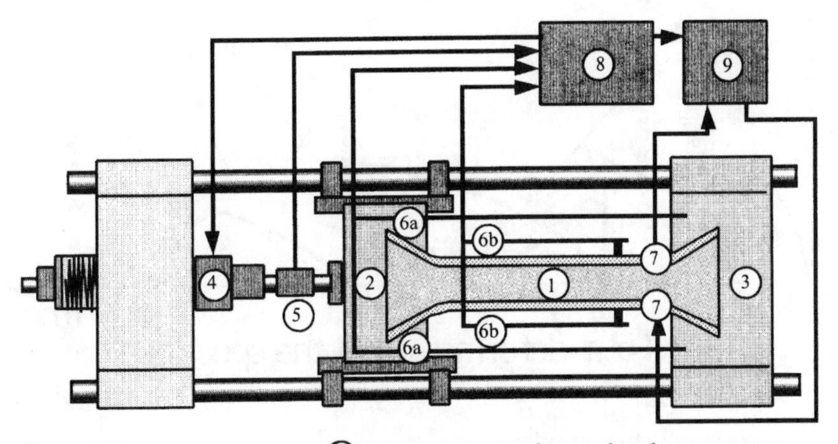

(1) Specimen
(2) Adjustable cross-head
(3) Fixed cross-head
(4) Step motor
(5) Load cell
(6a) Measurement of cross-head movements
(6b) Length measurement with carbon fibre bars
(7) Formwork with heating/cooling system
(8) PC for controlling and recording
(9) Cryostal for cooling/heating of the formwork

Figure 32. The temperature-stress testing machine.[94]

of the test is the thermal and autogeneous shrinkage stress curve. A disadvantage of the cracking frame is that the degree of restraint is significantly less than 100%. Furthermore, the level of restraint changes constantly with time because of both the hydration and the viscoelastic nature of concrete.

The first testing device specifically designed to study early age restrained autogeneous shrinkage was built at the Laboratoire Central des Ponts et Chaussées (LCPC) in France about ten years ago.[95] This modified version of the cracking frame, called by the authors the self-cracking apparatus, fully counteracts the volume changes. The lateral restraining members are replaced by a loading device attached to one of the two end grips, which is mobile, and equipped with a dynamometer. From the time of casting and throughout the test, the length of the specimen is maintained constant by means of a closed-loop system that monitors the strain and controls the loading device. During about the same period, at IBM (Germany), a similar device called the temperature-stress testing machine (TSTM) was built for the study of thermal cracking (see Fig. 32).

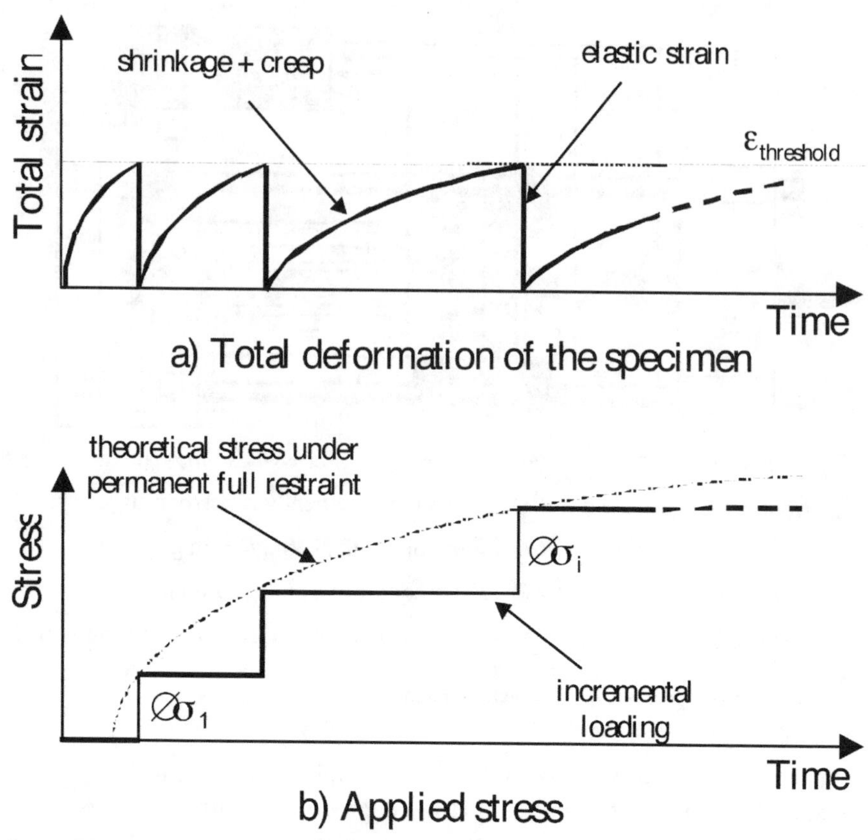

Figure 33. Principle of the discretized restrained shrinkage test.

More recently, some investigators[96–98] have built sophisticated systems on the basis of those developed at LCPC and IBM. Basically, they consist of a horizontal mold with enlarged heads in which the fresh concrete is placed directly after mixing. The mold is designed in such a way that shrinkage is theoretically unrestrained except by the enlarged heads. One of the heads is fixed, but the other can move and is linked to a load cell and to a closed-loop control system. When the shrinkage-induced deformation has reached a threshold value, an increment of load is applied to bring the specimen back to its original position. The operating principle is illustrated in Fig. 33. As the threshold strain value is reduced, the test approaches the conditions found in a purely restrained shrinkage test [Fig. 33(b)]. The load being increased step by step (instead of evolving in a quasi-continuous

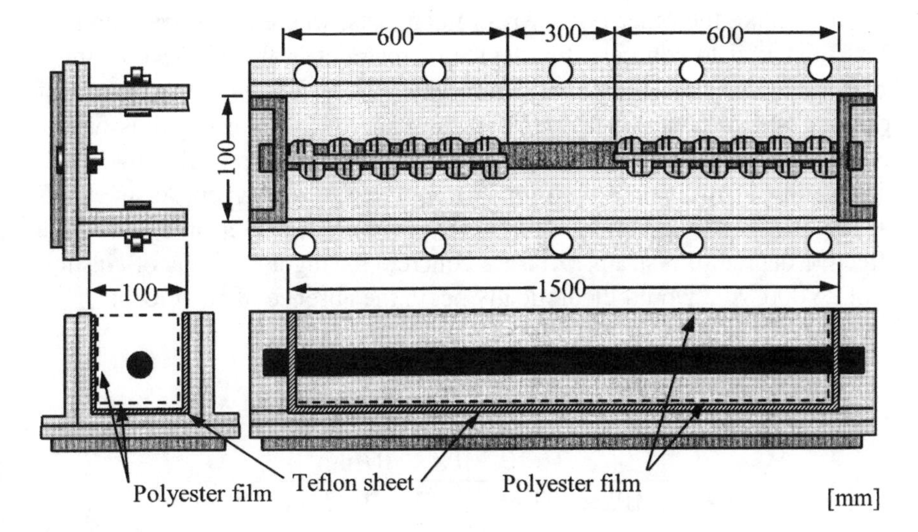

Figure 34. The JCI restrained shrinkage test: mold and specimen.[103]

fashion), this procedure is usually referred to as the discretized restrained shrinkage (DRS) test.

Using the free shrinkage measured on a companion specimen as a basis for the calculations, such a procedure can yield not only the stresses developed under restrained shrinkage, but also the relaxation capacity of the concrete.

Computers are now used for a better control of the loads and displacements, and some systems[94,99] also have temperature controls, which allow the phenomena to be tested at various temperatures and even under given temperature histories.

The other main type of restrained autogeneous shrinkage tests reported in the literature was used mostly in Japan.[100–102] Figure 34 shows the mold and the test specimen as described in the Japan Concrete Institute procedure.[103] The specimen basically consists of a concrete prismatic beam in which a deformed steel bar is embedded to provide longitudinal restraining reinforcement. In the central portion of the reinforcing bar, the steel surface is smooth and covered with a Teflon sheet over a sufficient length so as to ensure that there is no bond with concrete, thus producing a uniform stress distribution in both materials. Underneath the Teflon sheet, a strain gauge is applied on the steel surface for monitoring the steel strain during the experiment and determination of the stress induced by restrained shrinkage in the system. Unlike the other type of restrained tests, where the restraining system is external to concrete, the degree of restraint in the JCI setup cannot be adjusted during the test, nor can it reach 100%. The effective degree of restraint will in fact vary throughout the experiment as it depends on the relative stiffness of concrete to that of steel, and the concrete effective modulus significantly changes with time owing to aging and creep. As an example, given the configuration of the test specimen displayed in Fig. 34, the effective degree of restraint (λ) for a concrete having a modulus of elasticity of 25 000 MPa would theoretically be, in the absence of any creep:

$$\lambda = \frac{\varepsilon_c^{el}}{\varepsilon_c^{sh}} = \frac{E_s A_s}{E_s A_s + E_c A_c}$$
$$= \frac{2 \times 10^5 \text{MPa} \times 707\text{mm}^2}{\left(2 \times 10^5 \text{MPa} \times 707\text{mm}^2\right) + \left(25 \times 10^3 \text{MPa} \times 9293\text{mm}^2\right)}$$
$$= 0.38 \quad (28)$$

where ε_c^{el} is shrinkage-induced elastic strain in concrete, ε_c^{sh} is free shrinkage strain of concrete, E_c is Young's modulus of concrete, A_c is concrete area, E_s is Young's modulus of steel, A_s and is steel area (diameter of the bar in the central part is 30 mm).

According to the previous equation, the degree of restraint corresponds to the ratio of the induced elastic strain in concrete to the shrinkage strain. A full restraint would give a value of 1.0 and it would mean that shrinkage is completely offset by the elastic strain. In reality, concrete has a viscoelastic behavior. To account for this behavior, the elastic strain and the modulus of elasticity of concrete appearing in Eq. 28 should be replaced by the mechanical strain (instantaneous + creep strain) and the effective modulus of elasticity. For instance, if, at a given moment, the creep strain is equal to the elastic strain calculated in the previous example, provided that the elastic modulus has not changed, the effective modulus is half the elastic modulus value (E_{eff} = 12 500 MPa), and the effective degree of restraint (λ) becomes 0.55.

Like the ring test used by many investigators to study drying shrinkage cracking,[104] the JCI test is a simple and interesting tool to characterize the early age cracking behavior of cement-based materials. However, the fact that the material is only partially restrained and that the true degree of restraint at a given moment is not readily known complicates the analysis of the results. In addition, the effective degree of restraint depends on the properties of the material to be tested, such that comparisons between different concretes are not obvious.

Autogeneous Shrinkage Stress Development: Experimental Data

The effect of autogeneous shrinkage on stress development in a concrete structure is dependent on many factors, such as the kinetics and the intensity of the autogeneous strain, the effective degree of restraint, the material viscoelastic behavior, and the other volume changes occurring concurrently (e.g., thermal strain). The complex tests described previously are thus essential tools in the endeavor to understand, predict, and eventually control, at least to a certain extent, the early age behavior of cement systems. As mentioned previously, self-desiccation occurs shortly after the water/cement contact, but what about the corresponding induced stress? When does the restricted strain becomes harmful, or, at least, when does it generate significant internal stresses? One could go further and ask if it is necessary to generate significant or measurable stresses in such an early stage, when the material is in the transition between the liquid and solid phases, to cause significant damage to the developing microstructure. Unfortunately, these questions cannot be easily answered, but numerous valuable contributions published on the subject over the past decade have shed new light on the problem.

Among the few data available regarding the evolution of autogeneous shrinkage-induced stress in isothermal conditions, one of the earliest series was reported by Paillère et al.[95] In that study, ordinary and fiber-reinforced concrete mixtures prepared at various water/cement ratios (ranging between 0.26 and 0.44) were investigated using a uniaxial self-cracking test. Sealed prismatic samples (120 × 85 mm) were subjected to a fully restrained shrinkage test through the use of a closed-loop pneumatic loading system while, simultaneously, free shrinkage was measured on an identical unloaded specimen kept in the same environment. The results for the three unreinforced concrete mixtures are shown in Fig. 35. In Fig. 35(a), the autogeneous shrinkage results clearly illustrate the marked influence of the mixture water/cement ratio on its early volume stability. As it is reduced from 0.44 to 0.30, the initial expansion decreases slightly, but the shrinkage rate is much higher during the first few days. For the low water/cement ratio mixture, which contains silica fume in addition (15 wt%, actual water/binder ratio = 0.23), the initial expansion vanishes completely and

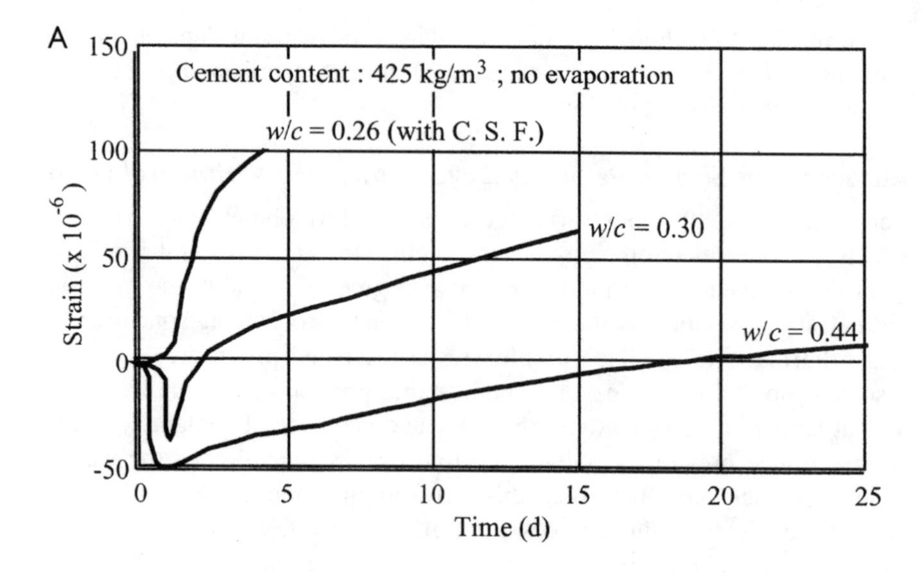

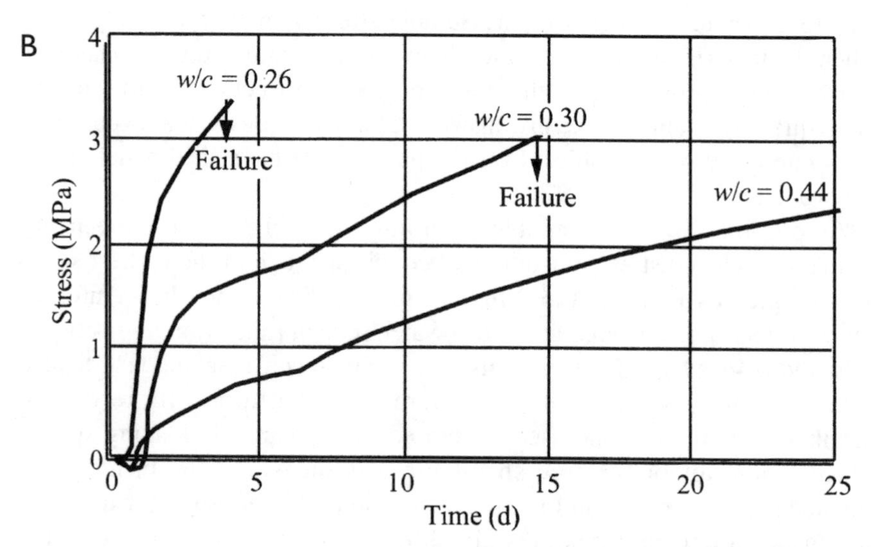

Figure 35. Results of (a) free and (b) restrained shrinkage in the self-cracking test apparatus on different high-strength concrete mixtures.[95]

shrinkage begins very shortly after casting at a higher rate than the 0.30 water/cement ratio mixture.

As can be seen in Fig. 35(b), the induced stress varies accordingly. The initial expansions of the 0.44 and 0.30 water/cement ratio mixtures do not induce any substantial stress, the materials having quite a low rigidity at

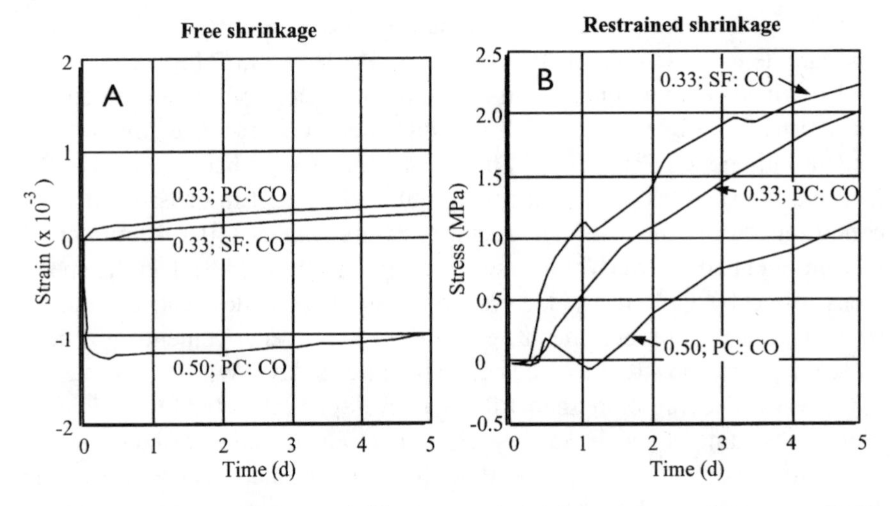

Figure 36. Results of (a) free and (b) restrained shrinkage tests performed in a DRS device on different high-strength concrete mixtures.[96]

that time, a low modulus of elasticity, and a high creep potential. Afterward, the increases in stress of all three mixtures are roughly proportional to shrinkage. In both the 0.26 and the 0.30 water/cement ratio mixtures, restrained shrinkage leads to cracking as the corresponding stress reaches 3.5 and 3.2 MPa after 4 and 15 days, respectively. In the 0.40 water/cement ratio mixture, shrinkage does not cause cracking within the duration of observation, but the induced stress is still quite significant, reaching more than 2.0 MPa at 25 days.

It thus appears that cracking could be induced by the sole effect of self-desiccation. These results should however be considered with caution, since the authors do not provide any information on reproducibility, as each curve corresponds presumably to a single test, nor is any information supplied concerning the location where the failure occurred in the test sample. In addition, the thermal conditions in which the tests were performed are not described.

In a subsequent study, using an experimental device directly inspired from that of Paillère et al.,[95] Bloom and Bentur[96] carried out restrained shrinkage tests with a 100% degree of restraint. Microconcrete mixtures (maximum aggregate size = 7 mm), made of portland cement (PC) and silica fume (SF), were prepared at various water/cement ratios (ranging from 0.33 to 0.50). Test results obtained on sealed specimens for the PC-0.50, PC-0.33, and SF-0.33 mixtures are shown in Fig. 36. As can be seen, the restrained shrinkage stress data obtained by Bloom and Bentur[96] are quite comparable to those previously reported by Paillère et al.[95] for the 0.30 and 0.44 portland cement concrete mixtures, both in terms of rate and intensity. This time, however, no cracking was observed for the low water/cement ratio concrete mixtures. This phenomenon is probably related to the fact that the samples of Bloom and Bentur[96] were kept in quasi-isothermal conditions during the tests. Their samples were smaller (40 × 40 mm cross section) and had a lower volume/surface ratio (10.0 vs. 24.9) than the specimens tested by Paillère et al.[95] Nevertheless, the self-desiccation induced stress is, once more, significant, even in the 0.50 water/cement ratio concrete for which the stress exceeds 1.0 MPa after 5 days. Within less than 24 h, a stress exceeding more than 1.0 MPa developed in the 0.33 water/binder ratio silica fume concrete. In the 0.33 water/cement ratio portland cement concrete, a stress of about half that figure is attained. These are by no means negligible with respect to the early development of strength and the risk of cracking.

Recently, some investigations were performed using testing devices similar to those of the previous studies, except that the testing procedure was discretized; that is, the strain compensation was imposed by constant steps instead of a quasi-continuous compensation. That kind of protocol, used by Kovler[97] to investigate the behavior of hardened concrete in drying conditions, allows the evaluation of the different deformation components, namely the elastic strain, the creep strain, and the shrinkage strain (see Fig. 33). Between each step where the specimen is brought back to its original length, the load is maintained constant at the level reached at the end of the compensation cycle.

Toma[80] ran such discretized restrained shrinkage (DRS) tests on both portland cement and silica fume concrete mixtures having water/cement ratios ranging from 0.25 to 0.45. The tests were conducted in semi-isothermal conditions (23°C) using sealed 50 × 50 × 1000 mm prismatic specimens. The compensation steps were fixed at 8×10^{-6} up to 24 h, and at 4×10^{-6} afterward. As can be seen in Figs. 37 and 38, for comparable water/cement ratios, the measured restrained shrinkage stresses after 5 days (168 h) were a little lower than the values reported by Paillère et al.[95] and Bloom and Bentur.[96] However, the free autogeneous shrinkage results stand within the range of values obtained in those previous studies. Although no failure was observed in any of the reported tests, these results confirm, once more, that autogeneous shrinkage-induced stresses are significant and that in both portland cement and silica fume low water/cement ratio concrete mixtures,

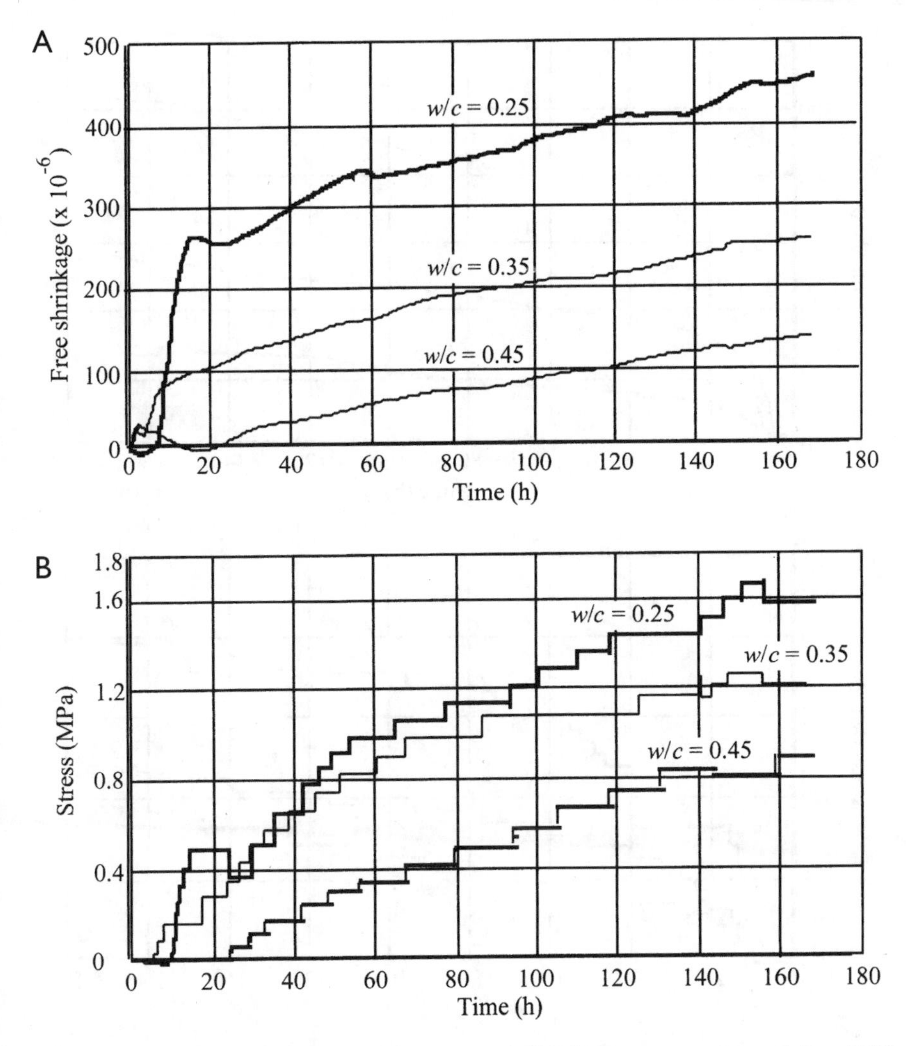

Figure 37. Results of (a) free and (b) restrained shrinkage tests performed in a DRS device on ordinary portland cement concrete mixtures with different w/c ratios.[80]

it can compete with the gain in strength during the first hours after casting. The results given in Figs. 35–37 clearly indicate that the concrete behavior goes through a transition when the water/cement ratio is decreased from the 0.40–0.50 range to values below 0.30. In low water/cement ratio systems, the free shrinkage and the induced stress increase sharply during setting, up to a period that presumably coincides with (or sets the limits of) the final

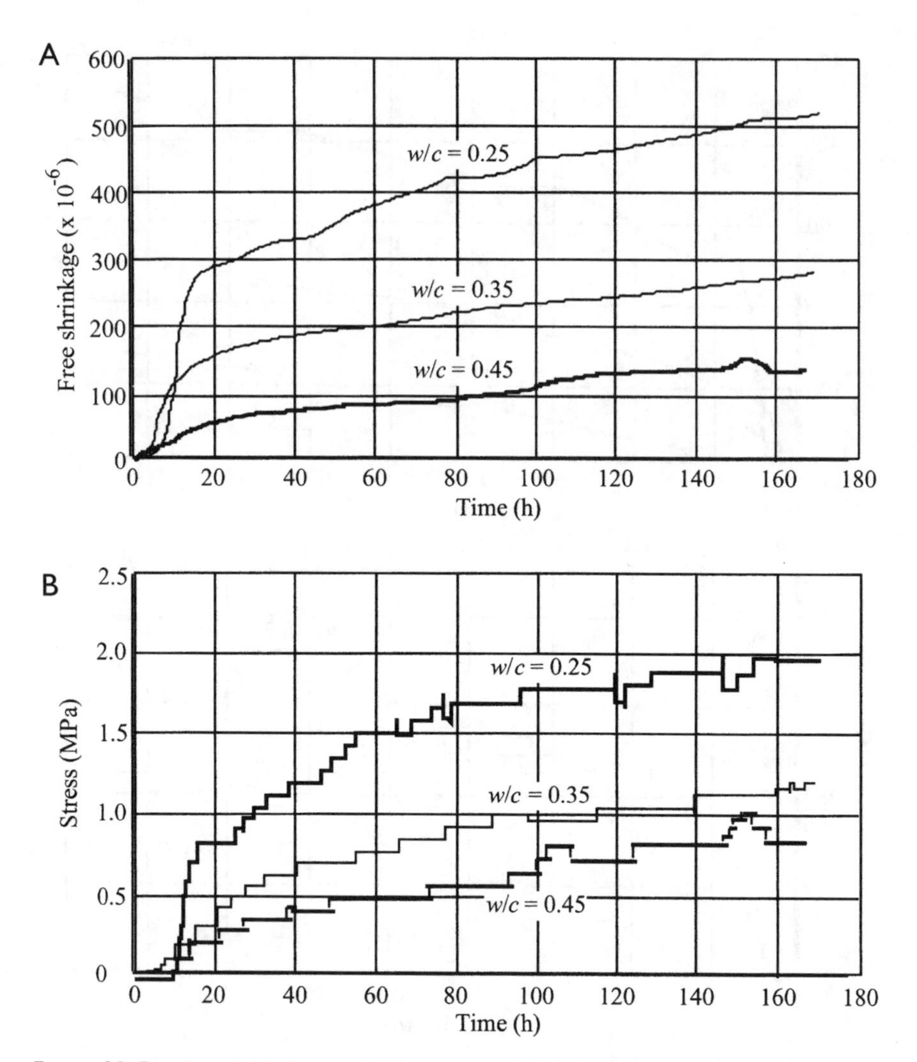

Figure 38. Results of (a) free and (b) restrained shrinkage tests performed in a DRS device on silica fume concrete mixtures with different w/c ratios.[80]

setting time. Beyond that point, the strain and stress rates become much like those encountered in higher water/cement ratio concretes.

DRS tests were also performed by Igarashi et al.[84] on cement pastes and concrete mixtures made of two types of cement (ASTM Type I and ASTM Type I + silica fume) and two water/cement ratios (0.33 and 0.25). The sealed specimens were exposed to an ambient temperature of 30°C and the

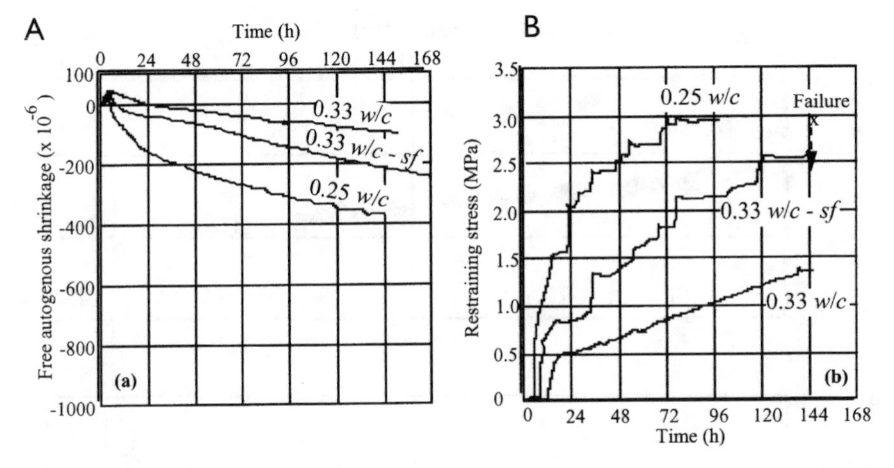

Figure 39. Results of (a) free and (b) restrained shrinkage tests performed in a DRS device on different high-strength concrete mixtures.[84]

compensation cycles were triggered at a strain of 5×10^{-6}. Although the free autogeneous shrinkage results for the concrete mixtures do not differ much from those reported by Toma[80] on similar concrete mixtures, except for the smoother transition after setting, the corresponding restraining stress was significantly higher in some cases (see Fig. 39). Both the specimens of the 0.33 water/binder ratio silica fume mixture and those of the 0.25 water/cement ratio portland cement mixture reached values on the order of 3.0 MPa at 7 days, and it can be seen on the graphs that the stress rates were still significant at that time.

Using a similar testing device but an experimental protocol where the strain compensation steps were significantly smaller (less than 1×10^{-6}), Bjøntegaard[99] obtained results that roughly match those of Toma,[80] provided that the small concrete composition differences are taken into account. Yet, no failure was recorded for the samples kept in isothermal conditions at 23°C.

Early age restrained shrinkage data on sealed low water/cement ratio concretes have also been reported lately by different investigators who have performed partially restrained tests[100–102] using either the previously discussed JCI testing method[103] or a slightly different version of this test. The results obtained by Sato et al.[100] on a 0.25 water/cement ratio concrete made with an ordinary portland cement containing 10% silica fume are presented in Fig. 40. The restraint being only partial, it is not surprising that

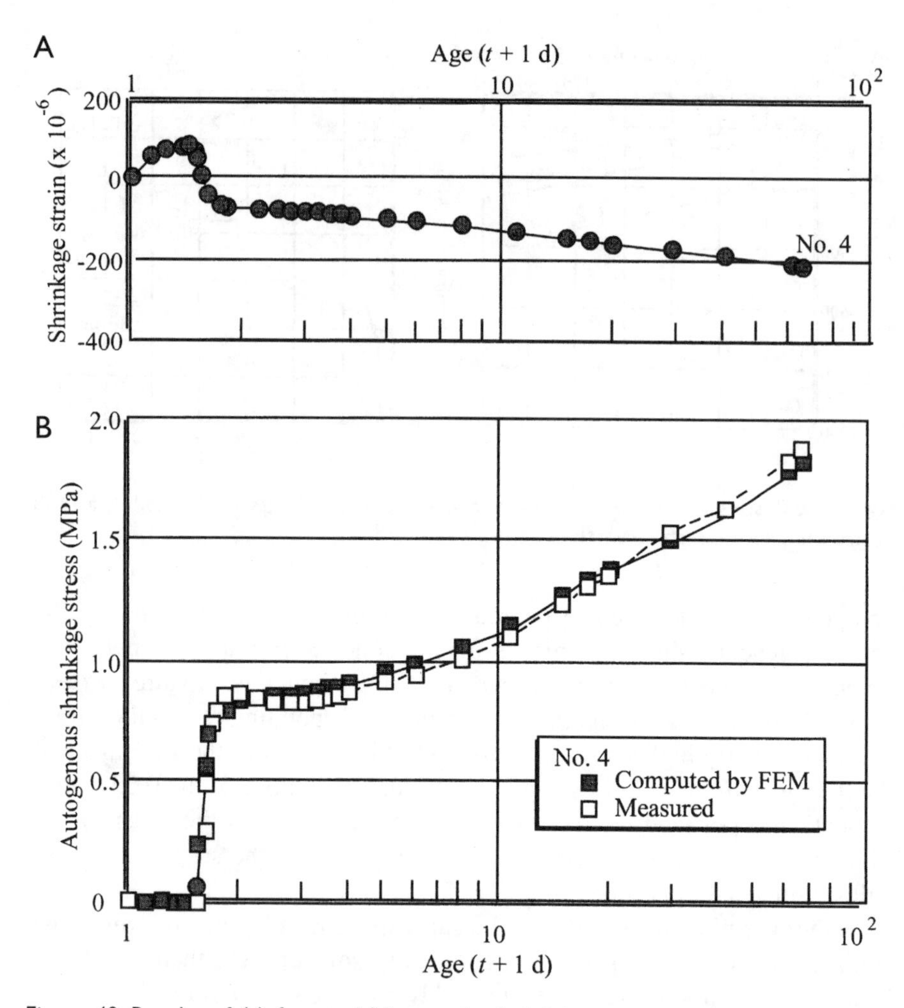

Figure 40. Results of (a) free and (b) restrained shrinkage tests performed on a 0.25 w/c concrete mixture in accordance with the JCI test procedure.[100]

the intensity of the shrinkage-induced stress is a little smaller than what was observed in the previous studies where a 100% restraint was achieved. Nevertheless, the value recorded within the first 24 h is quite significant, reaching almost 1.0 MPa at what is assumed to be approximately the end of the setting period. It must be stressed that with such a test configuration, the younger the concrete, the higher the effective degree of restraint, since it depends on the stiffness ratio between steel and concrete. Besides, allowing for the semi-log plots and the use of silica fume, it can be noted in Fig. 40

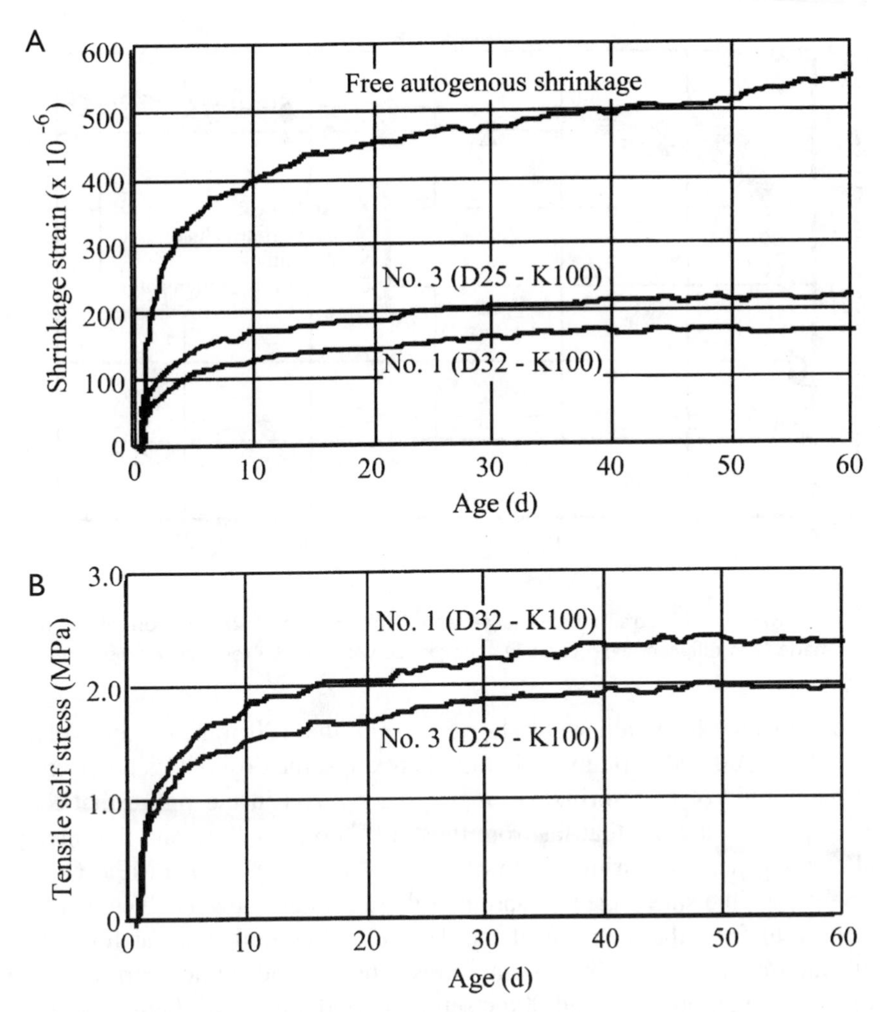

Figure 41. Results of (a) free and (b) restrained shrinkage tests performed on a 0.24 w/c concrete mixture using a modified JCI test procedure.[102]

that the shape of the stress curve is consistent with the behavior observed in fully restrained tests.

With a JCI test procedure slightly modified for different sample dimensions, Ohno and Nakagawa[102] performed a series of restrained shrinkage tests on a 0.24 OPC concrete. The restrained shrinkage results posted in Fig. 41 [Figs. 6(a) and 7(a) from Ohno and Nakagawa[102]] were obtained with JCI-like specimens, except for the overall length and the central

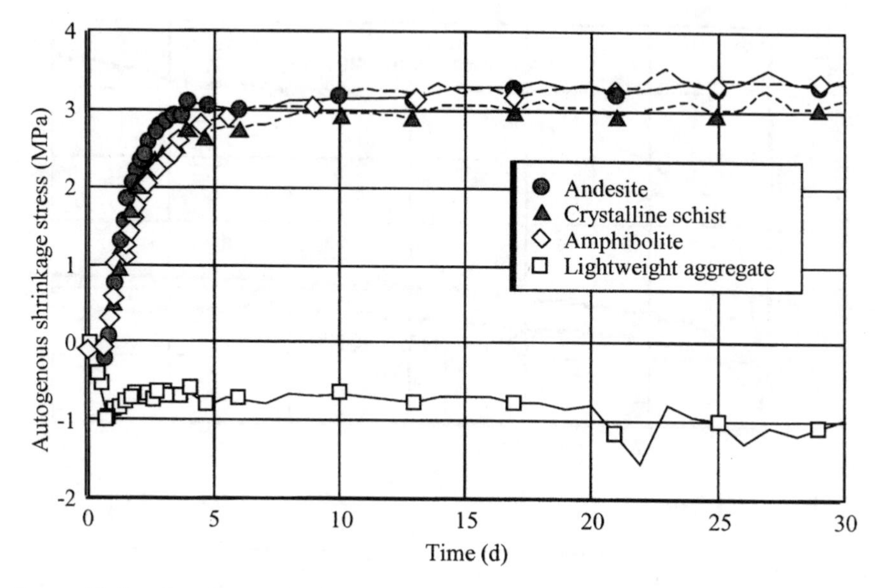

Figure 42. Results of restrained shrinkage tests performed on 0.28 w/c concrete mixtures made with different aggregates in accordance with the JCI test procedure.[101]

unbonded length, which were shorter (1000 and 100 mm, respectively, instead of 1500 and 300 mm). Though there are some differences regarding the magnitude of free shrinkage, it may be noticed in the figure that the self-stress data are not that far from those of Sato et al.[100] Again, the internal stress generated during the first day is of the order of 1.0 MPa. Conversely, both the shrinkage rate and the self-stress rate show a behavior fairly different from that of most of the data results compiled so far for very low water/cement ratio concrete mixtures. There is indeed no sharp transition occurring around the end of the setting period, the curves being characterized by a monotonically decreasing rate, just like drying shrinkage and creep. A relatively smooth transition was also observed by Matsuhita and Tsuruta,[101] who used the JCI test method to study the early age response of 0.28 water/cement ratio concrete mixtures made with different aggregates. Some of their results are displayed in Fig. 42. It can be seen that the stress still increases steadily well beyond the first day. Unfortunately, the corresponding free shrinkage data were not presented, making the analysis somewhat difficult.

From the above analysis, it is obvious that self-desiccation shrinkage can induce substantial tensile stresses in low water/cement ratio concrete mixtures at early ages. While cracking was not reached within the monitoring period in most of the reported tests, the measured induced stresses are not far from competing with the gain in tensile strength of concrete. It must be emphasized that the sole effect of autogeneous shrinkage in isothermal conditions is considered here, whereas in real conditions, potentially aggravating factors will add up, such as early age thermal stresses. Moreover, it should be stressed that in non-isothermal conditions, as encountered in many if not most real structures, the autogeneous shrinkage per se might be influenced by the thermal history, irrespective of the maturity.[105] In view of all this, it appears that the viscoelastic capacity of concrete could play an important role in the early age cracking outcome.

Stress Relaxation due to Creep

It has been stated previously that creep affects the magnitude of shrinkage right from the beginning at the microstructural level. The local stresses induced by self-desiccation promote viscoelastic compressive strains in the solid skeleton, the resulting apparent free contraction being significantly larger than what one would get in a purely elastic material.

At the meso and macro levels, according to the nature of the restraints, creep has a further influence on the development of shrinkage-induced stresses and ultimately on the occurrence of cracking. The internal state of stress caused by different forms of restraint to the autogeneous volume changes is obviously altered by the viscoelastic characteristics of the material. Nevertheless, determining the extent of such an effect is for many reasons a quite difficult task, especially in the very first hours in the life of concrete. While the creep of well-cured hardened concrete has been studied rather extensively over the years, there is not much data pertaining to the viscoelastic behavior when loading is applied within the first few days after casting. In fact, very few investigations have been dedicated in the first place to the elastic behavior of cement systems after a few hours of hydration.

Needless to say, from an experimental standpoint it is an extremely complex issue. Questions may arise also whether the response of the material to an imposed stress when undergoing the transition from a semi-liquid to solid state is truly viscoelastic.[106] Besides, restrained shrinkage causes tensile stresses and reliable tensile creep data are quite scarce. Most of the reported studies have been devoted primarily to the compressive behavior.

Nevertheless, along with the increasing attention paid to early behavior of concrete, the early age creep behavior has gradually begun to draw a little more interest. The tests can be classified into three categories: classical creep tests, relaxation tests, and discretized restrained shrinkage tests. In all cases, the creep strain is obtained with the assumption that the principle of strain superposition is applicable. Such an assumption is most likely inexact, but it is convenient from an engineering point of view and it has proven to give fairly acceptable predictions in the case of hardened concrete. It should be added that the interdependence between shrinkage and creep is overwhelmingly difficult to assess, especially at early age, since the phenomena cannot be totally isolated experimentally. As explained previously, the material undergoes creep at the microscale level right from the moment it starts to shrink, the latter being the results of stresses induced in the solid skeleton. At the meso level (or material level), the measurement of pure creep behavior would require the use of specimens that have reached hygrometric equilibrium and where hydration has stopped. These requirements are indeed impossible to fulfill at early age without altering the material to some extent.

In most studies concerned with the early viscoelastic behavior of concrete, compressive creep tests were performed. Such experiments were run by Okamoto[106] at two different stress levels (0.30 and 0.60) on very young ordinary concrete (water/cement ratio = 0.50), where the age upon loading varied between 2 and 7 h. It was observed that both the instantaneous and the viscoelastic responses evolved quite significantly with the age at loading. Within the period of time investigated, the elastic modulus measured at a load level of 0.35 increased from approximately 1 to 60 MPa. Unfortunately, the loads were sustained for only 30 min and no further information regarding the viscoelastic behavior is available up to and beyond the final setting time. The creep coefficients ($\phi = \varepsilon_c/\varepsilon_e$) at 30 min were on the order of 0.1 and a slight diminution of ϕ_{30} was observed with an increase in the age upon loading. Results from other investigations where the earliest loading was applied between 12 and 18 h after casting[107–109] show that the creep coefficient still varies considerably from that time up to a certain period after the final set. Figure 43 summarizes the results obtained by Persson[109] with high-strength silica fume concrete mixtures prepared at water/cement ratios ranging between 0.25 and 0.38. As can be seen in the figure, the compliance [$J(t,t')$], which is basically the sum of the unit elastic strain (ε_e/σ) and the unit creep strain (ε_c/σ), is extremely high for early-age concrete in comparison with the results obtained for mature concrete. The differences due to maturity are observed to be not only restricted to a limited duration

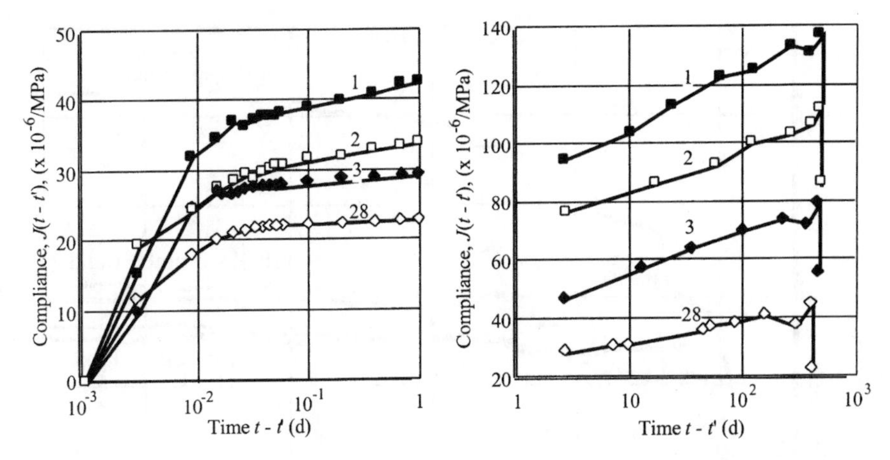

Figure 43. Early and long-term compliance functions of a 0.31 w/c concrete loaded in compression at different ages.[109]

after loading, as the creep rate was still higher after one year for early loaded concrete.

Some early age concrete creep tests have also been carried out in tension.[110–112] The earliest tests in these studies were started at 24 h. Globally, the reported specific creep data are of the same order of magnitude as the values recorded elsewhere in compression for similar concretes loaded at the same age. Hauggaard and Damkilde[110] have measured total basic creep strains on concrete exceeding 200×10^{-6} at a stress level of 0.8 (age upon loading was 24 h), as shown in Fig. 44. The magnitude of basic tensile creep is thus comparable to that of autogeneous shrinkage, which reinforces the idea according to which creep can significantly relieve self-desiccation stresses.

Strictly speaking, the problem of restrained shrinkage may be more a matter of relaxation than creep, since the stresses are caused by imposed strains. Unfortunately, data on relaxation are even rarer than those on tensile creep, given the additional degree of difficulty in performing the tests. In a recent study,[112] a series of early age creep and relaxation experiments in tension were conducted on cement pastes. Results of relaxation tests performed on a microsilica cement paste with a water/cement ratio of 0.25 are displayed in Fig. 45, along with the relaxation curves calculated on the basis of the corresponding creep results. The graph shows that just like the case for creep, the younger the material is, the higher the relaxation poten-

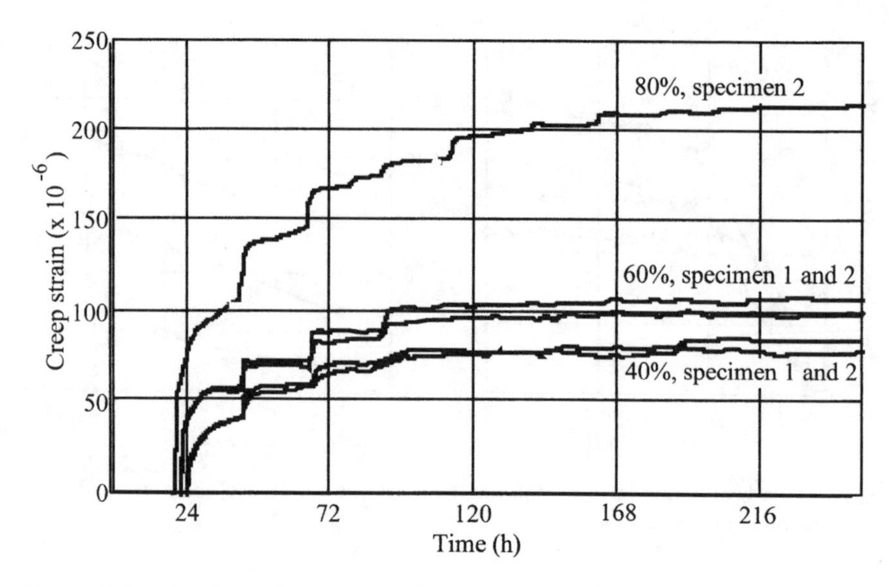

Figure 44. Results of tensile creep experiments performed at 1 day.[110]

tial is, provided that the comparison is made in reference to the relative stress. It can be readily observed that the initial induced stress is reduced by 50% and more within hours. Also, these results demonstrate that, at least for cement pastes, the relation between compliance $[J(t,t')]$ and relaxation $[R(t,t')]$ can quite satisfactorily be expressed as:

$$J(t,t') = 1 / R(t,t')$$
$$J(t,t') \times R(t,t') = 1 \qquad (29)$$

Although this equation is not exact for aging materials,[113] the predicted curves show relatively good agreement with the relaxation data, even for very early loading. The author in the case of a silica fume paste also observed similar agreement.

The type of tests that best simulate how concrete is self-stressed due to restricted early volume changes occurring during hydration is obviously a restrained shrinkage test. The only information one usually gets from such a test is however limited to the stress vs. time curve, the time to cracking, and perhaps, if the restraint is not perfect, the resulting contraction of the whole system (specimen + restraining device). Nevertheless, as explained previously, the different components of deformation, like creep, can be

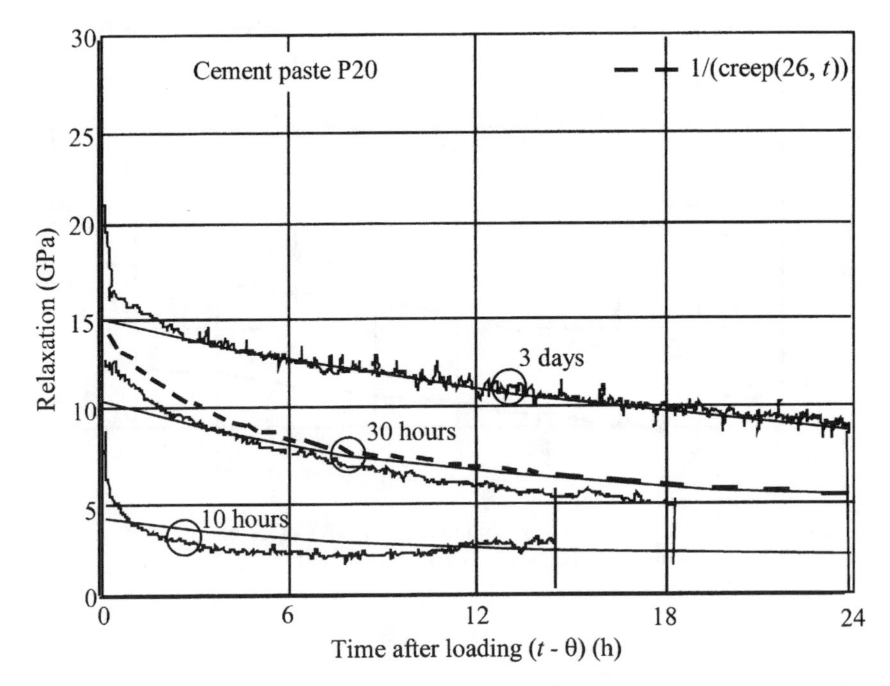

Figure 45. Results of relaxation tests in tension performed on a 0.25 w/c microsilica cement paste at different ages presented along with prediction curves calculated from tensile creep results.[112]

evaluated by using a discretized procedure of testing. Early age creep results obtained from experiments of the kind performed with DRS testing devices were reported in recent investigations[80,84] and showed similar trends. Figures 46 and 47 present the cumulative creep curves of portland cement and silica fume concrete mixtures, respectively, prepared at water/cement ratios ranging from 0.25 to 0.45. The corresponding free shrinkage and stress data were referred to previously and are summarized in Figs. 37 and 38. It appears clear that for all mixtures the creep rate and the shrinkage rate are closely related. In terms of magnitude, free autogeneous shrinkage is almost fully offset by cumulative creep. An important feature of these results is the observed influence of the water/cement ratio on the viscoelastic response. While conventional creep tests usually tell that creep decreases with the water/cement ratio, a quite different behavior was observed through these DRS tests. It is the viscoelastic response to the shrinkage induced stress that is observed, which does not necessarily give

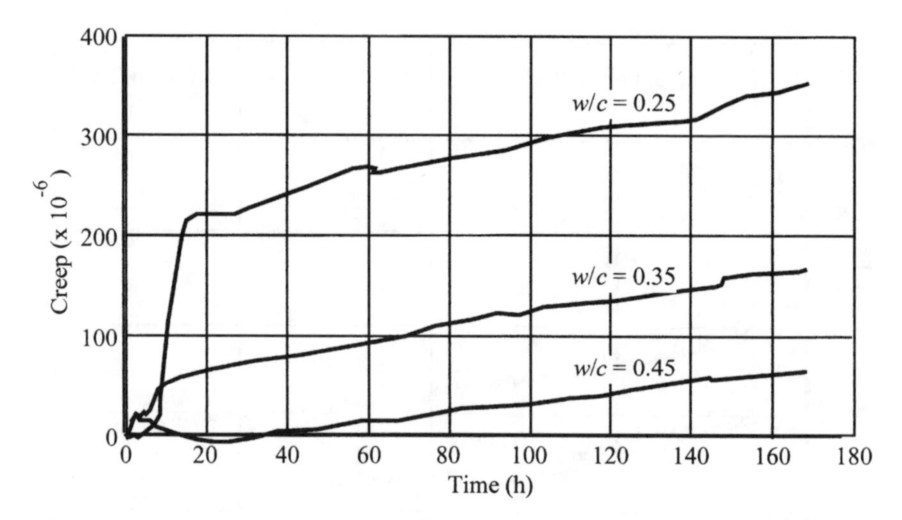

Figure 46. Cumulative creep curves obtained from DRS tests on ordinary portland cement concrete mixtures with different w/c ratios.[80]

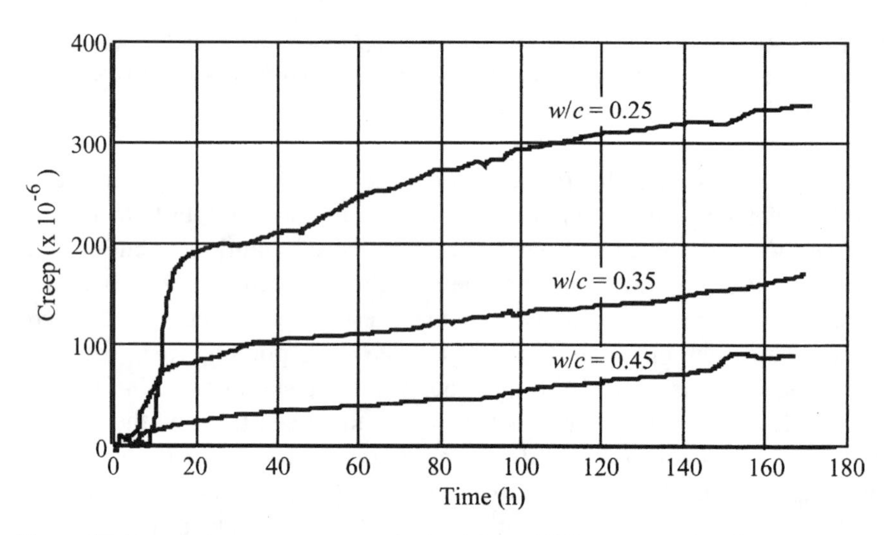

Figure 47. Cumulative creep curves obtained from DRS tests on silica fume concrete mixtures with different w/c ratios.[80]

an idea of the full creep potential of the material, but might reflect specific early age tendencies that are different from those encountered in more mature systems.[84] It stresses the paramount importance of investigating in greater depth the evolution of the properties that controls the behavior of concrete right from the time it starts to set.

From the previous discussion, it must be emphasized that creep relaxation plays a major role in relieving stresses induced by restrained shrinkage in concrete at early ages and that all the knowledge related to the creep of mature concrete that has been accumulated through the years is not necessarily applicable in that regard. For instance, the available models are generally not suitable for the prediction of the viscoelastic behavior at such an early stage of hydration.

Early Age Cracking Occurrence

It has been seen that the early age stress development in concrete of depends on different factors and phenomena. Once the level of stress is estimated, by using for instance Eq. 25, the risk of cracking can be evaluated by comparison with the stress the material can withstand at that particular time. The main approaches for that matter are based on the tensile strength or on fracture mechanics concepts.

Although relatively accurate prediction of cracking has been obtained in the case of restrained drying shrinkage of mature concrete[114] by using tensile strength as a criterion, it is not obvious whether it can be used satisfactorily in predicting early age cracking. In the latter case, the development of stress is competing with the gain in strength. Reliable data on the early strength development are thus required, especially for a few hours following the initial setting. Additionally, the stress and strain history of the material might affect its actual strength in such a way that it cannot be compared to the value obtained in a static test performed on a virgin companion specimen, because of some form of accumulated damage. It should be recalled that the time-dependent strength of concrete is less than that determined in a static test, a phenomenon commonly referred to as time failure.[115] The relevance of time failure in early age cracking problems is uncertain, but the fact that strength development might be affected in some way by the state of stress existing during hydration needs to be investigated. There is actually no data available on this issue.

A fracture mechanics criterion has been used at ACBM (USA) to predict shrinkage cracking in concrete.[93] Theoretically, it provides a more accurate method to account for stable crack growth, aggregate interlock, specimen size, and geometry. The approach that has been followed is based on two concepts, the *R* curve and the *G* curve, which respectively correspond to the curve of the material resistance to crack development and the curve of the energy rate applied to propagate the crack. These relations can be expressed as[93]:

$$R(\Delta a) = \beta_2 \psi (\Delta a)^{d^2} \tag{30}$$

$$G(\Delta a) = \sigma^2 \pi a F'(a/b) \tag{31}$$

where Δa is the length of crack growth, β_2, ψ, and d^2 are parameters that can be determined knowing experimentally measured material properties (K_{IC}, $CTOD_C$) and information relative to the specimen size and geometry, and $F'(a/b)$ is a geometric function that can be obtained from fracture mechanics handbooks. According to this approach, the stress level in the specimen will cause failure if the crack driving energy [$G(\Delta a)$], which depends on the stress level that develops due to the restraint, overcomes the fracture resistance of the material [$R(\Delta a)$]. While it seems to have provided an interesting correlation with experimental results obtained from the so-called ring tests (which mainly address the issue of drying shrinkage cracking of hardened concrete), no attempt appears to have been made toward an assessment in relation with early age cracking of concrete in the absence of moisture exchange. Again, the difficulty that arises is the lack of reliable data on the evolution of the relevant properties during the first hours and days of the life of concrete.

Whichever approach is used, further research is necessary to better understand and characterize the various phenomena involved in the hardening concrete, from the self-stressing driving forces to the characteristics of the material that shape its response to these driving forces. For instance, this could imply the investigation of rather complex (though maybe significant) issues, such as self-healing.

Concluding Remarks

With the recent development of new test procedures, engineers and scientists can now depend on a series of reliable tools to study the early age behavior of cement systems. A lot of work needs to be done on the viscoelastic behavior of concrete during the first days of hydration. Self-desiccation (and autogeneous shrinkage) clearly appears to be the driving force that initiates and controls the deformation of concrete. However, the early age behavior of a given concrete mixture cannot be predicted on the sole basis of free shrinkage measurements. As emphasized in the text, viscoelastic effects appear to have a very significant influence on the performance of the material.

Acknowledgments

The authors are grateful to the Natural Sciences and Engineering Research Council of Canada for its financial support for this project. The authors would also like to thank E.J. Garboczi and B. Zuber for their pertinent comments.

References

1. R.W. Burrows, *The Visible and Invisible Cracking of Concrete,* Monograph 11. American Concrete Institute, 1998.
2. D.B. MacDonald, P.D. Krauss, and E. Rogalla, "Early-age transverse deck cracking," *Concr. Int.,* **17** [5] 49–51 (1995).
3. R. Springenschmid, R. Breitenbucher, and M. Mangold, "Thermal cracking in concrete at early ages"; pp. 137–144 in *Thermal Cracking in Concrete at Early Ages.* Edited by R. Springenschmid. E&FN Spon, London, 1994.
4. D. Mitchell, A.A. Khan, and W.D. Cook, "Early-age properties for thermal stress analyses during hydration"; pp 265–306 in *Materials Science of Concrete V.* Edited by J.P. Skalny and S. Mindess. American Ceramic Society, Westerville, Ohio, 1998.
5. H.F.W. Taylor, *Cement Chemistry,* 2nd edition. Thomas Telford, London, 1997.
6. V.S. Ramachandran, R.F. Feldman, and J.J. Beaudoin, *Concrete Science — Treatise on Current Research.* Heyden, London, 1981.
7. J.M. Gaidis and E.M. Gartner, "Hydration mechanisms – Part II"; pp. 9-40 in *Materials Science of Concrete II.* American Ceramic Society, Westerville, Ohio, 1991.
8. C. Vernet and G. Cadoret, "Suivi en continu de l'évolution chimique et mécanique des bétons à hautes performances pendant les premiers jours"; in *Comptes-rendus du Colloque Voies Nouvelles du Béton.* Cachan, France, 1991. (In French.)
9. P. Paulini, "Reaction Mechanisms of Concrete Admixtures," *Cem. Concr. Res.,* **20** [6] 910–918 (1990).
10. H. Le Chatelier, *Recherches expérimentales sur la constitution des mortiers hydrauliques.* Dunod, Paris, 1904. (In French.)
11. H. Le Chatelier, "Sur les changements de volume qui accompagnent le durcissement des ciments," *Bull. Société pour l'Encouragement de l'Industrie Nationale,* **5** [5] 54–57 (1900). (In French.)
12. P. Paulini, "A weighing method for cement hydration"; pp. 248–254 in *9th International Congress on the Chemistry of Cement,* Vol. IV. 1992.

13. P. Paulini, "Chemical shrinkage as indicator for hydraulic bond strength"; in *10th International Congress on the Chemistry of Cement,* Vol. II. 1998.
14. P. Paulini and N. Gratl, "Stiffness formation of early age concrete"; pp. 63–70 in RILEM, Proceedings vol. 25. E&FN Spon, London, 1995.
15. D. Damidot, A. Nonat, and P. Barret, "Kinetics of tricalcium silicate hydration in diluted suspension by microcalorimetric measurements," *J. Am. Ceram. Soc.,* **73** [11] 3319–3322 (1990).
16. H. Justnes, E.J. Sellevold, B. Reyniers, D. Van Loo, A. Van Gemert, F. Verboven, and D. Van Gemert, "The influence of cement characteristics on chemical shrinkage"; Pp. 71–80 in *Autogeneous Shrinkage of Concrete.* E&FN Spon, London, 1999.
17. E. Tazawa, S. Miyazawa, and T. Kasai, "Chemical shrinkage and autogenous of hydrating cement paste," *Cem. Concr. Res.,* **25** [2] 288–292 (1995).
18. T.C. Powers, "Absorption of water by portland cement paste during the hardening process," *Ind. Eng. Chem.,* **27** [7] 790–794 (1935).
19. H. Justnes, A. Van Gemert, F. Verboven, and E. Sellevold, "Total and external chemical shrinkage of low w/c ratio cement pastes," *Adv. Cem. Res.,* **8** [31] 121–126 (1996).
20. S. Boivin, P. Acker, S. Rigaud, and B. Clavaud, "Experimental assessment of chemical shrinkage of hydrating cement pastes"; pp. 81–92 in *Autogeneous Shrinkage of Concrete.* E&FN Spon, London, 1999.
21. Japan Concrete Institute, *Report of the Technical Committee on Autogeneous Shrinkage of Concrete, in Autogeneous Shrinkage of Concrete.* E&FN Spon, London, 1999. Pp. 3–68.
22. M. Geiker, "Studies of Portland cement hydration: Measurements of chemical shrinkage and a systematic evaluation of hydration curves by means of the dispersion model," Ph.D. Thesis, Technical University of Denmark, Lyngby, Denmark, 1983.
23. M. Buil, "Contribution à l'étude du retrait de la pâte de ciment durcissante," Ph.D. Thesis, Rapport de recherche 92, École nationale des ponts et chaussées, Paris, 1979. (In French.)
24. M. Rey, *Nouvelle méthode de mesure de l'hydratation des liants hydrauliques,* Technical Publication 31. CERILH, Paris, 1979. (In French.)
25. R. Gagné, I. Aouad, J. Shen, and C. Poulin, "Development of a new experimental technique for the study of the autogeneous shrinkage of cement paste," *Mater. Struct.,* **32**, 635–642 (1999).
26. T. Knudsen, *Effect of Sample Size on the Hydration of Water-Cured Cement Pastes.* Institute of Mineral Industry, Technical University of Denmark, Lyngby, Denmark, 1987. Pp. 1–4.
27. B. Persson, "Chemical shrinkage and self-desiccation in portland cement based mortars," *Concr. Sci. Eng.,* **1**, 228–237 (1999).
28. H. Justnes, E. Sellevold, B. Reyniers, D. Van Loo, A. Van Gemert, F. Verboven, and D. Van Germert, "Chemical shrinkage of cementitious pastes with mineral additives"; pp. 73–84 in *Proceedings of the Second International Research Seminar on Self-Desiccation and Its Importance in Concrete Technology.* Edited by B. Persson and G. Fagerlund, Lund. 1999.
29. D. Bentz, O.M. Jensen, K.K. Hansen, J.F. Olesen, H. Stang, and C.J. Haecker, "Influence of cement particle size distribution on early-age autogeneous strains and stresses in cement-based materials," submitted to *J. Am. Ceram. Soc.*

30. J.M. Pommersheim, "Effect of particle size distribution on hydration kinetics"; pp. 301–306 in *MRS Symposium Proceedings,* Vol. 85. Materials Research Society, 1987.
31. T. Knudsen, "The dispersion model for the hydration of Portland cement – Part 1: General concepts," *Cem. Concr. Res.,* **14**, 622–630 (1984).
32. H. Justnes, B. Ardoullie, E. Hendrix, E.J. Sellevold, and D. Van Gemert, *The Chemical Shrinkage of Pozzolanic Reaction Products.* ACI Special Publication SP 179-11, 1998. Pp. 191–205.
33. S. Boivin, "Retrait au jeune age du béton: Dévelopement d'une méthode expérimentale et contribution à l'analyse physique du retrait endogène," Ph.D. Thesis, École Nationale des Ponts et Chaussées, Paris, 1999. (In French.)
34. J.H. Taplin, "A method for following the hydration reaction in Portland cement paste," Aust. J. Appl. Sci., **10**, 329–345 (1959).
35. D. Bentz, *CEMHYD3D: A Three-Dimensional Cement Hydration and Microstructure Development Modeling Package – Version 2.9, NISTIR.* National Institute of Standards and Technology (in press).
36. I. Jawed and J.P. Skalny, "Alkalies in cement: A Review — Part II : Effects of alkalies on hydration and performance of portland cement," *Cem. Concr. Res.,* **8** [1] 37–52 (1978).
37. A. Boumiz, "Etude comparée des évolutions mécaniques et chimiques des pâtes de ciment et mortiers à très jeune âge — Développement des techniques acoustiques," Ph.D. Thesis, Université Paris VII, 1995. (In French.)
38. A. Nonat, "Interaction between chemical evolution (hydration) and physical evolution (setting) in the case of tricalcium silicate," *Mater. Struct.,* **27**, 187–195. (1994),
39. S.P. Jiang, J.C. Mutin, and A. Nonat, "Studies on mechanism and physico-chemical parameters at the origin of cement setting: I. The fundamental processes involved during the cement setting," *Cem. Concr. Res.,* **25** [4] 779–789 (1995).
40. E. Sellevold, O. Bjontegaard, H. Justnes, and P.A. Dahl, "High Performance Concrete: Early Volume Change and Cracking Tendency"; pp. 229–236 in *Thermal Cracking in Concrete at Early Ages.* Edited by R. Springenschmid. E&FN Spon, London, 1994.
41. A. Radocea, *Water Pressure in Fresh and Young Cement Paste,* Publication No. 9. Nordic Concrete Research, 1990. Pp. 145–153.
42. J. Baron, "Les retraits de la pâte de ciment"; pp. 485–501 in *Le béton hydraulique.* Presses de l'Ecole Nationale des Ponts et Chaussées, 1982. (In French.)
43. T. Takashi, H. Nakata, K. Yoshida, and S. Goto, "Autogenous shrinkage of cement paste during hydration"; in *10th International Congress on the Chemistry of Cement, Vol. II.* 1997.
44. J.J. Beaudoin, V.S. Ramachandran, and R.F. Feldman, "Identification of hydration reactions through stresses induced by volume change — Part I: C_3A Systems," *Cem. Concr. Res.,* **21** [5] 809–818 (1991).
45. J.J. Beaudoin, V.S. Ramachandran, and R.F. Feldman, "Identification of hydration reactions through stresses induced by volume change — Part II: C_4AF Systems," *Cem. Concr. Res.,* **22** [1] 27–34 (1992).
46. C. Hua, P. Acker, and A. Ehrlacher, "Retrait d'autodessiccation du ciment: Analyse et modélisation," *Bulletin liaison Laboratoire des Ponts et Chaussées,* **196**, 79–89 (1995). (In French.)

47. T.C. Powers and T.L. Brownyard, "Studies of the physical properties of hardened cement paste — Part 9: General summary of findings on the properties of hardened portland cement paste," *Proc. Am. Concr. Inst.,* **43**, 971–1021 (1947).
48. T.C. Powers, "A discussion of cement hydration in relation to the curing of concrete," *Proc. Highway Res. Board,* **27**, 178–188 (1947).
49. G.R. Gause and J.R. Tucker Jr., "Method for determining the moisture condition in hardened concrete," *J. Res. National Bureau of Standards,* **25**, 403–415 (1940).
50. L.E. Copeland and R.H. Bragg, "Self-desiccation in portland cement pastes," *ASTM Bull.,* No. 204, 1–11 (1955).
51. V. Baroghel-Bouny, J. Godin, and J. Gawsewitch, "Microstructure and moisture properties of high performance concrete"; pp. 451–461 in *Fourth International Symposium on the Utilization of High Strength/High Performance Concrete.* 1996.
52. O.M. Jensen, "Thermodynamic limitation of self-desiccation," *Cem. Concr. Res.,* **25** [1] 157–164 (1995).
53. R.G. Patel, D.C. Killoh, L.J. Parrott, and W.A. Gutteridge, "Influence of curing at different relative humidities upon compound reactions and porosity in portland cement paste," *Mater. Struct.,* **21**, 192–203 (1988).
54. C. Hua, P. Acker, and A. Ehrlacher, "Analyses and models of the autogenous shrinkage of hardening cement paste — Part I: Modelling at macroscopic scale," *Cem. Concr. Res.,* **25** [10] 1457–1468 (1995).
55. C. Hua, A. Ehrlacher, and P. Acker, "Analyses and models of the autogenous shrinkage of hardening cement paste — Part II: Modelling at scale of hydrating grains," *Cem. Concr. Res.,* **27** [2] 245–258 (1997).
56. F.F. Radjy, "Swelling and Shrinkage," Laboratoriet for Bygningsmaterialer, Lecture Notes, Technical University of Denmark, 1971.
57. J.F. Young, Z.P. Bazant, J.W. Dougill, F.H. Wittmann, and T. Tsubaki, "Physical Mechanisms and their Mathematical Description, Creep and Shrinkage of Concrete: Mathematical Modeling"; in *Fourth RILEM International Symposium.* Edited by Z.P. Bazant. Evanston, Illinois, 1986. Pp. 44–78.
58. D.H. Bangham, "The Gibbs adsorption equation and adsorption on solids," *Trans. Faraday Soc.,* **83**, 805–811 (1937).
59. D.H. Bangham and F.A.P. Maggs, "The strength and elastic constants of coals in relation to their ultra-fine structure"; pp. 118–130 in *Proceedings of the Conference on Ultra-Fine Structure of Coals and Coke.* British Coal Utilization Research Association, 1943.
60. T.C. Powers, "Mechanisms of shrinkage and reversible creep of hardened cement paste"; pp. 319–344 in *International Conference on the Structure of Concrete.* Cement and Concrete Association, 1965.
61. C.F. Ferraris and F.H. Wittmann, "Shrinkage mechanisms of hardened cement paste," *Cem. Concr. Res.,* **17**, 453–464 (1987).
62. B.V. Derjaguin and N.V. Churaev, "Structural component of disjoining pressure," *J. Colloid Interface Sci.,* **49**, 249–255 (1974).
63. Z.P. Bazant, "Thermodynamics of interacting continua with surfaces and creep analysis of concrete structures," *Nuclear Eng. Res.,* **20**, 477–505 (1972).

64. F.H. Wittmann, "The structure of hardened cement paste: A basis for a better understanding of the material properties"; in *Proceedings of the Conference on Hydraulic Cement Pastes: Their structure and properties.* Sheffield, 1976.
65. D.P. Bentz, D.A. Quénard, V. Baroghel-Bouny, E.J. Garboczi, and H.M. Jennings, "Modelling drying shrinkage of cement paste and mortar — Part 1: Structural models from nanometers to millimeters," *Mater. Struct.,* **28**, 450–458(1995).
66. D.P. Bentz, E.J. Garboczi, C.J. Haecker, and O.J. Jensen, "Effects of cement particle size distribution on performance properties of portland cement-based materials," *Cem. Concr. Res.,* **29** [10] 1663–1671 (1999).
67. C. Hua, "Analyses et modélisations du retrait d'autodessiccation de la pâte de ciment durcissante," Ph.D. Thesis, École Nationale des Ponts et Chaussées, Paris, 1992. (In French.)
68. A.M. Neville, *Properties of Concrete,* 3rd ed. Pitman Publishing Limited, Marshfield . Massachusetts, and London, 1981.
69. Nachbaur et al., "Electrokinetic properties which control the coagulation of silicate cement suspensions during early-age hydration," *J. Colloid Interface Sci.,* **202** [2] 261–268 (1998).
70. T. Sato, T. Goto, and K. Sakai, "Mechanisms for reducing drying shrinkage of hardened cement by organic additives," *CAJ Rev.,* pp. 52–54, (1983).
71. T.A. Hammer, "Test methods for linear measurement of autogenous shrinkage before setting"; pp. 143–154 in *Autogeneous Shrinkage of Concrete.* E&FN Spon, London, 1999.
72. L. Barcelo, S. Boivin, S. Rigaud, P. Acker, B. Clavaud, and C. Boulay, "Linear vs volumetric autogeneous shrinkage measurement: Material behavior or experimental artefact"; pp. 109–126 in *Proceedings of the Second International Research Seminar on Self-Desiccation and Its importance in Concrete Technology.* Edited by B. Persson and G. Fagerlund. Lund, 1999.
73. H. Justnes, A. Van Gemert, F. Verboven, E. Sellevold, and D. Van Gemert, "Influence of measuring method on bleeding and chemical shrinkage values of cement pastes"; in *10th International Congress on the Chemistry of Cement, Vol. II.* 1997.
74. N. Setter and D.M. Roy, "Mechanical features of chemical shrinkage of cement paste," *Cem. Concr. Res.,* **8** [5] 623–634 (1978).
75. O.M. Jensen and P.F. Hansen, "Autogeneous relative humidity change in silica fume modified cement paste," *Adv. Cem. Res.,* **7** [25] 33–38 (1995).
76. E. Tazawa and S. Miyazawa, "Influence of cement and admixture on autogenous shrinkage of cement paste," *Cem. Concr. Res.,* **25** [2] 281–287 (1995).
77. E. Tazawa and S. Miyazawa, "Influence of constituents and composition on autogenous shrinkage of cementitious materials," *Mag. Concr. Res.,* **49** [178] 15–22 (1997).
78. D.P. Bentz and C.J. Haecker, "An argument for using coarse cements in high-performance concrete," *Cem. Concr. Res.,* **29** [4] 615–618 (1999).
79. O.M. Jensen and P.F. Hansen, "Autogenous deformation and change of the relative humidity in silica fume-modified cement paste," *ACI Mater. J.,* **93** [6] 539–543 (1996).
80. G. Toma, "Comportement au jeune âge des bétons," Ph.D. Thesis, Department of Civil Engineering, Laval University, 2000. (In French.)

81. S. Igarashi, A. Bentur, and K. Kovler, "Autogenous shrinkage and induced restraining stresses in high-strength concretes," *Cem. Concr. Res.,* (in press).
82. R.M. Edmeades, A.N. James, and J. Wheeler, "The formation of ettringite on aluminium surfaces," *J. Applied Chem.,* **16**, 361–368 (1966).
83. G. Toma, M. Pigeon, J. Marchand, B. Bissonnette, and A. Delagrave, "Early age autogenous restrained shrinkage: Stress build up and relaxation," submitted to *Concr. Sci. Eng.*
84. S. Igarashi, A. Bentur, and K. Kovler, "Stresses and Creep Relaxation Induced in Restrained Autogenous Shrinkage of High Strength Pastes and Concretes," *Adv. Cem. Res.,* **11** [4] 169–177 (1999).
85. G.D. de Haas, P.C. Kreijger, E.M.M.G. Niël, J.C. Slagter, H.N. Stein, E.M. Theissing, and M. van Wallendael, "The shrinkage of hardening cement paste and mortar," *Cem. Concr. Res.,* **5** [4] 295–320 (1975).
86. A. Bentur, S. Igarashi, and K. Kovler, "Internal curing of high strength concrete to prevent autogenous shrinkage and internal stresses by use of wet lightweight aggregates," *ACI Mater. J.,* (in press).
87. T.A. Hammer, O. Bjontegaard, and E. Sellevold, *Cracking Tendency of High Strength Lightweight Concrete at Early Ages.* 1999.
88. K. Van Breugel and J. de Vries, "Mixture optimization of low water/cement ratio high strength concretes in view of reduction of autogenous shrinkage"; pp. 365–375 in *International Symposium on High-Performance and Reactive Powder Concretes,* Vol 1. Sherbrooke, Canada, 1998.
89. D.P. Bentz and K.A. Snyder, "Protected paste volume in concrete: Extension to internal curing using saturated lightweight fine aggregate," *Cem. Concr. Res.,* **29** [11] 1863–1867 (1999).
90. S. Weber and H.W. Reinhardt, "A new generation of high performance concrete: Concrete with autogenous curing," *Adv. Cem. Based Mater.,* **6** [2] 59–68.(1997),
91. ACI Committee 207, "Mass concrete for dams and other massive concrete structures," *ACI Mater. J.,* **67** [4] 273–309 (1970).
92. M.E. Anderson, "Design and construction of concrete structures using temperature and stress calculations to evaluate early-age thermal effects,"; pp. 191–264 in *Materials Science of Concrete V.* American Ceramic Society, Westerville, Ohio, 1999.
93. J. Weiss, Wei Yang, and S.P. Shah, *Factors Influencing Durability and Early-Age Cracking in High-Strength Concrete Structures,* ACI Special Publication. American Concrete Institute, (in press).
94. R. Springenschmid, R. Breitenbücher, and M. Mangold, "Development of the cracking frame and the temperature-stress testing machine"; pp. 137–144 in *Thermal Cracking in Concrete at Early Ages.* Edited by R. Springenschmid. E&FN Spon, London, 1994.
95. A.M. Paillère, M. Buil, and J.J. Serrano, "Effect of fiber addition on the autogenous shrinkage of silica fume concrete," *ACI Mater. J.,* **86** [2] 139–144 (1989).
96. R. Bloom and A. Bentur, "Free and restrained shrinkage of normal and high-strength concretes," *ACI Mater. J.,* **92** [2] 211–217 (1995).
97. K. Kovler, "Testing system for determining the mechanical behaviour of early-age concrete under restrained and free uniaxial shrinkage," *Mater. Struct.,* **27** [170] 324–330 (1994).

98. M. Pigeon, G. Toma, A. Delagrave, B. Bissonnette, J. Marchand, and J.-C. Prince, "Equipment for the analysis of the behavior of concrete under restrained shrinkage at early ages," accepted by *Mag. Concr. Res.*
99. Ø. Bjøntegaard, "Thermal dilation and autogenous deformation as driving forces to self-induced stresses in high performance concrete," Ph.D. Thesis, NTNU Trondheim (Norway), 1999.
100. R. Sato, M. Xu, and Y. Yang, "Stresses due to autogenous shrinkage in high-strength concrete and its prediction"; pp. 327–338 in *Autogeneous Shrinkage of Concrete.* E&FN Spon, London, 1998.
101. H. Matsuhita and H. Tsuruta, "The influences of quality of coarse aggregate on the autogenous shrinkage stress in high-fluidity concrete"; pp. 339–350 in *Autogeneous Shrinkage of Concrete.* E&FN Spon, London, 1998.
102. Y. Ohno and T. Nakagawa, "Research of test method for autogenous shrinkage stress in concrete," pp. 351–358 in *Autogeneous Shrinkage of Concrete.* E&FN Spon, London, 1998.
103. Japan Concrete Institute JCI Research Report of Autogenous Shrinkage. 1996. Pp. 199–201.
104. S.P. Shah, C. Ouyang, S. Marikunte, W. Yang, and E. Becq-Giraudon, A method to predict shrinkage cracking of concrete, *ACI Mater. J.,* **95** [2] 339–346 (1998).
105. Ø. Bjøntegaard and E.J. Sellevold, Thermal dilation – Autogenous shrinkage: How to separate?"; pp. 233–244 in *Autogeneous Shrinkage of Concrete.* E&FN Spon, London, 1998.
106. H. Okamoto, "A study on creep properties of concrete at very early age," *Japan Cong. Mater. Res. Proc.,* **30**, 171–174 (1987).
107. D.M. Smith and M.I. Hammons, "Creep of mass concrete at early ages," *J. Mater. Civ. Eng.,* **5** [3] 411–417 (1993).
108. G. De Schutter and L. Taerwe, "Towards a more fundamental non-linear basic creep model for early-age concrete," *Mag. Concr. Res.,* **49** [180] 195–200 (1997).
109. B. Persson, "Basic Deformations of High-Performance Concrete"; pp. 59–74 in *Nordic Concrete Research Publication No. 20.* 1997.
110. A.B. Hauggaard and L. Damkilde, "Non-linearities in tensile creep of concrete at early ages"; pp. 16–28 in *Nordic Concrete Research Publication No. 20.* 1997.
111. B. Bissonnette and M. Pigeon, "Tensile creep at early ages of ordinary, silica fume and fiber reinforced concretes," *Cem. Concr. Res.,* **25** [5] 1075–1085 (1995).
112. B.F. Dela, "Eigenstresses in hardening concrete," Ph.D. Thesis, Technical University of Denmark, Lyngby, Denmark, 1999.
113. Z.P. Bazant, J. Dougill, C. Huet, T. Tsubaki, and F. Wittmann, "Material models for structural creep analysis, Creep and Shrinkage of Concrete: Mathematical Modeling,"; pp. 79–232 in *Preprints of the Fourth RILEM International Symposium.* Evanston, Illinois, 1986.
114. R.I. Gilbert, "Shrinkage cracking in fully restrained concrete members," *ACI Struct. J.,* **89** [2] 141–149 (1992).
115. M.A. Al-Kubaisy and A.G. Young, "Failure of concrete under sustained tension," *Mag. Concr. Res.,* **27** [92] 171–178 (1975).

Appendix: Chemical Shrinkage Calculations

Calculations of Justnes[11]

Gypsum

$$CaSO_4, 0.5H_2O + 1.5H_2O \rightarrow CaSO_4, 2H_2O$$

	$CaSO_4, 0.5H_2O$	$1.5H_2O$	$CaSO_4, 2H_2O$
Mass (g)	1	0.19	1.19
Molar weight (g/mol)	145.5	18.02	172.17
Number of moles (mmol)	6.89	10.33	6.89
Density (g/mL)	2.74	1.00	2.32
Volume (mL)	0.365	0.186	0.511

The volume change is then –0.040 mL per gram of gypsum.

C_3A

$$C_3A + 3C\bar{S}H_2 + 26H \rightarrow C_6A\bar{S}_3H_{32}$$

	C_3A	$3C\bar{S}H_2$	$26H$	$C_6A\bar{S}_3H_{32}$
Mass (g)	1	1.91	1.73	4.64
Molar weight (g/mol)	270.20	172.17	18.02	1255.26
Number of moles (mmol)	3.70	11.10	96.20	3.70
Density (g/mL)	3.03	2.32	0.998	1.78
Volume (mL)	0.330	0.823	1.733	2.607

The volume change is then –0.273 mL per gram of C_3A.

Early C_3S Hydration

$$C_3S + 5.3H \rightarrow C_{1.7}SH_4 + 1.3CH$$

	C_3S	$5.3H$	$C_{1.7}SH_4$	$1.3CH$
Mass (g)	1	0.418	0.995	0.422
Molar weight (g/mol)	228.32	18.02	227.2	74.09
Number of moles (mmol)	4.38	23.21	4.38	5.69
Density (g/mL)	3.15	0.998	2.01	2.24
Volume (mL)	0.317	0.419	0.495	0.188

The volume change is then –0.053 mL per gram of C_3S.

Late C_3S Hydration

$$2C_3S + 6.5H \rightarrow C_3S_2H_{3.5} + 3CH$$

	$2C_3S$	$6.5H$	$C_3S_2H_{3.5}$	$3CH$
Mass (g)	1	0.257	0.770	0.487
Molar weight (g/mol)	228.32	18.02	351.49	74.09
Number of moles (mmol)	4.38	14.24	2.19	6.57
Density (g/mL)	3.15	0.998	2.50	2.24
Volume (mL)	0.317	0.257	0.308	0.217

The volume change is then –0.049 mL per gram of C_3S.

Calculations of Paulini[12,13]

Reactions considered by the author:

$$2C_3S + 6H \rightarrow C_3S_2H_3 + 3CH$$

$$2C_2S + 4H \rightarrow C_3S_2H_3 + CH$$

$$C_3A + 6H \rightarrow C_3AH_6$$

$$C_4AF + 2CH + 10H \rightarrow C_3AH_6 + C_3FH_6$$

Reactant	Volume change, (cm^3/kg)	
	Paulini	Justness
C_3S	–53.2	–49
C_2S	–40	
$C_3A \rightarrow C_3AH_6$	–178.5	
$C_4AF \rightarrow C_3AH_6$	–111.3	
$C_3A \rightarrow$ Ettringite	–72	–273
$C_3A \rightarrow$ Monosulfate	–72.2	
Gypsum	–37.9 ~ –55.5	–40

Calculations of Tazawa[17]

$$2C_3S + 6H_2O \rightarrow C_3S_2H_3 + 3CH$$

$$2C_2S + 4H_2O \rightarrow C_3S_2H_3 + CH$$

$$C_3A + 3C\bar{S}H_2 + 26H_2O \rightarrow C_6A\bar{S}_3H_{32}$$

$$2C_3A + C_6A\bar{S}H_{32} + 4H_2O \rightarrow 3C_4A\bar{S}H_{12}$$

$$C_3A + CH + 12H_2O \rightarrow C_4AH_{13}$$

$$C_4AF + 3C\bar{S}H_2 + 27H_2O \rightarrow C_6AF\bar{S}_3H_{32} + CH$$

$$2C_4AF + C_6AF\bar{S}_3H_{32} + 6H_2O \rightarrow 3C_4AF\bar{S}H_{12} + 2CH$$

$$2C_4AF + 10H_2O + 2CH \rightarrow C_3AH_6 - C_3FH_6$$

The calculations are limited to the C_3S hydration. The densities of $C_3S_2H_3$ and CH are assumed to be equal to 2.71 and 2.40, respectively. According to the calculations, the reaction results in a decrease in absolute volume equal to 10.87%.

Transport Mechanisms and Damage: Current Issues in Permeation Characteristics of Concrete

Natalyia Hearn and John Figg
Univeristy of Windsor, Windsor, Ontario, Canada

"The artist would be well advised to keep his work to himself till it is completed . . . but the scientist is wiser not to withhold a single finding or a single conjecture from publicity."
— Johann Wolfgang von Goethe

The evolution of cementitious materials has been propelled by the need to improve and modify mechanical performance of concrete structures and to ensure sufficient durability to meet the specified design life. In order to ensure service life requirements, progressive durability characteristics of the concrete at various stages of maturity and deterioration must be assessed. Most studies have focused on the evaluation of cementitious microstructures based on their mix design and progressive maturity; very few studies, however, have focused on the effect of deterioration processes on the subsequent durability characteristics. This review evaluates current research correlating damage-induced cracking with permeation characteristics of deteriorating concrete.

Introduction

Permeation characteristics of concrete have been studied for over 100 years, with the first publication on the water permeability of concrete published in 1889 by Hyde and Smith.[1] The initial need for impermeability of concrete structures was for containment of water, oil (hot), liquid gas (cold), and gas. More recently, the issues of durability have resulted in the reevaluation of permeation characteristics of concrete, in order to prevent and control premature deterioration of concrete and corrosion of the embedded reinforcement.[2]

Currently, due to the diversity of concrete structures and their exposure, the issue of durability has become more significant than that of confinement. Moreover, if the durability is taken into consideration, the aspects of containment are usually satisfied. With durability being the primary focus, water permeability (water driven through concrete under a head of water) is no longer the most important transport mechanism, as capillary suction, ionic diffusion, and so on can be more significant in transporting aggressive agents from the environment into concrete.

The durability of concrete must be defined and quantified through testing. In general, durability can be defined as the ability of a material to resist aggressive environments. In order to quantify durability, tests must be conducted. In order to define durability, the test results must be analyzed in the context of the boundary conditions of the experimental setup. The tests must consider various transport mechanisms, while the analysis itself must consider the changes in the concrete's microstructure as the boundary conditions are altered.

Because of these factors, the study of durability is evolving into assessment of the interrelationship among various transport mechanisms and the effect of damage on the barrier characteristics of concrete. The chronological (historical) overview of the main driving forces in the development of permeability can be divided into three periods:

1. Containment analysis: Resulted in the "birth" of permeability testing and assessment of concrete (late 1800s to 1960s).
2. Durability: The most prominent issue driving development of barrier characteristics of concrete (1970–1990).
3. Damage assessment: Current area of testing and analysis in permeation assessment of concrete (present).

This chapter analyzes the most current issues in permeability dealing with the interrelationship among the various transport mechanisms and the sensitivity of permeability to various types of deterioration processes.

Transport Mechanisms

Definition

Historically the effectiveness of concrete containment structures was measured using Darcian permeability, where water flow is driven by an applied head. In assessment of concrete durability, not only Darcian flow, but all the possible means of water transport must be evaluated. Five primary transfer mechanisms can be defined: adsorption, vapor diffusion, liquid-assisted vapor transfer, saturated liquid flow, and ionic diffusion under saturated conditions. Figure 1 shows the various transport mechanisms and the driving forces behind them.

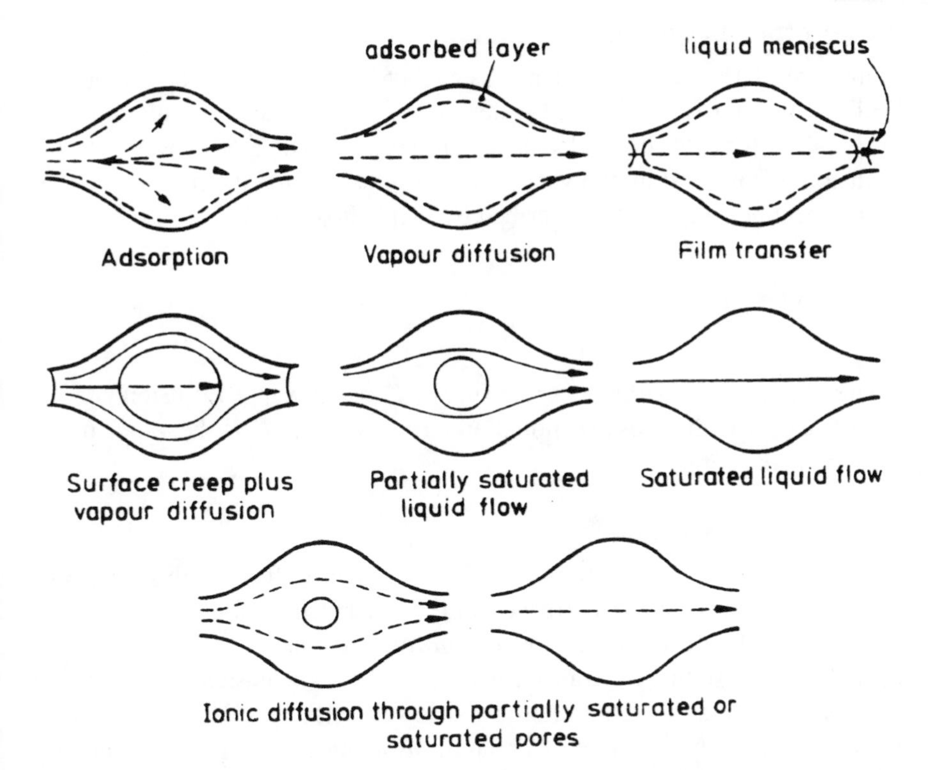

Figure 1. Idealized model of movement of water and ions within concrete (adapted from Ref. 3).

Adsorption

At relative humilities approaching zero, any external source of water will be initially adsorbed onto highly hydrophilic surfaces of the cement hydration products (Fig. 1). With increasing relative humidity, the energy of surface adsorption decreases as the subsequent water molecular layers are further removed from the hydration product surfaces; for example, the energy of surface adsorption of the first and fourth monolayers of water molecules are 1523 and 8 kJ/kg of H_2O, respectively. Adsorption is not only important at low relative humilities, but acts as a driving force along the pore walls, transmitting the flow ahead of the meniscus at higher relative humidity, and as part of the water movement at the initiation of bulk fluid flow in a dry pore.[4]

Water Vapor Diffusion

Below 45% RH, a capillary meniscus cannot be sustained inside the pores. With the thin layer of adsorbed water on the pore surfaces, the mass transfer is in the form of molecular diffusion[5] (Fig. 1). The rate of water vapor diffusion is governed by the vapor pressure gradient and the microstructure of the specimen and can be described by Fick's first law:

$$Q = -K_v A \frac{dp}{dx}$$

where Q is the vapor transport rate (m^3/s), K_v is the vapor diffusion coefficient (m/s), A is the cross-sectional area (m^2); and dp/dx is the vapor pressure gradient (m/m).

Liquid-Assisted Vapor Transfer

For relative humidity between 45 and 100%, menisci form in the pore system, with increasing RH corresponding to increasing pore sizes forming menisci. As the pore widths are not uniform, menisci are formed in the necks of pores, so that the transfer is propelled by condensation at the higher pressure side and evaporation at the lower pressure side (Fig. 1).[5] The result is that a short circuit is created, allowing rapid moisture transfer without coherent flow. This type of transfer has the same mechanism as wick action, except that wick action is measured on the bulk specimen, while liquid-assisted vapor transfer occurs on micro-scale within the hydrate structure.

Saturated Liquid Transfer

At saturation, the fluid transport is controlled by the viscosity of the fluid and the pressure gradient (Fig. 1). Darcian flow is a direct relationship between pressure gradient and resulting flow. Taking into account liquid properties, intrinsic permeability can be derived, which is solely a function of the pore structure:

$$D = \frac{\mu}{\rho g} K$$

where D is the intrinsic permeability (m^2), ρ is the density of the permeant (kg/m^3), μ is the viscosity (Pa·s), g is the gravitational acceleration (9.81 m/s^2), and K is the coefficient of permeability (m/s).

$$\frac{dq}{dt}\frac{1}{A} = K\frac{\Delta h}{L}$$

where dq/dt is the rate of flow (m^3/s), A is the cross-sectional area of the sample (m^3), Δh is the drop in hydraulic head across the sample (m), and L is the thickness of specimen (m).

Table I. Comparison of mass transfer of water through various OPC pastes using Darcian flow measured under high pressure 10 MPa (1000 psi) and wick action[6]

	Rate of water transmission (g/day)	
Mixes	Darcian permeability	Wick action (R^2)
0.3	$< 2.4 \times 10^{-3}$	2.32 (0.998)
0.4	4.1×10^{-2}	5.24 (0.998)
0.5	1.42	5.90 (0.999)

Ionic Diffusion

The presence of water within the cement structure in liquid form allows movement of dissolved ionic species driven by the molecular concentration gradient (Fig. 1). The rates of ionic migration are defined by Fick's law. Ionic diffusion is frequently the most significant transport phenomenon that determines the rates of physical and chemical deterioration.

In concrete, all of the transport mechanisms play a part in maintaining internal equilibrium. Durability parameters are mostly at risk under conditions where water is present in a liquid form rather than as vapor. In the liquid condition, dissolved species can migrate through concrete driven by the concentration gradient and following Fick's law of diffusion. If an evaporation front exists, the aqueous solution is pulled into the concrete through capillary suction, and at the evaporative front the crystallization of dissolved species can create tensile stresses in the pores. Under freezing conditions, the saturated pore system will undergo tensile stresses due to expansion of the forming ice.

A considerable amount of work has been done to measure permeation characteristics of concrete; however, very little work has been done to compare permeation rates obtained using various flow mechanisms. Conceptually, it is obvious that some transport mechanisms will be more effective than others at transmitting unit mass per unit time of liquid water, water vapor, or ionic species. In concrete, these rates of transmission are a function of the microstructural characteristics of the cementitious system. A study by Leong[6] compared transmission rates of OPC and blended pastes. For low-permeability materials, the wicking action was 1000 times greater than transmission driven by pressure head (Table I). Figure 2 relates the depth of penetration achieved through externally applied pressure and the size of the pore radius. Thus, the smaller the pore radius, the greater the

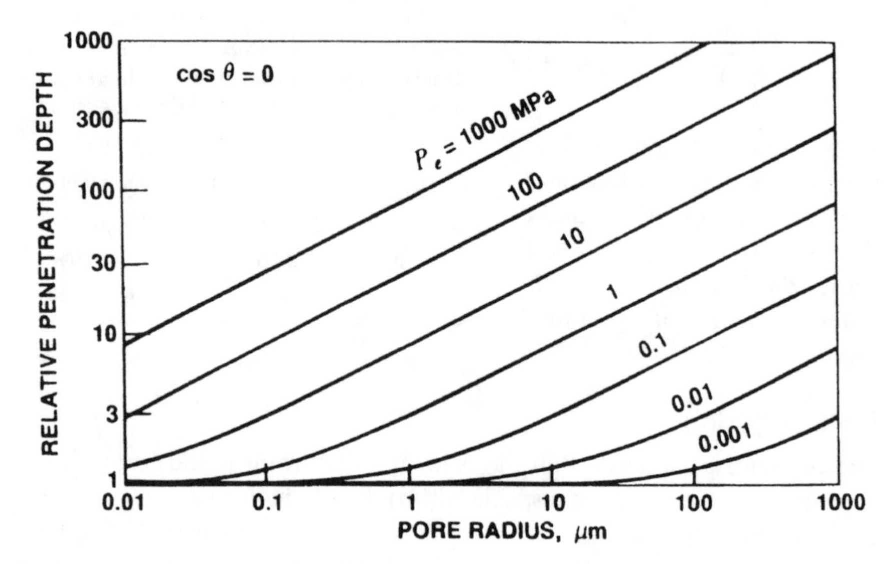

Figure 2. Influence of external pressure (P_e) on penetration (θ = contact angle of water).[7]

pressure required to achieve a given depth of penetration. In concrete, the transmission rates are further complicated by possible discontinuity of the pore structure, where both Darcian flow and capillary suction become "inactive," and any transmission is diffusion controlled.

In order to better comprehend the interaction between transmission rates and microstructural characteristics, key elements of the cementitious microstructure must be examined.

Microstructural Peculiarities of Concrete and Permeation Characteristics

All durability aspects of concrete are related to the movement of water and ionic species dissolved in it (Fig. 3). The rate of transport is a function of the pore structure, particularly the relative distribution of the various pore types in the cementitious matrix. In concrete, however, the distribution of the pore types, and therefore pore sizes, is not constant and is a function of the initial processing, the subsequent exposure history, and degree of saturation.

Concrete's microstructural characteristics have been described by numerous models, the three best-known models being those of Powers, Feldman and Sereda, and Munich. These models were developed in order

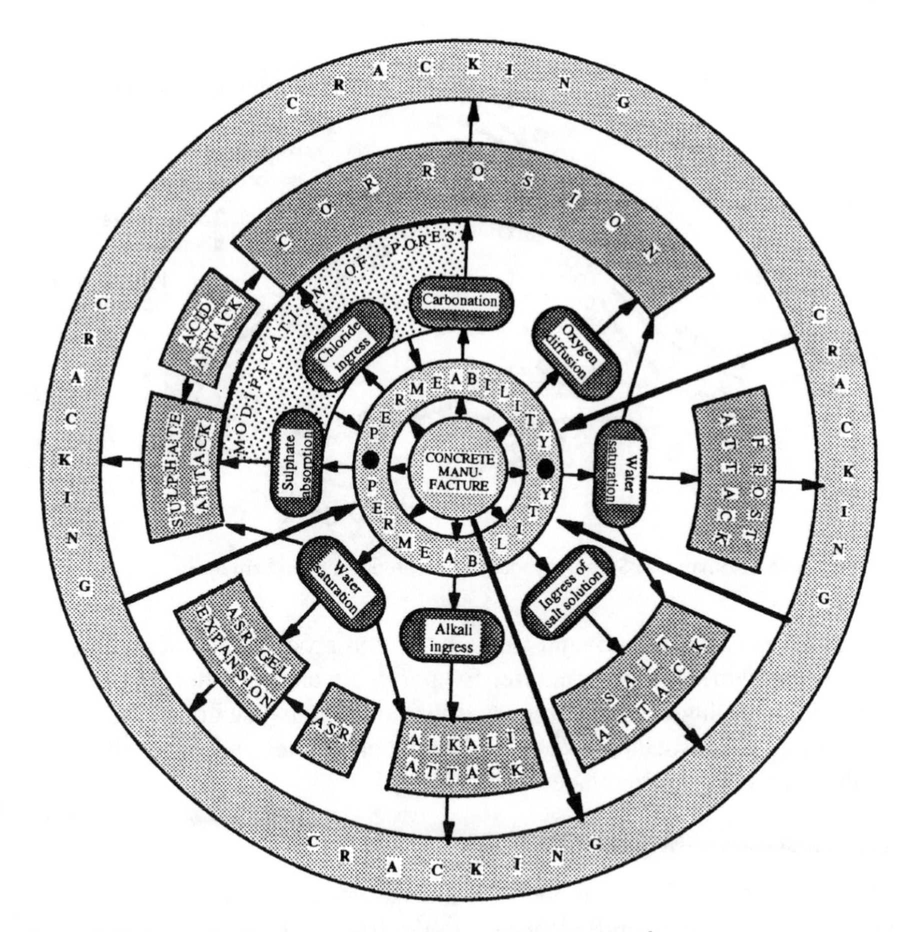

Figure 3. Relationship between permeability and deterioration.[8]

to explain concrete properties of technical importance (such as shrinkage, swelling, and creep mechanisms, etc.). In terms of permeation characteristics of concrete, these models are important in their definition of various types of water held in the cementitious structure and the changes in the structure as the water is added or removed. They classify the water held in the cementitious structure in the following ways.

The Powers Model

The Powers model classifies water in the cementitious matrix into two categories: non-evaporable and evaporable water. Non-evaporable water is

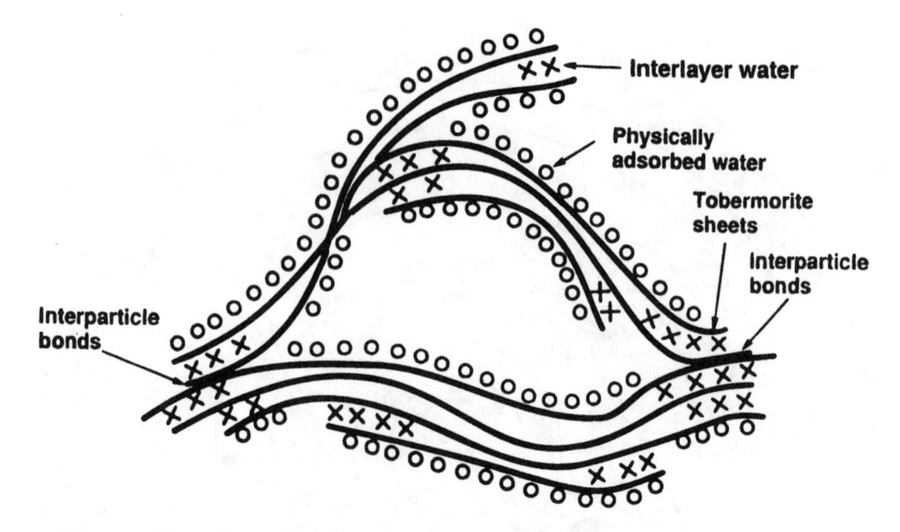

Figure 4. The Feldman and Sereda model of hydrated portland cement.[10]

water that is chemically bound to the hydration products and is an integral part of the cementitious structure. Evaporable water is water that can be removed by drying at some arbitrary drying condition (usually 105°C).[9] Evaporable water includes water held in the capillaries and water adsorbed on the poorly crystallized layers of C-S-H gel.

The Feldman and Sereda Model

The Feldman and Sereda model classifies water in a cementitious matrix into two categories: capillary water and gel water. Capillary water is the free water held in the cementitious structure. Gel water is divided into interlayer, hydrate (or zeolitic), and physically adsorbed water. The zeolitic water is the structural component of the C-S-H gel (Fig. 4). The volume of the physically adsorbed water (i.e., water in the gel pores) can be determined from the difference between the volume of evaporable water (excluding capillary water) and the zeolitic water. The classifications of the various types of water were achieved using scanning adsorption isotherms (Fig. 5), where water movement was plotted as a function of partial pressures.

The Munich Model

Munich model does not provide a specific classification of water,[12] except that all water is considered to be strongly attracted to the solid surfaces of

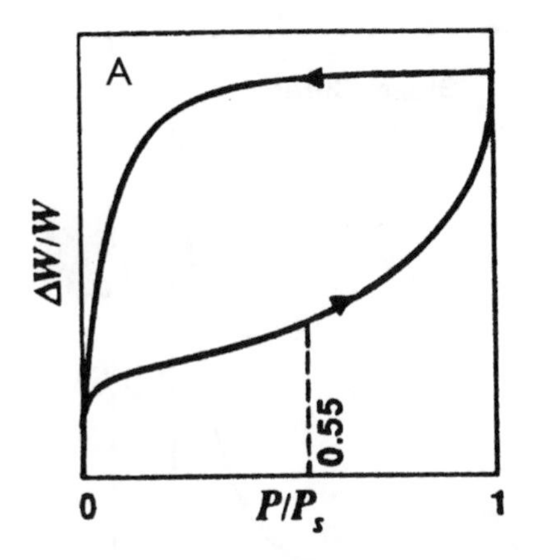

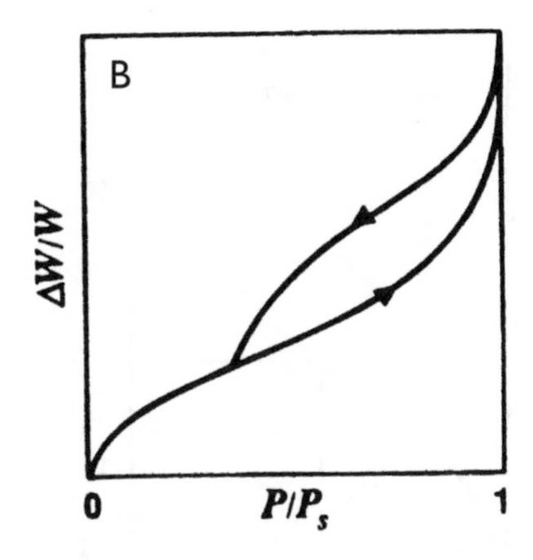

Figure 5. Sorption isotherms for (a) interlayer water and (b) adsorbed water.[11]

the C-S-H gel, so that all water to some degree is adsorbed water (Fig. 6).

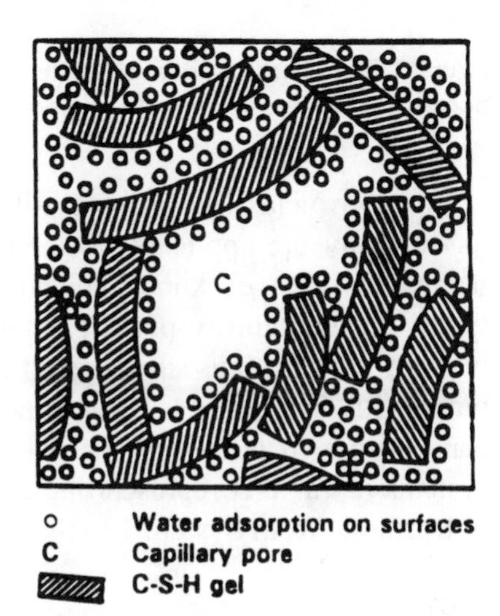

Figure 6. Schematic representation of C-S-H as described by the Munich model.[13]

Effect of Conditioning on Pore Structure

Conditioning of young and mature structures has a significant effect on the relative distribution of the gel and capillary water. For example, heat treatment of the young cement pastes through steam curing results in the significant decrease in the interlayer water and, thus, in the layered structure of the C-S-H gel, as shown in Fig. 7.

In mature systems, the degree of the swelling or collapse of the gel structure, which is dependent on the movement of interlayer water, which in turn is a function of the drying wetting routines, has a significant effect on the permeability. Previous work by Hearn[15] showed that if the interlayer water is removed from the gel structure, the capillary porosity increases due to the collapse of the gel pores, with an order of magnitude increase in the permeability.

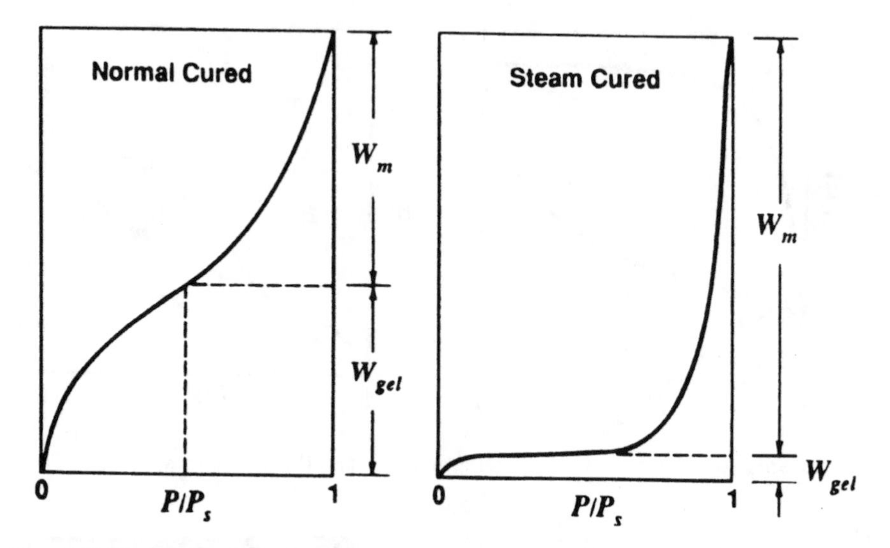

Figure 7. The effect of pore structure on the amounts of W_m (water in the capillary pores) and W_{gel} (gel water in the gel pores) (adapted from Ref. 14).

The above classifications of water in the cementitious structure, each in its own way, are applicable to the permeation analysis. Characteristic flow parameters are well known for porous structures. Figure 8 illustrates four types of permeability porosity models, while Fig. 1 shows flow mechanisms in cementitious structures. If the water classification models are combined with data in Figs. 1 and 8, porosity-permeation systems particular to cementitious materials can be developed, as shown in Table II. The three systems in Table II represent increasing cementitious content and/or levels of hydration from System 1 to System 3.

System 1

System 1 corresponds to high w/c ratios (above w/c = 0.7) and/or low levels of hydration, where high porosity is compounded by high connectivity. In this system, most of the water is free and unaffected by the surface forces of the hydration products. The discontinuity of the pore structure is never achieved, as discussed by Powers (Table III). Because the internal flow paths are unobstructed, the pressure- induced flow would provide the highest transmission rates through the pore structure.

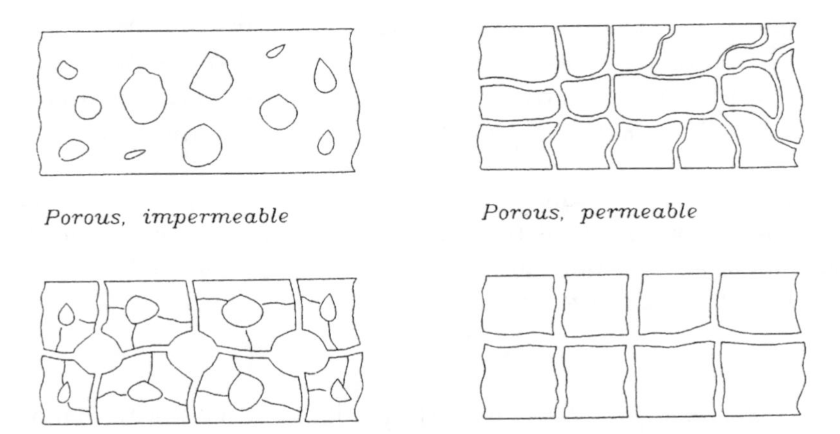

Figure 8. Illustration of the four porosity and permeability models.[3]

Table II. Three cementitious systems based on the porosity/permeation models

	System 1	System 2	System 3
Continuous pore radius	Large	Medium	Small
Pore sizes (μm)	10–100	0.1–10	0.001–0.1
Controlling transmission mechanisms	Darcian flow Ionic diffusion	Capillary suction Evaporation Ionic diffusion	Vapor diffusion

System 2

System 2 corresponds to w/c ratios of most structural concretes (w/c from 0.45 to 0.65). In this system, the existing porosity (as determined by the initial w/c) is divided into a multitude of finer pores and, under a proper curing regime, complete discontinuity of the pore system is possible (Table III). Under these conditions, the water is affected by the

Table III. Approximate age required to produce segmentation of the capillaries[9]

w/c	Time required to achieve discontinuity
0.40	3 days
0.45	7 days
0.50	14 days
0.60	6 months
0.70	1 year
> 0.70	Impossible

surface forces and the Darcian flow becomes less effective as a mass transfer mechanism. Higher pressures are required to achieve measurable flow. Work done by Leong[6] showed that the times required to reach zero flow conditions for 0.3, 0.4, and 0.5 w/c for hardened cement pastes tested at 1 MPa were 2, 5, and 17 days, respectively. Subsequent tests on these samples, using a higher pressure of 13 MPa, showed that only 0.3 w/c had achieved zero flow conditions (i.e., permeability less than 1×10^{-16} m/s), while 0.4 w/c and 0.5 w/c had permeabilities of $17 \pm 2 \times 10^{-16}$ m/s and $780 \pm 150 \times 10^{-16}$ m/s, respectively.

System 3

System 3 represents high-performance cementitious systems of w/c ratios below 0.45 that are well cured and often incorporate supplementary cementing materials. In this system the pore structure is mostly that of the internal C-S-H gel porosity, so permeability is a function of the transmission through gel. Darcian permeability of gel has been determined by Powers[16] to be 2×10^{-17} m/s, which was also taken to be the limiting permeability of the cementitious systems. This gel permeability value, however, is a theoretical determination. Experimentally, a number of researchers have reported zero flow or inability to measure Darcian flow.[17] This limiting value is a function of the applied pressure gradient during the experiment. For low-pressure systems (less than 1 MPa), the Darcian permeability was found to be 1×10^{-13} m/s, while under high-pressure testing (above 7 MPa) it was 1×10^{-16} m/s. Limiting permeability values, however, does not necessarily mean that the concrete at hand is completely impermeable to any aqueous transmission. Under these conditions, where gel segments the pore structure, the governing mass transfer mechanism becomes diffusion (Table I) of aqueous and ionic species instead of pressure-induced flow.

Porosity and Permeability Relationship of Systems 1, 2, and 3

As stated earlier, the relationship between porosity and permeability in a cementitious structure is primarily a function of the interconnectivity of the pore structure rather than of the total porosity. For all w/c ratios, the change in permeability during the hydration process can be as much as five orders of magnitude, while the decrease in the total porosity is one-tenth of the original value (Fig. 9).[18] The reason for such a difference is the segmentation of the pore structure.

Several theoretical methods have been established correlating permeability and pore structure.

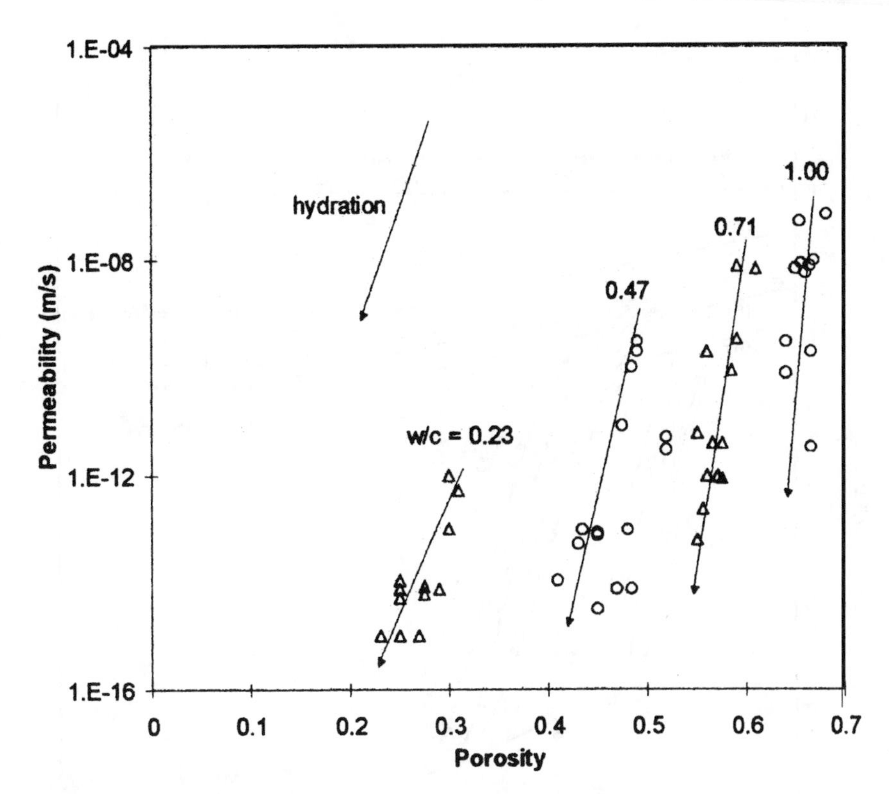

Figure 9. Relationship between permeability and porosity for various w/c ratios of hardened cement paste.[18]

The Powers Relationship

The Powers porosity/permeability relationship is shown in Fig. 10. The lines in this figure are defined by two boundary conditions:

1. Prior to hardening, as obtained from the rate of bleeding, using the following equation:

$$K_1 = \frac{Q_1 V_p}{(d_e - d_t)C}$$

where Q_1 is the rate of bleeding (cm/s), d_e is the density of the cement (g/cm^3), d_t is the density of the aqueous solution (g/cm^3), C is the weight of the cement (g), and V_p is the volume of the paste (cm^3).

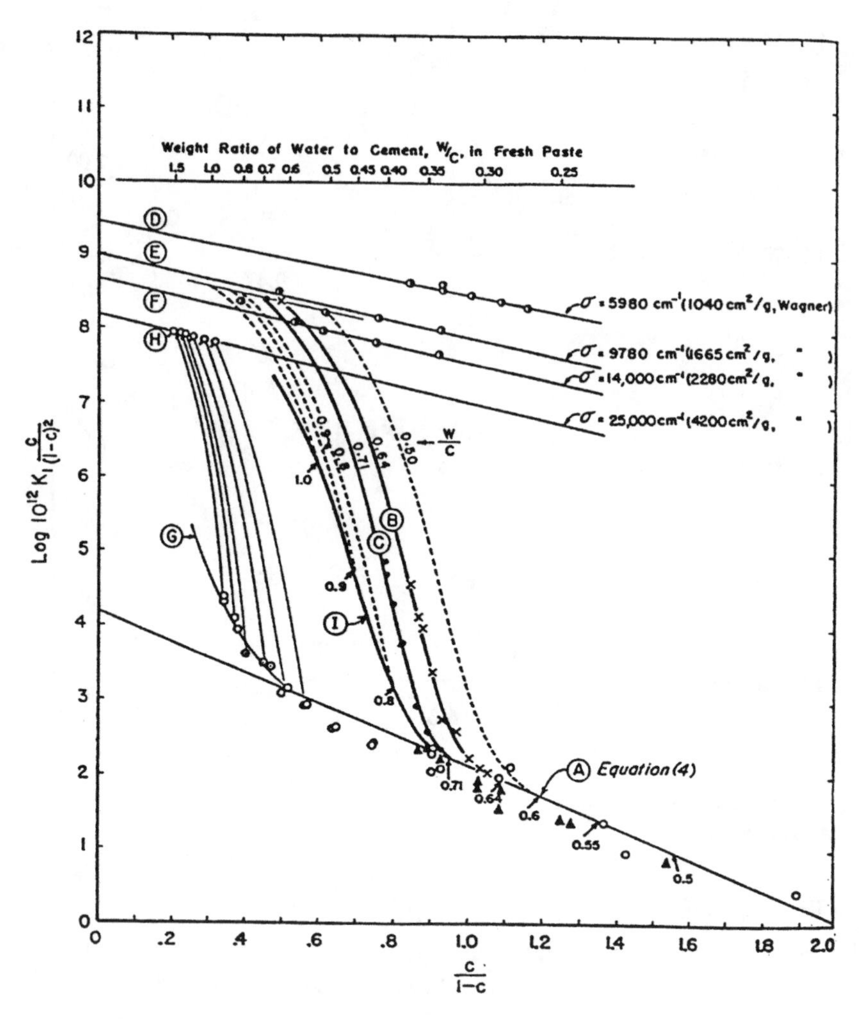

Figure 10. The Powers model of predicting permeability.[19]

2. Permeability of mature cement paste with discontinuous pore structure, defined by:

$$\log 10^{12} K_1 \frac{c}{(1-c)^2} = 4.2 - 2.097\left(\frac{c}{1-c}\right)$$

where K_1 is the coefficient of permeability (cm/s) and c is the volume of solids per unit volume of specimen.

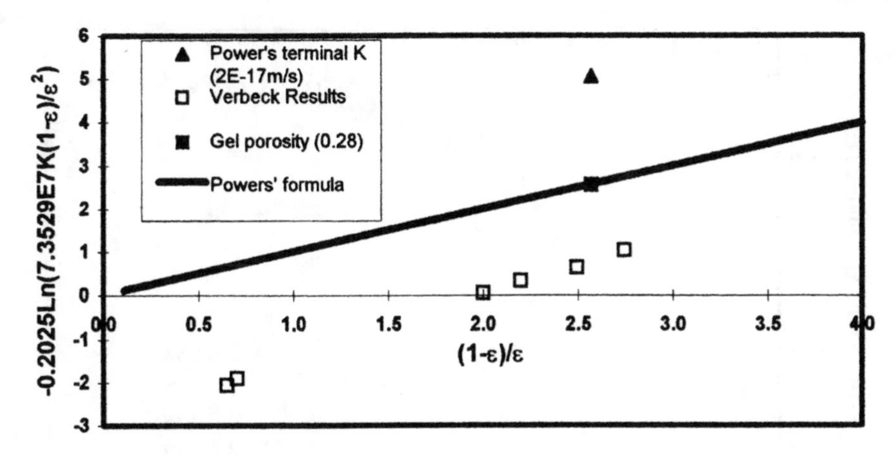

Figure 11. Permeability/porosity relationship using the Powers and Verbeck models.[6]

The two boundary conditions are connected through a series of curves, representing the changes in permeability with the hydration process, for the various w/c ratios.

The Verbeck Relationship

Powers's permeability of the mature pastes was modified by Verbeck to relate permeability with total porosity, resulting in the following equation[19]:

$$K = \frac{1.36 \times 10^{-10}}{\eta(T)} \frac{\varepsilon^2}{1-\varepsilon} \exp\left[-\left(\frac{1242}{T} + 0.7\right)\frac{1-\varepsilon}{\varepsilon}\right]$$

where K is the coefficient of permeability (cm^2), ε is the total porosity, η is the viscosity of water (poise), and T is the temperature (°K). In this equation, the porosity used was capillary porosity, as the gel pores were considered to be part of the solid phase and thus not contributing to the overall flow.

Graphical comparison of the Powers and Verbeck equations (Fig. 11) shows that the difference is a shift in the curves. Obviously, if equivalent porosity is capillary porosity, then the K value will increase. The main problem with this approach is the issue of how to deal with zero capillary systems or the systems where capillaries are completely segmented by the hydration products.

Comparison of the Powers postulated permeability of the C-S-H gel of $K = 2.05 \times 10^{-17}$ m/s and that obtained from his equation ($K = 4.52 \times 10^{-12}$

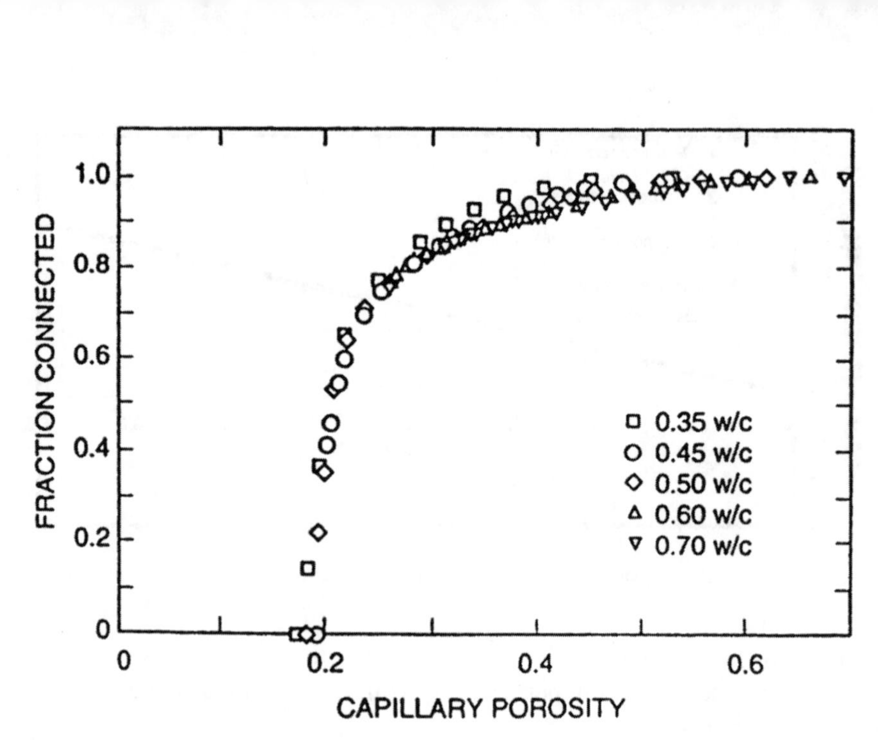

Figure 12. 3-D computer simulation of discontinuity of the pore structure at various levels of hydraton for different w/c ratios versus capillary porosity.[20]

m/s), shows five orders of magnitude difference (Fig. 11). Also, if $K = 2.05 \times 10^{-17}$ m/s is substituted into Verbeck's equation, the resulting capillary porosity is about 15%. The applicability of these equations at discontinuous porosity levels may be questionable. However, two interesting factors should be considered. First, the lowest experimentally measured flow in cementitious systems has been 1×10^{-16} m/s,[6,17] so that the Powers estimated gel permeability of 2.05×10^{-17} m/s fits reasonably well with the experimental data. And secondly, modeling of the hydration process by Garboczi and Bentz[20] showed that, as capillary porosity reaches ≈16%, the 3-D pore system becomes discontinuous (Fig. 12), which is close to the Verbeck 15% porosity discussed above.

Porosity and Pore Diameter Models

The above models assessed cementitious structure in terms of total pore volume. Considering that the internal mobility of water is highly dependent on the pore diameter, models were developed to include both total porosity and pore diameter effects. The approach taken by Hughes[21] summed the

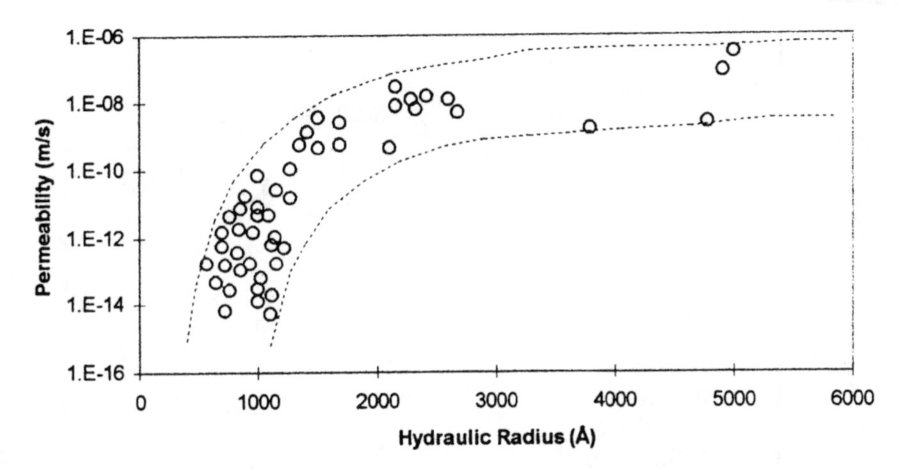

Figure 13. Relationship between permeability and hydraulic radius for hardened cement paste.[18]

flows through pores of the various pore radii by subdividing the pore size distribution into size components and considering an average radius within each component, using the Poiseuille tube model as the mathematical base.

A hydraulic radius relationship with permeability was used by Nyame and Illston,[18] where the hydraulic radius represents the ratio of the cross-sectional area to the wetted perimeter of the conduit, so that the higher the surface area, the lower the hydraulic radius and the greater the resistance to flow. However, the data shown in Fig. 13 do not provide limiting permeability values or reasonable correlation at low hydraulic radius values. At a hydraulic pore radius of 1000 Å, the permeability varies by four orders of magnitude.

The Katz-Thompson permeability theory incorporated the concept of the continuous pore radius and the diffusivity or conductivity of the fluid-saturated porous system, thus obtaining an estimate of the permeability. The maximum continuous pore radius has been defined as the minimum radius of the capillary pores that are geometrically continuous throughout all regions of the hydrated cement paste.[22]

Percolation Modeling

Garboczi and Bentz's computer simulations[20] show that the major reduction in the percolation through capillary porosity occurs at volume of capillaries below 40% and complete discontinuity was achieved at 16% (Fig. 12). The computer simulations also show the degree of hydration required to achieve

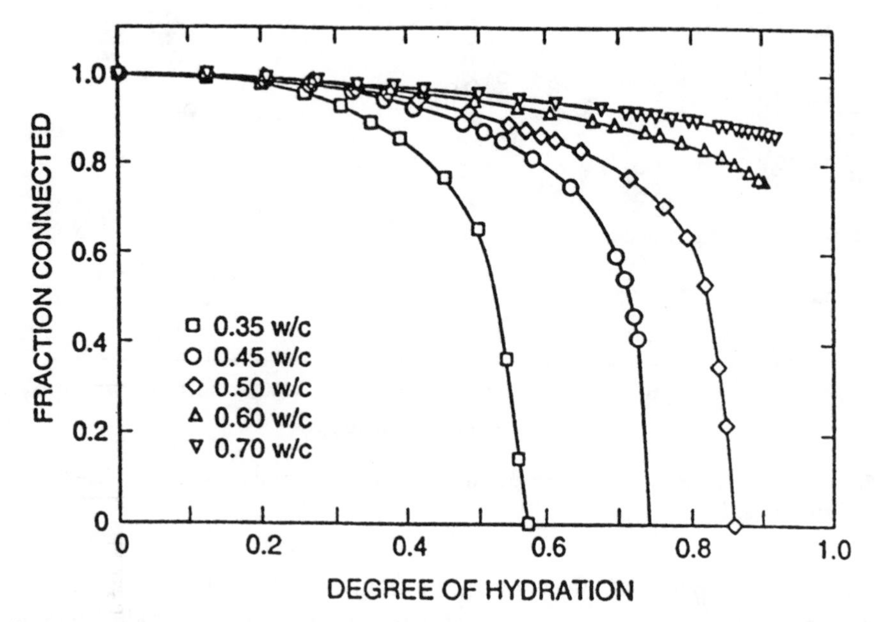

Figure 14. 3-D computer simulation of discontinuity of the pore structure at various levels of hydraton for different w/c ratios versus degree of hydration.[20]

discontinuity for various w/c ratios (Fig. 14). For w/c ratios above 0.6, the discontinuity of the pore structure is not possible, irrespective of the degree of curing and the resulting level of hydration. These results are similar to the Mills[23] data, where terminal levels of hydration were considered, while work by Powers (Table III) assumed 100% hydration. These results also supported earlier work by Bonzel[24] showing that pore discontinuity occurred faster at lower w/c ratios, and minimal changes in permeability were observed after a certain level of hydration was reached (Table IV).

Changes in Systems 1, 2 and 3 due to Deterioration

With the onset of a deterioration process, Systems 1, 2, and 3 can become interchangeable. For instance, extensive cracking of the highly impermeable System 3 would result in the flow characteristics being determined by the crack parameters rather than the original mix design. Therefore, the initially discontinuous pore structure turns into a continuously interconnected system that will easily support transmission through wick action and, under severe cracking, Darcian flow. In the same way, very permeable System 1, at the

Table IV. Length of wet curing past which only small changes in the barrier characteristics are possible[24]

w/c	Age (days)
0.45	7
0.60	28
0.70	90

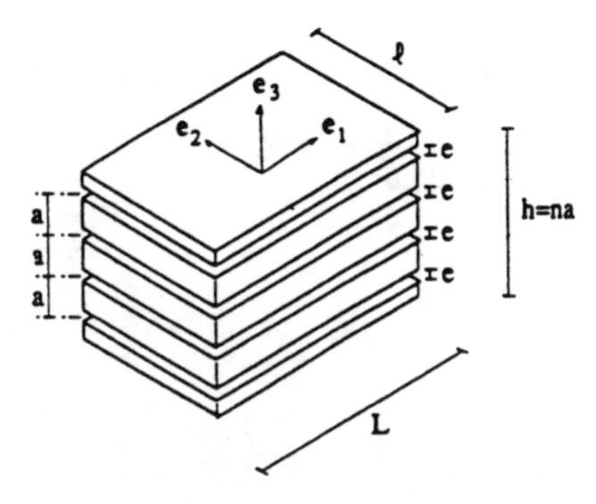

Figure 15. Model of a cracked medium proposed by Snow,[25] where parallel cracks are uniformally distributed and separated by a constant distance.

initial stages of chemical degradation processes, can become less permeable as the reaction products fill in the existing pore spaces (e.g., alkali aggregate reaction), or through self-healing of the cracks.

With the onset of the deterioration, the use of the permeation equations developed for the sound matrix become less applicable. In the initial stages of cracking, the cracks can be assessed as secondary porosity. Modeling of permeation characteristics of fractured material was summarized by Snow,[25] where the model was based on the analysis of parallel and uniformly distributed cracks, separated by a constant distance (Fig. 15). Based on this analysis, Fauchet[26] developed a model of cracked medium, where the total mass transfer was calculated from superimposition of the matrix permeability and crack permeability. Models by Sayers[27] and Bazant[28] assessed the permeability of the cracked matrix in terms of series of cracks of differing aperture (Fig. 16). Experimental results, however, have shown that actual permeation values were at least an order of magnitude lower than those predicted by the model. Bazant[28] explained this phenomenon by noting that in the real cracks there is an apparent and effective aperture, with the significant portion of the mass transfer occurring through the effective crack opening, which is narrower than the apparent one.

Very little work has been done to model the evolution of damage. Bourdarot[29] developed a relationship between permeability of intact and progressively cracked matrix:

$$k_i = k_{io}\left(\frac{k_{iu}}{k_{io}}\right)^{D^n}$$

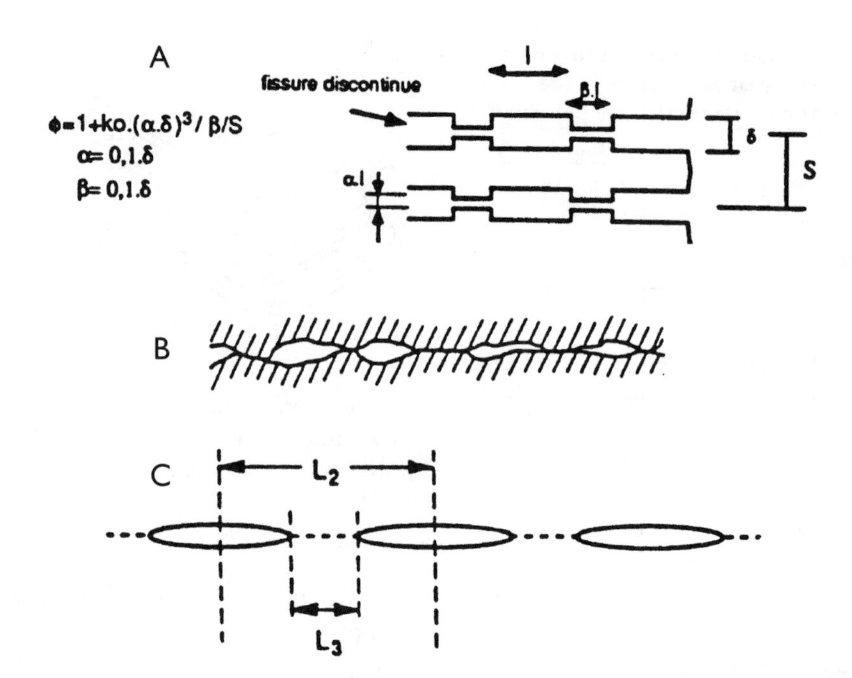

Figure 16. (a) Rugosity model proposed by Bazant et al., where ϕ is cracked medium apparent permeability, *ko* is the intitial permeability, *S* is the distance between two cracks, and *d* is apparent crack width.[28] (b) Sayers model for predicting the permeability of a cracked rock. A regular crack as shown in (b) and is modeled as a planar array of periodically spaced (L_3) fracture segments of equal length (c).[27]

where k_{io} is the initial permeability value of intact matrix, k_{iu} is the ultimate permeability value at ultimate level of damage, D is the damage coefficient ($0 \leq D \leq 1$), and $1 \leq n \leq 2.5$

Determination of the above empirical values makes this approach of limited use. It is intuitively obvious that cracking must modify the integrity of the intact matrix in terms of permeation characteristics and that at some level of cracking, crack characteristics become more significant than the original porosity of the intact matrix. It is not known, however, which crack parameters have the most effect on the mass transfer characteristics, and mathematical relationships relating the transition between intact and damaged cementitious matrices have not been developed.

Damage in Concrete

In service, concrete is exposed to a combination of physical and chemical processes. Some of these processes are beneficial to the welfare of the con-

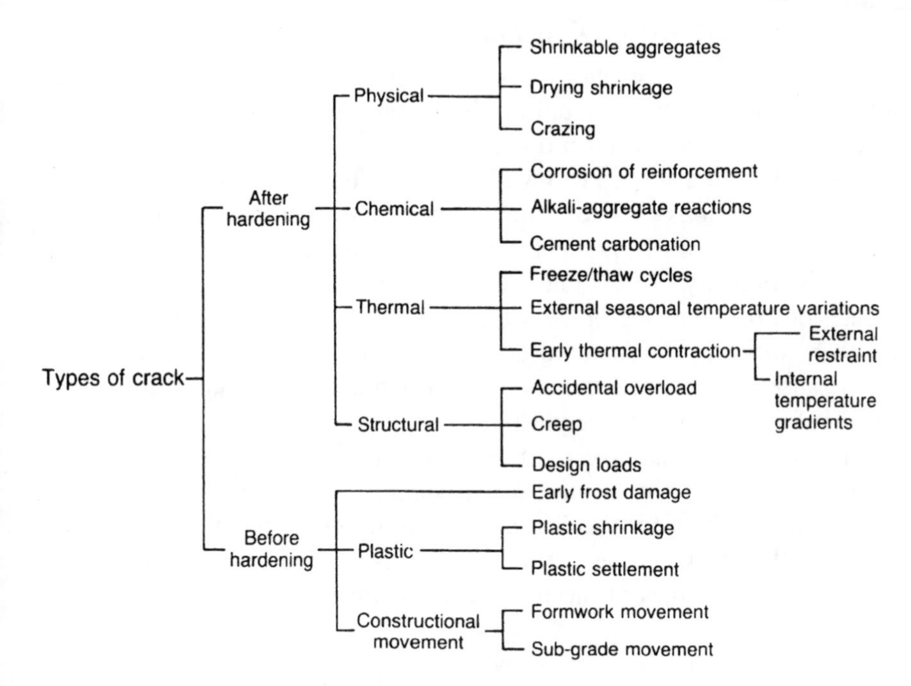

Figure 17. Crack classification based on the various deterioration mechanisms.[30]

crete, while others create irreversible damage. Beneficial processes are continuing hydration and the development of the microstructure long after the placement of concrete (e.g., long-term benefits of the supplementary cementing materials), and the related self-sealing characteristics of concrete. Most other processes damage the microstructural integrity of concrete. Shrinkage, mechanical loading, freeze-thaw, corrosion of reinforcement, and crystallization of salts induce physical stresses. Alkali-aggregate reaction, sulfate attack (including delayed ettringite formation), and acid attack chemically alter the binder component of the concrete and subsequently induce physical stresses in the concrete matrix. Figure 17 lists mechanisms that cause cracking in concrete.

Even though there are still some discussions regarding the cause and the microstructural modifications of the various deterioration processes, the physical signs of the progressive deterioration are well known.[30] What is not known, however, is how these deterioration processes affect the barrier characteristics and at what level of deterioration the barrier characteristics are affected.

The following sections assess available data on the interrelationship of the various damage mechanism and barrier characteristics of concrete.

Deterioration-Induced Cracking

Deterioration processes can cause two types of cracking: a single crack through a sound matrix or a uniformly distributed crack system. Clear[31] has discussed the mass transfer properties through a single crack and the effect of distribution of a single crack with a number of smaller cracks (one of the important functions of steel reinforcement and fibers). If, instead of one large crack of width W, n smaller cracks of width w form, so that $nw = W$, the proportionality of flow rates would be $Qnw : Qw = 1/n2$. Single cracks create localized damage or failure in the joint action of one element with another, however, the material components in themselves are intact. The main causes of localized single crack formation are mechanical loads, corrosion of reinforcement, and plastic shrinkage.[30]

In cases where damage is progressive and uniformly distributed, the resulting cracking creates an overall change in the characteristics of the system so that both mechanical and durability characteristics are globally affected. The main causes of uniformly distributed cracking are drying shrinkage, some types of mechanical loads, freeze/thaw, AAR, and sulfate attack.

The size relationships among various concrete elements are shown in Table V. The crack classification according to various crack parameters is given in Table VI and Fig. 17.

Physical Deterioration Processes

Shrinkage

Mechanisms

Shrinkage is controlled by the rate of removal of water from the pore structure of the cementitious system. Generally, two types of shrinkage cracking processes can be defined: shrinkage during the hardening process (on the macro-scale) and subsequent drying shrinkage of the hardened material (on the micro-scale).

The fracture mechanics mechanism of drying shrinkage is still disputed. Depending on which model of C-S-H gel is adopted (Powers, Feldman and Sereda, or Wittman), the explanation of drying shrinkage will vary. Taylor[33] summarized the findings of the various studies under the following four processes that may contribute to the shrinkage mechanism:

1. Capillary stress: Attractive force created by tension in the meniscus of evaporating water (important between 90 and 45% RH).

Table V. Size distribution of hydration products, pores, and cracks

Hydration products	Typical dimensions (μm)	Pore type	Typical dimensions of pores (μm)	Microcrack dimensions (μm)
C-S-H	1 × 0.1	Interparticle spacing between C-S-H sheets	0.001–0.03	1–60
CH	1 × 7	Capillary voids	0.01–50	
Ettringite	10 × 0.5	Entrained air bubbles	1–50	
Monosulfo-aluminate	1 × 1 × 0.1	Entrapped air voids	1000–3000	

Table VI. Typical size parameters of cracks in concrete[32]

Classification	Microcracks	Fine cracks	Large cracks
Width (mm)	<0.01	0.01–0.10	>0.10
Causes	Self-desiccation Hydration	Shrinkage Temperature damage	Mechanical loading
Crack depth (mm)	<2	<50	>100
Crack distance (mm)	5–40	10–200	30–200

2. Surface free energy: Energy due to the unsatisfied bonding forces on the surface of C-S-H layer, which is relaxed by removal once the last adsorbed water layer is removed (important below 20% RH).
3. Disjoining pressure: Repulsive force caused by the greater attraction of the water molecules to C-S-H layers than the attraction between the adjacent C-S-H layers. Drying reduces the disjoining pressure, thus bringing C-S-H layers close together.
4. Movement of interlayer water: Attractive force causing collapse to C-S-H gel layer, as the water molecules between the layers evaporate.

The relative importance of these process is still disputed.

On the macro-scale, rapid loss of moisture during the hardening period results in plastic shrinkage, due to the differential strains developed between the drying surface and the substrate concrete. Characteristic cracking of plastic shrinkage is either development of crazing, which is random shallow cracks formed on the surface, or deeper parallel cracks usually

BEFORE DRYING

E_1

E_2

$E_1 > E_2$

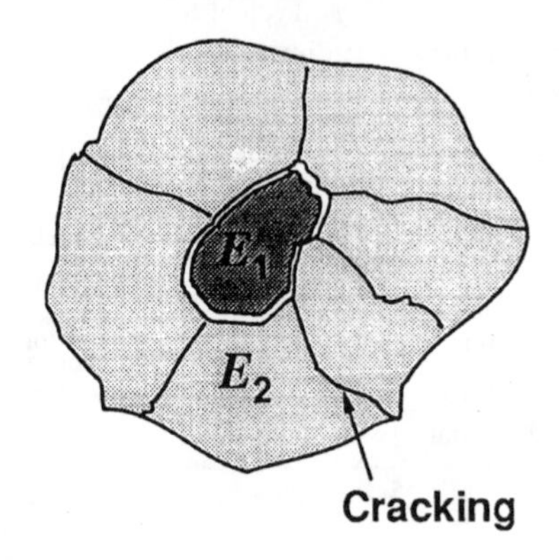

Figure 18. Effect of the difference in the elastic modulus on the cracking due to drying shrinkage.[34] E_1 is the elastic modulus of aggregate or unhydrated cement; E_2 is the elastic modulus of hydrated cement.

above the reinforcement. In the analysis of the effect of plastic shrinkage on the transport characteristics, the relative contributions of the transport through concrete and transport through cracks must be assessed.

On the micro-scale, drying of the cementitious microstructure results in the development of micro-cracks at any location where moduli of elasticity of two adjacent phases are different. In concrete, cracking due to drying and differential shrinkage can occur at all scales within the concrete structure. Figure 18 shows a system of a grain of high modulus of elasticity (E_1) surrounded by the solid of lower modulus of elasticity (E_2). The E_1 and E_2 system can be (a) an unhydrated cement particle surrounded by the hydration products, (b) a sand particle surrounded by the hardened cement paste, and (c) a coarse aggregate surrounded by the mortar. Moreover, because of the hydrophilic nature of cementitious pore surfaces and high surface tension in the capillary menisci, rapid removal of water results in the collapse of the finer pore system (80 Å).

Resulting Changes in Microstructure

Shrinkage drying can have a particularly severe effect on the cementitious microstructure for two reasons:

1. Removal of the interlayer water results in the collapse of the fine pore structure, causing widening of the capillary system.

Table VII. Effect of drying shrinkage cracks and self-sealing on permeability[15,17]

Sample ID	w/c	Virgin permeability*	Permeability after drying	Permeability after self-sealing
7/4	0.9	0.9×10^{-19}	300×10^{-19}	20×10^{-19}
12/2	0.9	0.7×10^{-19}	260×10^{-19}	20×10^{-19}
11/3	0.9	0.8×10^{-19}	110×10^{-19}	8.5×10^{-19}
9/1	0.9	0.4×10^{-19}	150×10^{-19}	2.5×10^{-19}

*Samples were water cured for 26 years.

2. Cracking across all scales of the cementitious microstructure results in the connectivity of the porosity.

Effect on Barrier Performance

Work by Hearn[15] showed that the widening of the capillaries due to the collapse of the finer pore structure resulted in an increase in the Darcian flow by one order of magnitude, while severe shrinkage cracks resulted in a three to four orders of magnitude increase in the permeability, as compared to never-dried samples (Table VII).[34] Similar results were obtained by Samaha and Hover[35] (Fig. 19).

One interesting by-product of severe drying shrinkage is the ability of concrete to self-seal once the water is introduced back into the pore structure. The self-sealing phenomenon is a function of the exposure, due to cracking, of the previously unexposed hydrated and unhydrated components, resulting in continuing hydration, and solution and redeposition of the dissolved and undissolved species in the cementitious pore structure.[17] The self-sealing has been shown to reduce permeability by several orders of magnitude (Table VII).

Mechanical Load

Mechanisms

Concrete is a non-ideal brittle material, as evidenced by the nonlinear response of the stress-strain curve. Under load the preference and extent of fracture propagation has been shown to be affected by the age of HCP,[36,37] the presence of water,[38] and concentration of $Ca(OH)_2$ in the pore solution.[39]

Certain characteristic properties of HCP and concrete, such as fracture brittleness, notch sensitivity, the proportion of tensile to compressive strengths, strength increase with the increased rate of loading, and presence

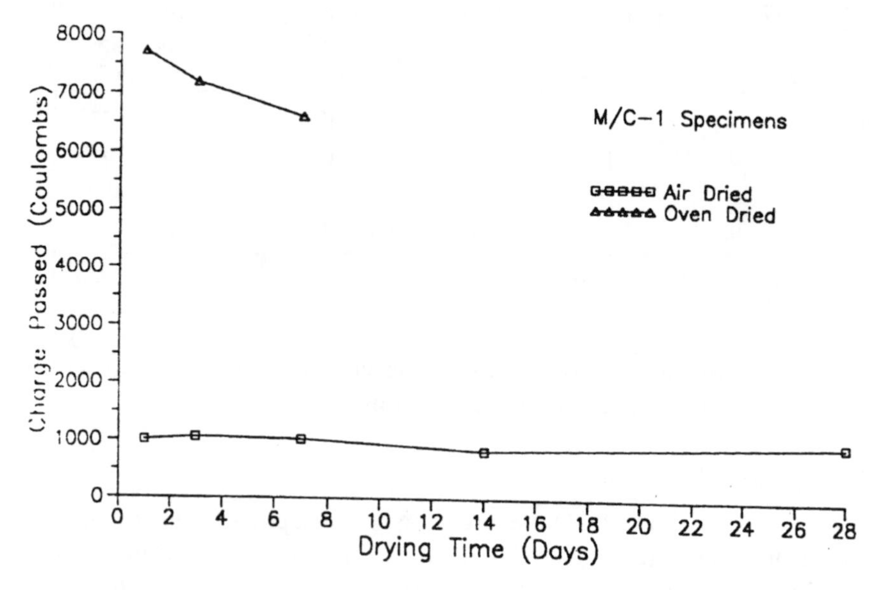

Figure 19. Relationship between total charge passed and drying time in the oven and air.[35]

of stress concentrations, suggests the possibility of application of Griffith theory to concrete.[40–42] Griffith postulated that the discrepancy between the actual strength of a brittle material and the theoretical strength based on the molecular bonds is due to the presence of flaws. Under load, these flaws result in high stress concentrations affecting small areas and thus causing microscopic fracture. Under a given stress level, critical crack length may be determined through the energy balance shown in Fig. 20(a). Crack propagation occurs once the slope of the parabolic energy release curve equals that of the linear energy requirement curve. In concrete, this energy balance is modified to allow for "crack arrestors," or high-strength areas, which demand a higher stress level to reactivate crack growth [Fig. 20(b)].[43,44] The Griffith theory of fracture and crack propagation, as applied to concrete, indicates failure through the largest flaws, suggesting a few major failure planes, rather than disintegration of the matrix through an extensive network of microcracks.

The main problem with such analysis is that the effect of subcritical crack propagation (due to non-ideal brittle behavior of concrete) and yield-

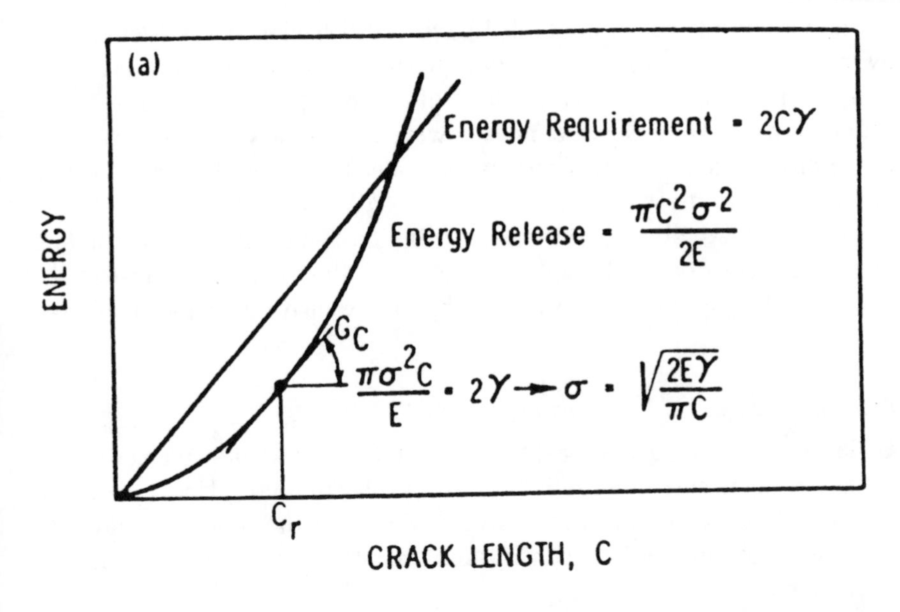

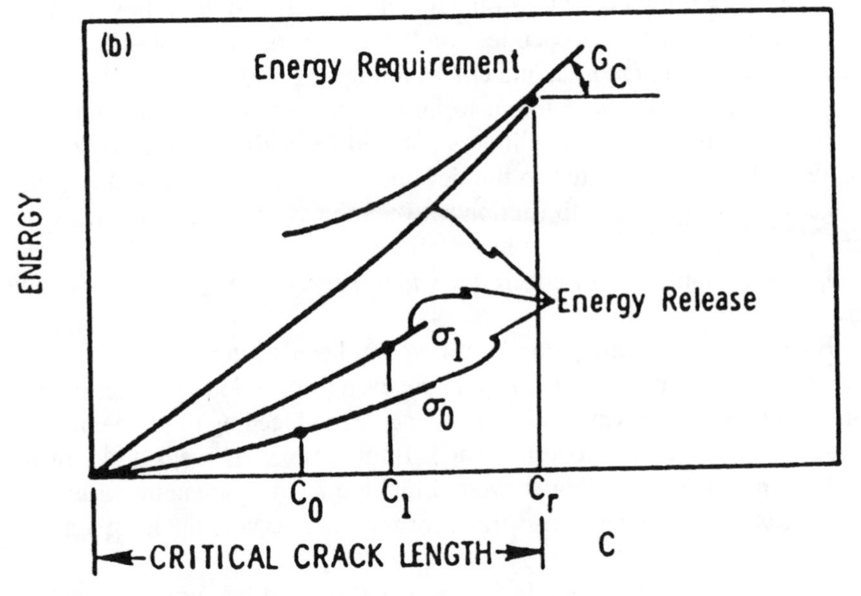

Figure 20. Crack growth (a) according to Griffith theory and (b) modified Griffith theory for concrete.[43]

ing is not considered on the global structural behavior. It has been shown by many researchers using a variety of testing techniques[45] that at the tip of every propagating crack exists a cloud of microcracks manifested by strain softening behavior. Bascoul et al.,[46] however, have shown that microcracks in the process zone, which were visible under load, closed after unloading of failure due to stress relaxation.

In summary, the literature indicates that although the catastrophic failure in concrete does occur through a few major failure planes, the rest of the matrix is also significantly weakened by the extensive network of microcracks[47,48]

Resulting Changes in Microstructure

Under most loading conditions, the concrete fracture occurs on clear failure planes, as in the case of tensile, flexural, and shear failures. Under compressive loading, however, progressive increase in the load correlates to increase in the microcrack density until, at prefailure loads, clear failure planes are established.

Early work on load-induced microcracking[49] defined the following three types of microcracks in concrete: bond cracks at the aggregate-paste interface, cracks through mortar, and cracks through aggregate.

Aggregate cracks were found to be rare at prefailure strains, although after failure the concrete cylinders showed extensive through-aggregate failure. This was attributed to the development of shear forces at the final stages of the failure, due to friction between the cylinder ends and the compressive frame. Through-aggregate failure is especially common in lightweight and high-strength concretes, where aggregate/paste bond strengths are high.

Bond cracks, formed in the high-porosity layer of the cement/aggregate interface, were found to exist in concrete even prior to loading. These could have been caused by segregation, settlement, and bleeding of fresh concrete, and drying or carbonation shrinkage. Bond cracks, as weak links in the cement matrix (Griffith flaws), were found to increase in length, width, and number with increasing stress/strain, forming first around the largest aggregates.

The development of mortar cracks was shown to be significant only at high strain levels (70–90% of the ultimate load and 1200–1800 $\mu\varepsilon$). Mortar crack propagation was mostly through the joining of bond cracks and linking of major voids, which together formed a continuous crack pattern. Extensive development of a continuous crack pattern leads to failure.

This type of classification is helpful in describing the various crack types and their formation. In terms of the contribution of the various cracks to the damage and ultimately failure, a more recent classification by Carrasquillo et al.[50] may be useful, too. The cracks were categorized into two types, simple bond cracks and combined cracks (Type I and Type II), with the stress-strain curve divided into three stages of crack development. Simple bond cracks occur first and refer to cracking at the mortar/coarse aggregate interface. These cracks are isolated and stable under constant load. As the load increases, the simple bond cracks are connected by mortar cracks, forming combined cracks. These cracks are grouped into Types I and II. Type I cracks are stable, consisting of one bond crack and mortar crack, or the combination of two bond cracks connected by a mortar crack. Type II combined cracks are unstable (i.e., they propagate spontaneously under constant load) consisting of at least two bond cracks and two mortar cracks. Propagation of Type II cracks leads to failure.

Carrasquillo et al.[50] found that under short-term load the development of Type II cracks for normal-strength concrete (35 MPa) occurred between 70 and 90% strain at peak stress, and for high-strength concrete (70 MPa), Type II cracks were insignificant even at 90% of strain. Similar results were also observed by Meyers et al.,[51] and Santiago and Hilsdorf[52] (Fig. 21).

Much field concrete is exposed to sustained loading. Under sustained loading conditions, the time-dependent component of deformation is related to mobility of water in wet concrete and proportion of pre-existing cracks in dry concrete. The long term stress-strain curve (Fig. 22) may also be divided into three stages of microcracking. In the first stage, up to 40% of f_c', the rate of growth of microcracks is very small. At about 75% of f_c', the effect of time-dependent deformation becomes considerable, but the crack propagation is mostly limited to the pre-existing bond cracks and few combined cracks connecting closely positioned bond cracks. In the last stage, at the stress level equal or above the long-term sustained strength (about 75% of f_c'), there is a sharp increase in bond and mortar cracking and propagation of combined cracks. The results of the various studies[54,55] indicate that the bond cracks are largely responsible for the time-dependent deformation at loads up to the long-term sustained strength. Above this level, extensive cracking is formed within a very short period of sustained loading either immediately, if the stress is high (90–95% of f_c'), or toward the end of failure time under sustained load, if the stress is just above the long-term strength.

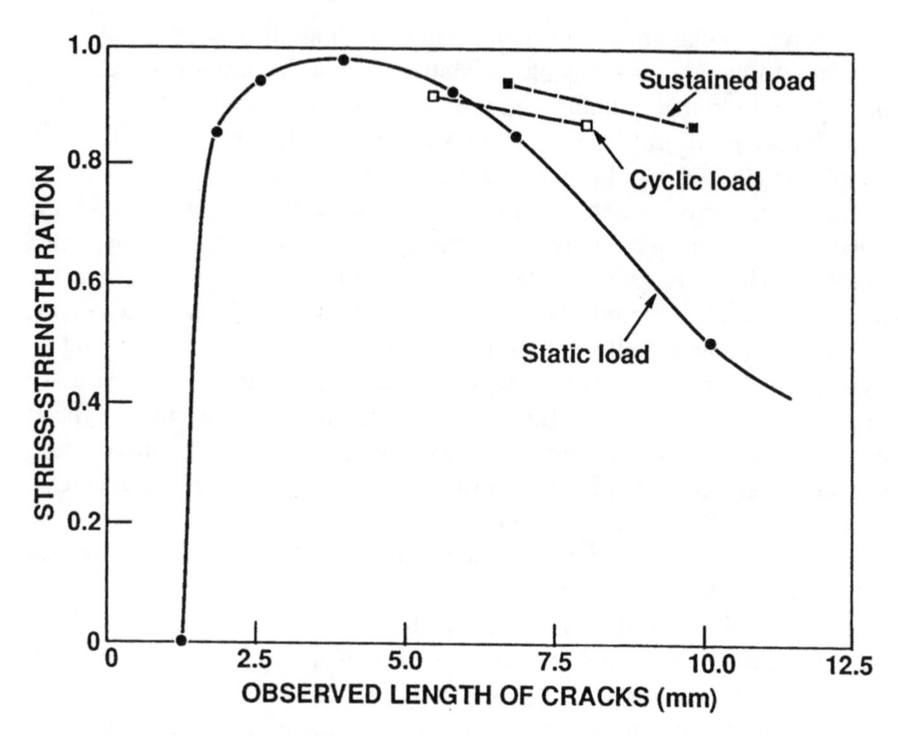

Figure 21. Observed length of cracks during compressive loading of concrete.[52]

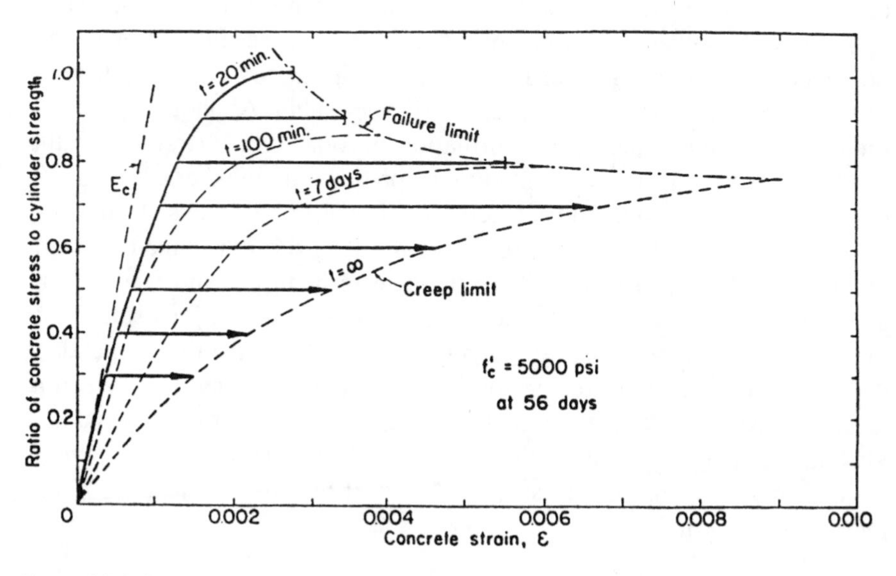

Figure 22. Influence of load intensity and duration on concrete strain.[53]

Effect on Barrier Performance

The fact that continuity of the cracks developed under compressive loading does not occur until prefailure loads is supported by the current research on the relationship between mechanically induced cracks and permeability. Only a few studies have addressed the issue of mechanically induced cracking and the changes in the transport characteristics of concrete. Research conducted on the compression induced microcracks has used loaded and unloaded specimens. Kermani[56] concluded that above 40% of the ultimate compressive strength, permeability sharply increases. On closer examination of his data, the sharp increase in the permeability did not occur until 70% of the ultimate load, as the water permeability results had a large standard deviation depending on the applied pressure and length of testing. Similar results were obtained by Hearn,[34] where a large standard deviation of the water permeability data masked the effects of the applied compressive load even at 80% of the ultimate (Fig. 23). Samara and Hover[35] similarly showed that increase in permeability did not occur until 75% of the ultimate loads. The permeability was assessed using the rapid chloride permeability test (ASTM C1202-94) and the observed increase above 75% of the ultimate was only in the order of 10–20% in the charge passed. Samara and Hover attempted to correlate the permeability changes to cumulative crack length change, as obtained by neutron radiography. No effect was observed for stress levels below 75% of f_c' (Fig. 24).

One of the main problems with the postmortem analysis is some closing of the cracks on unloading,[57] and the extent of damage is related to how long the peak load is maintained, especially at prefailure loads, where volumetric strain has a rapid rate of increase.[34,35] This observation has primary significance that in concrete the damage relationships may have better correlation with strain, rather than stress. Work by Pantazopoulou and Mills[58] on constitutive models of the interrelationships of the mechanical and microstructural characteristics of concrete has demonstrated this aspect of concrete performance. In terms of permeability, work by Hearn and Lok[59] on the correlation between volumetric expansion and permeability showed that the changes in permeation characteristics occurred once the volumetric expansion changed from negative to positive. This change occurred at stress level of about 70% of the ultimate (Fig. 25).

The strain relationship to permeability was clearly demonstrated by Shah[57,60] on the effect of crack width on the permeation characteristics. The increase in permeability occurring at crack opening displacement between 0.05 and 0.2 mm is shown in Fig. 26. Similar results were observed by

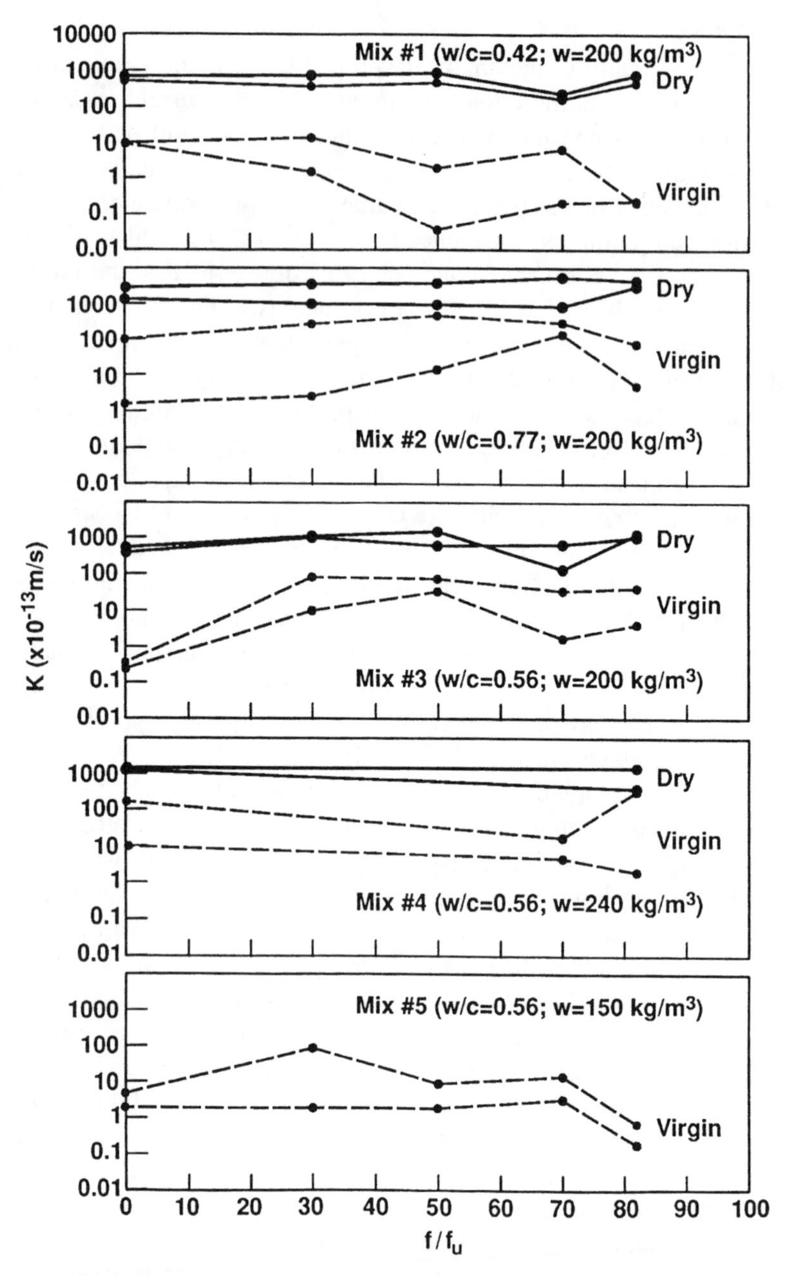

Figure 23. Permeability data showing the effect of compressive loading on permeability of concrete samples of various w/c ratios and before and after drying.[34]

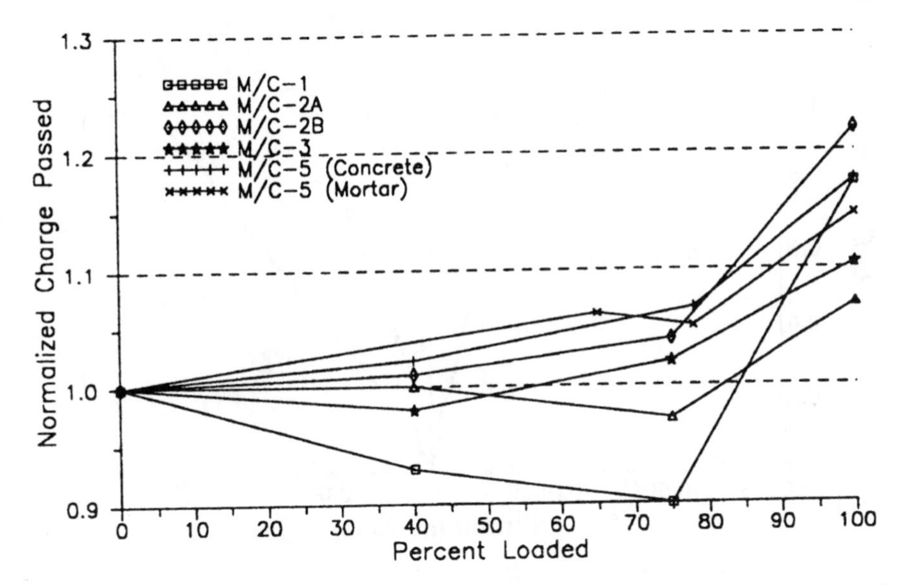

Figure 24. Normalized graph of charge passed versus compressive load level for all concrete and mortar samples.[35]

Mills,[61] where permeability increased exponentially with crack widths above 0.1 mm.

Discrete crack development is of particular importance in most reinforced concrete structures, particularly in connection with corrosion of the reinforcement. Discrete crack development and its effect on permeation characteristics will be discussed in the section on deterioration by corrosion mechanism.

Freeze/Thaw

Mechanisms

The main components of the physical process due to freezing and thawing of water in the cementitious pore structure are:

1. Nine percent expansion by volume of water upon freezing, thus creating tensile stresses within the pore system. Under saturated conditions the pressure is high enough to cause fracture.
2. Progressive depression of freezing point within smaller pore sizes (Fig. 27), with adsorbed water remaining unfrozen due to surplus energy at the surface.

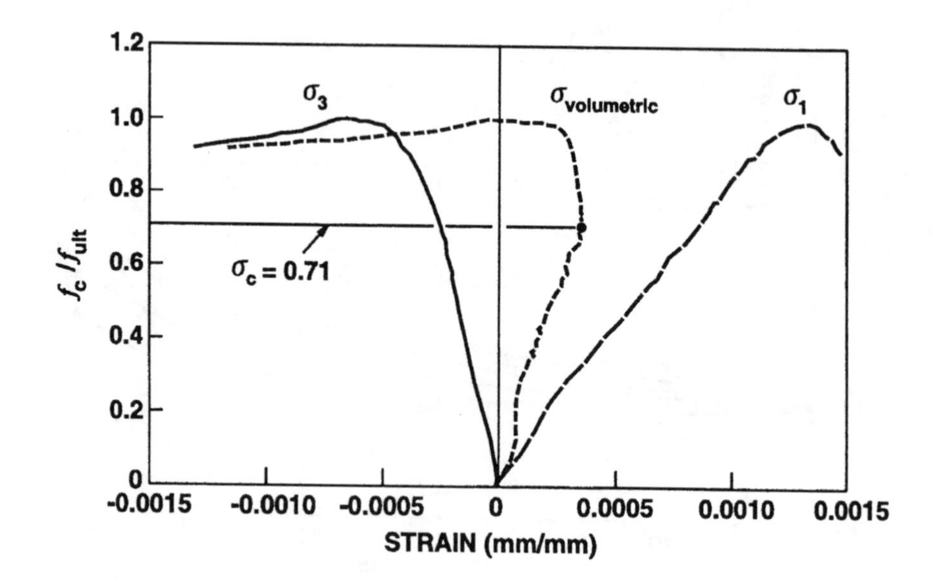

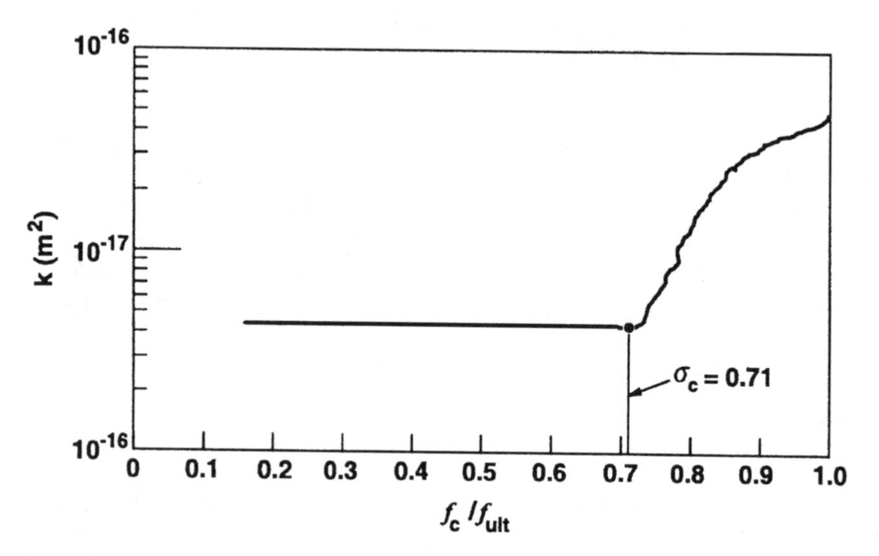

Figure 25. Relationship between changes in the gas permeability (K) and stress and volumetric strain of concrete cylinders loaded in compression. The gas permeability was tested under load.[59]

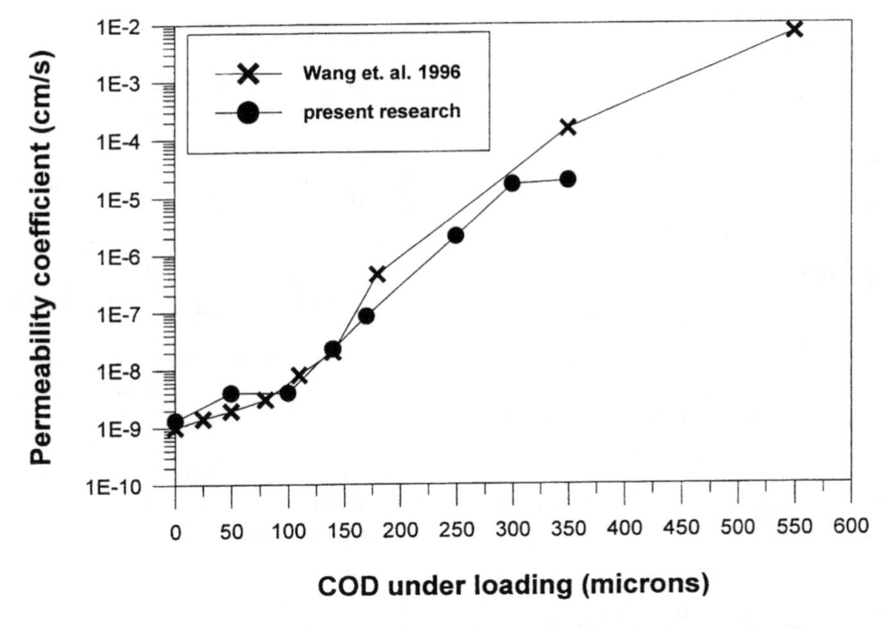

Figure 26. Correlation between crack opening displacement and permeability.[57]

3. In an unsaturated system, the change from water to ice will result in significant levels of evaporation.
4. Movement of water from smaller pores into larger pores, as the water in the larger pores freezes. This results in the development of hydraulic pressure within the pore system.
5. In the presence of de-icing agents, the freezing behavior of water is changed by lowering of the freezing point, which results in the temperature shock during the thawing process.

Resulting Changes in Microstructure

Freeze/thaw damage has the most severe effect on the cover of concrete. Compounded by the concentration gradient and depression of the freezing point in the presence of chlorides, freeze/thaw damage becomes particularly severe at the surface.

As volumetric increase due to ice formation starts in the larger pore system, the resulting fractures tend to bridge the larger pores, thus creating coarse interconnected pore system (as shown by System 1 in Table II).

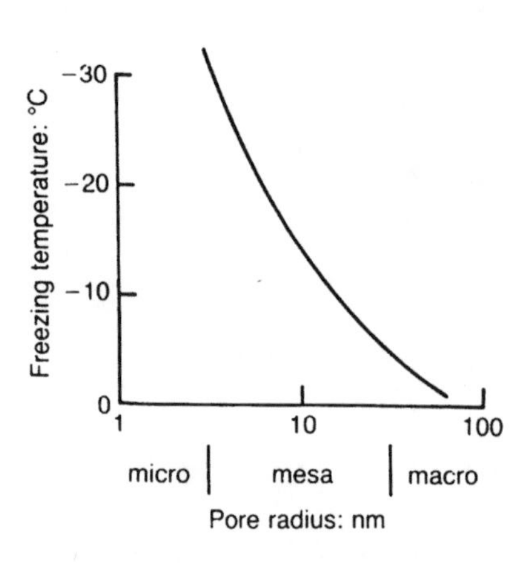

Figure 27. Depression of freezing point due to surface energy.[30]

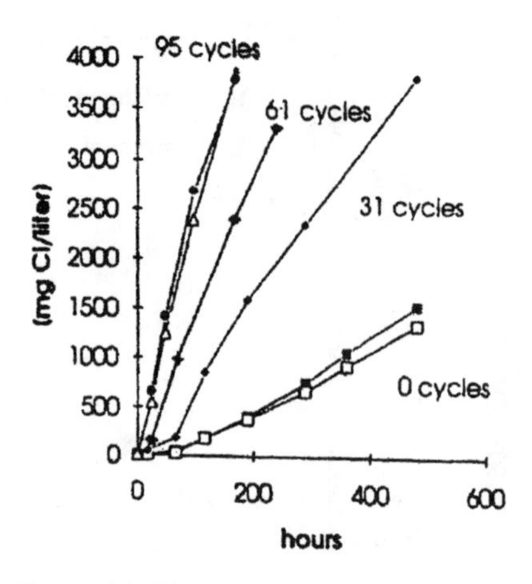

Figure 28. Effect of progressive cracking due to freezing and thawing on the chloride migration.[62]

Effect on Barrier Performance

Freeze/thaw studies have focused mostly on mix design optimization and the effect of cover damage on the reinforcement corrosion. The only information available on the effect of freeze/thaw damage on the permeation characteristic of concrete has been in connection with changes in the chloride penetration rates. Work by Jacobsen et al.[62] showed that progressive cyclical freeze/thaw damage results in increased chloride migration (Fig. 28).

Corrosion

Mechanisms

The onset of the corrosion process is dependent on the permeation characteristics of the host concrete. The two factors resulting in the depassivation of the "inert" steel in the concrete matrix are carbonation and chloride penetration, the rates of which are permeability dependent. Once general or local depassivation is achieved, the corrosion reaction is a function of the availability of oxygen and water, which again is permeability dependent. Progressive oxidation of the steel bars results in the volumetric expansion of the original steel into corrosion products, causing debonding and cracking of the host concrete.

Thus, in assessment of the corrosion process, two factors relating to cracking are of interest: how the existing cracks (due to shrinkage, loading, chemical deterioration, etc.) affect the depassivation rate and onset of the corrosion, and how corrosion-induced cracking affects the subsequent corrosion rate.

The existing cracks in concrete can be divided into three categories as shown in Table VI:

1. Uniformly distributed microcracks (due to drying, sulfate attack, AAR, etc.). Size will affect the global permeation characteristics of the host concrete.
2. Coincidental macrocracks, which occur along the length of the reinforcement and are usually due to plastic settlement and/or shrinkage, bond contraction and failure, or the corrosion process. The existence of the coincidental cracks is likely to lead to development of the general corrosion of the reinforcement.
3. Intersecting macrocracks, which occur across the reinforcement and are usually caused by mechanical loads (flexural or shear). The existence of the intersecting cracks may result in the development of the microcorrosion cells, where the anode is devolved at the crack tip, while the cathode is at the crack-free areas of the reinforcement.

In the presence of a crack the corrosion process can have two mechanisms (Fig. 29):

1. Microcell corrosion, where the anodic and cathodic process takes place at the tip of the crack, and the oxygen required for the cathodic reaction and the water required for the anodic reaction are supplied through the crack.
2. Macrocell corrosion, where reinforcement in the crack zone is depassivated and acts as an anode, while the passive steel between cracks acts as a cathode. The supply of oxygen through the uncracked concrete to the cathodic steel is usually the limiting step in the macrocorrosion reaction.[63]

Through this definition, it is likely that if all other factors remain equal, the permeation characteristics of cracks alone (not the host concrete) will be of greater significance in the microcell corrosion.

Resulting Changes in Microstructure

In connection with corrosion damage, the macrostructure, rather than microstructure, of the cementitious system is evaluated. Similarly, in examining the corrosion-induced cracking, only the effect of macrocracks is considered. In the cases of macrocracking the issue of the cementitious matrix

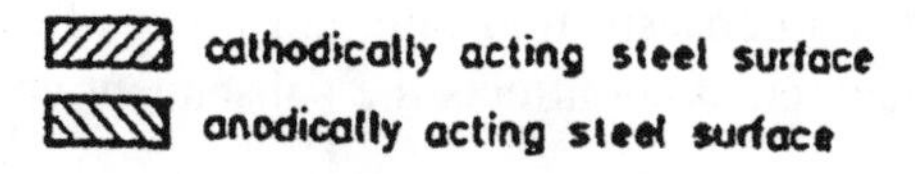

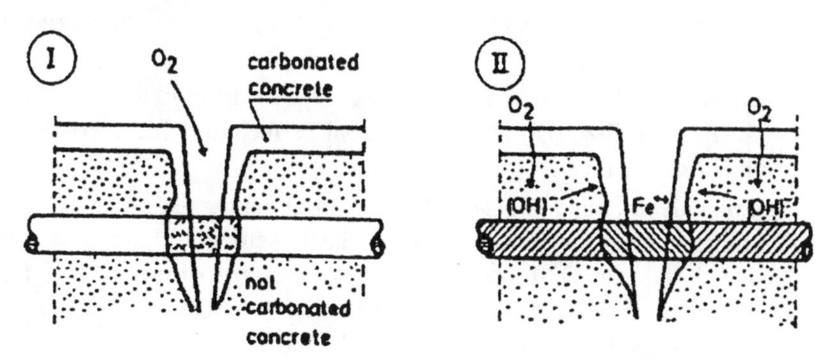

Figure 29. Schematic representation of two types of corrosion processes in the region of cracks.[63]

becomes secondary, as the permeation characteristics become largely governed by the permeation through cracks. It is of interest, however, to establish the critical crack parameters, so that concrete can be designed to regulate critical crack parameters (such as crack width, as discussed above). It can be argued, however, that crack parameters are only a function of the intensity and the duration of the exposure conditions, and under severe exposure, all crack parameters should be zero, thus the development of the high-performance concrete. High-performance concrete, however, is not crack-proof. Thus the key crack parameters with respect to corrosion should be defined: crack width at the surface, crack depth, crack width at the reinforcement, crack turtuosity, and crack locations (coincidental cracks and frequency of intersecting cracks).

The permissible crack widths are shown as a function of concrete cover in Table VIII.

Effect on Barrier Performance

In general, existence of cracks in concrete does increase its susceptibility to corrosion. Work by Jacobsen et al.[62] on the uniformly distributed cracks formed by freeze/thaw damage showed the increase in chloride concentration with the increased number of freeze/thaw cycles (Fig. 28). Discrete cracks have also been shown to increase the rate of mass transfer and thus

Table VIII. Permissible crack widths (mm) [64]

Rebar type		Environmental conditions due to vulnerability to corrosion reinforcement		
		Normal	Corrosive	Severely corrosive
Reinforced concrete	Deformed and round	0.05c	0.04c	0.035c
Prestressed concrete	Deformed	0.05c	0.04c	0.035c
	Prestressing steel	0.04c	0.035c	0.035c

affect corrosion resistance of the concrete. For example, Thaulow and Grelk[65] showed increased chloride concentrations around the cracks. Bazant et al.[66] demonstrated the effect of cracking on the drying rates, with the results showing that cracked specimens dried faster than noncracked concrete of the same mix. Lorentz and French[67] showed increased current density in cracked concrete in both coated and uncoated reinforcement (Fig. 30), thus indicating increased corrosion activity under cracked condition.

The permeation characteristics of the discrete cracks have been correlated to crack width. Petteson and Sandberg[68] produced correlation between chloride threshold and crack width, as measured on the surface. It has been shown, however, that unless the crack depth is greater than the 50% of the cover, the cover properties are not significantly impaired.[32] Moreover, the surface parameters of a crack may be very different from its parameters at the reinforcement.[69] Crack frequency has also been found to be a significant factor in the corrosion of reinforcement, where cumulative weight loss due to corrosion was shown to increase exponentially as the crack frequency was progressively increased from 0 to 20.[70] The variability of the crack width from the surface to the reinforcement and the crack frequency may explain the results of some studies, showing no correlation between crack widths and corrosion for crack widths up to 1 mm.[71,72]

In summary, the work on mechanically induced cracking and its effect on permeability and subsequent effect on corrosion rates shows some agreement on the importance of crack width, and the existence of critical of threshold crack parameters, beyond which permeation characteristics are affected. To date, the critical width is on the order of one-tenth of a millimeter; however, this value is based on very few studies conducted in this emerging area of research.

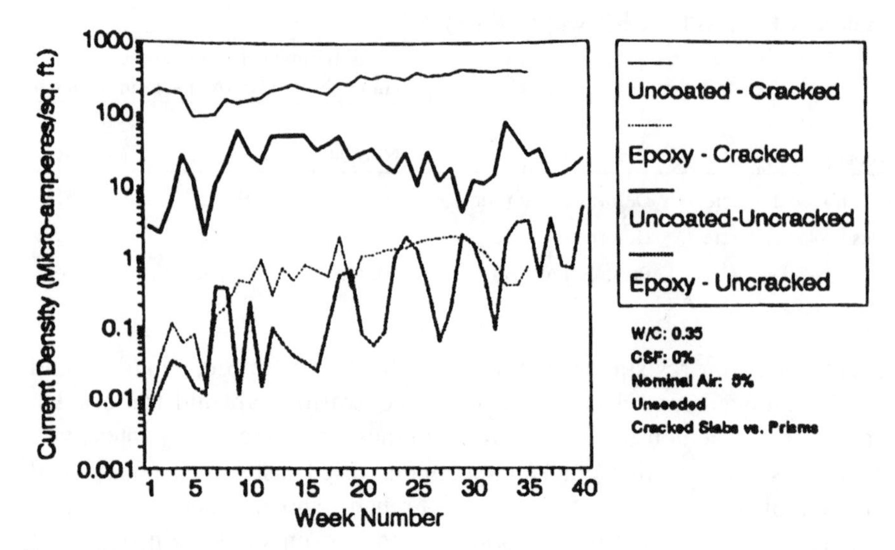

Figure 30. Effect of cracking on current density of coated and uncoated reinforcement.[67]

Chemical Deterioration

Alkali-Aggregate Reactions

Mechanisms

Alkali-aggregate reactions involve chemical interactions between the alkali hydroxides in the pore solution and certain types of reactive aggregates in concrete. The reaction products are either an alkali-silica gel or carbonate reaction products, which have a large capacity to imbibe water. The overall expansion is due to the expansion of the reactive aggregate particles and possibly swelling of the extruded reaction product (gel), resulting in the tensile stresses exerted on the surrounding cement matrix and subsequent cracking.

AAR has been classified into three types of reactions[73]:

1. Alkali-carbonate reaction (ACR) with argillaceous dolomitic limestone results in the formation of dark carbonate reaction products. These are found on the edges of the limestone aggregate particles.
2. Alkali-silicate/silica reaction (ASSR) occurs in alkali-rich environments in the presence of certain types of quartz-bearing aggregates. The silicate minerals in these aggregates absorb water and expand, causing disruption in the concrete's microstructure.

3. Alkali-silica reaction (ASR) involves reaction with the various forms of poorly crystalline reactive silica contained in certain types of aggregates, with the reaction product being alkali-silica gel (which differs from the products of ACR and ASSR).

ASR is the most common and most damaging of the three types of AAR. Diamond[74] has shown that the gel can expand as much as 80%, developing swelling pressure above 11 MPa, which is above the tensile stress limit of concrete. In order to sustain ASR, reactive silica in the aggregate, alkali hydroxides in the pore solution, a source of calcium (cement hydration products), and water are needed.

The overall effect of ASR is significant volumetric expansion, which causes visible cracking, sapling, and misalignment of concrete structural elements. While in the advanced stages of AAR the degradation signs are obvious; it is also of interest to determine whether the initial microcracking has effect on the fundamental properties of concrete (such as strength and durability).

Resulting Changes in Microstructure

In terms of transport characteristics of concrete, AAR creates two competing mechanisms. The gel formation causes partial blocking of the existing pore structure. Expansion creates internal microcracking, which subsequently may get filled with the expansive gel as the reaction proceeds. When considering the effect of AAR on the barrier characteristics of concrete, three factors must be considered: the permeability of the gel to the various transport mechanisms, the aging of the gel structure, and the level of expansion that corresponds to the changes in the barrier characteristics (if the gel is not present). These three factors can be used to assess relative importance of early and subsequent effects of AAR on the transport characteristics.

Effect on Barrier Performance

Very little research has been conducted on the effect of AAR on the barrier performance of concrete.

Work by Raphael et al.[75] has tested water permeability of concretes from eight dams affected by AAR and showed conflicting data between the extent of damage and the permeability levels, with permeability data ranging from 2×10^{-6} to 5×10^{-13} m/s. Thomas's research,[76] on the other hand, did find correlation between damage rating and percent expansion to the

Table IX. Effect of AAR damage on permeation characteristics[76]

Petrographic observations	Damage rating	RCPT (C)	O_2 permeability (m^2)
No significant damage due to AAR	<100	2 000–5 000	10^{-17}–10^{-16}
Minor cracking (width ~0.1 mm)	100–300	5 000–10 000	10^{-16}–10^{-15}
Extensive damage (width >0.5 mm)	>300	>10 000	10^{-15}–10^{-14}

changes in RCPT and gas permeability, for 10 hydraulic structures affected by AAR (Table IX).

The work by Hamada et al.[77] on AAR concrete exposed to chloride environment showed higher chloride content in AAR damaged concrete, thus showing increased level of chloride ion migration due to AAR damage.

Thomas's study[76] on the laboratory specimens found that changes in permeability were detectable at expansion levels above 0.10%, attributing that to the discontinuity in the formed cracks at lower expansion levels (0.04–0.60%). These expansion levels corresponded to >0.5 and <0.1 mm crack widths, respectively.

More recent work by Nelson[78] tested correlation of the effect of the expansion on RCPT, diffusion, and water and air permeability. The only clear relationship was found between the level of expansion and air permeability. For other types of permeation no clear increase was observed, even at expansion levels of 0.3% (Fig. 31). The sensitivity of air permeability to the level of expansion can be attributed to the dehydration of the ASR gel during the drying process, thus unblocking the flow passages. For the field concrete tested in the same study, the expansion levels did not correlate to the various transport mechanisms, including air permeability. This difference between young and old AAR samples may be due to the changes in the ASR gel structure, which becomes less sensitive to the presence of water.[79] In this study, no measurements were made of the crack widths.

Based on these very few studies, some preliminary conclusions can be made regarding effect of AAR damage on transport properties:

1. Early age expansion correlation with permeability may not be the same as those found in well-aged concretes (possibly due to changes in the gel[79]).
2. Transport through dry concrete is considerably higher than the AAR concrete in wet state, due to the hydrophilic nature of the gel.
3. Average crack width measurements, rather than the overall expansion levels, may provide better correlation with the various transport mechanisms.

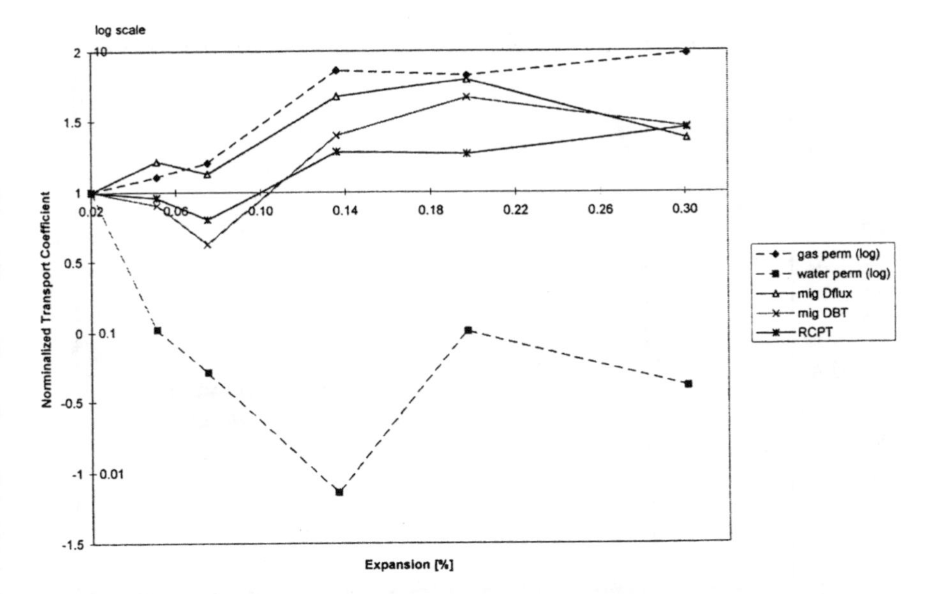

Figure 31. Data showing the relationship between percentage expansion of concrete samples due to AAR and various mass transport mechanisms.[78]

Sulfate Attack

Mechanisms

Sulfate attack is the result of complex reactions between sulfate ions and components of cement matrix. The sulfate may originate from internal sources (cement, aggregate, admixtures) or external sources (soils, ground water). The reactions with the cement components may lead to volumetric changes, increased porosity, destruction of the cementing properties of the C-S-H, and partial or complete removal of $Ca(OH)_2$ from the paste. As shown in Table V, $Ca(OH)_2$ crystals correspond in dimension to capillary porosity. Thus, removal of the $Ca(OH)_2$ will greatly increase the overall porosity and its connectivity. These findings have been supported by the modeling data from Garboczi and Bentz[80] (Fig. 32).

Resulting Changes in Microstructure

The mechanism of the chemical and microstructural change accompanying the sulfate attack will have an effect on the permeation characteristics, starting at the onset of the damaging reactions. For example, removal from the paste of $Ca(OH)_2$ and efflorescence (formation and leaching of sodium sul-

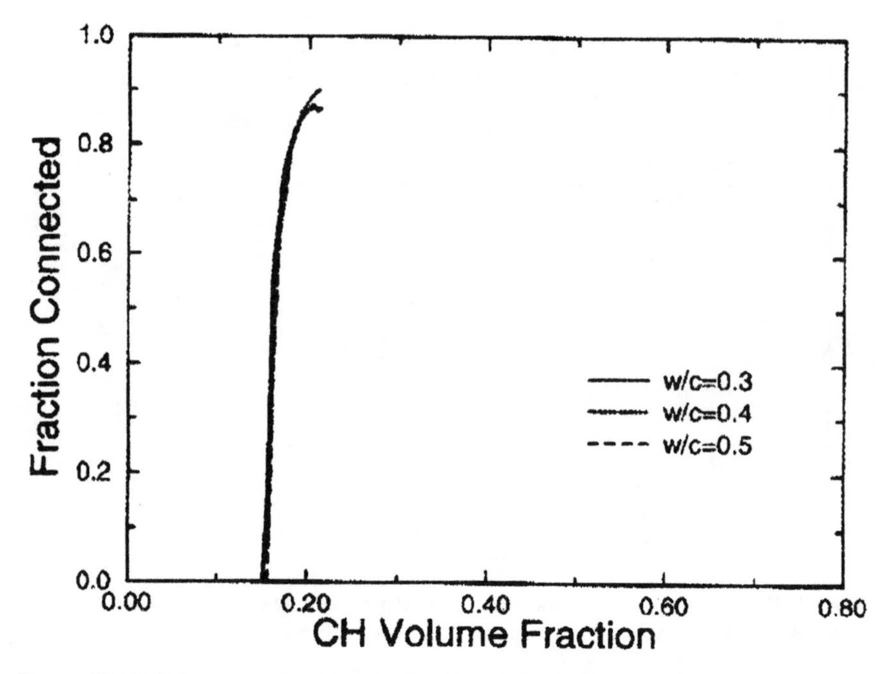

Figure 32. Modeling data showing the fraction of the CH phase that is part of a connected pathway as a function of the volume fraction of C-S-H for three water/cement ratio C_3S cement pastes.[80]

fate at exposed surfaces) may lead to increased porosity. At the same time, the formed reaction products may lead to temporary filling of the empty spaces within the matrix. Overall expansion of the paste, possibly due to formation of ettringite, thaumasite, or gypsum, eventually increases the crack width and density, thus allowing higher permeability.

Effect on Barrier Performance

The effect of progressive damage by sulfate attack and the subsequent changes in the permeation characteristics have not been addressed in the literature. A study by Katri et al.[81] examined the relative importance of the binder composition and permeability of the intact matrix. Progressive changes in the permeability with increasing levels of expansion were not investigated.

Conclusion

The correlation between deterioration processes and permeation characteristics is currently a hot issue in concrete science. In concrete, all durability issues are related to the changing interrelationship between evolution/deterioration of the microstructure and permeation characteristics. Understanding and quantifying these processes will provide needed information for the service-life prediction models and assessment of the degree of damage on the material properties.

As many of the degradation processes result in cracking of concrete microstructure, an evaluation of the critical crack parameters is required. To date, damage has been defined by degree of expansion, crack density, and crack width. Crack width seems to have the best correlation with the mass transfer characteristics. Many deterioration processes, however, do not produce discrete cracks, so that crack width determination in uniformly distributed crack system is difficult.

References

1. G.W. Hyde and W.J. Smith, "Results of experiments made to determine the permeability of cements and cement mortars," *J. Franklin Inst. Philadelphia,* pp. 199–207 (1889).
2. P. Bamforth, "Specifying and testing impermeability," BCA Course, Advances in Concrete Technology and Construction Practice, Walsall, Dec. 1994.
3. R.F.M. Bakker, *Permeability of Blended Cement Concretes,* ACI SP-79, Vol. 2. Edited by Malhotra. 1983. Pp. 589–605.
4. L.B.B. Peer, "Water flow into unsaturated concrete," Ph.D. Thesis, University of Cambridge, UK, 1990.
5. R.H. Mills, "Mass transfer of water vapour through concrete," *Cem. Concr. Res.,* **15**, 74–82 (1985).
6. K. Leong, "Discontinuity in pore structure and its effect on mass transfer through hardened cement paste," M.A.Sc. Thesis, University of Toronto, 1998.
7. H. W. Reinhardt, "Transport of chemicals through concrete"; pp. 209–241 in *Materials Science of Concrete III.* Edited by J.P. Skalny. American Ceramic Society, Westerville, Ohio, 1992.
8. A.E. Long, P.A.M. Basheer, and F.R. Montgomery, "In-situ permeability testing — A basis for service life prediction"; pp. 651–670 in *Advances in Concrete Technology,* Proceedings of the 3rd CANMET/ACI International Conference, Auckland, New Zealand, 1997.
9. T.C. Powers, "Structure and Physical Properties of Hardened Portland Cement Paste," *J. Am. Ceram. Soc.,* **41**, 1–6 (1958).
10. R.F. Feldman and P.J. Sereda, "A model for hydrated Portland Cement paste as deducted from sorption-length change and mechanical properties," *Mater. Struct.,* **1** [6] 509–520 (1968).

11. I. Soroka, *Portland Cement Paste and Concrete.* MacMillan Press Ltd., New York, 1979.
12. F.H. Wittman and G. Englert, "Water in hardened cement paste," *Mater. Struct.,* **1** [6] 535–546 (1968).
13. S. Mindess and J.F. Young, *Concrete.* Prentice-Hall Inc., New York, 1981.
14. T.C. Powers and T.L. Brownyard, *Studies of Physical Properties of Hardened Portland Cement Paste,* R&D Bulletin 22. Portland Cement Association, 1948.
15. N. Hearn, "Comparison of water and propan-2-ol permeability in mortar specimens," *Adv. Cem. Res.,* **8** [3] 81–86 (1996).
16. T.C. Powers, "The physical structure and engineering properties of concrete," R&D Bulletin 90. PCA, 1959. Pp. 1–28.
17. N. Hearn, "Saturated permeability of concrete as influenced by cracking and self-sealing," Ph.D. Thesis, University of Cambridge, Cambridge, UK, 1993.
18. B.K. Nyame and J.M. Illston, "Relationship between permeability and pore structure of hardened cement paste," *Mag. Concr. Res.,* **33** [116] 139–146 (1981).
19. T.C. Powers, L.E. Copeland, and H.M. Mann, "Capillary continuity or discontinuity in cement pastes," *J. PCA R&D Labs.,* **1** [2] 38–48 (1959).
20. E.J. Garboczi and D.P. Bentz, "Fundamental computer simulation models for cement-based materials" pp. 249–277 in *Materials Science of Concrete II.* Edited by J. Skalny and S. Mindess. American Ceramic Society, Westerville, Ohio, 1989.
21. D.C. Hughes, "Pore structure and permeability of hardened cement paste," *Mag. Concr. Res.,* **37** [133] 227–233 (1985).
22. D.N. Winslow and S. Diamond, "Mercury porosimetry study of the evolution of porosity in Portland Cement," *J. Materials,* **5**, 564–585 (1970).
23. R.H. Mills, "Factors influencing cessation of hydration in water cured cement pastes." Special Report 90. Highway Research Board, Washington, D.C., 1966.
24. J. Bonzel, "Der Einfluss des Zements, des W/Z Werts, des Alters und der Lagerung auf die Wasserundurchlässigkeit des Beton," *Beton,* No. 3, pp. 379–383 and No. 10, pp. 417–421 (1966).
25. D.T. Snow, "Anisotropic permeability of fractured media," *Water Resources Research,* **5** [6] 1273–1289 (1969).
26. B. Fauchet, "Analyse poroplastique des barrages en béton et de leurs fonation. Rôle de la pression interstitielle," Ph.D. Thesis, ENPC, 1991.
27. C.M. Sayers, "Fluid flow in a porous medium containing partially closed fractures," *Transport in Porous Media,* No. 6, 331–336. (1991)
28. Z.P. Bazant, S. Sener, and J.K. Kim, "Effect of cracking on drying permeability and diffusivity of concrete," *ACI Mater. J.,* **9-10**, 351–357 (1987).
29. E. Bourdarot, *Application of a porodamage model to analysis of concrete dams.* EDF/CNEH. 1991.
30. Comite Euro-International du Beton, *Durable Concrete Structures, Design Guide.* Thomas Telford, 1992. pp. 7–20.
31. C.A. Clear, "The effect of autogenous healing upon leakage of water through cracks in concrete." Technical Report 559. 1985
32. J.M. Frederiksen, L.O. Nilsson, P. Sandberg, and E. Poulsen, "A system for estimation of chloride ingress into concrete, theoretical background." Report No. 83. The Danish Road Directorate. 1997.

33. H.F.W. Taylor, *Cement Chemistry.* Academic Press Ltd., London, 1990.
34. N. Hearn, "Effect of shrinkage and load-induced cracking on water permeability of concrete," *ACI Mater. J.,* **96** [2] 234–241 (1999).
35. H.R. Samaha and K.C. Hover, "Influence of microcracking on the mass transport properties of concrete," *ACI Mater. J.,* **89** [4] 416–424 (1992).
36. R.L. Berger, *Science,* pp. 626–629 (1972).
37. R.L. Berger, F.V. Lawrence, and J.F. Young, *Cement and Concrete Research,* Vol. 3. 1974.
38. S. Mindess, J.S. Nadeau, and J.M. Hay, "Effects of different curing conditions on slow crack growth in cement paste"; pp. 953–965 in *Cement and Concrete Research,* Vol. 4. 1974.
39. J.E. Barrick and E.M. Krokosky, *J. Testing Eval.,* 4, 61–73 (1976).
40. A.M. Neville, *Civil Engineering.* pp. 1153–1156, 1308–1310, 1435–1439. London, 1959.
41. F.M. Kaplan, *J. Am. Concr. Inst.,* **58**, 591–610 (1961).
42. J. Glucklich, Proceedings of International Conference on Mechanical Behaviour of Materials, Kyoto, August 1971.
43. J. Glucklich, *ASCE Proc. J. Eng. Mech. Div.,* **89**, 127–138 (1963).
44. J.H. Brown, *Mag. Concr. Res.,* **24**, 185–196 (1972).
45. S. Mindess, "Fracture process zone detection"; in *Fracture Mechanics Test Methods for Concrete,* Report of Technical Com., 89-FMT. Edited by S.P. Shah and A. Carpinteri. Chapman & Hall, 1991.
46. A. Bascoul, C.H. Detriche, J.P. Olivier, and A. Turnatsinze, "Microscopical observation of the cracking propagation in fracture mechanics for concrete"; pp. 327–336 in *Fracture of Concrete and Rock: Recent Developments.* Edited by Shah, Swartz, Barr. Elsevier Applied Science, London, 1989.
47. A. Carpinteri and A.R. Ingraffea, *Fracture mechanics of concrete: Material characterization and testing.* Martinus Nijhoff Publ., 1984.
48. T.T. Hsu, F.O. Slate, G.M. Sturman, and G. Winter, "Microcracking of plain concrete and the shape of the stress-strain curve," *J. ACI,* pp. 209–223 (1963).
49. F.A. Blakely, "Some considerations of the cracking or fracture of concrete," *Civ. Eng. Public Works Rev.,* **50** (1955).
50. R.L. Carrasquillo, F.O. Slate, and A.H. Nilson, "Microcracking and behaviour of high strength concrete subjected to short-term loading," *ACI J.,* pp. 179–186 (1981).
51. B.L. Meyers, F.O. Slate, and G. Winter, "Relationship between time-dependent deformation and microcracking of plain concrete," *ACI J.,* pp. 60–68 (1969).
52. S.D. Santiago and H.K. Hilsdorf, "Fracture mechanism of concrete under compressive loads," *Cem. Concr. Res.,* **3**, 363–388 (1973).
53. H. Rusch, "Research toward a general flexural theory for structural concrete," *ACI J.,* pp. 1–29 (1961).
54. M.M. Samadi and F.O. Slate, "Microcracking of high and normal strength concrete under short- and long-term loading, *ACI Mater. J.,* pp. 117–127 (1989).
55. A.S. Ngab, F.O. Slate, and A.H. Nilson, "Microcracking and time-dependent strains in high strength concrete," *ACI J.,* pp. 262–268 (1981).
56. A. Kermani, "Permeability of stressed concrete," *Building Res. Info.,* **19** [16] 360–366 (1991).

57. C.M. Aldea, S.P. Shah, and A. Karr, "Permeability of cracked concrete," *Mater. Struct.,* pending publication.
58. S.J. Pantazopoulou and R.H. Mills, "Microstructural aspects of mechanical response of plain concrete," *ACI Mater. J.,* **92** [6] 605–616 (1995).
59. N. Hearn and G. Lok, "A method of measuring permeability of mortar under uniaxial compression," *ACI Mater. J.,* **95** [6] 691–694 (1998).
60. K. Wang, D. Jansen, S.P. Shah, and A. Karr, "Permeability study of cracked concrete," *Cem. Concr. Res.,* **27** [3] 381–393 (1997).
61. R.H. Mills, "Mass transfer of gas and water through concrete"; pp. 261–268 in *Concrete Durability,* ACI SP-100. Edited by Scarlon. 1981.
62. S. Jacobsen, J. Marchand, and B. Gerard, "Concrete cracks 1: Durability and self-healing – A review"; pp. 217–231 in *Proceedings of the Second International Conference on Concrete under Severe Conditions,* Tromso, Norway, June 1998.
63. P. Schiessl and M. Paupach, "Laboratory studies and calculations on the influence of crack width on chloride-induced corrosion of steel in concrete," *ACI Mater. J.,* **94** [1] 56–62 (1997).
64. A.D. Jensen and S. Chatterji, "State of the art report on micro-cracking and lifetime of concrete – Part 1," *Mater. Struct.,* **29** [Jan.] 3–8 (1996).
65. A. Neville, "Chloride attack of reinforced concrete: An overview," *Mater. Struct.,* **28**, 63–70 (1995).
66. Z.P. Bazant, S. Sener, and J.K. Kim, "Effect of cracking on drying permeability and diffusivity of concrete," *ACI Mater. J.,* [Sept-Oct.] 351–357 (1987).
67. T. Lorentz and C. French, "Corrosion of reinforced steel in concrete: effects of materials, mix composition and cracking," *ACI Mater. J.,* **28** [March-April] 181–190 (1995).
68. K. Petteson and P. Sandberg, "Chloride threshold levels and corrosion rates in cracked high performance concrete exposed in a marine environment"; presented at the 4th CANMET/ACI International Conference on Durability of Concrete, Sydney, Australia, August 17–22, 1997.
69. A.W. Beebly, "Corrosion of reinforces steel in concrete and its relation to cracking," *Struct. Eng.,* **56A** [3] 77–81 (1978).
70. C. Aray and F.K. Orfoi-Darco, "Influence of crack frequency on reinforcement corrosion in concrete," *Cem. Concr. Res.,* **26** [3] 345–353 (1996).
71. R. Francois and G. Arliguie, "Influence of service cracking on reinforcement steel corrosion," *J. Mater. Civ. Eng.,* **10** [1] 14–20 (1998).
72. N. Nilsen and B. Espolid, "Corrosion behaviour of reinforced concrete under dynamic loading," *Mater. Perform.,* [July] 44–50 (1985).
73. R.N. Swamy, *The Alkali-Silica Reaction in Concrete.* Blackie and Sons Ltd., 1992.
74. S. Diamond, "Alkali-reactions in concrete — Pore solution effects"; pp. 155–156 in Proceedings of the Sixth International Conference on Alkalis in Concrete. 1986.
75. S. Raphael, S. Sarkar, and P.C. Aitcin, "Alkali-aggregate reactivity — Is it always harmful?"; pp. 809–814 in *8th International Conference on AAR.* Edited by Okada, Nishibayashi, and Kawamura. 1997.
76. M.D.A. Thomas, "Synergy between alkali-aggregate reaction and other deterioration mechanisms in concrete"; pp. 571–586 in *Proceedings of 4th CANMET/ACI International Conference on Durability of Concrete,* Vol. 1. Edited by Malhotra. Sydney, Australia, 1997.

77. H. Hamada, N. Otsuki, and T. Fukute, "Properties of concrete specimens damaged by alkali-aggregate reaction, laumonite related reaction and chloride attack under marine environments"; pp. 603–608 in *8th International Conference on AAR.* Edited by Okada, Nishibayashi, and Kawamura. 1989.
78. N.K.C. Or, "Cracking due to alkali-silica reaction and its effect on durability properties of concrete," M.A.Sc. Thesis, University of Toronto, 1999.
79. T. Katayama and D. Bragg, "Alkali-aggregate reaction combined with freeze/thaw in Newfoundland, Canada — Petrography using EPMA"; pp. 243–250 in *10th International Conference on Alkali-Aggregate Reaction in Concrete.* Edited by A. Shayan. 1996.
80. E.J. Garboczi and D.P. Bentz, Material from the NIST modeling workshop. NIST, Gaithersburg, Maryland, 1999.
81. R.P. Katri, V. Sirivivatnon, and J.L. Yang, "Role of permeability in sulphate attack," *Cem. Concr. Res.,* **27** [8] 1179–1189 (1997)

The Use of Silica Fume to Control Expansion due to Alkali-Aggregate Reactivity in Concrete: A Review

M.D.A. Thomas and R.F. Bleszynski
University of Toronto, Toronto, Canada

> *". . . if you want to succeed in this world you don't have to be much cleverer than other people; you just have to be one day earlier . . ."*
> — Leo Szilard

There have been numerous studies on the effect of silica fume on alkali-silica reaction (ASR) in concrete, dating back more than 20 years. However, there is conflicting evidence concerning the long-term efficacy of silica fume for controlling the expansion of concrete containing reactive aggregates and the appropriate minimum "safe level" required for eliminating damaging expansion. Consequently, many national specifications do not give specific advice for using silica fume in this role or impose a blanket minimum level that may not be practical for other reasons. This review presents a synthesis of published data from laboratory and field studies for the purpose of developing a rational approach for specifying silica fume for controlling expansion due to ASR. It is clear from this study that the efficiency of silica fume is influenced to some degree by the nature of the reactive aggregate and to a much greater extent by the availability of alkalis in the concrete mix. Consequently, it would seem appropriate to base the minimum level of replacement on the availability of alkalis in the mix (and perhaps the type of reactive aggregate). Two approaches are presented in the paper; these are SF = 2.5 × alkali contributed by the OPC, where SF is the level of silica fume expressed a percentage of the mass of cementitious material and the alkali contributed by the Portland cement is expressed as kg of Na_2O_e per m^3 of concrete; and SF = 10 × alkali content of the blended cement, where the alkali content of the blended cement is expressed as percent Na_2O_e. Evidence is presented to demonstrate that these equations are consistent with published data from both expansion and pore solution studies. It is likely that similar approaches will be adopted by the latest versions of both the Canadian and British guidelines.

Introduction

Alkali-aggregate reaction (AAR) was first implicated as a cause of concrete deterioration almost sixty years ago (Stanton 1940). Since then, a number of approaches for avoiding the problem or, perhaps more accurately, "minimizing the risk of damage due to the reaction" have become accepted practice in the construction industry. These preventive methods include the avoidance of reactive aggregates, the use of low-alkali cement, controlling the alkali content of the concrete, and the use of supplementary cementing materials (SCM), including fly ash, slag, silica fume, and natural pozzolans.

Stanton (1940) reported the effective use of ground shale and pumice to control expansion due to AAR in his earliest work, and later demonstrated the efficacy of a wider range of natural pozzolans in this role (Stanton 1950). The first major use of a natural pozzolan to combat AAR was in the Davis Dam (Gilliland and Moran 1949), constructed over 40 years ago using an aggregate similar to that previously implicated in damaging AAR at the nearby Parker Dam. The 1950s saw the start of AAR research on fly ash and slag, since which time nearly 500 technical articles dealing with the effect of fly ash and slag on AAR have been published (Thomas 1996a). Interest in the possible use of silica fume in concrete began in the 1950s in Norway (Radjy et al. 1986). However, the first report on the potential use of silica fume to control AAR in concrete appeared in 1979 following research studies at the Building Research Institute in Iceland (Asgeirsson and Gudmundsson 1979). Much research has since been published to demonstrate that the incorporation of sufficient levels of silica fume in concrete is an effective means of suppressing AAR expansion. Although it is widely accepted that the level of silica fume required to prevent expansion is significantly less than the level of fly ash or slag required for the same purpose, there is no general consensus regarding the precise "safe level" of silica fume. This is likely to vary depending on a number of parameters, including the composition of the silica fume, the nature of the reactive aggregate, the alkali content of the concrete, and the type of exposure.

Although many countries now permit the use of SCMs, including silica fume, to control AAR expansion, there are differences regarding the method of use and specification (Nixon and Sims 1992). The minimum safe level of SCM varies between countries and there is considerable conflict regarding the alkalis in the SCM and the "availability" of these alkalis for reaction. The Canadian guidelines dealing with AAR in concrete (Canadian Standards Association 1994) give positive advice on the use of fly ash and slag with reactive aggregates, but intentionally avoid specific recommendations for using silica fume. This omission reflects certain reservations regarding the ability of silica fume to control AAR expansion in the long term when used at levels of replacement typically used in concrete practice, that is, less than 10 mass% of cementitious material (Bérubé and Duchesne 1991, 1992b; Chen et al. 1993; Durand 1993; Fournier et al. 1995).

This chapter presents a review of published research findings related to the effect of silica fume on AAR in concrete.

Alkali-Aggregate Reactions in Concrete

There are, essentially, two types of alkali-aggregate reaction: alkali-carbonate reaction (ACR) and alkali-silica reaction (ASR). ACR involves certain forms of dolomitic carbonate aggregates and its occurrence is restricted to a few remote locations around the world. ASR is far more widespread, and deterioration of concrete structures due to the reaction have been reported worldwide. Many forms of siliceous aggregate have been found to be reactive to some degree and sources of truly nonreactive aggregate are not readily available in some locations. Consequently, considerable effort has been made in the last five decades to find ways to control the reaction when reactive siliceous aggregates are used. This review focuses on the use of silica fume to control ASR in concrete.

Alkali-Silica Reaction

Alkali-silica reaction is actually a reaction between the hydroxyl ions in the pore solution and thermodynamically unstable silica in the aggregate; the silica is not directly attacked by the alkalis (Na + K). The alkalis contribute initially to the high concentration of hydroxyl ions in solution and later to the formation of an expansive alkali-silica gel. The extent or rate of dissolution is controlled by the alkalinity of the solution and the structure of the silica. The solubility of poorly crystalline or amorphous silica increases with pH; however, there have been few studies to determine the minimum hydroxide concentration of the pore solution required to initiate and sustain ASR in concrete. Diamond (1983a) reports that the threshold concentration is unlikely to be less than 0.25 M, and Kolleck et al. (1986) suggest a threshold of 0.20 M.

Well-crystallized or dense forms of silica, such as quartz, are relatively inaccessible to alkaline hydroxide solution and dissolution takes place only at the surface at a very slow rate. Although it is generally considered that quartz is not susceptible to alkali-silica reactions in concrete, Mather (1973) has shown that deleterious reaction can occur in concrete containing "relatively unstrained, megascopically crystalline quartz" after prolonged periods of alkaline exposure. This might suggest that no siliceous aggregate is truly inert with regard to ASR.

Despite general acceptance of the chemical reactions involved, a number of different expansion mechanisms have been proposed. Hansen's osmotic theory (1944) regards the cement paste surrounding reactive grains to behave like a semi-permeable membrane through which water (or pore solution) may pass but the larger complex silicate ions may not. The water is drawn into the reacting grain where its chemical potential is lowest. An osmotic pressure cell is formed and increasing hydrostatic pressure is exerted on the cement paste, inevitably leading to cracking of the surrounding mortar. Hansen's laboratory experiments confirmed that osmotic pressures were generated by sodium silicate solutions when separated from water by a cement paste membrane.

McGowan and Vivian (1952) disputed the classical osmotic theory on the basis that cracking of the surrounding cement paste "membrane" due to ASR would relieve hydraulic pressure and prevent further expansion. Their work showed that expansion was accompanied by the formation and widening of cracks due to mechanical (absorption) rather than hydraulic forces. They proposed an alternative mechanism based on the physical absorption of water by the alkali silica gel and subsequent swelling of the gel. Tang (1981) concurred with the water imbibition and swelling theory; his observations of polished sections suggested that expansion occurs before the reaction product becomes fluid. Thus, the presence and composition of a semi-permeable membrane is insignificant.

Powers and Steinour (1955a) proposed a compromise, suggesting that both osmotic and imbibition pressures may be generated, depending on whether the alkali-silicate complex is fluid or solid. Regardless of the mechanism, the fundamental cause of swelling is thermodynamically the same, that is, the entry of water into a region where the effect of a solute or of adsorption reduces its free energy. Diamond (1989) concurred, and regarded the distinction between the two mechanisms as purely formal. Dent Glasser (1979) regarded the gel itself to behave like a semi-permeable membrane as a result of its insolubility.

Alkali-Silicate Reaction

Alkali-silicate reaction was first proposed by Gillott et al. (1973) as the mechanism of expansion with various rock types such as greywackes, phyllites, and argillites. Expansion of concrete containing these rocks occurred with apparently relatively small amounts of gel compared with traditional alkali silica reactive rocks. It was proposed that the expansion was due to the exfoliation and swelling of certain layered silicates, particularly vermiculite, in the presence of alkali hydroxide solutions. Indeed, many of these rock types exhibit expansion in NaOH solution, whereas most reactive siliceous aggregates merely dissolve without expansion. However, other workers have attributed the expansion of greywacke-argillite rocks to reaction of microcrystalline quartz within the matrix (Blackwell and Pettifer 1992; Grattan-Bellew 1990), and others have challenged the existence of the alkali-silicate mechanism (Tang 1992).

Fournier and Bérubé (1991) suggest that "swelling-clay exfoliation" may contribute to concrete expansion but is unlikely to be the dominant mechanism with these rocks. They offer further supporting evidence of the role of swelling clay minerals in the expansion of siliceous limestone and argillaceous dolomitic limestone.

For the purposes of this review, it is assumed that alkali-silica reaction is the dominant mechanism in greywackes and argillites. The effect of silica fume on expansion due to alkali-silicate reaction (or clay exfoliation) is not considered. Indeed, to the authors' knowledge, there is no published data dealing with this issue.

Alkali-Carbonate Reaction

ACR was first discovered by Swenson (1957) as the cause of concrete deterioration in Canada and was subsequently implicated in structures in the United States (Hadley 1961). The reaction occurs between alkali hydroxides and certain argillaceous dolomitic limestones; these dolomites are characterized by a matrix of fine calcite and clay minerals with scattered dolomite rhombohedra. The reaction is characterized by the rapid expansion and extensive cracking of concrete, and structures affected by ACR usually show cracking within 5 years.

Although there is a lack of consensus regarding the precise mechanisms involved, it is generally agreed that the reaction is accompanied by a dedolomitization process. However, since this reaction results in a reduction in solid volume, the expansion must be attributed to an alternative mechanism. Furthermore, Grattan-Bellew and Lefebvre (1987) observed expansive ACR without any evidence of dedolomitization. Swenson and Gillot (1964) proposed that the adsorption of water and alkalis by anhydrous clay leads to volume increases, with dedolomitization developing access channels for moisture. Tang et al. (1987) suggested that the expansion is caused by water molecules migrating into restricted spaces originally occupied by dolomite crystals. Crystallization and rearrangement of brucite crystals, water imbibition by dry clay, and osmotic pressures are also considered to contribute to expansion.

The alkali carbonate produced in the dedolomitization reaction may react with lime in the cement paste, thereby "regenerating" alkalis for further reaction. Thus, provided sufficient alkali is available to initiate the reaction, the process may continue independently of the amount of alkalis available in the concrete. This could explain why low-alkali cements may not be effective in controlling damaging reaction in some cases (Fournier and Bérubé 1991).

Preventive Measures for AAR

Methods for minimizing the risk of damaging reaction include the following:

- Use of nonreactive aggregate.
- Use of low-alkali cement.
- Limiting the alkali content of the concrete.
- Use of SCM (fly ash, slag, silica fume, and natural pozzolans).

Limiting the alkali content of the cement or concrete, or the use of SCMs, has generally been found to be ineffective in controlling expansion due to ACR. Therefore, the reactive constituents must be avoided by selective quarrying, beneficiation, or selection of an alternative aggregate (Canadian Standards Association 1994). The use of nonreactive aggregates, though an obvious choice, is often not economically viable due to the unavailability of suitable sources. Thus, it is not unusual for reactive aggregates to be knowingly used in concrete with appropriate preventive measures. Furthermore, laboratory testing may fail to identify a deleteriously reactive aggregate; damage due to AAR has occurred in a number of cases where prior testing indicated the aggregates to be inert. Consequently, the adoption of suitable preventives is often required by specification even if the aggregate tests to be nonreactive.

Stanton's formative work (1940) on alkali-aggregate reaction indicated that expansive reaction is unlikely to occur (with alkali-silica reactive aggregates) when the alkali content of the cement is below 0.6% Na_2O_e. This value has become the accepted maximum limit for cement to be used with reactive aggregates in the United States and has been adopted by ASTM C150. This criterion takes no account of the cement content of the concrete, which, together with the cement alkali content, governs the total alkali content of concrete. Some national specifications take cognizance of this fact by specifying a maximum alkali level in the concrete; this limit is reported (Nixon and Sims 1992) to range from 2.5 to 4.5 kg/m^3 Na_2O_e; in some countries the limit may vary depending on the reactivity of the aggregate.

The use of low-alkali cement and limitation of the alkali content in concrete is not a sufficient safeguard in some cases. Damaging ASR has been detected in both field and laboratory concrete despite the use of low-alkali cements (Lane 1987; Stark 1980). Furthermore, the alkali content of concrete may increase during exposure as a result of alkali concentration caused by drying (or temperature) gradients, alkali release from aggregates, or the ingress of alkalis from external sources, for example, de-icing salts or sea water (Nixon et al. 1979, 1987; Oberholster 1992; Stark 1978, 1980; Stark and Bhatty 1986; Thomas et al. 1992; Van Aardt and Visser 1977; Way and Cole 1982; Zhongi and Hooton 1993).

Although many national specifications and guidelines for minimizing AAR permit the use of supplementary cementing materials (SCM), such as fly ash, slag, and silica fume, with potentially reactive aggregate, there is conflicting evidence regarding the efficacy of these materials in this role; this is reflected in the lack of consensus advice. Much of the controversy is centred around the alkalis in the SCMs and whether they are potentially available for reaction. A review of published studies concerned with the effect of silica fume on AAR is presented in the following sections of this report. A review of the effects of fly ash and slag has already been published (Thomas 1996a).

The Nature of Silica Fume

Silica fume is a by-product of metallurgical processes used in the production of silicon metal or ferrosilicon alloys. It is essentially composed of very fine spherical particles (average diameter of 0.1 mm, i.e., approximately 100 times smaller than average portland cement particles) of amorphous silica (typically >90% SiO_2) with a high surface area (~20 000 m^2/kg, compared with typical values in the range of 300–500 m^2/kg for portland cement). As such, silica fume is a highly reactive pozzolan and may be used to considerable advantage as a supplementary cementing material in portland cement concrete. A comprehensive review of the properties of silica fume concrete and guidelines for its production and use was published by ACI Committee 234 (ACI 1994).

Silica fume may be available to the consumer in different forms. ACI defines four types of product as follows:

1. As-produced silica fume powder, which has a very low bulk density and fine particle size, thus resulting in handling difficulties.
2. Slurried silica fume, which typically comprises a 50:50 mix of silica fume and water. Proprietary slurry products with various chemical admixtures are available.

3. Densified (or compacted) silica fume, which is agglomerated using compressed air or mechanical means. Chemical admixtures may be added. The interparticle forces holding the agglomerates together are easily overcome by concrete mixing.
4. Pelletized silica fume produced by mixing with a small quantity of water. Pelletized silica fume is not suitable as a direct addition to concrete but may be interground with portland cement.

Slurried or densified silica fume are the most commonly used product forms in North America.

Research on the use of silica fume in concrete first began in the 1950s at the Technical University of Norway, with the first documented use in structural concrete occurring in the same country in 1971 (Radjy et al. 1986). In the 1970s, investigations were carried out at the Icelandic Building Research Institute to determine the effect of silica fume on AAR (Asgeirsson and Gudmundsson 1979), and since 1979, the Icelandic state cement works has exclusively produced blended silica fume cement. Interest in North America began at about the same time with the first experimental project being constructed in Quebec in 1980 (Aitcin and Regourd 1985). This structure, a sidewalk at a ferrosilicon smelting plant, also contained highly reactive aggregate. By 1981 silica fume was being incorporated into ready-mix concrete in Montreal, and the following year saw the production of the first blended silica fume–portland cement in Canada (Isabelle 1986). The first specification for silica fume for use in concrete was published in Canada in 1985 (Isabelle 1986). The use of silica fume in the United States dates back to a similar time, with the U.S. Army Corps of Engineers developing a specification for and using high-strength silica fume concrete for erosion repairs to the stilling basin of the Kinzua Dam in 1983 (Holland et al. 1986). However, the first national specification did not appear until 10 years later (ASTM C 1240). A review of the development of ASTM C 1240 and other national specifications was published by Holland (1995).

The effects of silica fume on the properties of plastic and hardened concrete are now fairly well established (ACI 1994). If properly used, silica fume imparts significant improvement to the strength and durability of concrete, and the availability of this material together with high-range water reducers (superplasticizers) has been largely responsible for the development of high-strength and high-performance concretes. However, high dosages of silica fume may have undesirable effects on concrete properties, especially rheology, plastic shrinkage, and air void characteristics. Thus, specifications frequently restrict the level of silica fume that can be used in normal concrete practice. For instance, in Canada the maximum level of silica fume that can be added is limited to 10% of the total mass of cementitious material (Canadian Standards Association 1994). As mentioned previously, the ability of silica fume to suppress expansion due to ASR is widely accepted, but there is conflicting evidence concerning the quantity of silica fume required to completely ameliorate the deleterious effects of the reaction. In many cases, the quantity required has been reported to be in excess of that commonly used in concrete practice.

The beneficial effects of silica fume on concrete properties rely on adequate dispersion of the silica fume particles throughout the concrete mix. This can be achieved only if a suitable product form, sufficient levels of water-reducing admixtures, and suitable mix proportioning and mixing procedures are used. Furthermore the effect of silica fume on concrete will depend to some extent on its physical and chemical composition, and the properties of the other ingredients with which it is mixed. Unfortunately, details of the materials and procedures used are often omitted from published studies. Differences between materials may account for some of the discrepancy between various research programs.

The Effect of Silica Fume on Expansion due to ASR

Methods of Evaluation

Experimental methods for assessing the effect of silica fume on ASR expansion have included tests on mortars and concretes stored under laboratory conditions and concretes stored under field conditions. Field performance of actual structures obviously provides the most reliable indication of material behavior. However, there are too few well-documented cases (Aitcin and Regourd 1985; Olafsson 1989) of the use of silica fume with reactive aggregates in structures to permit conclusive recommendations to be made. Furthermore, silica fume does not have a long history of use and information on long-term field performance is not available.

A number of standard laboratory tests have been used to determine the efficacy of silica fume; the most commonly used include the ASTM C 227 Mortar Bar Test, the ASTM C 441 Pyrex Mortar Bar Test, and various forms of the concrete prism test, such as CSA A23.2 14A. More recently, interest has been shown in modified versions of the accelerated mortar bar test (CSA A13.2 25A and ASTM C 1260), which involves immersion of mortar specimens in 1 M NaOH at 80°C.

Pyrex has been used as a reference aggregate for assessing the performance of pozzolans and slags for many years. The potential for silica fume to reduce ASR expansion becomes immediately apparent if it is used at the specified 25 vol% replacement of ASTM C 441, shrinkage often being observed after the normal 14-day testing period (Popovic et al. 1984). Perry and Gillott (1985) used various levels of addition and found that 10% was effective in reducing the 14-day expansion by more than 75% compared to a control mix; this was the acceptance criteria at the time. Other workers have confirmed the ability of 10% silica fume (Hooton 1993; Rasheeduzzafar and Hussein 1991) or less (Bérubé and Duchesne 1992a) to meet this criterion. However, Perry and Gillott (1985) observed continued expansion of specimens containing silica fume beyond 14 days and questioned the reliability of short-term testing by this method. Since this time, other investigators have criticized the suitability of the test for assessing SCMs because results from the test do not correlate well with data from tests on concrete containing natural aggregates (Bérubé and Duchesne 1992a).

Studies in Iceland

Iceland has unique problems with regard to ASR since the only source of cement available on the island is low in silica and very high in alkali (~1.5% Na_2O_e), and many of the aggregates are deleteriously reactive. These problems and the solutions implemented by the Icelandic construction industry have been well documented (Asgeirsson 1986; Asgeirsson and Gudmundsson 1979; Gudmundsson and Asgeirsson 1975, 1983; Gudmundson and Olafsson 1996; Olafsson 1989) and present an interesting case history.

The Icelandic Building Research Institute began research into ASR in the early 1960s and discovered that many of the volcanic basalts used as aggregates for concrete were classified as reactive by ASTM C 227 (Mortar Bar Test) and ASTM C 289 (Quick Chemical Test) (Gudmundsson and Asgeirsson 1975). However, at this time ASR had not been found in Icelandic structures and it was felt that the low reactivity of the aggregate and low average temperature combined to minimize the risk of deleterious reaction in the field. It was concluded that no special provisions were required for minor construction, but that the use of imported low-alkali cement or proven local natural pozzolans should be used for major projects (e.g., dams and harbors).

During the late 1970s incidences of ASR in housing concrete were reported (Asgeirsson and Gudmundsson 1979), and had become a major

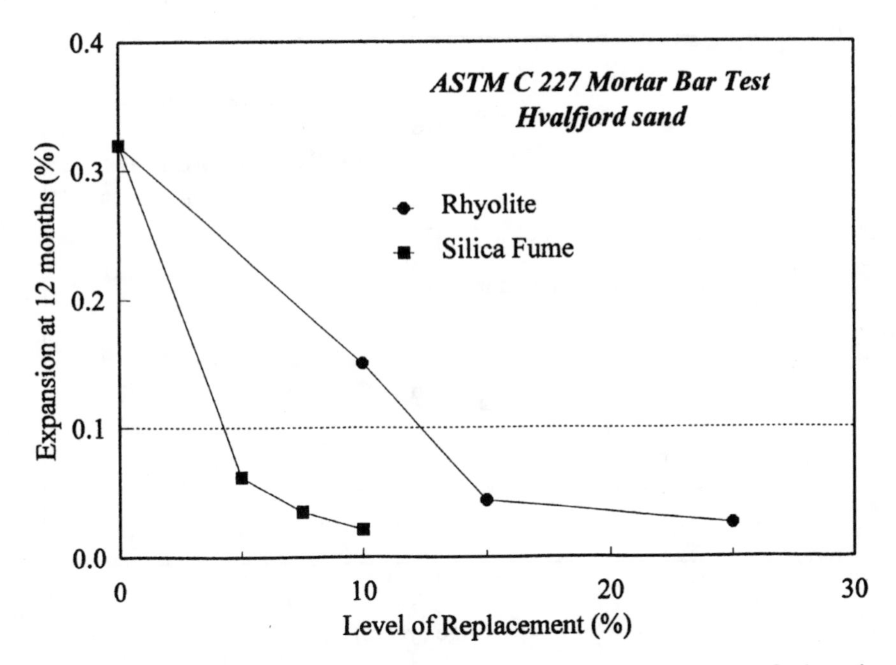

Figure 1. Comparison of preventative methods in Iceland (Asgeirsson and Gudmundsson 1979).

problem in the Reykjavik area by the end of the decade (Olafsson 1989). A contributing cause was the widespread use of an unwashed, sea-dredged sand (Hvalfjord sand) in housing concrete during the 1960s and 1970s (Olafsson 1989). The initial response to this problem was to intergrind rhyolite with the cement clinker at levels of 5–9% (Asgeirsson and Gudmundsson 1979). Research on silica fume began at the Building Research Institute in the early 1970s in response to the planned opening of a ferrosilicon plant close to the State Cement Works. Data published in 1979 (Asgeirsson and Gudmundsson 1979). showed that the by-product from the plant was more effective in controlling ASR expansion than rhyolite; early data are shown in Fig. 1.

In response to these early studies, the State Cement Works began to produce interground silica fume cement with 5% silica fume in 1979, and the level of silica fume was increased to 7.5% in 1983 (Olafsson 1989). Since 1979, virtually all concrete placed in Iceland has contained silica fume. Condition surveys have been carried out at regular intervals to monitor the performance of silica fume concrete in housing concrete and the results of

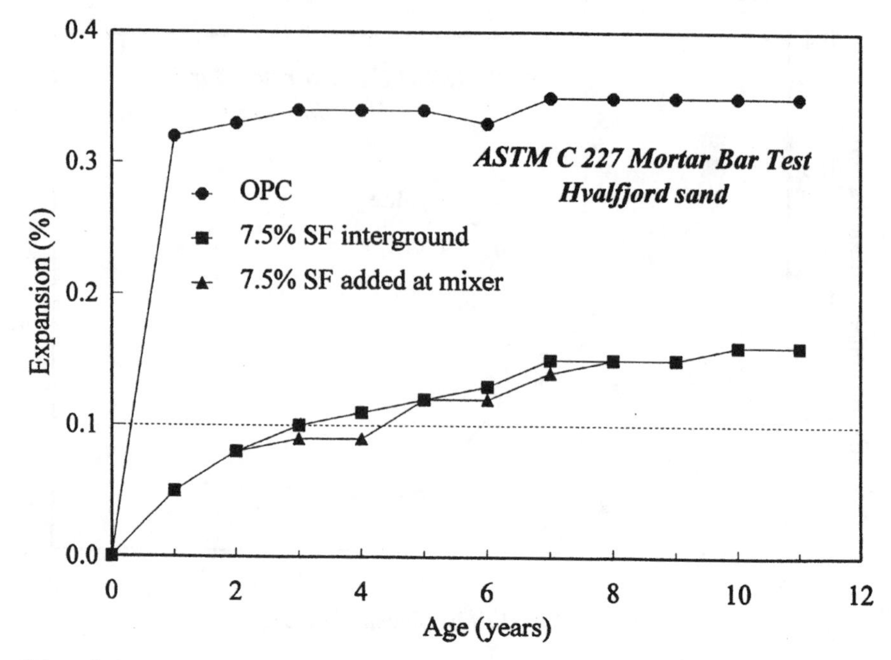

Figure 2. Long-term performance of silica fume in ASTM C277 (Gudmundsson and Olafsson 1996).

the survey have been published (Gudmundsson and Olafsson 1996; Olafsson 1989). Cores taken from the exterior walls of houses and stored at 38°C and 100% RH for 1 year have been examined by microscopy for signs of ASR. In 1989, Olafsson reported traces of ASR gel in cores taken from silica fume concrete walls but these were restricted to a few isolated cases. Incidences of ASR gel in silica fume concrete had not increased by 1996, although evidence of undispersed silica fume clumps was found (Gudmundsson and Olafsson 1996).

Long-term expansion results (ASTM C 227) have recently been reported for mortar bars containing Hvalfjord sand with silica fume, the fume being interground with the cement or added at the mixer. These results are shown in Fig. 2. The silica fume is effective in reducing the 12-month expansion to 50% of the allowable limit (0.10% at 12 months). However, this limiting value is reached after 4 years in the specimens containing silica fume, and the mortar bars continue to expand throughout the 11-year period.

Iceland now has over 18 years of field experience using silica fume with reactive aggregates in concrete. To date there are no reported cases of damage due to ASR in concrete containing either 5 or 7.5% silica fume, despite the use of cement that was recently reported to have an alkali content as high as 1.65% Na_2O_e (Gudmundsson and Olafsson 1996). However, the relevance of the field performance of silica fume concrete in Iceland to other countries with different reactive aggregates and climates must be interpreted with caution.

There are many reactive aggregates in Iceland; however, damage due to ASR has been almost entirely restricted to housing concrete constructed using unwashed, sea-dredged sand (Hvalfjord) during the period 1962 to 1979 (Olafsson 1989). There have been no reported occurrences of ASR damage in concrete constructed since the introduction of silica fume in 1979. However, this date also marks the preclusion of unwashed gravel materials in concrete and the absence of ASR damage may not solely be attributed to the presence of silica fume.

Experience has shown that the manifestation of ASR in field concrete takes a relatively long time in Iceland due to the low reactivity of the aggregates and the low average temperature. Indeed, ASR was not found in concrete containing the unwashed Hvalfjord sand until more than 10 years after its initial use. Laboratory testing has shown that silica fume considerably delays the expansion of mortar bars containing reactive Hvalfjord sand at 38°C, but may not completely suppress deleterious reaction in the long term. Thus it may well be too early to draw conclusions regarding the efficacy of the silica fume in field conditions.

Field Exposure Tests at the NBRI, South Africa

Investigations into the effect of silica fume on ASR in mortar bars were carried out at the National Building Research Institute (NBRI) in the Republic of South Africa as part of a wider program on the effect of SCMs (Oberholster and Westra 1981). In addition to replacing high-alkali portland cement with SCMs, tests were also carried out using low-alkali cement at various replacement levels to show the effect of dilution of the cement alkalis. The effect of diluting the high-alkali cement was to produce an almost proportional decrease in the mortar bar expansion at 555 days. The use of silica fume and fly ash produces proportionally greater reductions in expansion, suggesting that the SCMs have a positive role in suppressing ASR beyond mere dilution of cement alkalis. Only 10% silica fume was required to reduce the expansion to less than 0.10% at 555 days; this compares with 20% for fly ash and 30% for slag.

Table I. Results of NBRI field exposure tests with Malmesbury Hornfels aggregate

Cement content (kg/m^3)	Silica fume (mass%)	Active alkalis in concrete* (kg/m^3)	Reported expansion 1450 days†	Reported expansion 2700 days‡	Time to exp. > 0.05% (days)
350	0	3.85	0.17	0.32	950
345	3.5	3.87	0.51	0.26	1400
340	7	3.88	<0.01	<0.01	
450	0	4.95	0.17	0.28	870
444	3.5	4.97	0.08	0.15	1200
437	7	4.99	<0.01	0.09	2300

*Active alkalis includes available cement alkalis (ASTM C 311) plus alkali hydroxide but excludes alkalis in the silica fume.
†From Oberholster and Davies (1986).
‡From Oberholster (1989).

The NBRI began a series of field exposure tests using large beams (1000 × 450 × 300 mm) and cubes (300 mm) containing fly ash, slag, or silica fume (Oberholster 1989; Oberholster and Davies 1987). Details and performance of the silica fume specimens are given in Table I. Two series of mixes were cast with cementitious cement contents of approximately 350 and 450 kg/m^3. Within each series, 5 or 10% of the portland cement by mass was replaced with an equal volume of silica fume, resulting in silica fume levels of 3.5 and 7.0 mass%. The "active" alkali content was maintained at a constant level within a given series by addition of alkali hydroxide (using the same Na_2O/K_2O as the cement), assuming the silica fume to contribute no alkali.

The use of 3.5 or 7% silica fume delayed the onset of expansion and time to cracking in all cases. However, both mixes with 3.5% silica fume and the higher cement content mix with 7% silica fume eventually showed deleterious expansion and cracking. The expansion of control specimens with 450 kg/m^3 cementitious material exceeded 0.05% after 870 days, whereas specimens from the same mix series with 7% silica fume did not reach this value until 2300 days. However, once the onset of expansion was realized, the silica fume specimens appeared to exhibit similar expansion rates as the control specimens. Only the mix at the lower cement content with 7% silica fume failed to expand after 2700 days.

In South Africa silica fume replacement levels of 15% are recommended for preventing deleterious expansion (Oberholster 1994).

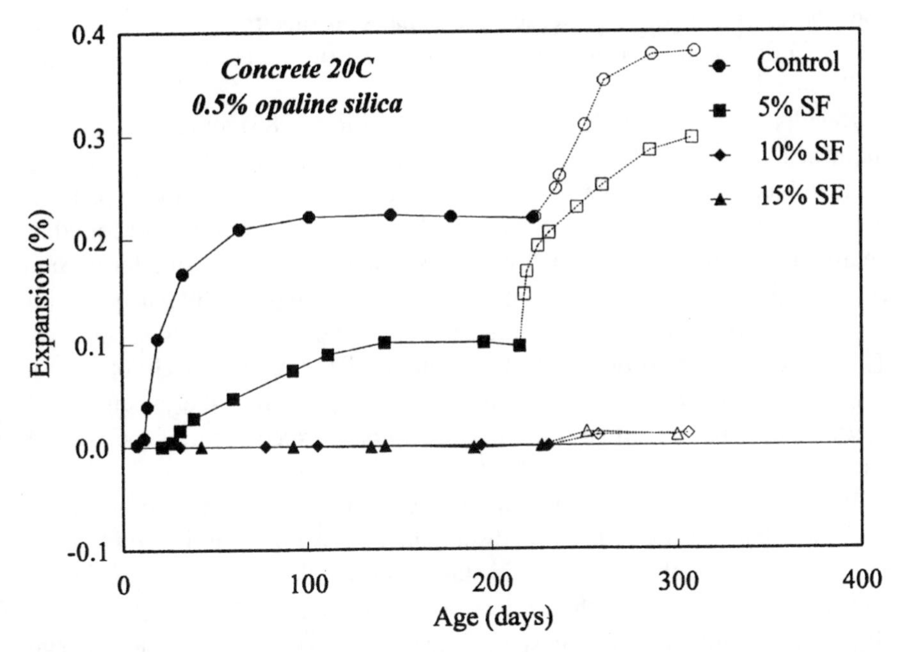

Figure 3. Effect of silica fume on expansion of concrete with opaline silica (Hobbs 1988).

Investigations with Opaline Silica

Various sources of opaline silica (such as Beltane opal from California) have frequently been used for ASR studies, presumably because the material reacts very quickly and studies can be completed in weeks or months (rather than years, as with slowly reacting aggregates). As such, a number of workers have questioned the validity of using opal for assessing slowly reacting pozzolans (such as fly ash) or slag (Nixon et al. 1986; Thomas 1996b). Furthermore, opal demonstrates a pessimum effect on the expansion of mortar bars or concrete. The expansion is very sensitive to the alkali-reactive silica ratio and other mix parameters (Hobbs 1988), and this often complicates the interpretation of results from mixes with and without SCMs (or any admixture with significant quantities of reactive silica or alkali, or both).

Hobbs (1988) produced data for concretes containing 0.5 or 1.0% of reactive silica by mass of aggregate, the mixes having relatively high alkali contents (~5 kg/m^3 Na_2O_e). The data are shown in Fig. 3 and demonstrate that 5% silica fume is insufficient, but higher levels (10%) are effective at least up to 220 days at 20°C. Exposure of the specimens to sodium chloride solution after 220 days increased the expansion of already damaged concretes (0 or 5% silica fume), but had no effect on the concrete with 10 or 15% silica fume.

Results of studies in Canada (Perry and Gillott 1985) and Japan (Kawamura et al. 1986, 1987) clearly demonstrate that rapid expansion may be achieved using opal. Silica fume showed a delaying effect on the expansion in both studies; however, levels of silica fume below 15% were not found to be effective in reducing expansion to less than 0.10% in the long term. Silica fume had a marked pessimum effect on the long-term expansion with levels of 5–10% actually increasing expansion above the control specimens. This effect was not observed in tests ran at 23°C (Perry and Gillot 1985), although the shape of the expansion curves suggests that the expansion of mortar bars with 5% silica fume may eventually exceed that of the control bars.

Results from studies using opaline silica to investigate the action of silica fume are at variance but these discrepancies have been assigned to pessimum effects by Hobbs (1988). His studies show that the expansion of mortars is clearly affected by the cement/opal ratio (or available Na_2O_e/reactive SiO_2) in the mix, and there may secondary effects due to the water/cement ratio. The absence of expansion at high reactive silica contents has been ascribed to rapid ASR and consequent alkali consumption while the concrete is in the plastic state. Once the concrete has hardened, insufficient alkali remains to promote further reaction and damaging expansion. It is for this reason that sufficient quantities of finely ground reactive aggregate can provide protection against expansion. The introduction of silica fume into the mix has much the same effect as increasing the opal or reactive silica content. Indeed, Wang and Gillott (1992) have described this phenomenon as a "competition" between silica fume and reactive siliceous aggregate to consume the alkalis. Hobbs (1988) estimated the quantity of amorphous silica required to suppress expansion in mortars with a high-alkali cement (1.04% Na_2O_e), and a range of opal contents and aggregate to cement ratios; the data are shown in Table II. The amorphous silica value can be equated to silica fume contents assuming the silica fume to comprise solely reactive SiO_2 and contribute no alkali to the mix, otherwise higher levels of silica fume are required.

The data in Table II imply that no expansion will occur provided the ratio of amorphous silica to available alkali is sufficiently high. The safe ratio appears to range from 12 for a w/c of 0.35 to 9 for a w/c of 0.59. Thus for a concrete with a w/c of 0.50 and a cement alkali level of 1.0% Na_2O_e, a

Table II. Amorphous silica required (mass% of cement) to prevent damaging expansion due to ASR in mortar bars with opaline silica

Mix details		Opaline silica content (mass% of aggregate)				
w/c	a/c	0.5	1	2	3	4
0.35	1	11.5	11	10	9	8
0.41	2	12	10	8	6	4
0.47	3	9.5	8	5	2	0
0.53	3.75	7	5	1.5	0	0
0.59	4.5	7	4.5	0	0	0

NOTE: The values in the last five columns refer to the amount of silica fume required to eliminate expansion assuming that the silica fume is 100% amorphous silica with a negligible alkali content and that the cement has an alkali level of 1.04% Na_2O_e. Data from Hobbs (1998).

silica fume content of 10% would be sufficient to eliminate deleterious reaction regardless of the opal content of the aggregate. Of course this assumes the fume to be 100% amorphous SiO_2, and higher quantities would be required as the SiO_2 content decreases and the Na_2O_e content increases. However, as either the aggregate/cement ratio or opal content increases, or the alkali content of the cement decreases, the level of silica fume required is reduced.

Hobbs (1988) predicted the level of silica fume required to suppress expansion in various studies, making adjustments for the various cement and silica fume compositions used by the individual investigators. Some of these results are compared in Table III with the observed values. There is fairly good agreement between the predicted and observed values, with the exception of the study by Perry and Gillott (1985). It may be significant that Perry and Gillott extended their tests beyond the ages reported by other workers.

Investigations with Cristobalite (Fused Silica)

Manufactured cristobalite (fused silica) has also been used as a source of reactive aggregate for assessing the effect of supplementary cementing materials such as fly ash (Hobbs 1994) and slag (Lumley 1993). Cristobalite aggregate has been proposed as a reference material for ASR studies as it can be manufactured under controlled conditions to produce a consistent and homogenous product (Lumley 1989). Hobbs (1994) selected the material for fly ash studies as it purportedly reacted at alkali levels close to the U.K. limit for use with reactive aggregates (3 kg/m^3 Na_2O_e). However,

Table III. Effect of mix proportions on silica fume level required to prevent damaging ASR expansion*

Authors	Silica fume SiO_2 (%)	Silica fume Na_2O_e (%)	Mix details w/c	Mix details a/c	Reactive silica % agg.	Cement alkali % Na_2O_e	Silica fume required to prevent exp. (%) Obs.	Silica fume required to prevent exp. (%) Pred.
Sprung & Adabian (1976)	95		0.45	2.25	4.0	High	4	4
Perry & Gillott (1985)	94	0.53		2.25	2.0	1.00	20	11
Kawamura et al. (1986)	91	2.38	0.40	0.75	5.0	0.76	20	16
Kawamura et al. (1987)	88	2.15						15
	89	1.97	0.40	0.75	5.0	0.76	15	15
	87	1.44						14
Hobbs (1988)	96	0.34	0.41	3.0	0.5	0.93	5–10	9–10
					0.1			7–9

*Modified from Hobbs (1988).

studies have shown that cristobalite may react at a much lower alkali contents, that is, closer to 2 kg/m^3 Na_2O_e (Lumley 1993), and Thomas et al. (1996) has suggested that results from studies using highly reactive cristobalite may not be applicable to concrete with less-reactive aggregates (i.e., those that react only at higher alkali levels).

Studies of the effect of silica fume on the expansion of concrete with cristobalite have been carried out at the Building Research Establishment (BRE) in the U.K. (Dunster et al. 1990) and at the University of Sheffield (Alasali 1989; Swamy 1990; Swamy and Alasali 1989).

In the BRE study, concrete specimens were cast with cristobalite contents of 5, 10, and 15% (by mass of aggregate) and the same proportions of silica fume (by mass of cementitious material). The silica fume used was supplied in slurried form; no chemical analysis is given but the SiO_2 is reported to be ≥ 85% and the Na_2O_e ≤ 2% (by mass of solids). Two portland cements with alkali levels of 0.7 and 1.0% Na_2O_e were blended to provide concretes with alkali contents of 2.5, 3.0, 3.5, and 4.0 kg/m^3 Na_2O_e, disregarding the alkalis from the silica fume (K_2SO_4 was added to the mixing water to supplement the alkalis where necessary) (Dunster et al. 1990).

Silica fume was effective in delaying the expansion of concrete, and the delay increased with fume content. For mixes with 15% silica fume, deleterious expansion (>0.04%) was not observed until 1 year. Although increasing levels of silica fume produced decreasing levels of expansion at 1 year, the data show that all the specimens (except the control specimens at lower alkali contents) are still expanding.

The results of the studies at Sheffield University have been adequately summarized by Alasali (1989). In that paper, results are reported for a range of concretes containing various SCMs (fly ash, slag, or silica fume) and cristobalite (at 15% by mass of aggregate). The silica fume used had a SiO_2 content of 98.51% and the alkali was reported to be less than 0.31% Na_2O_e. Concrete specimens were cast with a cementitious content of 300 kg/m^3 (using a high-alkali cement, 1.0% Na_2O_e), a w/c of 0.60, and silica fume content of 10 or 30%. A high-range water-reducing admixture was added to the silica fume concrete mixes. The alkali content was adjusted to provide a range from 3 to 6 kg/m^3 Na_2O_e by adding NaOH to the mixing water (the alkali in the silica fume was disregarded for the purpose of calculating the alkali level of the concrete).

The two control series gave essentially the same expansion at a given alkali loading regardless of the source of alkali (either by dosing with alkali salts or by using high-alkali cements). The expansion of concrete with 10% silica fume was similar to the control specimens at a given alkali level. This would imply that the silica fume (at 10% replacement) has no positive action on suppressing ASR expansion beyond acting as a diluent to the cement alkalis. Concrete with 30% silica fume effectively eliminated expansion regardless of the alkali content of the concrete mix.

Wei and Glasser (1989) reported expansion data for mortar bar tests (ASTM C 227) using cristobalite as part of a wider study on the role of silica fume in ASR. For mortars with high-alkali cement (1.24% Na_2O_e) and 10% silica fume (97.8% SiO_2 and 0.42% Na_2O_e), the expansion exceeded 0.10% after 180 days at 38°C. Mortar with 15% silica fume did not expand within the reported duration of the test (180 days).

Studies at CSIRO, Australia

Shayan et al. (1993a, 1993b) reported findings from investigations of steam-cured concrete containing silica fume. Expansion tests were performed on concrete specimens containing either a nonreactive basalt or a reactive granite aggregate. Specimens were subjected to steam curing at 75°C during the period from 3 to 11 hours after casting and subsequently stored at 40°C and 100% RH. The w/c was kept at 0.40 for all mixes and a

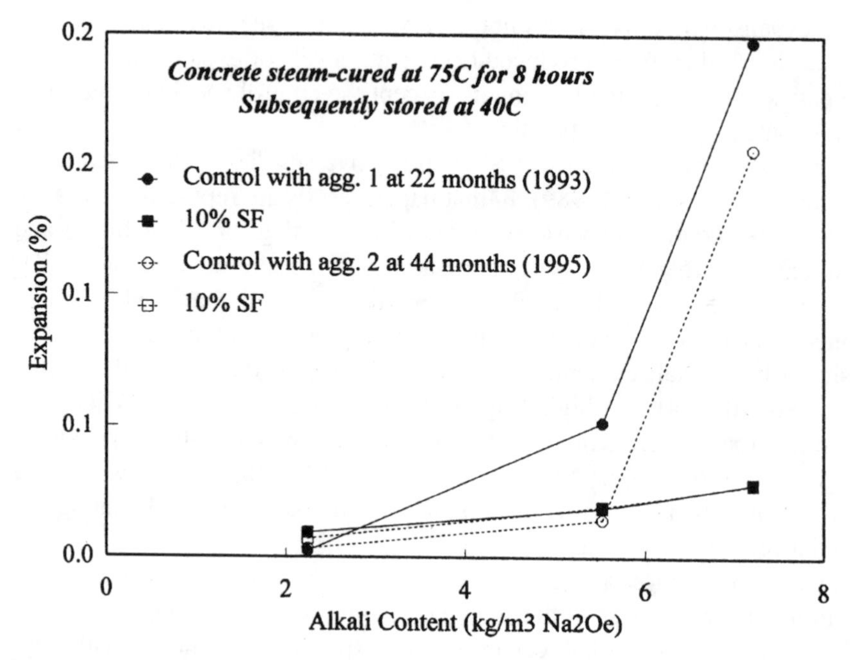

Figure 4. Effect of silica fume on expansion of steam-cured concrete (Shayan 1995a; Shayan et al. 1993a).

high-range water-reducer (napthelene-based formaldehyde) was used to maintain workability. The silica fume used was a densified product with an SiO_2 content of 93% and a Na_2O_e content of 0.28%. Concrete was cast with a cementitious material content of 400 kg/m^3 using a low-alkali portland cement (0.56% Na_2O_e). Alkali levels were increased to 1.38 and 1.80% Na_2O_e by mass of cement by adding NaOH to the mix. For the concrete with silica fume, 20 mass% of portland cement was replaced with 10% silica fume (e.g., 80 kg of cement replaced by 40 kg of silica fume).

Expansion data at 22 months are plotted in Fig. 4. Also shown are 44-month expansion data taken from a later paper (Shayan 1995a). The experimental procedures were the same for these studies, but it appears that the second study used a less reactive aggregate; however, this is not made clear by the authors. These results show that 10% silica fume is effective in preventing expansion even when the alkali level of the concrete is higher than that necessary to produce damage in the control specimens.

Studies with Canadian Aggregates

A number of expansion studies have been conducted using Canadian aggregates of varying reactivity. CANMET initiated studies on the effect of SCMs on AAR in the 1980s and have since carried out a number of testing programs in this area. An overview of the CANMET work was presented recently by Malhotra and Fournier (1995). A number of these studies have included silica fume (Chen et al. 1993; Fournier et al. 1994, 1995; Langley et al. 1995; Soles et al. 1987). Studies of the effect of silica fume (and other SCMs) on ASR expansion and reaction mechanisms, and of test methods for SCMs, have been carried out at Laval University, and the results have been published in a number of papers (Bérubé and Duchesne 1992a, 1992b; Bérubé et al. 1995; Duchesne and Bérubé 1992a, 1992b, 1994a, 1994b, 1994c). The authors of these papers also produced an overview on the effect of SCM on AAR in 1991 (Bérubé and Duchesne 1991). Durand (1993) has also produced data using Canadian aggregates.

A summary of 2-year concrete prism expansion results is given in Table IV. These studies used a modified version of the CSA test (except for Fournier et al. 1995). The current CSA Concrete Prism Test (A23.2 14A) specifies 420 kg/m^3 of cement raised to 1.25% Na_2O_e, which yields an alkali content of 5.25 kg/m^3 Na_2O_e. In all studies, the alkali content of the cement was raised to 1.25% Na_2O_e by addition of NaOH; however, different cement contents were used. Bérubé and Duchesne opted for a cement content of 350 kg/m^3, whereas Durand chose a cement content of 310 kg/m^3. The implication of these changes is a total equivalent alkali content less than 5.25% Na_2O_e (disregarding alkalis from the silica fume).

Fournier's results indicate that 7.5% and possibly 10% silica fume are insufficient for controlling expansion when the cement alkalis are 4.9 and 4.7 kg/m^3 Na_2O_e, respectively. An increased level of 12.5% silica fume is effective when the cement alkalis are 4.7 kg/m^3 Na_2O_e. However, inspection of the data reveals that concrete with 12.5% silica fume shows a trend of increasing expansion at 2 years. These tests are being continued and additional data will no doubt be reported in the future.

In Duchesne and Bérubé's study, the use of 5% silica fume delays the onset of expansion initially, but only for 3–6 months. Beyond this period, all the concrete with 5% silica fume expanded, and for specimens with rhyolitic tuff, the expansion actually exceeded that of the control. At this level of replacement, no discernible difference due to the nature of the silica

Table IV. Expansion of silica fume concrete prisms at 24 months: Canadian studies

Authors	Aggregate source	Cement alkalis (kg/m^3 Na_2O_e)	Silica fume % SiO_2	Silica fume % Na_2O_e	Silica fume level	Expansion at 24 months (%)
Fournier et al. (1995)	Greywacke	5.25	Control		0	0.267
		4.9	97.5	0.72	7.5	0.079
		4.7	97.5	0.72	10	0.039
		4.6	97.5	0.72	12.5	0.016
Bérubé & Duchesne (1992a); Duchesne & Bérubé (1994a)	Rhyolitic tuff	4.4	Control		0	0.245
		4.2	74.6	3.63	5	0.370
		3.9	74.6	3.63	10	0.177
		4.2	94.2	0.77	5	0.370
		3.9	94.2	0.77	10	0.033
	Siliceous limestone (Spratt)	4.4	Control		0	0.287
		4.2	74.6	3.63	5	0.239
		3.9	74.6	3.63	10	0.078
		4.2	94.2	0.77	5	0.275
		3.9	94.2	0.77	10	0.019
Durand (1993)	Chloritic schist	3.9	Control		0	0.077*
		3.7	91.85	0.64	5	0.023*
		3.5	91.85	0.64	10	0.009*
		3.3	91.85	0.64	15	0.001*
	Siliceous limestone (Spratt)	3.9	Control		0	0.137*
		3.7	91.85	0.64	5	0.047*
		3.5	91.85	0.64	10	0.016*
		3.3	91.85	0.64	15	0.006*
	Trois-Riviéres limestone	3.9	Control		0	0.175*
		3.7	91.85	0.64	5	0.044*
		3.5	91.65	0.64	10	0.000*

*These values have been estimated from figures presented in the original papers and are not absolute values as determined by experimentation.

fume was observed. The onset of expansion was delayed further with 10% silica fume and differences between the two silica fumes were observed. The expansion of concrete with the low-SiO_2, high-Na_2O_e silica fume exceeded the 0.04% expansion limit at 12 months with the tuff and 18 months with the limestone aggregate. The high-SiO_2, low-Na_2O_e fume was effective in suppressing expansion to below 0.04% at two years with both aggregates. However, there is an upward trend in the expansion data at this age.

All three aggregates in Durand's study (1993) produced deleterious expansion (>0.04%) in concrete with 0 or 5% silica fume. Higher levels of silica fume have been effective in suppressing expansion below 0.04%, although these tests were carried out at relatively low alkali contents (3.3–3.9 kg/m^3) compared with current CSA procedures.

Of particular interest when regarding these studies as a whole is the correlation of total alkali content of the concrete to the level of silica fume replacement to prevent deleterious expansion. In general, as the alkali content of the concrete increases, the amount of silica fume necessary to prevent expansion increases proportionately. Aggregate reactivity and silica fume composition, also being factors affecting replacement level, seem to be overshadowed by the effect of alkali content.

Long-Term Expansion of Concrete Prisms

The field studies reported by Oberholster (1989) emphasized the need for long-term data in ASR investigations, particularly in relation to silica fume, which may delay the onset of expansion beyond the scope of normal studies. The current Canadian Standard (CSA A23.1) recognizes this fact and specifies testing periods of two years to evaluate mixes containing fly ash or slag and undefined "longer testing periods" for silica fume concrete.

The implications of long-term expansion of silica fume concrete in these accelerated tests for concrete under normal service conditions are unclear. The concrete prism test uses raised alkali levels, although the alkali content in Durand's tests (3.5 kg/m^3 Na_2O_e from portland cement component) are not unrealistic. The storage conditions (38°C at 100% RH) undoubtedly accelerate expansion, although the relationship with field expansion rates is not clearly defined. The long-term data are clearly of concern if combinations of high-alkali cement and potentially reactive aggregate are to be used in structures that require extended service lives (e.g. >50 years); under such circumstances the ability of 10% silica fume to provide adequate protection is questionable.

Alkali Immersion Tests

Davies and Oberholster (1987) proposed the use of their accelerated mortar bar test (later adopted as CSA A13.2 25A and ASTM C 1260) for evaluating the effectiveness of supplementary cementing materials. They determined the minimum level of silica fume required to control expansion (<0.10% at 12 days) with a reactive greywacke/hornfels aggregate to be 13%. This value compared with minimum levels of 11 and 7% found by ASTM C 227 tests and field experience, respectively. However, later experience with the field tests showed 7% silica fume to be insufficient, and current advice in South Africa suggests a minimum of 15% (Oberholster 1994).

A number of workers have since carried out expansion tests on mortar or concrete specimens immersed in NaOH solutions at various temperatures (Berra et al. 1994, 1996; Bérubé and Duchesne 1992a; Bérube et al. 1995; Durand et al. 1990; Fournier et al. 1994, 1995; Langley et al. 1995). Reliable correlations have been found between the accelerated mortar bar test results and the expansion of concrete prism tests (Bérubé and Duchesne 1992a) and mortar bars (Berra et al. 1994) stored at 38°C and 100% RH. It has been suggested that the effectiveness of SCMs in controlling the expansion of mortar bars stored in 1M NaOH is related to their ability to control the alkalinity of the pore solution during the test (Berra et al. 1994; Bérubé et al. 1995). As such, the results are highly sensitive to the test duration, since alkalis will inevitably penetrate the specimens from the host solution.

Silica fume (7.5–12.5%) was found to delay the onset of expansion of concrete containing reactive aggregate and stored in 1 M NaOH at 38°C; however, expansions in excess of 0.04% eventually occurred after extended periods of time (up to 2 years) regardless of the silica fume content of the concrete (Fournier et al. 1995; Langley et al. 1995).

The results from expansion tests are difficult to interpret with regard to the effect of silica fume on ASR expansion. Since the necessary alkalis for reaction are abundantly available in the host solution, such tests may provide a measure of the resistance of the concrete to ion ingress and little else.

The Effect of Silica Fume on Expansion due to ACR

Comparatively few studies on the effect of silica fume on alkali-carbonate reaction have been carried out. Incidences of concrete damage due to ACR are limited, since there are few reactive carbonate sources that are otherwise suitable as concrete aggregates. Accepted measures for controlling ASR, such as the use of low-alkali cement, fly ash, or slag (at normal high levels of replacement), have been found to be ineffective in controlling expansion due to ACR (Chen et al. 1993; Rogers and Hooton 1992). Consequently, the Canadian Standard (Candadian Standards Association 1994) prohibits the use of reactive carbonate rocks in concrete.

Concrete prisms stored at 20°C and cast with reactive dolomitic limestone from Kingston, Ontario, and a high alkali cement (1.1% Na_2O_e)

blended with various amounts of silica fume were studied by from Soles et al. (1987). A high-SiO_2 fume with 95.4% SiO_2 and 0.51% Na_2O_e, and a low-SiO_2 fume with 74.7% SiO_2 and 2.60% Na_2O_e were used (Chen et al. 1993). The results show that 15% silica fume has little effect in ameliorating the deleterious effects of ACR. Indeed, the use of the low-silica, high-alkali silica fume actually exacerbated expansion.

Expansion data from studies by Deng and Tang using concrete microbars (10 × 10 × 40 mm) cast again with Kingston dolomite, a low-alkali cement (0.42% Na_2O_e) and a silica fume from Nanjing (88.3% SiO_2 and 0.95% Na_2O_e). Microbars were manufactured with 15 vol% of aggregate (5–8 mm) and a w/cm ranging from 0.27 (0% silica fume) to 0.52 (30% silica fume), and subsequently stored at 40°C and 100% RH. The microbar data show little improvement with 10% silica fume but significantly reduced expansion at levels of 20 and 30%. The authors ascribe this effect to the substantially lower pore solution pH when high levels of silica fume were present, and observed similar behavior with high levels of fly ash (70%) and slag (90%).

Silica Fume as the Source of Reactive Aggregate

Pettersson (1992) made a serendipitous discovery of cracking in mortars with 10% granulated silica fume that were immersed in 1 M NaCl for corrosion studies. Further testing showed that the use of granulated silica fume could actually promote ASR expansion due to the presence of agglomerations of silica fume, which behaved as reactive aggregate particles. The morphology and composition of the reaction product reported by Pettersson appears to be typical of ASR gel.

Bonen and Diamond (1992a, 1992b) found similar evidence of "reacted" silica fume lumps (up to 80 μm) in 1-year-old cement pastes immersed in 2% $MgSO_4$ solution. The presence of agglomerations in this study is somewhat surprising since uncompacted silica fume from a commercial source was used, superplasticizer was incorporated in the mix, and a high–shear action mixer was used. The authors likened the reaction to ASR, although the reaction products were not expansive. However, the authors suggest that expansion might have resulted had the cement alkalis been higher.

Several workers have since implicated silica fume agglomerations as the cause of alkali-silica reaction in a number of field and laboratory cases (Marusin and Shotwell 1995; Shayan et al. 1993b, 1994; Lagerblad and Utkin 1995). Although these are relatively isolated incidents, material suppliers and concrete producers should be aware of the potential for this phenomenon to occur. Such cases further demonstrate the need to ensure adequate dispersion of silica fume during mixing. This is equally important for both laboratory studies and field use of silica fume concrete. Unfortunately, many laboratory studies have not used appropriate admixtures (high-range water reducers) in the manufacture of mortar or concrete for expansion studies or paste specimens for pore solution studies. Furthermore, the product form of the silica fume and the mixing processes used may not lend themselves to sufficient dispersion. In such cases, the true effect of silica fume may not be realized in the test program and, indeed, the presence of poorly dispersed lumps may actually exacerbate any ASR expansion. Prudence dictates "post-mortem" examination of laboratory specimens to ascertain the true cause of deterioration in expansion tests. Such precautions are recommended in concrete prism testing according to CSA A23.2-14A (Canadaian Standards Association 1994).

The Role of Silica Fume in ASR

The use of silica fume affects many of the chemical and physical properties of fresh and hardened concrete. Some of these changes have a direct impact on the rate, extent, and manifestation of alkali-silica reaction. These include:

- Reduced pore solution alkalinity.
- Reduced ionic diffusion and water permeability.
- Consumption of $Ca(OH)_2$.
- Improved $Ca(OH)_2$ distribution at the aggregate/cement interface.

The reduced alkalinity of the pore solution has become generally accepted as the predominant mechanism by which silica fume suppresses ASR. Reduced ionic and water mobility may have a significant effect on retarding the rate of reaction and subsequent expansion (water uptake) of the reaction product. The role of $Ca(OH)_2$ in ASR expansion has received relatively little attention with regard to silica fume. Other mechanisms, such as self-desiccation, may affect the development of damaging reaction especially at low water-cement ratios.

Pore Solution Effects

The solubility of thermodynamically unstable silica increases with pH and thus the risk of damaging ASR may be expected to decrease as the pH decreases. There is no universally accepted threshold pH value below which ASR is thought to cease (or reduce to an insignificant rate), although significant reaction below the range pH 13.3–13.4 is unlikely (Diamond 1983a; Kolleck et al. 1986). This range equates to hydroxyl ion concentrations of 0.20–0.25 M.

Pore Solution Evolution in Portland Cement Pastes

After about 24 h of hydration at room temperature, the pore solution of normal portland cement paste is composed almost entirely of alkali hydroxides (Na^+, K^+, OH^-) with minor amounts of calcium and other ions (e.g., silica and sulfate) (Diamond 1983a; Penko 1983). At later ages the alkalinity of the pore solution is controlled by the alkali content of the cement, the water/cement ratio, and the degree of cement hydration. At 28 days the hydroxyl ion concentration (in mol/L) has been found to be approximately equal to 0.7 times the alkali content of the cement (in % Na_2O_e) for a paste with w/c = 0.5 (Nixon and Page 1987; Diamond 1989). The OH^- concentration is higher at lower w/c (and vice-versa) and increases with hydration due to the progressive removal of free water by the hydration process.

Effect of Reactive Aggregate

When reactive aggregate is incorporated into mortar or concrete, alkali and hydroxyl ions are removed from solution due to reaction with the amorphous (or microcrystalline) silica. The effect of Beltane opal on the hydroxyl ion concentration of mortars stored at 20°C was studied by Diamond et al (1981). Alkalis are rapidly removed by reaction with the opal and a steady-state condition at approximately 0.28 M OH^- is reached after 1 month. At 40°C there was a rapid acceleration of the reaction and a steady-state condition was attained after just 3 days (Diamond 1981). Similar results were obtained by Thomas (1996b) for concrete containing a reactive flint sand from the United Kingdom. Although there is an initial rapid reduction of OH^- ions in solution, the reaction generally proceeds more slowly with a steady concentration of approximately 0.27 M OH^- being attained after 1 year. Other workers have shown lower alkali levels (~0.2 M OH^-) in concrete containing reactive siltstone (Sibbick et al. 1996). The attainment of steady-state conditions does not necessarily signify the completion of the reaction. It merely means that there is an equilibrium between the alkalis in the liquid and solid phases. If the reaction continues, alkalis consumed from the pore solution may be replaced by alkalis released from the hydrates. Alternatively, alkalis may be regenerated from existing reaction product with calcium taking its place (cation exchange).

Table V. Results from pore solution studies of paste specimens

Authors	OPC (% Na_2O_e)	w/c	Temp. (°C)	Age (days)	% SiO_2	% Na_2O_e	Level	OH^- (mM)
Diamond (1983a, 1993b)	0.71	0.50	22	80	Control		0	520
		0.50	22	80	94	0.94	5	290
		0.50	22	80	94	0.94	10	150
		0.50	22	80	94	0.94	20	80
		0.50	22	65	94	0.94	30	16
		0.4	22	174	Control		0	590
		0.4	22	174	94	0.94	30	44
Page and Vennesland (1983)	1.19	0.50	22	84	Control		0	743
					96	0.39	10	228
					96	0.39	20	78
					96	0.39	30	9.8
Glasser and Marr (1985)	0.50	0.60	19	420	Control		0	263*
						0.68	15	71*
Kawamura et al. (1987)	0.79	0.40†	38	70	Control		0	645
				50	91	2.39	5	510
				50	91	2.39	10	415
				60	91	2.39	15	300
				50	88	2.15	5	490
				70	88	2.15	10	410
				30	88	2.15	15	305
Andersson et al. (1989)	0.92	0.50†		300	Control		0	251‡
		0.35†			96	0.44	20	25‡
Yilmaz and Glasser (1990)	0.55§	0.35†	20	180	Control		0	501‡
					98	0.42	15	100‡
Rasheeduzzafar and Hussain (1991)	0.65¶	0.60	20	180	Control		0	348
					97	0.30	10	105
					97	0.30	20	66
	0.9¶	0.60	20	180	Control		0	510
					97	0.30	10	273
					97	0.30	20	67
	1.2¶	0.60	20	180	Control		0	735
					97	0.30	10	403
					97	0.30	20	69
	1.5¶	0.60	20	180	Control		0	946
					97	0.30	10	533
					97	0.30	20	70
Duchesne and Bérubé (1992b, 1994c)	1.05	0.50	38	365	Control		0	857*
					94	0.77	5	681*
					94	0.77	10	420*
					75	3.63	5	895*
					75	3.63	10	617*

Table V, continued

Authors	OPC (% Na_2O_e)	w/c	Temp. (°C)	Age (days)	% SiO_2	% Na_2O_e	Level	OH^- (mM)
Durand et al. (1990)	0.92	0.50†	38	180	Control		0	405
					92	0.64	5	320
					92	0.64	10	160
					92	0.64	15	80
Shayan et al. (1993)	0.56	0.50	23**	56	Control		0	280
					93	0.28	10	30
Wiens et al. (1995)		0.60†	20	730	Control		0	360
					91	0.81	25	76
Nagataki and Wu (1995)	0.65	0.50	20	28	Control		0	794‡
					87	1.33	10	316‡

*Calculated from the sum of ($Na^+ + K^+$).
†High-range water reducer (superplasticizer) used.
‡Calculated from pH.
§Oil-well cement.
¶Alkali from OPC + silica fume + NaOH.
**Concretes steam cured for 8 h (from 3 to 11 h after mixing).
NOTE: Many of the OH^- values in this table have been estimated from figures presented in the original papers and are not absolute values as determined by experimentation.

Effect of Silica Fume on Pore Solution Composition

The effects of silica fume on pore solution composition were first reported in 1983 (Diamond 1983b; Page and Vennesland 1983). Since then there have been a number of studies, details of which are summarized in Table V. These studies differ in the materials used, mix proportions, temperature of storage, age at testing, method of solution analysis, and reporting format. In all cases the method of solution extraction was based on the techniques developed by Longuet et al. (1973), although the applied pressures and duration of loading may vary. A number of workers have monitored the change in solution composition with time, but the data reported in Table V are restricted to the final solution concentration reported by the investigators. The pore solution data in Table V are reported as mM of OH^-, however, in some cases these values have been calculated as the sum of alkali ions ($Na^+ + K^+$) or from pH data. The assumption that electrical neutrality is maintained by equal number of alkali and hydroxyl ions (i.e., $Na^+ + K^+ = OH^-$) in the pore solution of mature cement paste is reasonable since the solution is composed almost entirely of these ions. However, as the OH^- concentration falls below 0.1 M, there is an increased likelihood of other ions coming into solution (e.g., Ca2+).

Page and Vennesland (1983) demonstrated that there is a decrease in the OH^- (and associated Na^+ and K^+) concentration beyond mere dilution of the cement alkalis during the first 7 days; this continues up to 28 days. Beyond 28 days the pore solution composition remains relatively stable up to the end of the test (84 days). There is a significant reduction in alkalinity with increasing silica fume content. At 20% silica fume the OH^- concentration falls below 0.1 M (i.e., pH < 13.0) and at 30% OH^- is less than 0.01 M (pH < 12.0) as compared to 0.74 M (pH < 13.9) for the control. The proportion of alkali ions in the pore solution (as a percentage the total alkalis from the portland cement and silica fume) was calculated from solution analyses and evaporable water contents, and found to vary from 84% for portland cement to 9% for paste with 30% silica fume.

Similar pore solution data were reported by Diamond (1983b). Diamond, however, found that the effect of silica fume was less pronounced at lower w/cm (he compared pastes at w/cm = 0.50 and 0.40) and suggested that this may have implications for the typical silica fume concrete mixtures used in practice.

Duchesne and Bérubé (1992b, 1994c) showed very different trends for more typical levels of silica fume (5 and 10%) over longer time periods. Their data show a reduction in alkalinity during the first 7–28 days, but beyond this period the alkalinity starts to increase again, reaching values in excess of 0.5 M OH^- (pH > 13.7) in pastes with 10% silica fume and 0.7 M (pH > 13.8) in pastes with just 5% fume, as compared to 0.92 M (pH > 13.9) for the OPC control specimen. This study differed from those of Page and Vennesland (1983) and Diamond (1983) in that the pastes were stored at a higher temperature (38°C as compared with 22°C). It is likely that the higher pH values observed for specimens stored at elevated temperatures are due to a change in reaction kinetics as well as a decreased capacity of the hydrates to bind alkalis.

Figure 5 shows the OH^- concentration (at the latest reported age) plotted against the silica fume content for all the studies listed in Table V. The wide scatter of data on the vertical axis (0% silica fume) reflects the differences in the portland cement used, particularly the alkali content, which ranged from 0.5 to 1.5% Na_2O_e (including added NaOH). As the level of silica fume is increased, the OH^- concentration is reduced, the one exception being when 5% of a low-silica (75% SiO_2), high-alkali (3.63% Na_2O_e) silica fume was used (Duchesne and Bérubé 1992b, 1994c). At high levels of silica fume (20–30%), the alkalinity of the pore solution is very low (pH <

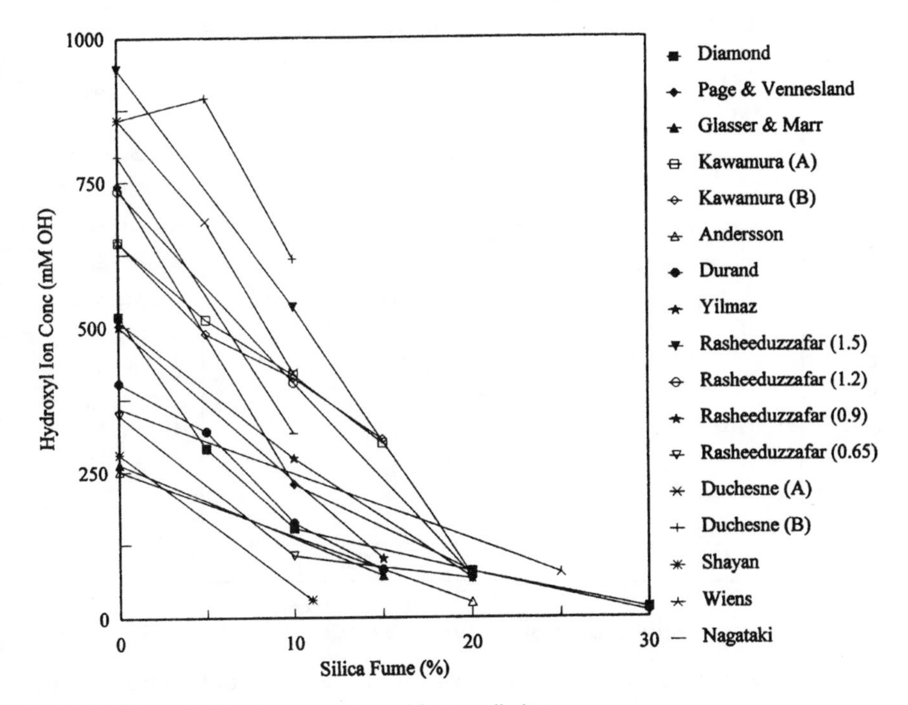

Figure 5. Effect of silica fume on pore solution alkalinity.

13.0) in all cases, regardless of the alkali level of the portland cement. However, at 10% silica fume, the OH^- concentration covers a very broad range from approximately 0.1 to 0.6 M (13.0 < pH < 13.8).

Rasheeduzzafar and Hussain (1991) studied a series of pastes in which the alkali content (from cement + silica fume + NaOH added to the mix) was varied in the range of 0.65 to 1.5% Na_2O_e (by mass of cement + silica fume). The pore solution results (at 6 months), shown in Fig. 6, demonstrate that the alkalinity of the pore solution in both control pastes and pastes with 10% silica fume was controlled by the total amount of alkali in the mix. However, at 20% silica fume, the pore solution was found have a relatively constant OH^- concentration (~0.07 M) regardless of the alkalis present in the mix constituents. Data from other studies are plotted on the same figure and, while there appears to be no single relationship linking all the studies, the same trends are observed.

Hobbs (1988) has suggested that results of pore solution studies may not be applicable to concretes containing reactive aggregates. The removal of

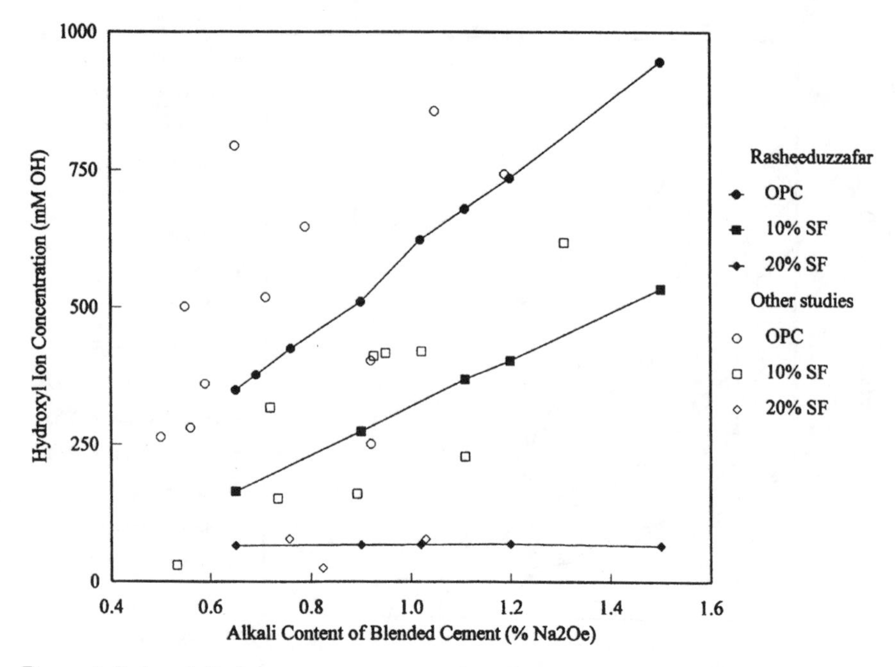

Figure 6. Role of alkali content on pore solution alkalinity of silica fume pastes (data from Table V).

alkalis by reaction with the aggregate lowers the pore solution alkalinity and changes the mass balance. Hobbs contends that this may lead to further alkali release from the hydrates. Wei and Glasser (1989) and Durand (1990) investigated the combined effect of silica fume and reactive aggregate on the pore solution alkalinity of mortars and concrete, respectively. Mortars with reactive cristobalite eventually attain a pore solution alkalinity in the range of 0.20–0.25 M, regardless of the presence of silica fume (10 or 15%). However, mortars with 10% silica fume (and cristobalite) exhibit unusual behavior in that the alkalinity of the pore solution increases during the first 2–3 months and becomes considerably higher than companion specimens without silica fume. The same behavior was observed with Pyrex glass as a reactive aggregate (Wei and Glasser 1989). Mortars with 10% silica fume and cristobalite eventually suffered deleterious expansion, although this was delayed considerably compared with the control. The use of 15% silica fume effectively inhibited expansion during the reported test period (180 days at 38°C). The authors ascribe the beneficial effect of silica fume to self-desiccation and conclude that considerable potential for expansion remains in the silica fume mortars should moisture availability be increased. Durand et al. (1990) show that the pore solution alkalinity of concrete with reactive siliceous limestone (Spratt) is lower than that for companion paste specimens with the same silica fume content. However, the differences at 10% silica fume are small and may be an artifact of the higher w/cm (0.70) used in the concrete compared with the paste (0.50). Expansion data for concrete specimens with the same aggregate, but a lower w/cm (0.50), show 10 and 15% silica fume to be effective in suppressing deleterious expansion up to 3.5 years, although the long-term performance may be questionable. The difference between the pore solution alkalinity of paste and concrete is significant for the control specimens and those containing 5% silica fume. Deleterious expansion was observed for both types of concrete.

Duchesne and Bérubé (1992b, 1994c) compared the performance of a number of SCMs (3 fly ashes, 1 slag, and 2 silica fumes) in pore solution studies on pastes and concrete prism expansion tests. The authors concluded that SCMs, which were effective in lowering the alkali concentration of the pore solution to below 0.65M ($Na^+ + K^+$), were also able to limit the expansion of concrete prisms to 0.04%.

Rasheeduzzafar and Hussein (1991) suggested a "broad correlation" between expansion and pore solution data for SCMs. However, they proposed that the threshold alkalinity of the pore solution for damaging ASR (with Pyrex glass) is in excess of 0.4 M OH^- for silica fume mortars, whereas lower thresholds may exist for mortars without SCM.

Effect of Silica Fume on Hydrate Composition

The reduction in pore solution alkalinity in the presence of silica fume is a corollary of the increased alkali-binding capacity of the hydrates (specifically CSH). Bhatty and Greening (1978) found that CSH with a low Ca/Si ratio was able to retain more alkali (Na + K) compared to hydrates of higher lime to silica ratios. This was later explained by Glasser and Marr (1985) on the basis of the surface charge on the CSH, which is dependent on the Ca/Si ratio. At high ratios, the charge is positive and the CSH tends to absorb cations. As the Ca/Si ratio decreases, the positive charge reduces, becoming negative at low Ca/Si ratios, for example, less than 1.3 (Glasser 1992). Negatively charged CSH has an increased capacity to sorb cations, especially alkalis.

Table VI. Results from X-ray microanalysis studies

Authors	OPC	w/c	Temp.	Age	Silica fume			Analysis		
	% Na_2O_e		(°C)	(days)	% SiO_2	% Na_2O_e	Level	Ca/Si	% Na_2O	% K_2O
Regourd et al. (1993)*	0.80	0.50	20	28	89	2.24	0	1.70		
							5	1.43		
Regourd et al. (1982)†	1.15	0.58		200	94	0.47	15	1.3		0.40
		0.83					20	0.9		0.40
Uchikawa et al. (1989)‡	0.59	0.50	20	91	84	1.52	0	1.8	0.03	0.10
							10	1.3	0.10	0.25

*Analysis of CSH (around clinker grains) on fracture surface of mortar bars. Mortars also contain 30 or 25% of "inert" ground crystalline slag.
†Analysis of CSH (around clinker grains) of field concretes from SKW plant in Bécancour, Québec [see Aitcin and Regourd (1985)].
‡Analysis of CSH in hardened concrete.

Results from X-ray microanalysis studies of silica fume cement systems are presented in Table VI. The Ca/Si ratio was found to vary from 1.43 to 0.90 for specimens with silica fume contents of 5–20%, which compares with expected ratios of 1.60–1.80 for portland cement of the same age. There appears to be a concomitant increase in the alkali content of the hydrates due to the presence of silica fume.

Durand et al. (1990) analyzed the residual solid phases following pore solution tests. Details of the mixes used for pore solution studies are given in Table V. The residual material was crushed, washed in distilled water for 24 h, and dried prior to analysis. The authors report that the alkali content of these phases is increased by 25% for paste with 5% silica fume and 275% for paste with 15% silica fume compared with the solid residue of plain portland cement pastes.

Nagataki and Wu (1995) showed a reliable correlation between the alkali concentration of the pore solution and the CaO/SiO_2 ratio of the cementitious material for a range of blended cements with different levels of slag and silica fume (including ternary blends). Bhatty and Greening (1987) observed a similar relationship between the amount of alkali retained in paste specimens after 14 years in a fog room and the Ca/Si ratio of the CSH for blended cements with various pozzolans (opal, calcined shale, and fly ash).

Tenoutasse and Marion (1987) observed morphological and compositional similarities between hydrates formed in blended silica fume cement pastes and "conventional" alkali-silica gel (i.e., ASR reaction product). Wang and Gillott (1992) found that the expansive properties (free swelling and water absorption) of the reaction products from alkali-silica fume and alkali-opal systems were also similar. However, provided the silica fume (or opal) is sufficiently fine and adequately dispersed, the expansion will not manifest itself in concrete.

These data provide support for the concept of a "competition" between silica fume and reactive siliceous aggregate (Wang and Gillot 1992). Provided there is a sufficient quantity of silica fume adequately dispersed through the paste, alkalis will be removed from solution by direct reaction with the silica fume or incorporation into the CSH gel, thus preventing deleterious reaction with the aggregate. The efficacy of the silica fume will depend to some extent on the relative rates of reaction between the silica fume and the aggregate. The competition becomes more intense when highly reactive forms of aggregate such as Beltane opal are present. In such cases, the aggregate may react at similar rates to the silica fume, and higher levels of silica fume may be required to lower the alkalinity before deleterious ASR can occur.

Diffusion Effects

Bakker (1981) explained the beneficial effects of slag in suppressing ASR on the basis of reduced ionic mobility and water permeability. The diffusion of Na^+ and K^+ ions was found to decrease by 15 times at 14 days in pastes with slag (75%) compared to control pastes; this effect would be expected to retard the rate of ASR in mortar or concrete. Similar reductions in permeability (10–100 times) are thought to further reduce expansion by retarding the rate of water imbibition by any ASR gel that does form (Bakker 1981). He suggested that alkali sorption by CSH was not a significant factor and that the alkali level in concretes with high slag levels (>65%) is of no importance with regard to alkali-silica reaction.

Uchikawa et al. (1989) measured the effect of 10% silica fume on the diffusion of Na^+ ions in cement pastes, mortars, and concretes. Silica fume had little effect on the diffusion coefficient measured at 1, 28, or 91 days in cement pastes or mortars. However, there was a significant reduction in diffusion in concrete due to the presence of silica fume, the reduction reaching approximately 5 times at 91 days. The authors ascribed this behavior to the reduced porosity [and $Ca(OH)_2$] of the transition zone in silica fume concrete due to pozzolanic reaction. Such effects would not be apparent in cement pastes; however, it is surprising that they were not observed in mortar specimens.

The authors are not aware of any other studies of alkali ion diffusion in silica fume cement pastes, mortars, or concretes. However, the effects of silica fume on pore structure, chloride ion diffusion, and water permeability are well documented. The addition of silica fume produces little alteration to the total porosity of cement paste, but has a pronounced effect on the pore size distribution. The volume of larger capillary pores is reduced in silica fume pastes and there is a concomitant increase in finer gel porosity (Mehta and Gjørv 1982). The pore structure refinement is attributed to both a physical filler effect and the pozzolanic reaction, and results in reduced fluid permeability. Permeability reductions due to silica fume may be more pronounced in concrete owing to the further microstructural improvement of the transition zone. It is not the purpose of this report to review the literature pertaining to the permeability of silica fume concrete; it is sufficient to observe that the use of 10% silica fume may be expected to reduce the coefficient of permeability by approximately one order of magnitude compared to portland cement concrete of the same w/cm (ACI 1994). Similar reductions in chloride diffusion coefficients have been observed with silica fume (Byfors 1987).

Bakker (1981) concluded that the restrictions to ionic and water movement imposed by high slag levels (>65%) were sufficient to eliminate ASR expansion. Lower levels of silica fume (i.e., 10–20%) are likely to have similar effects on the transport properties. However, it is possible that such effects only retard the onset and subsequent rate of reaction rather than eliminate ASR entirely. Such retardation may be satisfactory provided deleterious expansion is delayed beyond the required service life of the concrete. It is also possible that the mere slowing of the reaction may allow reaction products to be accommodated within the concrete, thus reducing stresses. Furthermore, if stresses are developed more slowly, they are more likely to be relieved by creep effects.

Role of Calcium

The mechanisms of expansion proposed by Hansen (1944) and by McGowan and Vivian (1952) were consistent in that they did not consider calcium in the primary mechanism of reaction and expansion. Dent Glasser (1979) also ignored the role of calcium in her description of gel structure. However, Hansen (1944) postulated that although $Ca(OH)_2$ was not involved in the initial reaction, it may serve to fuel the reaction through "alkali regeneration" by reacting with the alkali-silica gel and releasing some of the alkali. Powers and Steinour (1955a, 1955b) postulated that the ratio of lime to alkali determined the type of reaction product formed. At low concentrations of $Ca(OH)_2$, an expansive or "unsafe" alkali-silica product is formed, but if sufficient $Ca(OH)_2$ is present, a nonexpansive or "safe" calcium-alkali-silica complex is formed.

Struble and Diamond (1981a, 1981b) suggested that this explanation was an oversimplification. They studied the swelling properties of a series of synthetic alkali-silicate (Na_2O-SiO_2-H2O) and lime-alkali-silicate (CaO-Na_2O-SiO_2-H_2O) gels of varying composition. No consistent relationship between the composition of the gel and either the free expansion or the swelling pressure was found. Furthermore, the presence of calcium had no significant effect on the free expansion or the swelling pressure generated by the gel. Calcium gels behaved in a similar manner to gels of the same

Na_2O/SiO_2 ratio without calcium. However, calcium gels were observed to remain solid after testing, whereas the sodium silicate gels generally became fluid. The molecular structure of some of these synthetic gels was found to be unstable, and the gel became fluid to an extent that it was able to pass through the semi-permeable membrane used to constrain gel movement.

In a further study of opal in model pore solutions, Struble (1987) demonstrated that significant quantities of silica (2 M) can remain in solution with alkali hydroxides in the absence of cement hydration products (calcium) without the formation of alkali-silica gel. However, in mortars containing cement together with opal, the silica content of the pore solution was more than 2000 times lower (<0.001 M), and gel was observed to form. This would imply that silica simply dissolves in alkali hydroxide solution and does not form alkali-silicate gel in the absence of calcium (Diamond 1989).

A number of workers have suggested that the presence of $Ca(OH)_2$ is required for damaging reaction to occur (Chatterji 1979, 1988, 1989; Chatterji and Clausson-Kass 1984; Diamond 1989; Tang and Han 1980; Thomas 1995; Thomas et al. 1991). The dependence on $Ca(OH)_2$ for the promotion of damaging AAR is not a recent phenomenon. Conrow (1952) suggested that the expansion of concretes containing a siliceous sand-gravel may be related to the quantity of $Ca(OH)_2$ produced by the cement and that the beneficial effect of pozzolan is related to its ability to react with $Ca(OH)_2$. In a discussion of Conrow's paper, Mather stated an observation that concrete that has undergone AAR is characterized by "materially reduced quantities of crystalline calcium hydroxide." He suggested that $Ca(OH)_2$ may be consumed by the reaction and that "the mere consumption of calcium hydroxide by reaction with a pozzolan is sufficient to explain the beneficial effects of pozzolans in preventing abnormal expansion."

Chatterji (1979) showed that the expansion of mortar bars containing opaline silica, immersed in saturated NaCl at 50°C, could be eliminated if $Ca(OH)_2$ was removed prior to exposure. Lime removal was achieved by first leaching the bars in concentrated $CaCl_2$ or by the addition of a highly siliceous pozzolan (diatomite). In addition, mortar bars with very low cement contents (aggregate/cement = 19) did not expand; this is attributed to insufficient $Ca(OH)_2$ in these specimens. Significant quantities of silica gel were observed to precipitate on the surfaces of mortar bars with low cement content. Chatterji (1988, 1989) has proposed an ASR expansion mechanism with the role of $Ca(OH)_2$ clearly defined. Chatterji suggested that expansion occurs in concrete when the net amount of material entering a reactive silica grain (K^+, Na^+, Ca^{2+}, OH^-) exceeds the amount of material leaving the grain (Si^{4+}). The calcium concentration in the pore solution surrounding the grain controls the rate of diffusion of silica away from the reactive site and at high levels of calcium the migration of silica is prevented, leading to expansion of the reactive grain.

Tang et al. (1983) suggested that the preventative effect of mineral admixtures is related to their acidic oxide content (SiO_2 + Al_2O_3 + Fe_2O_3) and their ability to remove $Ca(OH)_2$ and lower the basicity of the system. He found that adding 10% CaO to mortar mixes containing opal and various admixtures increased the expansion, and explained this on the basis of the increased basicity [$CaO/(SiO_2 + Al_2O_3 + Fe_2O_3)$] of the cement. Similar behavior was reported by Thomas (1995); the addition of $Ca(OH)_2$ to fly ash concretes containing reactive flint was found to substantially increase the expansion of specimens stored in alkaline solution. The absence of expansion in fly ash concretes despite evidence of reaction has been attributed to the formation of a fluid or soluble reaction product in the absence of $Ca(OH)_2$ (Thomas 1995; Thomas et al. 1991).

Figure 7 shows the effect of adding $Ca(OH)_2$ (9%) on the expansion of mortars containing opal (Wang and Gillot 1991). In the absence of lime, 20% silica fume effectively suppressed expansion throughout the reported duration of the test (550 days). The addition of $Ca(OH)_2$ resulted in increased expansion for mortars with and without silica fume, however, the effect was more marked for the silica fume mortar.

Wei and Glasser (1989) measured calcium hydroxide contents in mortars with and without reactive aggregate (Pyrex or cristobalite). Their results, shown in Fig. 8, indicate that calcium hydroxide is consumed in the alkali-silica reaction.

The complete consumption of $Ca(OH)_2$ by pozzolanic reaction with silica fume is only achieved at relatively high levels of replacement (~25%). This is demonstrated in Fig. 8, which shows the relative $Ca(OH)_2$ content in pastes and mortars containing various amounts of silica fume. The data have been drawn from a range of studies using different materials, mix proportions, storage conditions, and methods of $Ca(OH)_2$ determination. Duchesne and Bérubé (1994b) demonstrate that most of the consumption occurs in the first 7 days.

With 10% silica fume, the reduction in $Ca(OH)_2$ is typically around 50% (compared with plain OPC), although there is significant variation. Conse-

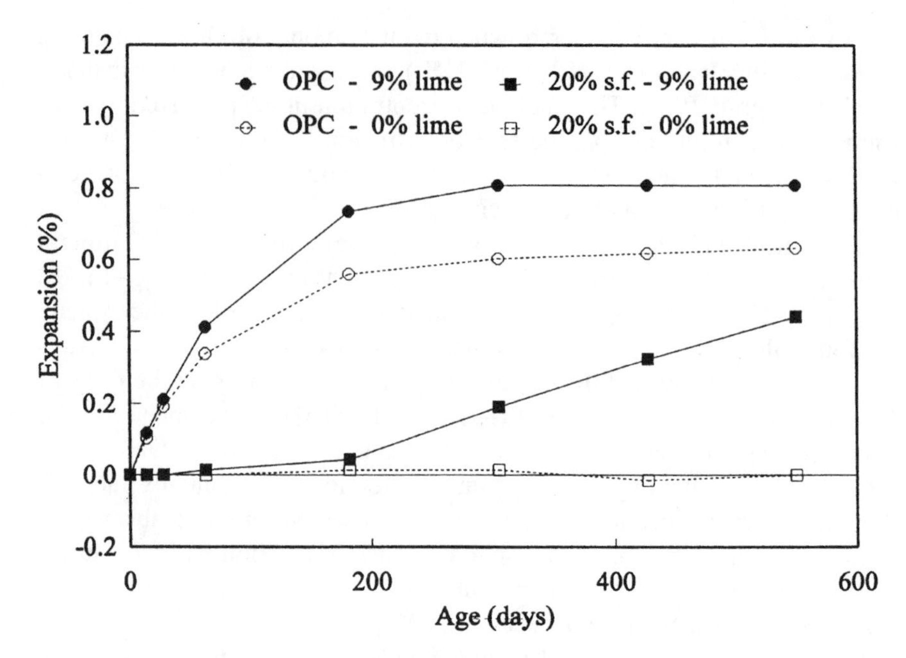

Figure 7. Effect of lime on expansion due to ASR (Wang and Gillot 1991b).

quently, it is hard to accept the consumption of $Ca(OH)_2$ as the only mechanism responsible for reducing AAR expansion in concretes containing silica fume. However, the use of SCMs can produce localized $Ca(OH)_2$ depletion and increased density of hydration products in the interfacial zone. These changes decrease the buffer of OH^- ions and reduce ionic mobility in this region, and it has been suggested (Larbi and Bijen 1992) that these effects are of paramount importance for reducing the alkali-silica reaction in concrete. Furthermore, the complete removal of $Ca(OH)_2$ may not be necessary to prevent damaging expansion. It is possible that there is a limiting $Ca(OH)_2$ content below which further reaction or expansion does not occur.

Summary and Recommendations

Early research data indicated silica fume to be highly effective in controlling expansion of concrete due to ASR, with levels of 10% or less being sufficient to eliminate damaging expansion in mortar bars with reactive aggregate from Iceland (Asgeirsson and Gudmundsson 1979) and South

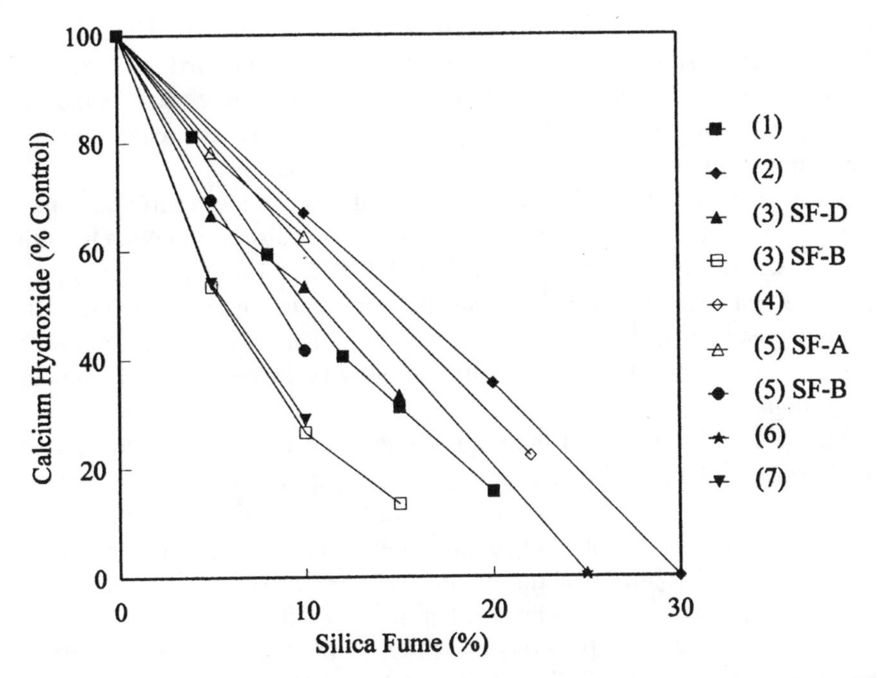

Figure 8. Consumption of lime by silica fume. (1) Sellevold and Nielsen (1987); (2) Glasser et al. (1988); (3) Kawamura et al. (1987); (4) Mehta and Gjørv (1982); (5) Duchesne and Bérubé (1994); (6) Wiens et al. (1995); (7) Nagataki et al. (1995).

Africa (Oberholster and Westra 1981). A few years later, pore solution studies revealed the marked effect of silica fume on pore solution composition, with pH values below 13.0 being attained with 20% silica fume (Diamond 1983b; Page and Vennesland 1983). Lower levels of replacement (5 and 10%) also produced a considerable reduction in pore solution alkalinity. With the courage of their convictions, the Icelanders have been using silica fume with high-alkali cement (~1.5% Na_2O_e) and highly reactive aggregate for housing concrete since 1979; to date there are no reported incidences of ASR in such concrete (Gudmundsson and Olafsson 1996).

In the last decade, the efficacy of lower levels of silica fume in controlling expansion in the long term has been questioned by a number of workers. Perry and Gillott (1985) showed 10% silica fume to substantially delay but not eliminate expansion with opaline aggregate. Kawamura et al. (1987) showed that 15% replacement may not be enough for some silica fumes used with opal. Work with cristobalite also showed that substantially higher levels of silica fume than 10% may be required to eliminate expansion in the long term (Alasali 1989; Dunster et al. 1990). Opal exhibits marked pessimum behavior and cristobalite reacts at relatively low alkali contents; findings from studies using such aggregates may not be applicable to concrete containing less-reactive forms of silica (Bérubé 1991).

More recent studies in Canada with natural reactive aggregates have further demonstrated the potential for deleterious expansion in concrete with 10% silica fume (Duchesne and Bérubé 1992a, 1994b; Fournier et al. 1995). Silica fume may provide a respite from damaging reaction for periods of 2 years or more for concrete prisms stored at 38°C, but the evidence suggests that the limit of 0.04% will eventually be exceeded with a number of aggregates.

The relevance of typical laboratory expansion tests to the behavior of field concrete has not been adequately investigated. In most laboratory tests, elevated temperatures and alkali contents are used to accelerate reaction and specimens are maintained at high humidity. The only reported case of damaging ASR in silica fume concrete stored under field conditions comes from South Africa, where concrete with 7% silica fume exhibited cracking after 3–7 years (Oberholster 1989; Oberholster and Davies 1987). These blocks had high alkali contents (~5 kg/m^3 Na_2O_e) and were exposed to a relatively warm climate. Blocks with lower alkali content (~4 kg/m^3 Na_2O_e) and 7% silica fume had not expanded significantly after 7.5 years. The absence of deleterious reaction in silica fume concrete in Iceland may be the result of the relatively cold climate. Silica fume at 7.5% replacement has not proven to be effective in controlling long-term expansion in mortar bars with Icelandic aggregate (see Fig. 2), with expansions in excess of 0.1% being attained after 4 years at 38°C. A further 20 years or so of exposure in Icelandic conditions may be required before a reliable evaluation of the long-term field performance can be made.

A wide range of data is available from expansion studies, but synthesis of the reported findings is complicated by the considerable variation in test methods and materials used by various investigators. Sources of variation include:

- Properties of silica fume used.
- Nature of reactive aggregate.
- Alkali content of portland cement.
- Mix proportions.
- Type of specimen.

- Storage conditions.
- Duration of testing.

Despite these differences, a number of general trends can be discerned from the literature:

- Expansion decreases with increasing silica fume content, although pessimum behavior has been observed in some studies at levels of replacement of 5% silica fume.
- There are insufficient data to allow the effect of silica fume composition to be determined. However, silica fume with atypically low SiO_2 or high Na_2O_e are not as effective in controlling expansion.
- At normal levels of silica fume (5–10%), expansion increases with the alkali content of the cement or the total alkali content of the concrete.
- The efficacy of the silica fume is influenced by the nature of the reactive aggregate and appears to be less able to control expansion with highly reactive cristobalite or opal.
- Silica fume retards the rate of expansion, the effect becoming more marked at higher replacement levels.

Duchesne and Bérubé (1994c) suggested that the long-term expansion of silica fume concrete may be due to the release of alkalis from the hydrates; they reported increasing alkali concentrations with time for solutions extracted from silica fume pastes. It has been suggested that, provided some $Ca(OH)_2$ remains, calcium may replace alkalis in the hydrates, thereby releasing Na^+ and K^+ ions to the pore solution (Bérubé and Duchesne 1991). It is not clear why this mechanism operates in silica fume concrete and not in concrete with fly ash or slag, or no SCM. In the authors' opinion, the delaying effect of silica fume is more likely a consequence of the reduced ion and water mobility.

The pore solution alkalinity of cement paste containing 10% silica fume is clearly related to the original alkali content of the blended materials (see Figs. 5 and 6) and alkali levels may still exceed 0.5 M OH^- when 10% silica fume is present. There is no generally agreed-upon threshold alkalinity above which ASR is sustained and below which it ceases. Diamond (1983a) and Kolleck et al. (1986) have suggested that reaction is unlikely below 0.25 and 0.20 M, respectively. Other workers have suggested much higher concentrations such as 0.65 M (Duchesne and Bérubé 1994c). The value of 0.65 M is clearly not tenable, as the results in Fig. 5 show that this is only exceeded with 10% silica fume once in all the studies reviewed; on this occasion the alkali content of the blend was 1.5% Na_2O_e. It is possible that the threshold level varies depending on a multitude of parameters, especially aggregate type and temperature. Regardless of the actual value, the risk of ASR obviously increases as the alkalinity of the pore solution increases.

Specifications take cognizance of the increased risk at higher alkali levels by imposing a limit either on the alkali content of the cement or, more commonly, on the total alkali content of the concrete (Nixon and Sims 1992). The current Canadian standard requires the alkali content of the concrete to be limited to 3 kg/m^3 Na_2O_e when reactive aggregates are used. Although there are a number of other sources of alkali (mixing water, admixtures, and aggregates), the chief contributor of alkali is portland cement. However, SCMs may contain appreciable quantities of alkali, often exceeding that of the portland cement they are combined with (especially fly ash). There are different approaches for calculating the alkali contribution from the SCM. In the Canadian standard, fly ash and slag can be used to control the alkali content of the concrete, provided minimum replacement levels of 20% (Class F ash) or 50% (slag) are used. In such cases, the ash and slag are considered to contribute no alkali to the concrete. There is no equivalent advice for silica fume.

The results of tests using Canadian aggregates show that for a given OPC alkali content (i.e., disregarding alkalis in silica fume) expansion is always reduced by the use of 10% silica fume, but 5% silica fume may increase expansion. Furthermore, concretes with 10% silica fume have exhibited only deleterious expansion in 2-year laboratory tests when the OPC alkalis have exceeded 4 kg/m^3 Na_2O_e, the exception being the unusually low-SiO_2 and high-Na_2O_e silica fume used by Duchesne and Bérubé (1994b). Similar behavior was been observed in the field tests reported by Oberholster (1989) for concrete with 7% silica fume. Consequently, it appears reasonable to endorse silica fume with the same status as fly ash and slag, and to permit it to be combined with portland cement for use with reactive aggregates, provided the alkali content of the concrete does not exceed 3 kg/m^3 Na_2O_e. As with fly ash and slag, the alkalis in the silica fume may be disregarded, provided the silica fume is used at a level of replacement exceeding 7% (to avoid pessimum effects). It would also be prudent to adopt compositional controls for the silica fume by specifying minimum silica contents (e.g. >85–90% SiO_2) and maximum alkali contents (e.g. <1.0–1.5% Na_2O_e).

Such a specification, though limited, would at least permit the use of silica fume with reactive aggregates. However, since silica fume is used at relatively small replacement levels, it would not be possible to achieve the same level of portland cement dilution (and hence alkali reduction) that can be achieved with 20% or more fly ash or 50% slag. Alternative uses of silica fume could take advantage of the increased effectiveness in suppressing expansion and pore solution alkalinity at higher levels of replacement. For example, the minimum level of silica fume might be specified as

$$SF = 2.5 \times AL$$

where SF is the silica fume content (% minimum) and AL is the alkali content in concrete from OPC (kg/m^3 Na_2O_e). This would result in a minimum replacement level of 12.5% silica fume when the alkalis in the concrete were 5 kg/m^3 Na_2O_e, which is in close agreement with the observations of Fournier et al. (1995) for concrete containing greywacke.

An alternative specification might be based on the alkali content of the blended cement rather than total alkali content of the concrete. Suitable relationships may be developed from pore solution studies, as shown in Figs. 9 and 10; these figures were plotted using the published data presented in Table V and Fig. 5. In Fig. 9, the silica fume content of pastes is plotted against the alkali content of the blended cement. Symbols reflect the hydroxyl ion concentration of the pore solution extracted from the paste and are classified as low (<250 mM OH^-), medium (250 to 500 mM OH^-), and high (>500 mM OH^-). The hashed line shown on the figure is drawn to separate pastes of low and high alkalinity and could form the basis for specifying minimum silica fume contents. When the alkali content of the cement is below 0.6% Na_2O_e, silica fume is not required. When the cement alkali exceeds 0.6% Na_2O_e, silica fume is required and the level increases with the alkali content of the blend (i.e., cement + silica fume alkalis). The line is constructed such that silica fume levels below 6% are not permitted in order to avoid pessimum effects. The relationship shown actually yields silica fume replacement levels that are 10 times the alkali content of the blend expressed as percentage Na_2O_e.

In Fig. 10, the silica fume content required to reduce the pore solution alkalinity to 250 mM OH^- was estimated from Fig. 5 and then plotted against the alkali content of the blended cement. There is a reasonable fit between these data, with the exception of the point representing the data of Andersson et al. (1989). These investigators reported an OH^- concentration

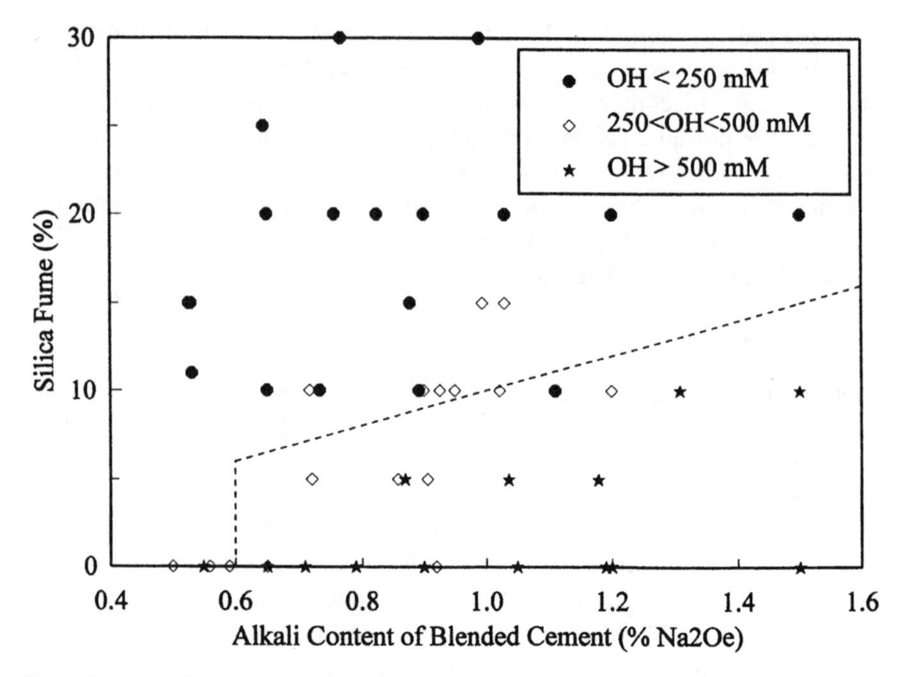

Figure 9. Pore solution alkalinity in silica fume pastes (data from Table V).

of just 251 mM for solution extracted from a 300-day-old paste produced using high-alkali cement (0.92% Na_2O_e); such a result appears to be erroneous in the light of accepted relationships between pore solution data and OPC alkalis (Nixon and Page 1987). Specifying 10% silica fume for every 1.0% Na_2O_e in the blend would appear to be appropriate to ensure a "safe" level of alkalinity.

The concept of increasing levels of silica fume with higher alkali contents is supported by data from expansion studies using opaline silica. Damaging expansion with opal is prevented provided the ratio of amorphous silica to available alkali is maintained sufficiently above the pessimum. A safe SiO_2/Na_2O_e ratio of 10 appears to be appropriate from the data in Table II. Assuming silica fume to be composed of 90% amorphous silica and 1% Na_2O_e, this ratio translates to approximately 11% silica fume for every 1% available Na_2O_e (from cement and silica fume). This would appear to coincide with the level of silica fume required to maintain a "safe" level of pore solution alkalinity (i.e., ~0.25 M OH^-).

The actual values suggested in the previous discussion related to specifications are not intended to be absolute and are given merely to illustrate the

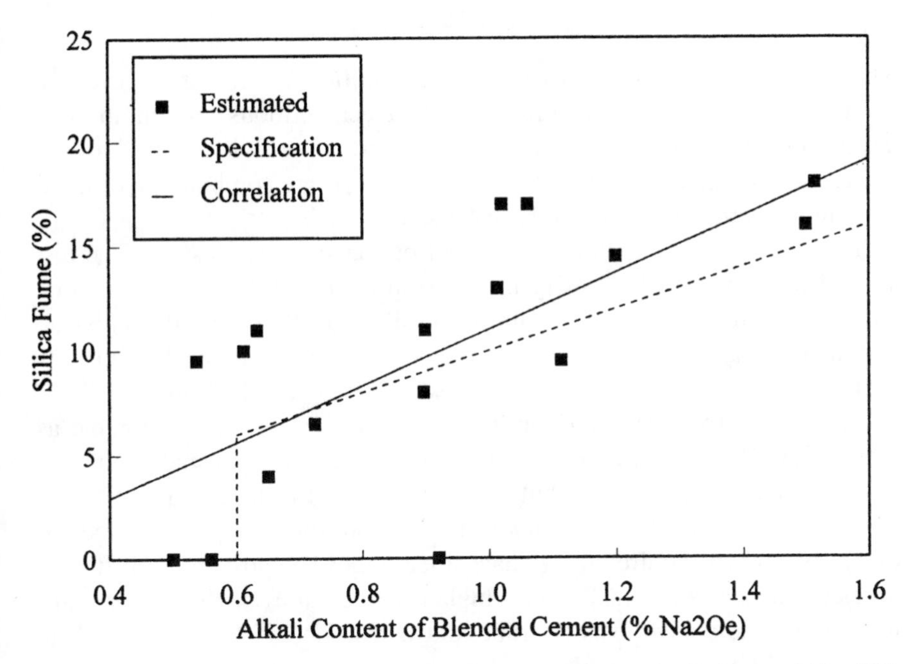

Figure 10. Level of silica fume required for pore solutions with alkalinities OH < 250 mM (estimated from Fig. 5).

concept of specifying the silica fume content on the basis of the available alkalis.

The level of silica fume may be restricted for other practical considerations or by specification, and it may not be possible to attain the level required for combating ASR. Similar problems are encountered with fly ash and slag, which are required to be used at levels of 20% (Class F ash) and 50%, respectively, when reactive aggregates are present (Canadian Standards Association 1994). These levels are, however, in excess of the maximum levels permitted by some regional specifications. The Ontario Ministry of Transport, for instance, limits the maximum amount of fly ash and slag to 10 and 25%, respectively. There is a perception that higher levels will lead to poor resistance to de-icer salt scaling (Afrani and Rogers 1994). It may be possible to overcome these difficulties by using ternary (or even quaternary) blends of portland cement, silica fume, fly ash, or slag. Writing a specification for all the possible combinations may be difficult without appropriate performance tests. However, one approach might be to use a formula similar to

$$(FA / 20) + (SG / 50) + (SF / 15) \geq 1$$

where FA is Class F fly ash level (% cementitious), SG is slag level (% cementitious), and SF is silica fume level (% cementitious). Again, the values presented are not intended as absolute. Further work is required to elucidate the synergistic effects when different SCMs are combined in order to determine appropriate levels of replacement.

Ideally, specification should be based on performance testing. The test should be capable of identifying the appropriate level of silica fume required to prevent expansion and be sensitive to the nature of aggregate and alkali content (and quantity) of the cement. As such, accelerated tests such as the Accelerated Mortar Bar Test (CSA A23.2 25A and ASTM C 1260) and the Pyrex Mortar Bar Test (ASTM C 441) are not suitable as they stand. The concrete prism test (CSA A23.2 14A) is usually run at a fixed cement alkali loading, but may be modified to assess the effect of alkali content. Unfortunately, this test takes too long, with test periods in excess of 2 years required to evaluate silica fume. The possibility of further accelerating this test (e.g., using a higher temperature) or the use of high-temperature autoclave tests (Bérube and Duchesne 1992a; Tang and Han 1983) should be investigated.

References

ACI Committee 234. 1994. "Guide for the use of silica fume in concrete." American Concrete Institute, Detroit.

Afrani, I., and C.A. Rogers. 1994. "The effects of different cementing materials and curing on concrete scaling," *Cem. Concr. Aggr.,* **16**, 132–139.

Aitcin, P.-C., and M. Regourd. 1985. "The use of condensed silica fume to control alkali-silica reaction — A field case study," *Cem. Concr. Res.,* **15** [4] 711–719.

Alasali, M.M. 1989. "Alkali-aggregate reaction in concrete: Investigations of concrete expansion from alkali contributed by pozzolans or slag"; pp. 41–451 in Proceedings of the Third International Conference on the Use of Fly Ash, Silica Fume, Slag and Natural Pozzolans in Concrete, ACI SP-114, Volume 1. Edited by V.M. Malhotra. American Concrete Institute, Detroit.

Andersson, K., B. Allard, M. Bengtsson, and B. Magnusson. 1989. "Chemical composition of cement pore solutions," *Cem. Concr. Res.,* **19** [3] 327–332.

Asgeirsson, H. 1986. "Silica fume in cement and silane for counteracting of alkali-silica reactions in Iceland," *Cem. Concr. Res.,* **16** [3] 423–428.

Asgeirsson, H., and G. Gudmundsson. 1979. "Pozzolanic activity of silica dust," *Cem. Concr. Res.,* **9** [2] 249–252.

Bakker, R.F.M. 1981. "About the cause of the resistance of blastfurnace cement concrete to the alkali-silica reaction"; paper S252/29 in *Proceedings of the 5th International Conference on Alkali-Aggregate Reaction,* Cape Town.

Berra, M., T. Mangialardi, and A.E. Paolini, 1994. "Application of the NaOH bath test method for assessing the effectiveness of mineral admixtures against reaction of alkali with artificial siliceous aggregate," *Cem. Concr. Compos.*, **16**, 207–218.

Berra, M., G. De Casa, and T. Mangialardi. 1996. "Evolution of chemical and physical parameters of blended cement mortars subjected to the NaOH bath test"; pp. 483–491 in *Proceedings of 10th International Conference on Alkali-Aggregate Reaction in Concrete,* Melbourne. Edited by A. Shayan.

Bérubé, M.-A., and J. Duchesne. 1991. "The effectiveness of silica fume in suppressing expansion due to alkali-aggregate reaction"; presented at CANMET/ACI Workshop on the Use of Silica Fume in Concrete, Washington, D.C.

Bérubé, M.-A., and J. Duchesne. 1992a. "Evaluation of testing methods used for assessing the effectiveness of mineral admixtures in suppressing expansion due to alkali-aggregate reaction"; pp. 549–575 in *Proceedings of the 4th International Conference on Fly Ash, Silica Fume, Slag and Natural Pozzolans in Concrete,* ACI SP132, Vol. 1. Edited by V.M. Malhotra. American Concrete Institute, Detroit.

Bérubé, M.-A., and J. Duchesne. 1992b. "Does silica fume merely postpone expansion due to alkali-aggregate reactivity?"; pp. 71–80 in *Proceedings of the 9th International Conference on Alkali-Aggregate Reaction in Concrete,* Vol. 1. The Concrete Society, Slough, U.K.

Bérubé, M.-A., J. Duchesne, and D. Chouinard. 1995. "Why the accelerated mortar bar method ASTM C 1260 is reliable for evaluating the effectiveness of supplementary cementing materials in suppressing expansion due to alkali-silica reactivity," *Cem. Concr. Aggregates,* **17** [1] 26–34.

Bhatty, M.S.Y., and N.R. Greening. 1978. "Interaction of alkalis with hydrating and hydrated calcium silicates"; pp. 87–112 in *Proceedings of the Fourth International Conference on the Effects of Alkalis in Cement and Concrete,* Purdue.

Blackwell, B.Q., and K. Pettifer. 1992. "Alkali-reactivity of greywacke aggregates in Maentwrog Dam (North Wales)," *Mag. Concr. Res.,* **44** [161] 255–264.

Bonen, D., and S. Diamond. 1992a. "Investigations on the coarse fraction of a commercial silica fume"; pp. 103–113 in *Proceedings of the 14th International Conference on Cement Microscopy.* ICMA, Duncanville, Texas.

Bonen, D., and S. Diamond. 1992b. "Occurrence of large silica fume–derived particles in hydrated cement paste," *Cem. Concr. Res.,* **22** [6] 1059–1066.

Byfors, K. 1987. "Influence of silica fume and fly ash on chloride diffusion and pH values in cement paste," *Cem. Conr. Res.,* **17** [1] 115–130.

Canadian Standards Association, 1994. "A23.1-94 Concrete Materials and Methods of Concrete Construction and A23.2-94 Methods of Test for Concrete." CSA, Rexdale, Ontario, Canada.

Chatterji, S. 1979. "The role of $Ca(OH)_2$ in the breakdown of portland cement concrete due to alkali-silica reaction," *Cem. Concr. Res.,* **9** [2] 185–188.

Chatterji, S. 1988. "Alkali-silica reaction and mineral admixtures"; pp. 125–131 in *Durability of Concrete: Aspects of Admixtures and Industrial By-Products,* International Seminar, April 1986, Stockholm. Swedish Council for Building Research.

Chatterji, S. 1989. "Mechanisms of alkali-silica reaction and expansion"; pp. 101–105 in *Proceedings of the 8th International Conference on Alkali-Aggregate Reaction,* Kyoto. Edited by K. Okada, S. Nishibayashi, and M. Kawamura.

Chatterji, S., and N.F. Clausson-Kass. 1984. "Prevention of alkali-silica expansion by using slag-slag-portland cement," *Cem. Concr. Res.,* **14** [6] 816–818.

Chen, H., J.A. Soles, and V.M. Malhotra. 1993. "Investigations of supplementary cementing materials for reducing alkali-aggregate reactions." *Cem. Concr. Compos.,* **15** [1/2] 75–84.

Conrow, A.D. 1952. "Studies of abnormal expansion of Portland cement concrete"; pp. 1205–1227 in *Proceedings of the American Society for Testing and Materials,* Vol. 52.

Davies, G., and R.E. Oberholster. 1987. "Use of the NBRI accelerated test to evaluate the effectiveness of mineral admixtures in preventing the alkali-silica reaction," *Cem. Concr. Res.,* **17** [1] 97–107.

Deng, Min, and Tang Mingshu. 1993. "Measures to inhibit alkali-dolomite reaction," *Cem. Concr. Res.,* **23** [5] 1115–1120.

Dent Glasser, L.S. 1979. "Osmotic pressure and the swelling of gels," *Cem. Concr. Res.,* **9**, 515–517.

Diamond, S. 1983a. "Alkali reactions in concrete — Pore solution effects"; pp. 155–166 in Proceedings of the 6th International Conference on Alkalis in Concrete. Edited by G.M. Idorn and Steen Rostam. Danish Concrete Association, Copenhagen.

Diamond, S. 1983b. "Effects of microsilica (silica fume) on pore-solution chemistry of cement pastes," *Commun. Am. Ceram. Soc.,* no. 5, 82–84.

Diamond, S. 1989. "ASR — Another look at mechanisms"; pp. 83–94 in Proceedings of the 8th International Conference on Alkali-Aggregate Reaction, Kyoto. Edited by K. Okada, S. Nishibayashi, and M. Kawamura.

Diamond, S., R.S. Barneyback, and L.J. Struble. 1981. "On the physics and chemistry of alkali-silica reactions"; paper S252/22 in *Proceedings of the 5th International Conference on Alkali-Aggregate Reaction,* Cape Town.

Duchesne, J., and M.-A. Bérubé. 1992a. "An autoclave mortar bar test for assessing the effectiveness of mineral admixtures in suppressing expansion due to AAR"; pp. 279–286 in *Proceedings of the 9th International Conference on Alkali-Aggregate Reaction in Concrete,* Vol. 1. The Concrete Society, Slough U.K.

Duchesne, J., and M.-A. Bérubé. 1992b. "Relationships between portlandite depletion, available alkalies and expansion of concrete made with mineral admixtures"; pp. 287–297 in *Proceedings of the 9th International Conference on Alkali-Aggregate Reaction in Concrete,* London, Vol. 1. The Concrete Society, Slough, U.K.

Duchesne, J., and M.-A. Bérubé, 1994a. "Available alkalies from supplementary cementing materials," *ACI Mater. J.,* **91** [3] 289–299.

Duchesne, J., and M.-A. Bérubé. 1994b. "The effectiveness of supplementary cementing materials in suppressing expansion due to ASR: Another look at reaction mechanisms. Part 1: concrete expansion and portlandite depletion," *Cem. Concr. Res.,* **24** [1] 73–82.

Duchesne, J., and M.-A. Bérubé. 1994c. "The effectiveness of supplementary cementing materials in suppressing expansion due to ASR: Another look at reaction mechanisms. Part 2: Pore solution chemistry," *Cem. Concr. Res.,* **24** [2] 221–230.

Dunster, A.M., H. Kawan, and P.J. Nixon. 1990. "The effect of silica fume to reduce alkali-silica reaction in concrete"; pp. 193–199 in Proceedings of the Fifth International Conference on Durability of Building Materials and Components. Edited by J.M. Baker, P.J. Nixon, A.J. Majumdar, and H. Davies. E&FN Spon, London.

Durand, B. 1993. "Preventive Measures Against Alkali-Aggregate Reactions in Concrete." IREQ-93-067. Institut de Recherche d'Hydro-Quebec, Varennes.

Durand, B., J. Berard, and R. Roux. 1990. "Alkali-silica reaction: The relation between pore solution characteristics and expansion test results," *Cem. Concr. Res.,* **20** [3[419–428.

Fournier, B., and M.-A. Berube. 1991. "General notions on alkali-aggregate reactions"; pp. 7–69 in *Petrography and Alkali-Aggregate Reactivity.* CANMET, Ottawa.

Fournier, B., A. Bilodeau, and V.M. Malhotra. 1995. "CANMET/industry research consortium on alkali-aggregate reactivity"; pp. 169–180 in *CANMET/ACI International Workshop on Alkali-Aggregate Reactions in Concrete.* Natural Resources Canada.

Fournier, B., V.M. Malhotra, W.S. Langley, and G.C. Hoff. 1994. "Alkali-aggregate reactivity (AAR) potential of selected Canadian aggregates for use in offshore structures"; pp. 657–686 in Third International Conference on the Durability of Concrete, ACI SP-145. Edited by V.M. Malhotra. American Concrete Institute, Detroit.

Gilliland, J.L., and W.T. Moran. 1949. "Siliceous admixture for the Davis Dam," *Eng. News Record,* 3 February, p. 62.

Gillott, J.E., M.A.G. Duncan, and E.G. Swenson. 1973. "Alkali-aggregate reaction in Nova Scotia. IV Character of the reaction," *Cem. Concr. Res.,* **3**, 521–535.

Glasser, F.P. 1992. "Chemistry of the alkali-aggregate reaction"; pp. 96–121 in *The Alkali-Silica Reaction in Concrete.* Edited by R.N. Swamy. Blackie, London, 1992.

Glasser, F.P., K. Luke, and M.J. Angus. 1988. "Modification of cement pore fluid compositions by pozzolanic additives," *Cem. Concr. Res.,* **18** [2] 165–178.

Glasser, F.P., and J. Marr. 1985. "The alkali binding potential of OPC and blended cements," *Il Cemento,* **82**, 85–94.

Grattan-Bellew, P.E. 1990. "Canadian experience with the mortar bar accelerated test for alkali-aggregate reactivity"; pp. 17–34 in *Canadian Developments in Testing Concrete Aggregates for Alkali-Aggregate Reactivity,* Report EM-92. Edited by C.A. Rogers. Ministry of Transportation, Downsview, Ontario, 1990.

Grattan-Bellew, P.E., and P.J. Lefebvre. 1987. "Effect of confinement on deterioration of concrete made with alkali-carbonate reactive aggregate"; pp. 280–285 in *Proceedings of the 7th International Conference on Concrete Alkali-Aggregate Reactions.* Edited by P.E. Grattan-Bellew. Noyes Publications, New Jersey.

Gudmundsson, G., and H. Asgeirsson. 1975. "Some investigations of alkali aggregate reaction," *Cem. Concr. Res.,* **5** [3] 211–220.

Gudmundsson, G., and H. Asgeirsson. 1983. "Factors affecting alkali expansion in Icelandic concretes"; pp. 217–221 in *Proceedings of the 6th International Conference on Alkalis in Concrete.* Edited by G.M. Idorn and Steen Rostam. Danish Concrete Association, Copenhagen.

Gudmundsson, G., and H. Olafsson. 1996. "Silica fume in concrete — 16 years of experience in Iceland"; pp. 562–569 in *Proceedings of the loth International Conference on Alkali-Aggregate Reaction in Concrete,* Melbourne. Edited by A. Shayan.

Hadley, D.W. 1961. "Alkali reactivity of carbonate rocks — Expansion and dedolomitization"; pp. 462–474 in *Proceedings of the Highway Research Board,* Vol. 40.

Hansen, W.C. 1944. "Studies relating to the mechanism by which the alkali-aggregate reaction proceeds in concrete," *J. Am. Concr. Inst.,* **15** [3] 213–227.

Hobbs, D.W. 1988. *Alkali-Silica Reaction in Concrete.* Thomas Telford, London.

Hobbs, D.W. 1994. "The effectiveness of fly ash in reducing the risk of cracking due to ASR in concretes containing cristobalite." *Mag. Concr. Res.,* **46** [168] 167–175.

Holland, T.C. 1995. "Specification for silica fume for use in concrete"; pp. 607–638 in *Second CANMET/ACI International Symposium on Advances in Concrete Technology,* ACI SP-154. Edited by V.M. Malhotra. American Concrete Institute, Detroit.

Holland, T.C., A. Krysa, M.D. Luther, and T.C. Liu. 1986. "Use of silica-fume concrete to repair abrasion-erosion damage in the Kinzua dam stilling basin"; pp. 841–863 in *Proceedings of the 2nd International Conference on Fly Ash, Silica Fume, Slag and Natural Pozzolans in Concrete,* ACI SP91, Vol. 2. Edited by V.M. Malhotra. American Concrete Institute, Detroit.

Hooton, R.D. 1993. "Influence of silica fume replacement of cement on physical properties and resistance to sulfate attack, freezing and thawing, and alkali-silica reactivity," *ACI Mater. J.,* **90** [2] 143–151.

Isabelle, H.L. 1986. "Development of a Canadian specification for silica fume"; pp. 1577–1587 in *Proceedings of the 2nd International Conference on Fly Ash, Silica Fume, Slag and Natural Pozzolans in Concrete,* ACI SP91, Vol. 2. Edited by V.M. Malhotra. American Concrete Institute, Detroit, pp. 1577-1587.

Kawamura, M., K. Takemoto, and S. Hasaba. 1986. "Effect of silica fume on alkali-silica expansion in mortars"; pp. 999–1012 in *Proceedings of the 2nd International Conference on Fly Ash, Silica Fume, Slag and Natural Pozzolans in Concrete,* ACI SP91, Vol. 2. Edited by V.M. Malhotra. American Concrete Institute, Detroit.

Kawamura, M., K. Takemoto, and S. Hasaba. 1987. "Effectiveness of various silica fumes in preventing alkali-silica expansion"; pp. 1809–1819 in *Concrete Durability, Katherine and Bryant Mather International Conference,* ACI SP-100, Vol. 2. Edited by J.M. Scanlon. American Concrete Institute, Detroit.

Kollek, J.J., S.P. Varma, and C. Zaris. 1986. "Measurement of OH^- concentrations of pore fluids and expansion due to alkali-silica reaction in composite cement mortars"; pp. 183–189 in *Proceedings of the 8th International Congress on the Chemistry of Cement,* Vol. 3, Rio de Janeiro.

Lagerblad, B., and P. Utkin. 1995. "Undispersed granulated silica fume in concrete — Chemical system and durability problems"; pp. 89–97 in *Materials Research Society Symposium Proceedings,* Vol. 370.

Lane, D.S. 1987. "Long-term mortar-bar expansion tests for potential alkali-aggregate reactivity"; pp. 336–341 in Proceedings of the 7th International Conference on Concrete Alkali-Aggregate Reactions. Edited by P.E. Grattan-Bellew. Noyes Publications, New Jersey.

Langley, W.S., B. Fournier, and V.M. Malhotra. 1995. "Effectiveness of silica fume in reducing expansion of concrete made with reactive aggregates from New Brunswick, Canada"; pp. 165–195 in *Real World Concrete.* Edited by G. Singh.

Larbi, J.A., and J.M. Bijen. 1992. "Effect of mineral admixtures on the cement paste-aggregate interface"; pp. 655–669 in *Proceedings of the 4th International Conference on Fly Ash, Silica Fume, Slag and Natural Pozzolans in Concrete,* ACI SP132, Vol. 1. Edited by V.M. Malhotra. American Concrete Institute, Detroit.

Longuet, P., L. Burglen, and A. Zelwer. 1973. "La phase liquide du cement hydrate," Revue Materiaux de Construction et travaux Publics, **676**, 35–41.

Lumley, J.S. 1989. “Synthetic cristobalite as a reference reactive aggregate”; pp. 561–566 in Proceedings of the 8th International Conference on Alkali-Aggregate Reaction, Kyoto. Edited by K. Okada, S. Nishibayashi, and M. Kawamura.

Lumley, J.S. 1993. “The ASR expansion of concrete prisms made from cements partially replaced by ground granulated blastfurnace slag,” *Const. Build. Mater.,* **7** [2] 95–99.

Malhotra, V.M., and B. Fournier. 1995. “Overview of research in alkali-aggregate reactions in concrete at CANMET”; pp. 1–45 in *CANMET/ACI International Workshop on Alkali-Aggregate Reactions in Concrete.* Natural Resources Canada.

Marusin, S.L., and L.B. Shotwell. 1995. “Alkali-silica reaction in concrete caused by densified silica fume lumps. A case study”; pp. 45–59 in *Fifth CANMET/ACI International Conference on Fly Ash, Silica Fume, Slag and Natural Pozzolans in Concrete, Supplementary Papers.*

Mather, K. 1973. “Examination of Cores from Four Highway Bridges in Georgia.” Miscellaneous Paper C-73-11, U.S. Army Engineering Waterways Experimental Station, Vicksburgh, Mississippi.

McGowan, J.K., and H.E. Vivian. 1952. “Studies in cement-aggregate reaction. XX. The correlation between crack development and expansion of mortar,” *Austral. J. Applied Sci.,* **3**, 228–232.

Mehta, P.K., and O.E. Gjørv. 1982. “Properties of portland cement concrete containing fly ash and condensed silica-fume,” *Cem. Concr. Res.,* **12** [5] 597–595.

Nagataki, S., and C. Wu. 1995. “A study of the properties of portland cement incorporating silica fume and blast furnace slag”; pp. 1051–1068 in *Fifth CANMET/ACI International Conference on Fly Ash, Silica Fume, Slag and Natural Pozzolans in Concrete,* ACI SP 153, Vol. 2. Edited by V.M. Malhotra. American Concrete Institute, Detroit.

Nixon, P.J., and C.L. Page. 1987. “Pore solution chemistry and alkali aggregate reaction”; pp. 1833–1862 in *Concrete Durability, Katherine and Bryant Mather International Conference,* ACI SP-100, Vol. 2. Edited by J.M. Scanlon. American Concrete Institute, Detroit.

Nixon, P.J., and I. Sims. 1992. “RILEM TC106 alkali aggregate reaction — Accelerated tests interim report and summary of national specifications”; pp. 731–738 in *Proceedings of the 9th International Conference on Alkali-Aggregate Reaction in Concrete,* London, Vol. 2. Concrete Society, Slough, U.K.

Nixon, P.J., R.J. Collins, and P.L. Rayment. 1979. “The concentration of alkalies by moisture migration in concrete — A factor influencing alkali aggregate reaction,” *Cem. Concr. Res.,* **9** [4] 417–423.

Nixon, P.J., C.L. Page, R. Bollinghaus, and I. Canham. 1986. “The effect of a PFA with a high total alkali content on pore solution composition and alkali silica reaction,” *Mag. Concr. Res.,* **38** [134] 30–35.

Nixon, P.J., I. Canham, C.L. Page, and R. Bollinghaus. 1987. “Sodium chloride and alkali-aggregate reaction”; pp. 110–114 in *Proceedings of the 7th International Conference on Concrete Alkali-Aggregate Reactions.* Edited by P.E. Grattan-Bellew. Noyes Publications, New Jersey.

Oberholster, R.E. 1989. “Alkali-aggregate reaction in South Africa. Some recent developments in research”; pp. 77–82 in *Proceedings of the 8th International Conference on Alkali-Aggregate Reaction,* Kyoto. Edited by K. Okada, S. Nishibayashi, and M. Kawamura).

Oberholster, R.E. 1992. "The effect of different outdoor exposure conditions on the expansion due to alkali-silica reaction"; pp. 623–628 in *Proceedings of the 9th International Conference on Alkali-Aggregate Reaction in Concrete,* Vol. 2. The Concrete Society, Slough, U.K.

Oberholster, R.E. 1994. "Alkali-silica reaction"; pp. 181–207 in *Concrete Technology.* Edited by Fulton.

Oberholster, R.E., and G. Davies. 1987. "The effect of mineral admixtures on the alkali-aggregate expansion of concrete under outdoor exposure conditions"; pp. 60–65 in *Proceedings of the 7th International Conference on Concrete Alkali-Aggregate Reactions.* Edited by P.E. Grattan-Bellew. Noyes Publications, New Jersey.

Oberholster, R.E., and W.B. Westra. 1981. "The effectiveness of mineral admixtures in reducing expansion due to alkali-aggregate reaction with Malmesbury group aggregates"; paper S252/31 in *Proceedings of the 5th International Conference on Alkali-Aggregate Reaction,* Cape Town.

Olafsson, H. 1989. "AAR problems in Iceland — Present state"; pp. 65–70 in *Proceedings of the 8th International Conference on Alkali-Aggregate Reaction,* Kyoto. Edited by K. Okada, S. Nishibayashi, and M. Kawamura.

Page, C.L., and Ø. Vennesland. 1983. "Pore solution composition and chloride binding capacity of silica-fume cement pastes," *Mater. Struct.,* **16** [91] 19–25.

Penko, M. 1983. "Some Early Hydration Processes in Cement Paste as Monitored by Liquid Phase Composition Measurements," Ph.D. Thesis, Purdue University.

Perry, C., and J.E. Gillott. 1985. "The feasibility of using silica fume to control concrete expansion due to alkali-aggregate reactions," *Durability Build. Mater.,* **3**, 133–146.

Petterson, K. 1992. "Effects of silica fume on alkali-silica expansion in mortar specimens," *Cem. Concr. Res.,* **22** [1] 15–22.

Popovic, K., V. Ukraincik, and A. Djurekovic. 1984. "Improvement of mortar and concrete durability by the use of condensed silica fume," *Durability Build. Mater.,* **2**, 171–186.

Powers, T.C., and H.H. Steinour. 1955a. "An investigation of some published researches on alkali-aggregate reaction. I. The chemical reactions and mechanism of expansion," *J. Am. Concr. Inst.,* **26** [6] 497–516.

Powers, T.C., and H.H. Steinour. 1955b. "An interpretation of some published researches on the alkali-aggregate reaction. Part 2: A hypothesis concerning safe and unsafe reactions with reactive silica in concrete," *J. Am. Concr. Inst.,* **26** [8] 785–811.

Radjy, F.F., T. Bogen, E.J. Sellevold, and K.E. Loeland. 1986. "A review of experiences with condensed silica-fume concretes and products"; pp. 1135–1152 in Proceedings of the 2nd International Conference on Fly Ash, Silica Fume, Slag and Natural Pozzolans in Concrete, ACI SP91, Vol. 2. Edited by V.M. Malhotra. American Concrete Institute, Detroit.

Rasheeduzzafar and S.E. Hussain. 1991. "Effect of microsilica and blast furnace slag on pore solution composition and alkali-silica reaction," *Cem. Concr. Compos.,* **13**, 219–225.

Regourd, M., B. Mortureux, P.-C. Aitcin, and P. Pinsonneault. 1982. "Microstructure of field concretes containing silica fume"; pp. 249–260 in *Fourth International Conference on Cement Microscopy.*

Regourd, M., B. Mortureux, and H. Hornain. 1983. "Use of condensed silica fume as filler in blended cements"; pp. 847–865 in *Proceedings of the First International Conference on Fly Ash, Silica Fume, Slag and Natural Pozzolans in Concrete,* ACI SP79, Vol. II. Edited by V.M. Malhotra. American Concrete Institute, Detroit.

Rogers, C.A., and R.D. Hooton. 1992. "Comparison between laboratory and field expansion of alkali-carbonate reactive aggregates"; pp. 877–884 in *Proceedings of the 9th International Conference on Alkali-Aggregate Reaction in Concrete,* Vol. 2. The Concrete Society, Slough, U.K.

Sellevold, E.J., and T. Nielsen. 1987. "Condensed silica fume in concrete: A world review"; pp. 167–229 in Supplementary Cementing Materials for Concrete. Edited by V.M. Malhotra. CANMET.

Shayan, A. 1995a. "Behaviour of precast prestressed concrete railway sleepers affected by AAR"; pp. 35–56 in *Real World Concrete.* Edited by G. Singh.

Shayan, A. 1995b. "Developments in testing for AAR in Australia"; pp. 152–169 in *CANMET/ACI International Workshop on Alkali-Aggregate Reactions in Concrete.* Natural Resources Canada.

Shayan, A., G.W. Quick, and C.J. Lancucki. 1993a. "Reactions of silica fume and alkali in steam-cured concrete"; pp. 399–410 in *Proceedings of the 16th International Conference on Cement Microscopy.* ICMA, Duncanville, Texas.

Shayan, A., G.W. Quick, and C.J. Lancucki. 1993b. "Morphological, mineralogical and chemical features of steam-cured concretes containing densified silica fume and various alkali levels," *Adv. Cem. Res.,* **5** [4] 151–162.

Shayan, A., G.W. Quick, and C.J. Lanucki. 1994. "Reactions of silica fume and alkali in steam-cured concrete"; pp. 399–410 in *Proceedings of the 16th International Conference on Cement Microscopy.* ICMA, Duncanville, Texas.

Sibbick, R.G., and C.L. Page. 1995. "Effects of pulverized fuel ash on alkali-silica reaction in concrete." *Constr. Build. Mater.,* **9** [5] 289–293.

Soles, J.A., V.M. Malhotra, and R.W. Suderman. 1987. "The role of supplementary cementing materials in reducing the effects of alkali-aggregate reactivity CANMET investigations"; pp. 79–84 in *Proceedings of the 7th International Conference on Concrete Alkali-Aggregate Reactions.* Edited by P.E. Grattan-Bellew. Noyes Publications, New Jersey.

Sprung, S., and M. Adabian. 1976. "The effect of admixtures on alkali-aggregate reaction in concrete"; pp. 125–137 in *Proceedings of the 3rd International Conference on the Effect of Alkalies on the Properties of Concrete,* London.

Stanton, T.E. 1940. "Expansion of concrete through reaction between cement and aggregate," *Proc. Am. Soc. Civ. Eng.,* **66** [10] 1781–1811.

Stanton, T.E. 1950. "Studies of use of pozzolans for counteracting excessive concrete expansion resulting from reaction between aggregates and the alkalies in cement"; pp. 178–203 in *Pozzolanic Materials in Mortars and Concretes,* ASTM STP 99. American Society for Testing and Materials, Philidelphia.

Stark, D. 1978. "Alkali-silica reactivity in the Rocky Mountain region"; pp. 235–243 in *Proceedings of the 4th International Conference on Effects of Alkalies in Cement and Concrete,* Publication No. CE-MAT-1-78. Purdue University, West Lafayette, Indiana.

Stark, D. 1980. "Alkali-silica reactivity: Some reconsiderations," *Cem. Concr. Aggr.,* **2** [2] 92–94.

Stark, D., and M.S.Y. Bhatty. 1986. "Alkali-silica reactivity: Effect of alkali in aggregate on expansion"; pp. 16–30 in *Alkalies in Concrete,* ASTM STP 930. American Society for Testing and Materials, Philadelphia.

Struble, L.J. 1987. "The Influence of Cement Pore Solution on Alkali Silica Reaction," Ph.D. Thesis, Purdue University, Indiana.

Struble, L.J., and S. Diamond. 1981a. "Swelling properties of synthetic alkali silica gels," *J. Am. Ceram. Soc.,* **64** [11] 652–656.

Struble, L.J., and S. Diamond. 1981b. "Unstable swelling behaviour of alkali silica gels," *Cem. Concr. Res.,* **11** [4] 611–617.

Swamy, R.N. 1990. "Role and effectiveness of mineral admixtures in relation to alkali silica reaction"; pp. 219–254 in *Proceedings of the G.M. Idorn International Symposium on the Durability of Concrete,* ACI SP-131. American Concrete Institute, Detroit.

Swamy, R.N., and M.M. Al-Asali. 1989. "Effectiveness of mineral admixtures in controlling ASR expansion"; pp. 205–210 in *Proceedings of the 8th International Conference on Alkali-Aggregate Reaction,* Kyoto. Edited by K. Okada, S. Nishibayashi, and M. Kawamura. pp. 205–210.

Swenson, E.G. 1957. "A reactive aggregate undetected by ASTM tests"; pp. 48–51 in *Proceedings of American Society for Testing and Materials,* Vol. 57.

Swenson, E.G., and J.E. Gillott. 1964. "Alkali-carbonate reaction," *Highway Res. Record,* [45] 21–40.

Tang, M. 1981. "Some remarks about alkali-silica reactions," *Cem. Concr. Res.,* **11**, 477–478.

Tang, M. 1992. "Classification of alkali-aggregate reaction"; pp. 648–653 in *Proceedings of the 9th International Conference on Alkali-Aggregate Reaction in Concrete,* London, Vol. 2. The Concrete Society, Slough, U.K.

Tang, M., and S. Han. 1980. "Effect of $Ca(OH)_2$ on alkali-silica reaction"; p. 94 in *Proceedings of the 7th International Conference on the Chemistry of Cement,* Vol. II.

Tang, M., and S. Han. 1983. "Rapid methods for determining the preventive effect of mineral admixtures on alkali-silica reaction"; pp. 383–386 in *Proceedings of the 6th International Conference on Alkalis in Concrete.* Edited by G.M. Idorn and Steen Rostam. Danish Concrete Association (Dansk Betonforening, DBF), Copenhagen.

Tang, M., Y.F. Ye, M.Q. Yuan, and S.H. Zheng. 1983. "The preventive effect of mineral admixtures on alkali-silica reaction and its mechanism," *Cem. Concr. Res.,* **13**, 171–176.

Tang, M., Z. Liu, and A. Han. 1987. "Mechanism of alkali-carbonate reaction"; pp. 275–279 in *Proceedings of the 7th International Conference on Concrete Alkali-Aggregate Reactions.* Edited by P.E. Grattan-Bellew. Noyes Publications, New Jersey.

Tenoutasse, N., and A.M. Marion. 1987. "The influence of silica fume in alkali-aggregate reactions"; pp. 71–75 in *Proceedings of the 7th International Conference on Concrete Alkali-Aggregate Reactions.* Edited by P.E. Grattan-Bellew. Noyes Publications, New Jersey.

Thomas, M.D.A. 1995. "Microstructural studies of ASR in fly ash concrete"; in *Proceedings of the 17th International Conference on Cement Microscopy.*

Thomas, M.D.A. 1996a. "Review of the effect of fly ash and slag on alkali-aggregate reaction in concrete." Report BR314, Building Research Establishment, Construction Research Communications Ltd., Watford, U.K.

Thomas, M.D.A. 1996b. "The effect of fly ash on alkali-aggregate reaction in concrete." BRE Report, Building Research Establishment, (in press).

Thomas, M.D.A., P.J. Nixon, and K. Pettifer. 1991. "The effect of pulverized fuel ash with a high total alkali content on alkali silica reaction in concrete containing natural U.K. aggregate"; pp. 919–940 in *Proceedings of the 2nd CANMET/ACI International Conference on Durability of Concrete,* Vol. 2. Edited by V.M. Malhotra. American Concrete Institute.

Thomas, M.D.A., B.Q. Blackwell, and K. Pettifer. 1992. "Suppression of damage from alkali silica reaction by fly ash in concrete dams"; pp. 1059–1066 in *Proceedings of the 9th International Conference on Alkali-Aggregate Reaction in Concrete,* Vol. 2. Concrete Society, Slough, U.K.

Thomas, M.D.A., B.Q. Blackwell, and P.J. Nixon. 1996. "Estimating the alkali contribution from fly ash to expansion due to alkali-aggregate reaction in concrete," *Mag. Concr. Res.,* **49** [177] 251–264.

Uchikawa, H., S. Uchida, and S. Hanehara. 1989. "Relationship between structure and penetrability of Na ion in hardened blended cement paste mortar and concrete"; pp. 121–128 in *Proceedings of the 8th International Conference on Alkali-Aggregate Reaction,* Kyoto. Edited by K. Okada, S. Nishibayashi, and M. Kawamura.

Van Aardt, J.H.P., and S. Visser. 1977. "Calcium hydroxide attack of feldspars and clays: Possible relevance to cement-aggregate reactions," *Cem. Concr. Res.,* **7**, 643–648.

Wang, H., and J.E. Gillott. 1991. "Mechanism of alkali-silica reaction and the significance of calcium hydroxide," *Cem. Concr. Res.,* **21** [4] 647–654.

Wang, H., and J.E. Gillott. 1992. "Competitive nature of alkali-silica fume and alkali-aggregate (silica) reaction," *Mag. Concr. Res.,* **44** [161] 235–239.

Way, S.J., and W.F. Cole. 1982. "Calcium hydroxide attack on rocks," *Cem. Concr. Res.,* **12**, 611–617.

Wei, X., and F.P. Glasser. 1989. "The role of microsilica in the alkali-aggregate reaction," *Adv. Cem. Res.,* **2** [8] 159–169.

Wiens, U., W. Breit, and P. Schiessl. 1995. "Influence of high silica fume and high fly ash contents on alkalinity of pore solution and protection of steel against corrosion"; pp. 741–761 in *Fifth CANMET/ACI International Conference on Fly Ash, Silica Fume, Slag and Natural Pozzolans in Concrete,* ACI SP 153, Vol. 2. Edited by V.M. Malhotra. American Concrete Institute, Detroit.

Yilmaz, V.T., and F.P. Glasser. 1990. "Reaction of alkali-resistant glass fibres with cement. Part 2. Durability in cement matrices conditioned with silica fume," *Glass Technol.,* **32** [4] 138–147.

Zhongi, X., and R.D. Hooton. 1993. "Migration of alkali ions in mortar due to several mechanisms," *Cem. Concr. Res.,* **23** [4] 951–961.

Delayed Ettringite Formation in Concrete: Recent Developments and Future Directions

Michael Thomas
Department of Civil Engineering, University of Toronto

"It is a capital mistake to theorize before one has data. Insensibly one begins to twist facts to suit theories, instead of theories to suit facts."
— Sherlock Holmes

Summary

This report reviews recent developments concerning the problem of delayed ettringite formation* (DEF) in concrete and offers some suggestions for future directions with regard to minimizing the risk of damage in new construction and prioritizing research needs. Considerable controversy surrounds the phenomenon of DEF, much of it stemming from litigation, and this has prompted a great debate in the literature. Copious technical publications have surfaced recently, their number seemingly out of proportion with the limited extent of scientific study in the area and, indeed, the relative scarcity of this problem in the field. Unfortunately this has led to discord in the industry, mainly between the producers of cement and the manufacturers of precast concrete, with consultants and experts generally taking sides with one side or the other. While one side believes that recent changes in cement manufacturing practices are responsible, the other maintains that poorly regulated heat curing processes in precast plants lead to the problem. However, it is this author's opinion that DEF is a "concrete problem" that affects and must be dealt with by the industry as a whole. For instance, one could argue that it is the demand to increase productivity that has led precasters to intensify the heat treatment process and cement producers to

---

*The term "delayed ettringite formation" is used here to describe the phenomenon whereby the ettringite that normally (and harmlessly) forms during the early periods of cement hydration is prevented from doing so, for example, because of excessive heat curing, and forms subsequently in the mature concrete. Under certain conditions this delayed formation of ettringite may be accompanied by expansion of the cement paste, resulting in cracking and deterioration of the concrete. This phenomenon should be distinguished from other forms of internal sulfate attack, which may occur due to excessive quantities of sulfate in the concrete, for example, from sulfate-contaminated aggregate.

develop more reactive and finer ground cements, and that the combination of these factors has led to increased occurrences of DEF.

One of the more disturbing claims made in recent years is that concrete that has not been exposed to elevated temperatures may suffer deterioration due to so-called ambient-temperature DEF. The main culprit in this case is apparently the high sulfur content of some modern-day cement clinkers, reportedly the result of using high-sulfate and waste fuels, which result in a slow or late release of sulfate over time. This school of thought seems to have been developed to explain the damage to non-heat-cured railway ties in a recent litigation case and has gained momentum ever since. All of the evidence to support this viewpoint comes from investigations of damaged non-heat-cured concrete structures. It is very difficult to substantiate these claims for the following reasons:

1. It is difficult to confirm the role that DEF has played in deterioration (other mechanisms such as ASR may have occurred).
2. It is possible that the internal concrete temperature exceeded 70°C due to the heat of hydration of cement (thus ambient temperature was not maintained).
3. Details of the cement used are invariably missing from these cases (i.e., the composition of the cement or clinker cannot be confirmed).

Carefully controlled laboratory studies, on the other hand, have indicated that expansion due to DEF cannot occur unless the mortar or concrete is heat cured above some threshold temperature (generally $\geq$ 70°C). Furthermore, it has been shown that clinker sulfates are generally available in the same time frame as gypsum. Thus, the balance of scientific evidence seems to refute the ambient-temperature theory.

The composition of the cement certainly has an influence on the expansive behavior of mortars and concretes that have been cured at elevated temperatures. Generally, for a given curing regime the extent of expansion is higher for cements with high levels of sulfate, alkali, C_3A, and fineness. Such properties are generally characteristic of high–early strength (Type III) cements. It is certainly ironic that the cements most sensitive to elevated temperature are those that are most likely to be heat cured.

A number of investigators have presented complex relationships between DEF expansion and cement chemistry (and fineness) and it is clear from these that DEF cannot be prevented by simply limiting a single component of the cement (e.g., SO_3). Any prescriptive measure to control DEF by cement chemistry would require limits on at least three properties of the cement, which would probably preclude many of the Type III cements currently in production. The prevention of DEF by controlling the cement would best be facilitated by the development of a suitable performance test to evaluate the DEF susceptibility of a particular cement. However, no such test exists at this time.

The association between DEF and damage due to alkali-silica reaction (ASR) and freeze-thaw is also reviewed. In many cases, the coexistence of ettringite with other deterioration mechanisms is simply a consequence of secondary ettringite recrystallizing into previously formed cracks and voids. Such a process is not a primary deterioration process and very probably does not cause any damage at all. However, there is evidence that high alkali concentrations in the pore solution may hinder the early formation of ettringite and that ettringite may form later if there is a subsequent reduction in the pore solution alkalinity (e.g., due to ASR). Under such circumstances it has been demonstrated that, in heat-cured concrete, DEF may contribute to the expansion initiated by ASR. This highlights the need to pay particular attention to the testing and selection of aggregates for use in heat-cured concrete. The role of ettringite in "blocking" air bubbles, and thereby compromising the air-void system and hence the freeze-thaw resistance of concrete, is not clear. However, the accumulation of ettringite in air voids, whether a damaging phenomenon or not, occurs as a consequence of secondary ettringite recrystallization and is not associated with "classical DEF."

At first sight, eliminating the risk of DEF in precast concrete production would appear to require the manufacturer to either closely regulate the heat-curing cycle (particularly controlling the maximum concrete temperature to <70°C) or use cements with low reactivity. Both approaches would probably result in reduced early age concrete strength and, consequently, lower productivity. However, there are other approaches for minimizing the risk of DEF, although they are less well defined; these include the use of pozzolans or slag, selection of suitable aggregates, and incorporation of sufficient quantities of entrained air. A manufacturer also has other options (in addition to heat and cement reactivity) for enhancing early age strength; these include the use of finely divided pozzolans (silica fume or metakaolin), chemical accelerators, or superplasticizers (to lower the water/cementitious material ratio). The performance of the final product depends on the interaction between all the various materials and manufacturing processes that make up the system and not just the cement or heat-curing temperature. Optimization of this system requires the means to quantify the performance of the final product both in terms of strength (for productivity) and durability (e.g., DEF susceptibility). Unfortunately, this is not currently possible, as there is no generally accepted performance test for evaluating the risk of

Table I.

	German	CSA A23.4
Minimum holding period and maximum concrete temperature during holding	3 h at 30°C 4 h at 40°C	
Maximum rate of temperature increase (°C/h)	20	
Maximum concrete temperature (°C)	60*	70
Maximum cooling rate (°C/h)		10

*Individual values may be 5°C higher.

DEF. In the author's view, the need to develop a suitable test for determining the risk of DEF for a particular combination of materials and production methods is the most important issue in resolving problems related to DEF.

Many European countries and Canada have adopted heat-curing practices similar to those introduced in Germany 20 years ago. The German and Canadian recommendations are summarized in Table I.

Where these recommendations have been enforced, they are reported to have been effective in reducing or eliminating the incidences of DEF regardless of the composition of the cement. In view of the uncertainty regarding the role of cement composition and the absence at this time of accepted test methods for evaluating the potential for DEF, it may be prudent to adopt similar regulations where DEF is a concern.

Introduction

Delayed ettringite formation was first implicated as a primary cause of deterioration in heat-cured precast concrete in the early 1980s in Germany. The decade that followed saw a substantial effort to confirm the validity of DEF as a deterioration process, elucidate the mechanisms of formation and expansion, determine the conditions required for its occurrence, and evaluate the role of cement composition. Day (1992) produced a comprehensive review of the literature published up to 1990, with the objective of determining the potential for the problem in North America. He concluded that the composition of some high–early strength cements and production practices of some precasters could combine to produce a DEF problem, although he was not aware of the occurrence of any such problem in North America. Soon after publication of Day's report, certain investigators claimed that DEF was the primary cause of deterioration in concrete railway ties (Marusin 1993a; Mielenz et al. 1995). Later publications have laid claim to the occurrence of DEF in non-heat-cured concrete (Hime 1996a, 1996b; Diamond 1996). The railway tie case, and the associated litigation, involved many concrete experts worldwide and sparked a great debate in North America regarding DEF. The fallout from the case, which is still being felt, has included a sharp growth in the number of technical papers, presentations, and even symposia dealing with DEF, and a book (Wolter 1997), which, unfortunately, have not been matched by a similar increase in the scientific study of the phenomenon.

Currently there is a great deal of controversy surrounding DEF, particularly with regard to its correct diagnosis in field concrete and the role of heat curing and cement (or clinker) composition. Much of this controversy has been fueled by the opinions of experts and different sectors of the industry, which (in this author's opinion) have been somewhat polarized through litigation. Such conflict has made it difficult for cement producers and precast concrete manufacturers to reach a consensus regarding the best solution to the problem. In regions where DEF is alleged to have occurred (e.g., Iowa and Texas), the lack of consensus guidelines from the concrete industry as a whole has led to reduced user confidence. This in turn may increase the attractiveness of alternative materials (e.g., asphaltic concrete for pavements or steel for bridge girders) or lead users to introduce drastic changes to their own cement and concrete specifications in an attempt to control the risk of DEF.

The purpose of this report is to review some of the recent developments related to the problem of DEF in concrete. The report is not intended to be a review of all the pertinent literature as, since Day's report (1992), a number of detailed and authoritative reviews have been published either as technical papers (Taylor 1994a; Lawrence 1995a) or as part of Ph.D. theses (Fu 1996; Lewis 1996). Instead, this report attempts to look at the problem of DEF from the perspective of the concrete industry by identifying the immediate concerns, establishing a viable approach to minimize the risk of DEF in new construction, and prioritizing research needs. Particular emphasis is placed on the following areas:

- Role of heat curing.
- Potential for ambient-temperature DEF.
- Role of cement composition (e.g., cement and clinker sulfates).
- Means for mitigating the damage due to DEF (e.g., blended cements).
- Relationship with ASR.
- Research needs.

Mechanisms of DEF and Associated Expansion

The term "delayed ettringite formation" has been used to describe a process in concrete whereby the normal early formation of ettringite is postponed, predominantly as a consequence of a deliberate or adventitious excursion to high temperature (typically >70°C). Ettringite forms during later storage at normal temperatures and high humidity, and this delayed formation may be accompanied by deleterious expansion of the concrete. The process of DEF is differentiated from normal sulfate attack in that there is no requirement for an external source of sulfate. The necessary ingredients (calcium, alumina, and sulfate) are already present when the concrete is mixed and placed, although the process does appear to require an external supply of moisture to produce damaging expansion. Glasser (1996) refers to DEF as an "isochemical" process in that the bulk composition of the system remains constant and the deterioration is the result of "spontaneous mineralogical changes."

Numerous studies have shown that ettringite cannot be detected (e.g., by X-ray diffraction) in cement paste, mortar, or concrete samples immediately after heat treatment at temperatures above about 70°C. This might indicate that ettringite is unstable at such temperature. However, Glasser (1996) explains that the disappearance of ettringite at elevated temperature is due to its increased solubility and that provided sufficient SO_4^{2-} is dissolved in the pore solution, ettringite is stable up to at least 100°C. This is shown in the experiments of Heinz and Ludwig (1987), who demonstrated that ettringite was still present immediately after heat treatment at 90°C in mortars with sufficient quantities of added anhydrite (equivalent to 8.6% SO_3 by mass of cement). Increased pore solution alkalinity will also increase the solubility of sulfate, leading to higher SO_4^{2-} concentrations in solutions at elevated temperature; this has been shown experimentally (Wieker and Herr 1989b) and theoretically (Glasser et al. 1995).

Some of the SO_4^{2-} remains in solution after heat treatment and some appears to enter the C-S-H gel, which is produced rapidly during the heating process. The evidence for residual dissolved SO_4^{2-} after heat curing comes from pore solution studies (Wieker and Herr 1989b; Glasser 1996) and can be explained by the absence of sufficient alumina and calcium in solution to promote the precipitation of ettringite or other sulfate forms (Glasser et al. 1995, Glasser 1996). The increased amounts of sulfate compared with alumina and calcium in solution is a result of the incongruent dissolution of ettringite at high temperature (Glasser 1996). It has been proposed that most of the sulfate and nearly all of the alumina that dissolve are absorbed by the C-S-H (Taylor 1994b). The effect of cement composition and heat curing conditions on the composition of the C-S-H has been studied extensively at Imperial College, London, using SEM/EDS techniques (Scrivener and Taylor 1993; Lewis et al. 1995; Lewis 1996; Scrivener and Lewis 1997). Microanalysis shows that the sulfate and aluminate content of the "inner product" C-S-H increases with curing temperature and that the composition of the C-S-H (particularly the sulfur/aluminum atomic ratio, S/Al) immediately after heat curing provides a reliable indication of the future potential for DEF expansion.

Analysis (e.g., X-ray diffraction) indicates that ettringite starts to form shortly after heat curing during subsequent moist storage at normal temperature (Ghorab et al. 1980; Sylla 1988; Scrivener and Taylor 1993). The reformation of ettringite is often accompanied by expansion; indeed it has been shown (Lewis et al. 1995) that much of the ettringite forms during the period of expansion (although the magnitude of the expansion is not directly related to the quantity of ettringite formed). This association suggests that the formation of ettringite is the direct cause of expansion. However, this has been contested in a recent paper by Scrivener, Damidot, and Famy (1997), since it was shown that the pattern of ettringite formation may be similar in heat-cured mortars that do not expand, which indicates that the crystalline ettringite detected by XRD is not the direct cause of expansion.

A common observation in concrete or mortar that has supposedly undergone expansion due to DEF is the presence of gaps around aggregate particles, which are frequently partially or completely filled with macroscopic ettringite. This led early workers (Heinz and Ludwig 1987) to conclude that the formation of ettringite at the cement/aggregate interface is the "origin of the degradation." Strong support for this argument is the apparent absence of expansion in steam-cured cement pastes (Lawrence 1993). However, more recent data have shown that paste specimens may expand, although the time to the onset of expansion is greatly delayed and the ultimate expansion is much less than in samples containing aggregate (Odler and Chen 1995, 1996; Lawrence 1995b; Yang et al. 1996). In one study (Yang et al. 1996), no expansion was observed until after 2 years of storage in water at 20°C for cement pastes initially cured at 100°C for 3 h. Other investigators have suggested that pre-existing cracks, caused by other deterioration mechanisms such as ASR or thermal expansion, are required for damaging DEF to occur in cement pastes (Sylla 1988; Fu et al. 1994; Fu and Beaudoin 1996).

It has been argued that the conditions necessary for crystal pressure growth, for example, supersaturation of the concrete pore solution with respect to ettringite, are unlikely to be maintained in mortar or concrete (Johansen et al. 1993). Assuming a degree of supersaturation of $K/K_s = 2.4$, Taylor (1994b) calculated that the maximum pressure exerted by a growing crystal would be 3 MPa according to the Kelvin equation. Using a much higher degree of supersaturation ($K/K_s = 109.74$), Deng and Tang (1994) have suggested that pressures in excess of 50 MPa may be a thermodynamic possibility due to ettringite growth in portland cement paste. A more elaborate analysis involving thermodynamics and fracture energy has been presented by Fu et al. (1994). They showed that ettringite would nucleate preferentially in cracks where relatively small crystallization pressures (and degrees of supersaturation) would be required to cause crack growth. Further support for this theory is presented by Diamond (1996).

Johansen et al. (1993) proposed an alternative mechanism for DEF expansion, which involves a uniform expansion of the cement paste (see also Skalny, Johansen, et al. 1996). Evidence for this phenomenon is the apparent growth of the cement paste away from the aggregate particles, resulting in gaps, the size of which are approximately proportional to the size of the aggregate. Some of these gaps are subsequently filled with ettringite due to recrystallization or Otswald ripening, a process that describes the tendency of small, confined crystals to dissolve and reprecipitate elsewhere as larger crystals in cracks and voids (Johansen et al. 1993). The observation that not all of the gaps are filled with ettringite supports the argument that the gaps open first and do not require the growth of ettringite at the cement/aggregate interface.

Microstructural studies at Imperial College lend support to the paste expansion theory. Microanalysis of "undesignated product" (Scrivener and Taylor 1993) and "inner product" C-S-H (Scrivener and Lewis 1997) show that sulfate and aluminate are intimately mixed with the C-S-H immediately after heat curing and that ettringite forms within the C-S-H gel during subsequent storage in water at room temperature. This formation appears to coincide with the onset of expansion, whereas the appearance of ettringite in voids within the paste occurs subsequently as a result of recrystallization.

Glasser et al. (1995) also support the paste expansion theory. However, they hypothesize that the ettringite forms mainly as a result of sulfates in the pore solution reacting with calcium and aluminate in the solid phases (e.g., maybe as monosulfate). When this occurs in pores and voids, well-developed, macrocrystalline ettringite forms and its growth is innocuous. However, when sulfates diffuse into the dense C-S-H gel and encounter calcium and aluminate, the ettringite that forms does so in a confined space, resulting in expansion of the matrix.

Diamond (1996) has criticized the paste expansion theory mainly on the basis of the observation that the width of gaps that form around aggregate particles are not necessarily proportional to the aggregate size, nor do they form around all of the aggregate particles within an expanded mortar or concrete.

Provided the aggregate is volumetrically stable, it is indisputable that gaps around aggregate particles indicate that the paste has expanded. To expect such gaps to form in a consistent manner implies a degree of homogeneity in the expansion process that is somewhat inconsistent with the heterogeneous nature of concrete.

Field Studies

Difficulties Interpreting Field Data

The presence of ettringite in cracks and voids of exposed concrete is not indicative of a damaging process involving sulfates from either external or internal sources. Primary ettringite, formed by the normal hydration of portland cement, tends to subsequently dissolve and "grow" by successive dissolution of small crystals in confined spaces and recrystallization of macroscopic crystals into larger spaces, a process known as Otswald ripening that tends to reduce the specific surface of the crystals. The process is sometimes referred to as secondary ettringite formation and is not harmful to concrete. Any exposed concrete that is cracked (e.g., due to loading, freeze/thaw, alkali-silica reaction, etc.) is likely to contain some quantity of crystalline secondary ettringite lining or filling voids and cracks. This reprecipitation of ettringite into existing cracks is not a mechanism of deterioration, although in this author's experience it is frequently taken as one. It is generally agreed that the presence of gaps around aggregate particles, which may be filled with crystalline ettringite, is a reliably diagnostic feature of DEF expansion. However, other processes may cause expansion of the paste, producing the same feature (e.g., unsoundness).

In almost all reported field cases of damage due to DEF evidence of some other deterioration process, frequently alkali-silica reaction (ASR), has been present. In these circumstances it is difficult to deconvolute the evidence and determine the initial cause of damage, the direct contribution of DEF to the observed damage, and whether damage would have occurred due to DEF in the absence of the other mechanism.

Field studies can be very useful in studying the long-term performance of materials exposed to environmental loads. However, the interpretation of data must be done with a good deal of caution. There is rarely accurate information available regarding the composition of the concreting materials, concrete proportions, production methods, or placing conditions. It is usually not possible to establish the precise curing conditions and determine the extent of any early-age excursion to elevated temperature (due to applied or autogenous heat). Thus it is usually necessary to build up a database of field cases, each one meticulously documented, in an attempt to represent all the material and environmental parameters. Isolated cases rarely yield useful information and can often lead to a distorted view of "typical behavior."

DEF in Concrete Railway Ties

While DEF has been implicated as a contributory cause of deterioration in a number of concrete structures, it is perhaps significant that it has been more closely associated with premature damage in heat-cured railway ties than any other form of construction. DEF has reportedly occurred in railway ties in Germany (Heinz and Ludwig 1986; Sylla 1988), Finland (Tepponen and Eriksson 1987), the former Czechoslovakia (Vitousova 1991), the United States (Mielenz et al. 1995; Scrivener 1996; Scrivener, Johansen, and Skalny 1998), South Africa (Oberholster et al. 1992), and Australia (Shayan and Quick 1992). The mass production of concrete railway ties invariably involves the use of high–early strength cements and accelerated curing; indeed, the rate of production has a strong impact on the economics of the process. The finished elements are also subjected to relatively severe in-service loads and often to aggressive environmental conditions. It is also significant that in nearly all of the reported cases of DEF in railway ties at least one other cause of deterioration has been identified. In most cases, the other form of deterioration has been ASR. The relationship between ASR and DEF in field and laboratory investigations is discussed in detail later.

Some controversy has arisen over the role of DEF in concrete railway ties in the northeastern United States. Some of the damaged ties could be traced back to being produced on a Friday, which is significant because some of these ties were not steam cured when the production schedule allowed for them to be left in the casting beds at ambient temperature over the weekend. This seems to provide strong support for the argument of Scrivener and others (Scrivener 1996; Scrivener, Johansen, and Skalny 1998) that another mechanism, for example, ASR, was primarily responsible for deterioration. These investigators found evidence that DEF had contributed to damage only in ties that could be traced back to "hot spots" in the casting beds (i.e., locations that could be subjected to excessively high temperature). However, to explain the phenomenon of "Friday ties," proponents of the DEF school suggested that expansion due to late formation of ettringite may be caused by the slow release of sulfate from high-sulfate clinker, and that this process may occur without elevated temperature curing (Mielenz et al. 1995; Hime 1996a, 1996b). No evidence has been presented to support this hypothesis, but this appears to be the origin of ambient temperature DEF, which is discussed in more detail later.

Problems with railway ties have led to changes in production practices in some countries. Tepponen and Eriksson (1987) reported that the use of heat treatment was abandoned in Finland in the 1980s. Non-heat-cured ties made with high–early strength cements from Finland and Germany have been in service since 1982. An investigation made after 5 years indicated improved field performance (compared to heat-cured ties) and no evidence of DEF. The author of this report is not aware of any investigation of these ties since 1987. Skalny and Locher (1997) reported that recommendations for controlling heat curing for railway ties were implemented as early as 1977 in Germany and were later adopted for all concrete. These recommendations included a maximum temperature limit for the concrete of 60°C. Skalny and Locher report that no further incidences of damage to concrete railway ties have been observed in West Germany since these practices were adopted. However, it should be noted that details regarding other changes in the materials and procedures used in Finland and Germany are not available, and it is not possible to ascribe the improved performance solely to the elimination of excessive temperatures. A detailed and well-documented investigation of the long-term performance and production practices of concrete railway ties is required if advantage is to be taken from the European experience.

DEF in Cast-in-Place Concrete

In 1990 the British Cement Association (BCA) reported isolated occurrences of damage due to DEF in concrete beams that were believed to have been constructed with rapid-hardening portland cement (RHPC) and using steam curing (Lawrence et al. 1990). The apparent absence of DEF in cast-in-place concrete led the authors to suggest that the application of external heat might be a prerequisite for damaging DEF. However, the BCA's view has since changed and Hobbs (1997) recently presented details of field cases of DEF in cast-in-place concrete structures in the United Kingdom. These concretes were of large section and high cement content (~500 kg/m^3), and Hobbs estimated that the early peak temperature was probably in excess of 85°C in all cases. He reports no incidence of DEF in concrete that has not been subjected to elevated temperature.

BCA's investigations highlight an important point. If the process of DEF expansion does require an excursion to some threshold temperature (e.g., >70°C), as indicated by laboratory studies, it is probably irrelevant whether this occurs either deliberately or adventitiously. It is certainly not unheard of for the internal temperature of concrete to exceed 80°C due to the exothermic hydration of cement. For instance, the autogenous temperature rise in a large concrete section (e.g., minimum dimension >1.0 m) containing 500 kg/m^3 of portland cement can be expected to be on the order of 60°C above the initial placing temperature. It should not surprise us then if damage due to DEF is not restricted to heat-cured precast concrete.

Experiences in Texas

Damage attributed to DEF has been reported in precast concrete bridge beams and other cast-in-place elements (e.g., light-pole foundations) in Texas (Lawrence et al. 1997a, 1997b). In one case, 56 out of 69 precast concrete box beams, manufactured in 1991, were found to be deteriorated to the extent that they were unusable. These beams were subjected to an extensive investigation involving a number of different laboratories that have failed to agree on a single diagnosis. Extensive map-cracking of these elements occurred within 1 year of manufacture. The Texas Department of Transport concluded that DEF is the main cause of deterioration due to the presence of ettringite "nests" with radiating cracks in the paste and gaps around some aggregate particles. Some investigators diagnosed ASR as the main cause of deterioration on the basis of the presence of reactive aggregate, cracked aggregate particles, and alkali-silica gel. In addition, in terms of composition the 56 damaged beams were differentiated from the other 13 beams only by the use of cement with a higher alkali content (similar sulfate content).

Some of the damaged beams were not steam cured (Lawrence et al. 1997b). However, this may not be significant in view of the composition of the concrete and the ambient conditions during fabrication. In order to meet the very high release strength requirements, it is necessary to use relatively high contents of cement with high fineness. This fact, coupled with the high ambient temperatures during casting, that is, >35°C (Lawrence et al. 1997b), is likely to lead to internal concrete temperatures in excess of 80ºC in thick sections.

No details regarding the deteriorated cast-in-place concrete have been made available to this author. However, there seems to be an inclination to extend the findings of the forensic investigations on the precast beams to these other structures. In other words, the parties involved in the evaluation of these cases seem prepared to accept only one mechanism of deterioration to explain the cracking in all the structures, although it is possible (and even probable) that some elements have undergone ASR, some DEF, and some both.

Field Evidence of Ambient-Temperature DEF

As mentioned above, the ambient-temperature DEF school seems to have its roots in the concrete railway tie litigation in northeastern United States. To explain the apparent damage of Friday ties that were not heat cured, Mielenz et al. (1995) reported the following:

> The distress to most ties was largely or completely due to delayed ettringite formation (DEF). The DEF, in turn, was primarily caused by the presence of large amounts of sulfate in the cement used in the mix, and also to the very slow solubility of that sulfate because of its source: the clinker. Sulfate formed in the clinker was probably present as anhydrite and as a component of the silicate phases. Such sulfate is only slowly soluble. Our finding that DEF in precast concrete can occur without steam curing, and due to cement composition, appears not to have been previously reported in the literature.

The clinker sulfate level in the cement used for much of the production of these ties was approximately 1.2% SO_3. However, the investigation of Mielenz and his coworkers did not establish a causal link between the clinker sulfates and damage to the concrete. Indeed, the role of clinker sulfates is pure conjecture and is certainly not a scientific “finding” of the investigation as suggested by the authors.

Shortly after the publication of this paper, Hime (1996a) published further discussion on the role of clinker sulfates. In this paper, Hime suggests that the recent emergence of DEF as a durability problem may be directly linked to changes in cement production, which have led to substantial increases in the sulfate content of clinkers and finished cements. It is stated that the sulfate in clinker may be as high as 4%, although such levels are denied by cement producers on the basis that such values are not possible in normal kiln operations. Hime attributes the increased sulfate levels to the burning of high-sulfur fuels and wastes (e.g., tires) in cement kilns. In fact, studies have shown that the burning of tires in cement kilns generally does not increase the SO_3 content of the clinker, because the tires have a similar sulfur content to the coal they replace (Schrama et al. 1995). In addition to the railway ties, Hime cites a number of other cases where DEF has apparently caused deterioration of the concrete. Only vague descriptions of the affected structures are given, whereas no information is given on the exposure conditions, the nature of the distress, and the details of the investigation carried out to confirm the role of DEF. Furthermore, details of the materials and methods used to produce the concrete are not provided. In spite of the absence of such data, Hime concludes that

> High sulfate clinkers containing slowly soluble sulfate may result in production of ettringite in concrete months or years after casting, leading to destructive expansion of the concrete product or structure. High temperature curing may exacerbate the occurrence, but is not necessary for it.

There is no evidence presented in the paper that any of the structures supposedly damaged by DEF were produced using cement with a high-sulfate clinker, nor that any of the structures were damaged by DEF. The absence of real data is highlighted by the following statement by the author (Hime 1996a):

> Although cement production and concrete curing records have not yet been made available, it has been reported that cement manufacturers in the area were burning fuels and wastes that contained high concentrations of sulphur.

This is pure conjecture and, like the rest of the information in the paper, does not make a case for the existence of ambient-temperature DEF or role of clinker sulfates. Other papers with essentially the same theme have been published by the same author (Hime 1996b; Marusin and Hime 1997).

Diamond (1996) recently provided further support to the delayed formation of ettringite in non-heat-cured concrete due to the slow release of sulfate from the clinker. The evidence is based on four cases reported in the literature (Pettifer and Nixon 1980; Jones and Poole 1987; Larive and Louarn 1992; Saloman et al. 1992) and "personal experience." These cases warrant closer examination.

Pettifer and Nixon (1980) examined cores from five damaged structures in the United Kingdom and one in South Africa. All of these concretes showed evidence of deterioration due to alkali-silica reaction and the authors observed that ettringite was frequently associated with the reaction product in cracks and voids. They proposed that the reaction of calcium and sulfate with hydrated calcium aluminates ($C_4A{\cdot}19H$) precipitates ettringite while releasing hydroxyl ions to the pore solution, thereby fueling ASR. The origin of the sulfate required for this process was not dealt with by the authors. Jones and Poole (1987) examined cores from three structures in the United Kingdom, all of which were damaged by ASR. In one of the concretes, macrocrystalline ettringite was found in cracks often associated with alkali-silica gel or accumulations of portlandite. The authors dismissed the mechanism proposed by Pettifer and Nixon (1980) on the basis that the ettringite formation must have occurred after ASR had reached an advanced stage. They concluded that the ettringite recrystallized from primary ettringite, a process that is favored by the microcracking caused by ASR and the presence of alkali silica gel, which provides a medium for crystal nucleation and growth. It is suggested that the growth of this secondary ettringite in cracks may exacerbate the expansion and damage (Jones and Poole 1987).

Larive and Louarn (1992) reported on the examination of cores taken from 19 bridges in France built between 1936 and 1981. Evidence of damage due to ASR was found in all the structures. Various forms of ettringite were also observed in these structures, but secondary ettringite recrystallization was significantly less advanced in the six structures built since 1980 (approximately 10 years old). Although not stated by the authors, this might indicate that some level of damage due to ASR is required before the secondary ettringite formation can occur. This conclusion was reached by Saloman et al. (1992) as a result of their studies of cores from four structures. These investigators found evidence of secondary ettringite only in samples that were severely damaged by ASR.

These field reports do not lend support to ambient-temperature DEF. First, it is not possible to determine whether ettringite formation has even contributed to damage, let alone whether it is the primary cause of destruction. In all of these cases ASR damage was present, and three of the papers suggest that the cracks were initially formed by this reaction and provided the space for secondary ettringite precipitation (which may or may not have exacerbated damage). Second, the temperature history of the structures is not known and the maintenance of ambient temperature cannot be confirmed. Many of these cases cited in the four papers are massive elements (e.g., dams or foundations) that are likely to experience significant autogenous temperature rise.

Diamond's "personal experiences" include cases of DEF in the absence of ASR in field concrete that has not been steam cured (Diamond 1996). Some of these have apparently been associated with cements high in sulfate (4–5% SO_3). The only information provided in the paper related to these cases is an electron micrograph showing a cracking characteristic of DEF in non-steam-cured concrete. Once again, no details are given regarding the type of structure or its exposure, age, curing history, or material composition. There are no details of the cement used.

The latest link between high-sulfate clinker and ambient-temperature DEF was reported by Collepardi (1997). In this paper, Collepardi makes the case for DEF caused by the late release of sulfate from high-sulfate clinker on the basis of previously reported "proved cases" of DEF in non-steam-cured concrete and cites the work of Mielenz et al. (1995), Hime (1996a), and Diamond (1996). As discussed above, these reports fail to prove the role of clinker sulfates or the occurrence of damage due to DEF in concrete that has not been exposed to elevated temperature. Collepardi also provides support for ambient-temperature DEF based on his own field experiences. In addition to confirming the findings of Mielenz et al. (1995) for concrete ties, Collepardi reports evidence of DEF in cast-in-place transmission tower footings and precast roofing sheets. The roofing sheets are of interest as they apparently suffered DEF damage in the absence of any aggregate, but since they were steam-cured "at high temperatures," the case is not relevant here. The cast-in-place footings were located near Acona, Italy. They contained reactive aggregate, and were originally diagnosed as suffering from ASR. Later analysis detected ettringite in the cracks and relatively high sulfate levels in the concrete. Based on the sulfate content in the concrete, Collepardi (1997) reports that the sulfate content of the cement may have been as high as 4.4% SO_3 and further reports that the clinker sulfate may have been as high as 2% SO_3. Chemical analyses of the cement or the clinker were not available. No information is given regarding the potential for high internal temperatures during curing.

Taken collectively, the published information regarding DEF damage in non-steam-cured concrete appears formidable. However, when the cases are examined individually, there is no evidence that unequivocally demonstrates that damage due to DEF can occur in concrete that has not been exposed to elevated temperature (e.g. >70°C) or that the late release of sulfates from clinker contributes to the process. This statement is made for the following reasons:

- In nearly all of the cases reported, it is not possible to eliminate the possibility of an excursion to high temperature due to the heat released by the cementitious components of the concrete. Thus the maintenance of an ambient temperature cannot be confirmed.
- In most of the reported cases, cracking due to ASR is also reported and insufficient information is available to discern whether or not ettringite formation actually contributed to damage.
- There are insufficient details given about some of the investigations to allow any kind of objective opinion to be reached by the reader.
- Information regarding the cement and clinker composition is conspicuously absent from all the field cases that supposedly support the role of late sulfate release from cement clinker.

Indeed it is this author's opinion that the case for ambient-temperature DEF and the role of clinker sulfates is largely based on conjecture and hearsay. However, the absence of unequivocal evidence does not mean that these issues are not a cause for concern. The field cases discussed above present valuable information. It is certainly reasonable to develop working hypotheses on the basis of the field data; these might include the potential for late sulfate release and DEF expansion in concrete that has not been exposed to elevated temperature. However, the next stage in the scientific process is to test any such hypothesis in experiments that permit the control of exposure conditions (e.g., temperature), the elimination of other processes (e.g., ASR), and the use of materials of known composition. Carefully controlled laboratory testing has been carried out or is underway to elucidate the role of clinker composition and curing temperature. The results from these studies are reported below. It is perhaps ironic that the proponents of ambient-temperature DEF have reported no such testing.

Laboratory Studies

Role of Elevated Temperature

In 1992, Day presented a comprehensive analysis of more than 300 published articles dealing with DEF or related subjects. One of Day's major conclusions was that "there is no evidence that secondary (delayed) ettringite formation has been a principal cause of deterioration in non-heat-treated concrete."

Since this review, data from a number of laboratory studies have been reported, and the one general consensus that can be drawn from the different investigations is that it is not possible to induce damaging expansion due to DEF in the laboratory without heat treatment, regardless of cement composition (Lawrence 1995b; Lewis 1996; Kelham 1996; Fu 1996; Michaud and Suderman 1997b). Although there does not appear to be any single threshold temperature, values in excess of 70°C are invariably required to promote expansion in the laboratory.

Lawrence (1995b) reported data for 55 cements with sulfate contents in the range of 2.06–4.40% SO_3; unfortunately, clinker compositions were not reported. Mortar bars were heat cured at temperatures in the range of 60–100°C for various periods of time. For mortar bars heat cured for 3 h at 100°C, 36 of the 55 cements tested showed significant expansion. Data were presented only for 22 cements heat cured at 65°C, and none of these showed any significant expansion after storage at this temperature, even though 20 of the 22 cements cured at 65°C had expanded when cured at 100°C.

Kelham (1996) examined five clinkers with sulfate contents in the range of 0.3–1.6% SO_3 and sulfate/alkali molar ratios, $[SO_3]/[Na_2O_e]$, as high as 3.29. These clinkers were ground with different forms of sulfate to give target fineness values of 350 or 450 m^2/kg and sulfate levels of 4 or 5% SO_3 in the finished cement. Mortar bars were heat cured at 20, 70, 80, or 90°C for 12 h and subsequently stored in water at 20°C. Only mortars cured at 90°C showed significant expansion due to DEF. Some mortars with K_2SO_4 or KOH added to the mixing water did expand when cured at 80°C, but not when cured at 70°C. In a later paper, Kelham (1997) included data for a mineralized clinker with 2.6% SO_3 and $[SO_3]/[Na_2O_e] = 4.75$. Mortar bars made with cement produced from this clinker also failed to show expansion unless heat cured at 90°C.

Effect of Cement Composition

Heinz and coworkers have suggested that the ratio of sulfate to alumina in the cement is a key indicator of DEF susceptibility. In one series of tests (Heinz and Ludwig 1987), mortars were cast with PZ55 cement (German high–early strength portland cement) and anhydrite added to produce sulfate contents in the range of 2.0–8.6% SO_3 (by mass of cement). The expansion of bars initially heat cured at 90°C for 12 h was reported to vary depending on the sulfate/alumina ratio of the cement (plus anhydrite). When $\bar{S}/A \leq 0.67$ (molar ratio), no significant expansion was observed after 2 years, whereas all mortars with $\bar{S}/A \geq 0.67$ showed expansion greater than 0.40% after 1 year of storage in water at 20°C. In a later study (Heinz et al. 1989), the results for the PZ55 cement plus anhydrite additions were compared with similar data for a range of different cements (including composite cements with silica fume, slag, and fly ash). In this analysis the oxide ratio $\bar{S}^2/A$ (where A represents the "active" alumina in the C_3A only and excludes that in the C_4AF and any mineral admixture) was considered to be the most influential parameter and a critical value of $\bar{S}^2/A = 2.0$ was proposed. Cementitious systems in which $\bar{S}^2/A < 2.0$ did not exhibit any expansion even after "intensive heat treatment" (Heinz et al, 1989). Most of the cements with $\bar{S}^2/A > 2.0$ showed some level of expansion, the principal exception being PZ45HS (sulfate-resisting cement), which did not expand despite an obviously high sulfate/active alumina ratio, even when the SO_3 level was raised to 4% (by the addition of calcium sulfate) and the curing temperature was raised to 100°C.

Lawrence (1995b) reported data for 55 cements with a wide range of compositions. The bulk of the expansion tests were carried out on mortar bars initially cured at 100°C for 3 h and subsequently stored in water at 20°C. Significant expansion was observed for 37 of the 55 cements tested in this manner. Correlation coefficients between mortar bar expansion values at different ages and various compositional parameters of the cement have been presented, and Lawrence (1995b) reported statistically significant coefficients for expansion at 1200 days, given in Table II.

In a different paper, Lawrence (1995a) also proposed the fineness as having a significant positive correlation with the expansion at 800 days, although the coefficient reported was only 0.244. In this paper, the following equation was presented as the best prediction of expansion at 800 days:

$$Exp_{800} = 9.51 + 0.304 \cdot SO_3 + 0.00085 \cdot \mathit{fineness} \\ + 1.728 \cdot \mathit{comb.}Na_2O_e - 0.162 \cdot CaO - 0.040 \cdot C_3A$$

Figures 1 and 2 show Lawrence's expansion data on a plot of Al_2O_3 versus SO_3 and active Al_2O_3 vs. $(SO_3)_2$, respectively. The limits proposed by Heinz and coworkers (1987, 1989) — $[SO_3]/[Al_2O_3] = 0.7$ (molar ratio) and $(SO_3)_2/Al_2O_3 = 2.0$ (oxide ratio) — are also shown. It is clear from these figures that Heinz's empirically derived relationships cannot be extended to Lawrence's data.

Table II.

Parameter	Correlation coefficient
SO_3	0.593
MgO	0.679
Na_2O_e	0.335
Na_2O combined*	0.687
K_2O soluble	0.358
Na_2O_e combined*	0.449
SrO	0.702
CaO	–0.695
"Free" CaO	–0.389
BaO	–0.391
Mn_2O_3	–0.361
C_3S (by Bogue)	–0.463
C_3A (by XRD)	–0.407

*Alkalis remaining in cement after shaking with water for 1 h.

Kelham (1996, 1997) tested seven clinkers ground with various amounts and forms of sulfate to different target fineness values. Mortar bars were heat cured at various temperatures and subsequently stored in water at 20°C. Only mortars heat cured at 90°C generally showed significant expansion, and the long-term expansion (>600 days) was found to increase with the fineness of the cement. The sulfate content of the finished cement was found to have a pessimum effect, with the maximum expansion occurring at around 4.0% SO_3. Kelham also found that adding the sulfate in the form of K_2SO_4 (dissolved in the mix water) increased the expansion of mortars compared with those containing the same sulfate content in the form of $CaSO_4$. Increased expansion was also observed when KOH was added to the mix water. Kelham proposed the following relationship between long-term expansion of mortars initially heat cured at 90°C and cement composition:

$$
\begin{aligned}
Exp = {} & 0.00474 \cdot SSA + 0.0768 \cdot \mathrm{MgO} + 0.217 \cdot \mathrm{C_3A} \\
& + 0.0942 \cdot \mathrm{C_3S} + 1.267 \cdot \mathrm{Na_2O_e} \\
& - 0.737 \cdot ABS(\mathrm{SO_3} - 3.7 - 1.02\mathrm{Na_2O_e}) - 10.1
\end{aligned}
$$

The term *ABS(X)* is the absolute value of the expression *X* and accounts for the pessimum sulfate content and the dependence of this value on the alkali content. The fineness or specific surface area of the cement is represented by *SSA* in the above equation.

Odler and Chen (1995) studied the behavior of four cements with C_3A contents in the range of 6.6–10.6% and SO_3 contents between 3.4 and

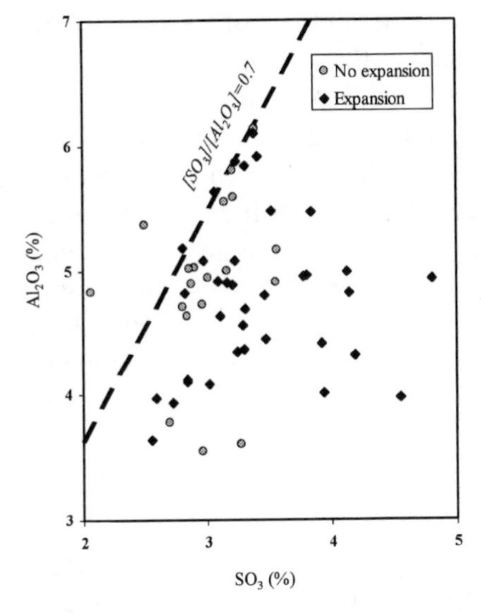

Figure 1. Lawrence's (1995b) expansion data: Al_2O_3 vs. SO_3.

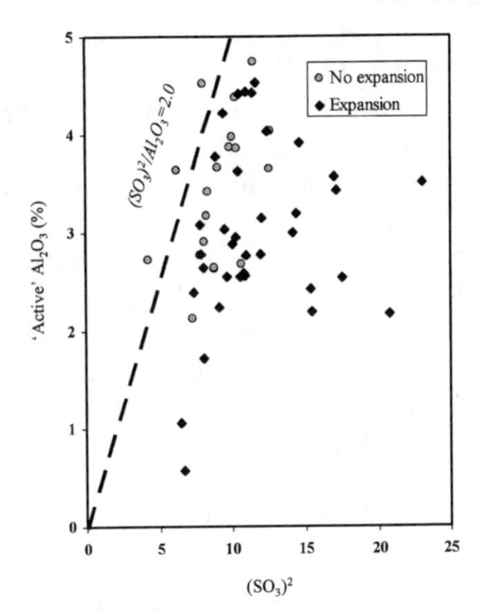

Figure 2. Lawrence's (1995b) expansion data: Active Al_2O_3 vs. $(SO_3)^2$.

5.0%. Significant expansion was observed only in paste specimens initially heat cured at 90°C and with both high C_3A (10.2%) and high SO_3 (5.0%) contents. Similar dependence of the expansion of pastes on high SO_3 and C_3A contents has been reported by Hempel et al. (1992).

Taylor (1994a) suggested threshold values for certain components of portland cement based on earlier published work by Lawrence (1993); these values were $Na_2O_e < 0.83\%$, $Na_2O < 0.22\%$, $SO_3 < 3.6\%$, and MgO < 1.6 ± 0.2%. Taylor proposed that cements that exceeded two or more of these limits had a very high likelihood of showing DEF expansion when heat cured, whereas cements that "passed" all of the above criteria would probably not expand.

All of these analyses show that the sulfate content of the cement is an important parameter in determining the risk (and extent) of DEF expansion. This may be expected since the amount of ettringite that can be formed in a closed system is limited by the sulfate availability. However, the data in Fig. 1 clearly show that there is no single relationship between sulfate content and expansion, nor is there a single level of sulfate below which damage due to DEF will not occur.

The analyses of Heinz and Ludwig (1987), Heinz et al. (1989), and Lawrence (1995b) suggest that the expansion decreases as the alumina (or active alumina) content increases. However, the findings of Kelham (1996, 1997) and Odler and Chen (1995) have shown that the highest expansions occur with the cements with the highest SO_3 and C_3A contents.

Taking the results collectively, it appears that cements produced from high-reactivity clinker (e.g., high C_3A, C_3S, Na_2O_e, and fineness) with a high SO_3 content have the greatest susceptibility to DEF expansion when heat cured. These parameters characterize high–early strength cements, and Kelham (1996) showed a clear correlation between the 2-day compressive strength of mortar and the expansion of the same mortar subsequent to heat curing at 90°C. It is ironic that the cements that are the most susceptible to the adverse effects of elevated temperature are also those that are most likely to be heat cured.

Role of Clinker Sulfates

Field evidence supporting the phenomenon of ambient-temperature DEF due to the delayed release of sulfates from clinker compounds was discussed above. This section reviews laboratory studies aimed at elucidating the role of clinker sulfates and their potential for contributing to DEF. Recent experimental studies in this area seem to be restricted to the cement industry (Miller and Tang 1996; Klemm and Miller 1997; Michaud and Suderman 1997a, 1997b; Herfort et al. 1997; Tennis et al. 1997) although an excellent treatise was recently presented by Taylor (1996). It is incongruous that no experimental data have been presented by the proponents of ambient-temperature DEF.

One of the most comprehensive studies of clinker sulfates and their role in DEF was recently conducted at the Portland Cement Association (PCA) in Chicago (Miller and Tang 1996; Klemm and Miller 1997). Laboratory investigations have been carried out on 33 clinker samples with SO_3 contents in the range of 0.03–3.00% and, perhaps more significantly, $[SO_3]/[Na_2O_e]$ molar ratios from 0.06 to 2.54. High sulfate/alkali ratios (e.g., $[SO_3]/[Na_2O_e] > 1.0$) indicate that the sulfate is not adequately balanced by alkalis and will likely exist as less soluble forms (Taylor 1990). This was confirmed by Klemm and Miller (1997) using selective dissolution techniques. XRD analysis of residues after extraction of the calcium silicates and free lime showed the main sulfate phases in clinkers with $[SO_3]/[Na_2O_e] < 1.0$ to be readily soluble arcanite (K_2SO_4) and aphthitalite ($3K_2SO_4 \cdot Na_2SO_4$). As the $[SO_3]/[Na_2O_e]$ molar ratio increased above 1.0, both calcium langbeinite ($2CaSO_4 \cdot K_2SO_4$) and soluble anhydrite (γ-$CaSO_4$) were found in increasing quantities. A later paper confirmed that the anhydrite precipitated from SO_3 leached from the calcium silicates during the salicylic acid-methanol extraction (Tennis et al. 1997), indicating that the anhydrite may not be present in the clinker prior to treatment. The level of sulfate detected in the silicates was determined to range from 0.07 to 0.83% SO_3 (Miller and Tang 1996). No evidence of insoluble β-$CaSO_4$ was found in any of the clinkers studied. Chemical analysis of residues after KOH sugar extraction showed that the sulfate in the silicate phases is balanced by sufficient alumina to ensure that the resulting hydrate phase is monosulfate rather than ettringite (i.e., $[Al_2O_3]/[SO_3] > 1.0$).

The fate of the sulfate during hydration was monitored for one of the clinkers (2.27% SO_3 and $[SO_3]/[Na_2O_e] = 1.74$) used in the PCA study (Klemm and Miller 1997). The analysis showed that after 24 h of hydration, ettringite accounted for all of the SO_3 originally present as aphthitalite, calcium langbeinite, or anhydrite, or substituted in the silicates.

Five of these clinkers were later used to study the effect of clinker sulfates on the expansion of heat-cured mortars (Tennis et al. 1997). These clinkers had SO_3 contents in the range of 0.07–2.27% and $[SO_3]/[Na_2O_e]$ molar ratios between 0.24 and 1.74, and were used to produce cements with SO_3 contents of 2.58–4.45%. Mortar bars fabricated from these cements were heat cured at 70 or 80°C prior to long-term storage in water at 23°C. Results to 28 weeks were reported by Tennis and coworkers, and no expansion was observed for any of the specimens at that time. These tests are obviously not sufficiently advanced to permit conclusions to be drawn. These tests will be continued to at least 3 years and the program has recently been extended to include heat curing at 90°C and cements ground to higher fineness (i.e., typical of ASTM Type III).

Sulfate solubility studies carried out by Michaud of Lafarge Corporation have reached broadly similar conclusions regarding the dissolution kinetics of anhydrite and calcium langbeinite. In an initial study using pure phases of C_3A, $Ca(OH)_2$, calcium langbeinite, and anhydrite, Michaud and Suderman (1997a) demonstrated that calcium langbeinite is rapidly soluble and may be expected to behave in a similar manner to interground gypsum. "Dead-burned" anhydrite (i.e., burned at 1400°C for 2 h) showed reduced solubility. However, most of it dissolved after 2 days, and all of it after 4 days, to be replaced by ettringite and later by monosulfate.

In a second study, Michaud and Suderman (1997b) prepared laboratory clinkers with high sulfate contents and sulfate/alkali ratios; the composition of the clinkers was reported as given in Table III.

Table III.

Clinker	SO_3	K_2O	Na_2O	$[SO_3]/[K_2O]$	Sulfate phases (by XRD)
1	3.05	1.03	0.07	3.5	$2CaSO_4{\cdot}K_2SO_4$ only
2	3.84	0.86	0.08	5.2	$2CaSO_4{\cdot}K_2SO_4$ & $CaSO_4$

Anhydrite was produced in the clinker only with an extremely high level of sulfate, and this material is certainly not representative of industrially produced cement clinker. Quantitative analysis of the sulfate phases in Clinker 2 showed that 2.1% SO_3 was present as calcium langbeinite, 0.7% was in the silicate, and 0.65% was present as anhydrite. This clinker was hydrated (water/clinker = 0.50) and the evolution of the sulfate phases followed using quantitative X-ray diffraction. The results show that most of the anhydrite dissolves quickly — within a few hours — and nearly all of it has gone after a few days.

Two cements were prepared from Clinker 2 with sulfate contents of 3.84 (i.e., pure clinker) and 4.5%. Three cements were prepared from Clinker 1 with 3.05 (pure clinker), 4.0, and 4.5% SO_3. The potential for ambient-temperature DEF expansion was determined using mortar bars cured at either 20 or 38°C in various environments. All measured expansions were below 0.03% after 12 months of storage, suggesting that there is no potential for expansion due to DEF at ambient temperature. These mortar bars will continue to be monitored.

Further work was recently presented by Aalborg Portland of Denmark (Herfort et al. 1997). They studied the distribution of sulfates in 10 clinkers. These included three industrially produced clinkers (including one mineralized clinker), five clinkers produced in a laboratory scale rotary kiln, and two clinkers produced under laboratory conditions in a stationary kiln. Sulfate contents of the clinkers ranged from 0.27 to 6.48% SO_3 and the molar $[SO_3]/[Na_2O_e]$ ratio was between 1.03 and 10.31. Anhydrite was found only in three clinkers, all of which had sulfate/alkali molar ratios in excess of 4.0. Two of these were laboratory clinkers, and the third was produced in a full-scale kiln during a limited production run using deliberately enhanced SO_3 levels to promote anhydrite formation.

The sulfate in the silicate phases was found to correlate well with both the total SO_3 and $[SO_3]/[Na_2O_e]$ in the clinker and was reported to be as high as 1.45% SO_3. However, it was also found that the Al_2O_3 content of the silicates increased with the SO_3 content and that the molar ratio of alumina to sulfate was in excess of 1 (i.e., $[Al_2O_3]/[SO_3] > 1.0$). The authors contend that the alumina is likely to be released at the same rate as the sulfate and that the molar ratio favors the formation of monosulfate rather than ettringite (Herfort et al. 1997).

The fate of the sulfates during hydration (water/clinker = 0.50) was monitored by XRD for a clinker containing 1.5% anhydrite. The results show a substantial dissolution of anhydrite in the first 10 min, and that the XRD peak corresponding to anhydrite was absent after 7 h (Herfort et al. 1997). Long-term expansion studies are being carried out using some of these clinkers.

Kelham (1996, 1997) carried out expansion studies on heat-cured mortars using a range of cements produced from seven clinkers. The initial study (Kelham 1996) included a clinker with a sulfate content of 1.6% and an $[SO_3]/[Na_2O_e]$ ratio of 3.3. A mineralized clinker with 2.6% SO_3 and $[SO_3]/[Na_2O_e] = 4.75$ was included later (Kelham 1997). None of the cements produced from these high-sulfate clinkers expanded in mortars heat cured at temperatures of 70°C or lower, even when the sulfate content of the finished cement was as high as 5% SO_3. Only mortars with added K_2SO_4 or KOH (in the mixing water) exhibited expansion when heat cured at 80°C. Expansion was observed for most of the cements when tested in mortars heat cured at 90°C, the exception being the cements produced from sulfate-resisting cement clinker (C_3A = 0.6%). While acknowledging the influence of the sulfate in the finished cement, Kelham (1997) states that the source of that sulfate (i.e., as clinker sulfate or interground gypsum) has little impact on the behavior of mortars.

The data from these experimental studies do not support the hypothesis that the late release of sulfate from high-sulfate clinkers may lead to expansion due to DEF in non-heat-cured concrete. Indeed it is probable that the clinker sulfates will not behave very differently from interground gypsum, even when the concrete is heat cured. These experimental observations are consistent with theoretical considerations based on cement chemistry as put forward by Taylor (1996).

Hime (1996a) postulates that in clinkers with high sulfate/alkali ratios (i.e., $[SO_3]/[Na_2O_e] > 1.0$), the sulfates are not adequately balanced by the alkalis and may be encapsulated in the silicate phases, becoming available at later ages for the delayed formation of ettringite. In the abstract of the same paper he proposes that imposing a limit of 1% SO_3 on the clinker may prevent damage due to DEF in non-steam-cured concrete (this statement is not elaborated on in the body of his paper). No scientific evidence to support these claims is offered by Hime, and the balance of evidence from experimental studies and theoretical considerations are in direct contrast to Hime's statements. Further discussion of these issues was recently published by Skalny, Johsansen, and Miller (1997).

DEF and Alkali-Silica Reaction

Numerous field investigations have indicated a possible association between ASR and DEF. In some of these cases it has been concluded that ASR is the primary cause of cracking and deterioration, and that the subsequent precipitation of ettringite into the cracks is a secondary effect that may or may not cause further damage (Shayan and Quick 1991/2, 1992; Oberholster et al. 1992; Meland et al. 1997). Other workers have suggested that the chemical changes resulting from ASR may promote the formation of ettringite (Brown and Bothe 1993). In contrast, it has been postulated that the formation of ettringite leads to increased pore solution alkalinity, thereby enhancing the chances of ASR (Pettifer and Nixon 1980). It is the opinion of some workers that DEF is commonly misdiagnosed as ASR due to the apparently similar appearance of alkali-silica gel and amorphous ettringite when examined by optical microscopy (Marusin 1993b, 1994a, 1994b). On the other hand, Shayan and Quick (1994) demonstrated that ASR was a major cause of deterioration of railway ties in Finland that had previously been determined as suffering from DEF and freeze-thaw action (Tepponen and Eriksson 1987).

The harmless recrystallization of "secondary ettringite" into cracks and voids was discussed earlier in this paper. This is a common phenomenon and may occur in concrete subjected to moisture fluctuations regardless of whether or not the concrete has been steam cured. In a study of damaged railway ties in Australia, Shayan and Quick (1992) observed only symptoms of DEF in concrete that had significant deterioration due to ASR. The subordinate role of DEF was confirmed in a series of laboratory tests conducted on mortar bars and concretes with different aggregates (Shayan and Quick 1991/2; Shayan and Ivanusec 1996). Expansion and cracking was not observed in heat-cured (up to 80°C) specimens without reactive aggregate, even when the gypsum content of the mix was raised. The addition of gypsum did increase the expansion of concrete containing reactive aggregate when heat cured at 80°C, but not when cured at 40°C (Shayan and Ivanusec 1996). The authors concluded that DEF contributes to expansion only in heat-cured concrete containing reactive aggregates.

Similar results have been reported by Diamond and Ong (1994). Heat-cured (10 hours at 95°C) mortars containing limestone sand did not expand significantly during subsequent storage at 23°C and 100% RH. However, mortars containing reactive silica (Beltane opal or cristobalite) showed considerable expansion and cracking, which, during the early stages (i.e., up to 28 days), was accompanied by the formation of alkali-silica gel. However, expansion continued in the long term with no gel being produced, and the authors concluded that the later expansion may be due to ettringite formation. The role of DEF was further confirmed by the observation that significantly increased amounts of ettringite (i.e., approximately 50% more) were formed in mortars containing reactive aggregate, compared with the mortar with nonreactive limestone.

The results of these experimental studies are consistent with the field observations of Oberholster et al. (1992), who determined that ASR was a prerequisite for damaging ettringite formation in concrete railway ties, and with the hypothesis of Fu and coworkers (1994, 1996) regarding the role of cracking as a precursor to DEF. However, there may also exist a chemical relationship between ASR and DEF as suggested by Brown and Bothe (1993). They studied the hydration of relatively pure phases of C_3S, C_3A, and gypsum in potassium hydroxide solutions of varying concentration (0.0–2.0 M KOH) and at various temperatures (25–80°C). Their results show that ettringite formation is inhibited by alkali, especially in the range of 0.5–1.0 M KOH. The authors contend that the process of ASR may, by reducing the pore solution alkalinity, promote the formation of ettringite. Such an effect may be expected to occur where local concentrations of alkali are particularly low, such as at the interface between the cement paste and reactive silica. Formation of ettringite in the interfacial zone may also be promoted by the increased availability of calcium hydroxide at such locations (Brown and Bothe 1993).

It has also been proposed that ASR may be promoted by the formation of ettringite. Pettifer and Nixon (1980) observed ettringite in association with alkali-silica gel in a number of field concretes. The reaction of sulfate with calcium aluminates also consumes calcium from calcium hydroxide, thereby releasing more hydroxyl ions to the pore solution according to the following sequence of reactions (Pettifer and Nixon 1980):

$$Ca(OH)_2 \rightarrow Ca^{2+} + 2OH^-$$

$$4CaO{\cdot}Al_2O_3{\cdot}19H_2O + 3Ca^{2+} + 3SO_4^{2-} + 13H_2O \rightarrow 3CaO{\cdot}Al_2O_3{\cdot}3CaSO_4{\cdot}31H_2O + Ca(OH)_2$$

The increased alkali hydroxide concentration increases the chances of alkali-silica reaction; a similar mechanism has been proposed by Regourd et al. (1981). However, it is difficult to reconcile this argument, since the hydroxyl ion concentration in concrete pore solutions is controlled by the availability of the alkali ions and requirements for electroneutrality, which is maintained only if the source of sulfate ions in the second reaction is alkali sulfate.

Michaud et al. (1997) showed that hydrated calcium sulfoaluminates may not be stable in alkali-silicate solutions. In the vicinity of reacting aggregate particles, amorphous or semi-crystalline products containing calcium, silica, alkali, and some alumina may form, but the product does not incorporate sulfate. Thus, sulfate ions diffuse to areas low in silica where, if calcium and alumina are available, ettringite is precipitated. In this manner ASR facilitates secondary ettringite formation.

Marusin (1993b, 1994a, 1994b) claims that many investigators have mistaken ettringite for alkali-silica gel, both in the petrographic examination of thin sections and SEM study of polished samples. She has proposed a methodology for distinguishing the two products using SEMs equipped with energy-dispersive X-ray analysis (Marusin 1994b). However, as pointed out by Johansen et al. (1993), the diagnostic process does not merely rely on the identification of reaction products, but requires a more in-depth interpretation of the symptoms.

DEF and Freeze-Thaw Damage

It has been suggested that the premature failure of a number of concrete pavements in the midwestern United States may be a consequence of ettringite filling entrained air voids and thereby reducing the resistance of the concrete to cyclic freezing and thawing (Marks and Dubberke 1996; Ouyang and Lane 1997). Observations of air voids filled or lined with ettringite in deteriorated field concrete have been made using low-vacuum scanning electron microscopy and energy-dispersive X-ray analysis (Marks and Dubberke 1996). Laboratory studies have indicated that the freeze-thaw resistance of concrete made with some cements (and fly ash) may be reduced after extended periods of moist curing (Ouyang and Lane 1997). The reduced resistance of the concrete is associated with a reduction in its "effective air void content," which discounts voids filled with ettringite.

Similar findings were recently published by Stark and Bollmann (1997), who studied the effect of alternating temperature and moisture conditions on concrete performance. Concrete beams were sawed from a pavement 2 weeks after placement. The concrete was typical of pavement quality with a cement content of 340 kg/m^3 and an air content of 5.5%. The cement had a sulfate content of 3.2% SO_3 and an alkali content of 0.79% Na_2O_e. These beams were subjected to three levels of treatment in the laboratory. The most aggressive treatment involved a 2-week drying period at 60°C, 8 weeks in water at 20°C, and a 7-h freezing cycle to –20°C. This treatment resulted in microcracking and an accumulation of ettringite in the cracks and air voids that was conspicuous by optical and scanning electron microscopy after only 2 cycles. The authors detected ettringite growth on the surface of air bubbles after 2 cycles of the most benign treatment used (2 weeks at 20°C and 65% RH and 8 weeks in water at 20°C), and these "bundles" of ettringite got larger with each cycle. The concrete was reported to be "undamaged" after 4 cycles of the different treatments, there being no reduction in strength, elastic modulus, or pulse velocity. However, after 4 cycles the concretes were tested using the RILEM CDF frost/de-icing salt resistance test, and a severe reduction in performance was observed for the concrete that had been through the most severe treatment prior to testing. This reduced performance was manifested as increased scaling and poor transmission of ultrasound, indicating a internal damage to the structure of the concrete (Stark and Bollmann 1997).

A recent study was conducted by Detwiler and Powers-Couche (1997) to investigate the same phenomenon. Concrete specimens were produced using typical pavement-quality concrete with target air contents of 2, 4, or 6%. Three cements were used: a Type I cement with 12% C_3A and 2.03% SO_3, the same cement with added gypsum to bring the SO_3 content to 3.14%, and a Type II cement with 5% C_3A and 2.72% SO_3. Specimens were moist cured for 3 days and stored in laboratory air for 25 days prior to testing using modified versions of the ASTM test for freezing and thawing in water (ASTM C 666 Procedure A). The modifications included using 3% NaCl solution instead of water (in some cases the NaCl was contaminated with 5% gypsum) and the inclusion of a drying cycle over every weekend during the test period. Not surprisingly, the performance of the concrete was found to be dependent on its air content. There was no deterioration at all in specimens with 6% air. Concrete with 2% air deteriorated after anywhere between 170 cycles (for Type I cement + gypsum) to 360 cycles (for Type I cement). Ettringite was not observed in the voids of any of these concretes. The behavior of the specimens with 4% air is the most interesting. Ettringite was observed to form in the air bubbles over a period of

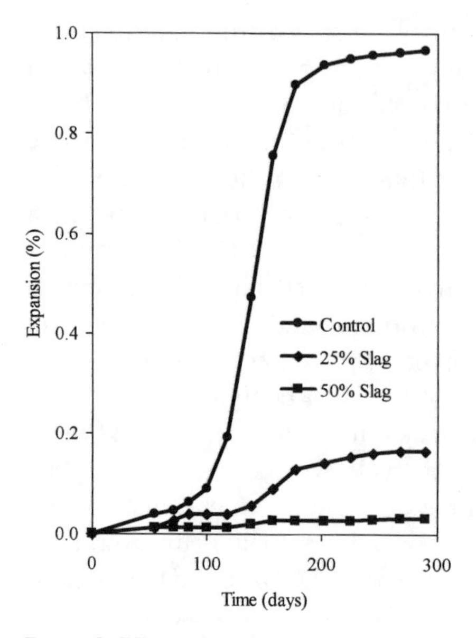

Figure 3. Effect of slag on DEF expansion (Ghorab et al. 1980).

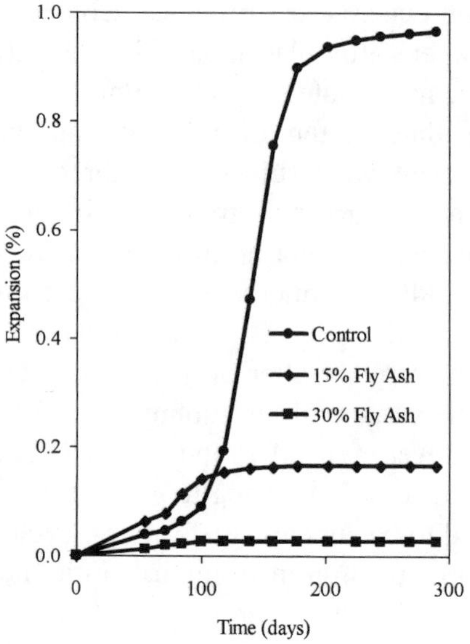

Figure 4. Effect of fly ash on DEF expansion (Ghorab et al. 1980).

time, but the authors conclude that frost damage occurs first, and that secondary ettringite formation is a consequence of the damage and not a cause.

There appears to be conflicting evidence from the various published studies in this area. A systematic study of pavements in service is required to resolve this issue. This study should include pavements with both good and poor field performance rather than concentrating on the "failures," as has been the case in the past.

Effect of Supplementary Cementing Materials

One of the first reported laboratory studies of expansion due to DEF included data indicating the potential ameliorating action of slag, fly ash, and a natural pozzolan, Bavarian trass (Ghorab et al. 1980). Mortar bars, produced using a high–early strength cement (German PZ55 cement), were subjected to a 2-day heat treatment with a maximum temperature of 100°C, prior to storage in water at laboratory temperature. Figures 3 and 4 show the effect of partially replacing the PZ55 with slag and fly ash, respectively, and similar results were obtained for the trass. The results to 290 days show that the use of 30% fly ash or Bavarian trass, or 50% slag, is effective in suppressing deleterious expansion. In a later paper (Heinz et al. 1989) it is reported that there was no damage to these mortars after 10 years of storage. This later paper also reports that 5–10% silica fume may reduce expansion, but actual expansion data and test details are not given.

In spite of these encouraging results, relatively little work has been carried out since to confirm the beneficial effects and better understand the mechanisms involved when mineral admixtures are present. Heinz and coworkers (Heinz et al. 1989) suggested that the "active" alumina present in these materials may be responsible for reducing DEF, and this is an extension of their relationship between expansion and the $\bar{S}/A$ (molar ratio) of the cement. In a recent presentation, Kalde and Ludwig (1997) confirmed the importance of the available Al_2O_3 in the mineral admixture for controlling DEF expansion.

Lewis (1996) confirmed the ability of 50% slag to control the expansion of heat-cured (up to 90°C) mortar bars produced using a high–early strength clinker with up to 4% SO_3. Microanalysis of the inner C-S-H product after heat curing showed substantially increased alumina contents in the presence of slag, such that the SO_3/Al_2O_3 ratio would be consistent with the formation of monosulfate rather than ettringite.

Fu (1996) studied the effect of slag, fly ash, silica fume, and natural zeolite on the expansion of mortars. Mortars were heat cured at 90°C for 12 h and subsequently oven dried at 85°C prior to long-term storage in water at 23°C. Figure 5 summarizes the expansion of specimens up to 91 days. The use of 30% slag or 30% low-lime (3.4% CaO) Class F fly ash was effective in controlling expansion (i.e., < 0.1%) throughout the period reported. Silica fume, at 15% replacement, was slightly less effective. The natural zeolite and higher-lime (12.3% CaO) Class C fly ash, both at 30% replacement, delayed the onset of expansion compared with the control, but still exhibited significant expansion after 91 days. Fu's study did not extend to examining the role of the mineral admixtures, although he suggests that the beneficial effect may be related to the ability of products of the pozzolanic reaction to absorb sulfate.

Lagerblad and Utkin (1994) examined the effects of water/cement ratio, superplasticizers, and silica fume on expansion due to DEF. Figure 6 shows the expansion of some of the mortar bars heat cured at 95°C prior to storage in water at 20°C. The addition of 10% silica fume clearly retards the rate of expansion, but does not eliminate expansion. Further reduction in the rate of expansion was observed when the water/cementitious material ratio was reduced through the use of a superplasticizer. The authors suggest

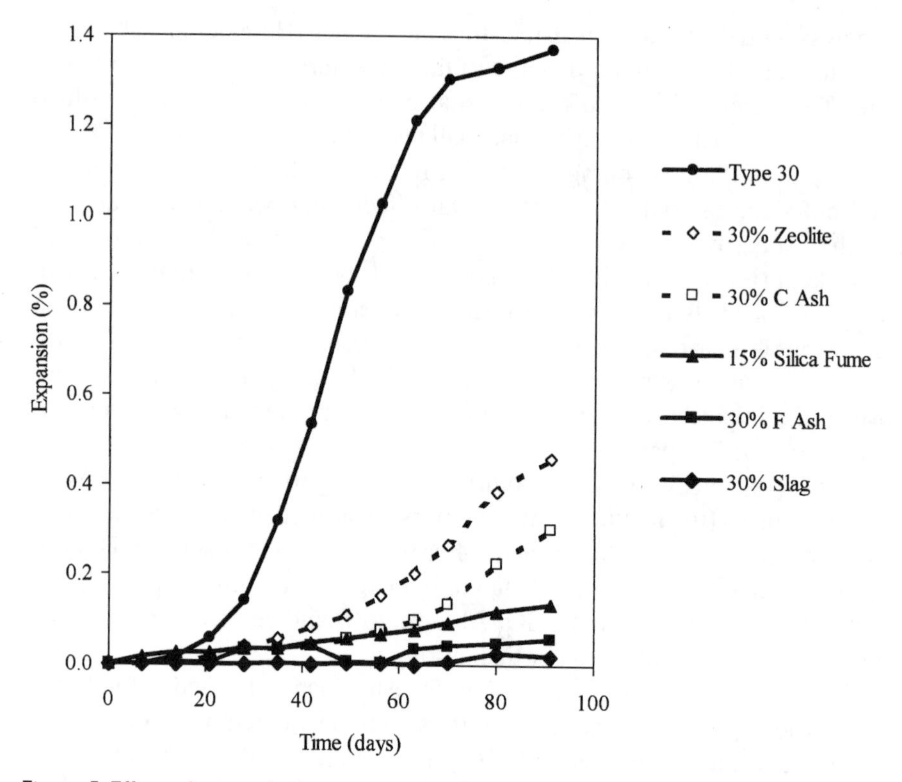

Figure 5. Effect of mineral admixtures on DEF expansion (Fu 1996).

that these effects are due to reduction in permeability, which slows down the rate of delayed ettringite formation and associated expansion, but does not prevent it.

Meland et al. (1997) measured volumetric expansion (using water displacement) of concrete cylinders initially heat cured at 85°C and subsequently stored in water at 20°C. Control specimens showed a volumetric expansion of more than 4% after 1 year, whereas specimens containing 8% silica fume showed no significant expansion after 2 years. Examination by SEM/EDS indicated ettringite in cracks between the paste and aggregate of the heat-cured concrete without silica fume. Sulfoaluminate was also found in the matrix of non-heat-cured concrete and heat-cured concrete with silica fume, but the SO_3/Al_2O_3 ratio was consistent with monosulfate. These findings are in contrast to those of Shayan et al. (1993), who found that the addition of silica fume promoted the formation of ettringite in heat-cured concrete, possibly as a direct result of reducing the pore solution alkalinity.

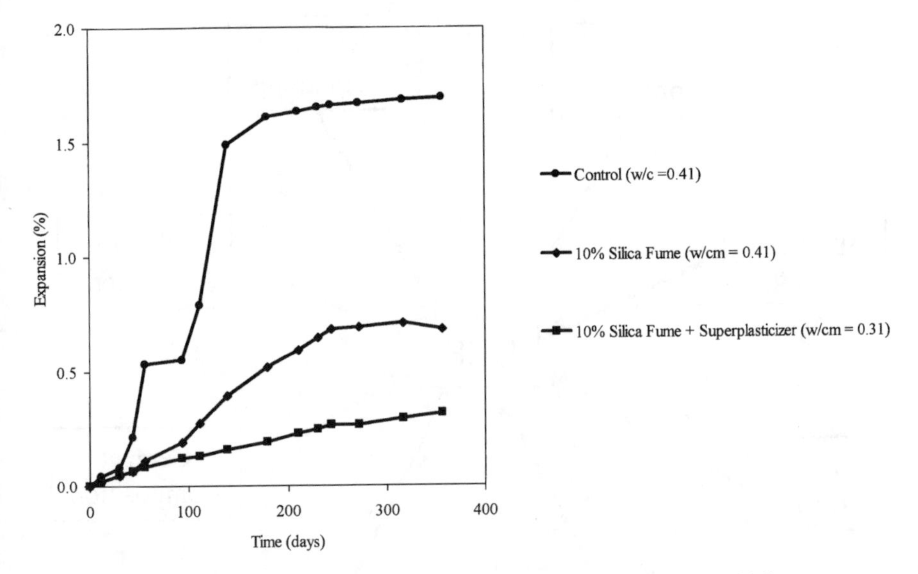

Figure 6. Effect of silica fume on DEF expansion (Lagerblad and Utkin 1994).

These workers suggested that silica fume may be effective in suppressing DEF by stabilizing the early formation of ettringite.

More work is clearly required in this area to develop a fuller understanding of the role of mineral admixtures in controlling DEF.

Minimizing the Risk of DEF

A number of different factors have been reported to influence the risk of damage due to DEF (Fig. 7), including:

- Exposure to elevated temperature at early ages (whether due to heat curing or autogenous temperature rise).
- Manufacturing processes (preset time, rate of heating, maximum internal temperature, holding time, cooling rate, and possibly other factors, such as prestressing release times).
- Composition of the finished cement (particularly fineness, SO_3, Na_2O_e, and C_3A).
- Composition of the cement clinker (particularly SO_3 or $[SO_3]/[Na_2O_e]$).
- Use of pozzolans and slag (sufficient quantities may reduce or eliminate expansion).

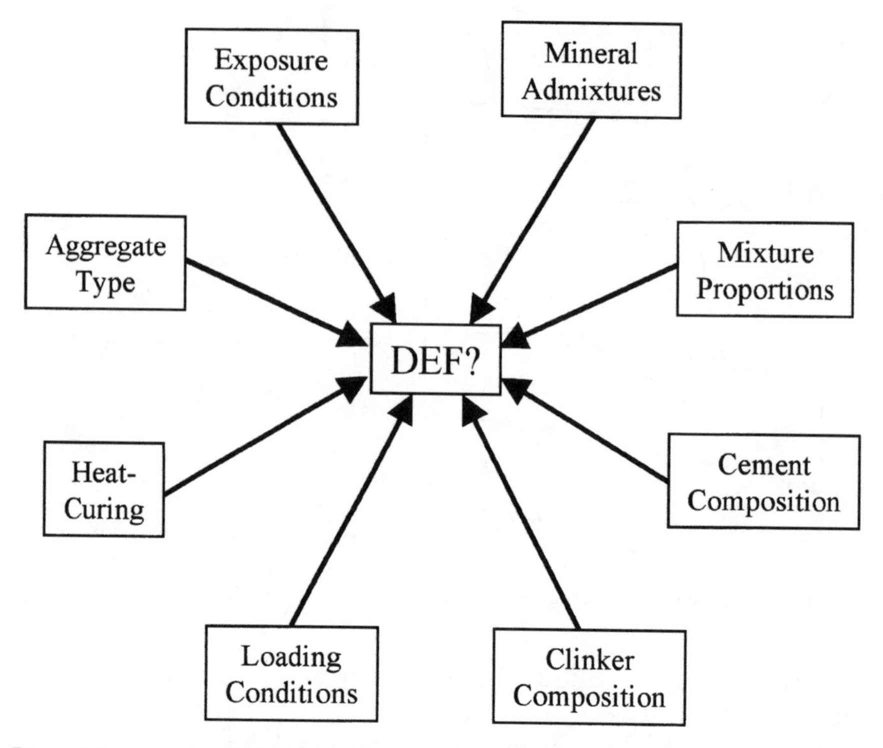

Figure 7. Factors affecting the risk of damage due to DEF.

- Nature of the aggregate (mortars containing limestone tend to expand to a lesser extent than those containing siliceous aggregates; alkali-reactive aggregates may exacerbate DEF).
- Mixture proportions (high cement contents lead to increased temperature rise and higher alkali concentrations in the pore solution, whereas air entrainment appears to reduce damage).
- Size of element (higher internal temperatures are likely to be reached in larger elements, whereas small elements have greater accessibility to external moisture).
- Exposure conditions (damaging DEF requires excess moisture and may be exacerbated by wetting and drying, or freezing and thawing).
- Loading conditions (internal fracturing due to excessive stress can lead to increased expansion due to DEF).

Thus the risk of damage due to DEF may be minimized by the control of one or more of these conditions. Obviously, some of these factors have a greater influence on DEF than other factors, and some are, to a great extent, outside of the control of the concrete producer. For example, the use of limestone may reduce the rate and extent of expansion, whereas the use of low-C_3A cement (e.g., ASTM Type V) has been shown to be effective in eliminating damage. And while a producer can control the environment during manufacture, he has little or any control on the exposure conditions that the concrete will ultimately be exposed to.

In most circumstances, the concrete producer's main options for controlling damage due to DEF are restricted to the regulation of the manufacturing process and the selection of materials. For all practical purposes, this translates to controlling the temperature history of the concrete or the composition of the cement, or both. It is unfortunate that the control of one or both of these conditions is likely to reduce the rate of strength gain at very early ages and, consequently, adversely affect the rate of production.

Limits for Portland Cement

The effects of cement and clinker composition on laboratory studies of DEF were discussed earlier in this paper. From the published literature it is apparent that the composition of the clinker has a subordinate role compared to that of the cement. Although studies have shown relationships between DEF expansion and various compositional parameters of the cement, there is no single parameter that can be used to reliably predict the performance of a particular cement. For example, due to the role of sulfate in DEF, it may be tempting to impose a limit on the SO_3 content of the cement used for heat-cured concrete. However, such an action would not be sufficient to guarantee satisfactory performance, as both Lawrence (1995b) and Kelham (1996, 1997) have demonstrated damaging expansion in heat-cured mortar bars produced with cements of relatively low sulfate content ($\leq$ 3.0% SO_3). As suggested by Taylor (1994a), it would probably be necessary to limit other components in addition to SO_3; such limits might include some combination of Na_2O, Na_2O_e, MgO, C_3A, and fineness.

An alternative to placing prescriptive limits on cements is to develop a performance test for evaluating the susceptibility of a particular cement to DEF. Many different mortar bar expansion tests have been used to study DEF and one (or more) of these may provide the basis for a standard test method. However, as with any acceptance test, it is essential that any DEF test is correlated with field performance or, at least, long-term tests on concretes, as will be discussed later.

Use of Pozzolans or Slag

The use of pozzolans or slag as a means of reducing the risk of damage due to DEF was discussed earlier in this paper. It is clear that the use of a sufficient quantity of suitable fly ash or slag is an effective means of eliminating damaging expansion in heat-cured mortars or concrete. However, there has been comparatively little work in this area and it is not possible to give firm guidance on how these materials should be used in this role (i.e., minimum levels of replacement required). The development of a suitable performance test for DEF would provide the basis for assessing the ability of blended cements to resist DEF.

Controlling Temperature History

A number of recommendations for heat-treated concrete have appeared in recent years. The German Committee for Reinforced Concrete was perhaps the first to offer specific advice to avoid DEF; their specification for concrete exposed to "damp" conditions is summarized in Table I; also shown are the limits in the Canadian Standards Association specification for precast concrete (CSA A23.4-1994). Other authorities offer similar advice, although the absolute values of limits may differ slightly. Cooling rates are also specified in some cases and some specifications require the concrete to be protected from moisture loss.

These limits are consistent with the results from laboratory studies, which show unequivocally that expansion due to DEF occurs only when mortars or concretes are heat cured above some threshold temperature (i.e., generally above 70°C). This statement applies regardless of the composition of the cement and clinker (within the limits reported).

Experience with the German recommendations dates back to 1977, and there is evidence that these and similar practices have been successful in eliminating damage due to DEF (Skalny and Locher 1997). Indeed, the following statement was made in a recent paper by Stark et al. (1998):

> Experience so far indicates that compliance with the rules for the warm-curing of concrete given in the guidelines issued by the DAfStb (German Committee for Reinforced Concrete) ensures that no harmful ettringite formation is to be expected as a result of warm curing.

However, as mentioned earlier, a detailed review of the changes in production practices, materials, and field performance in Germany and other countries where similar practices have been adopted would reveal the true benefits of implementing such controls.

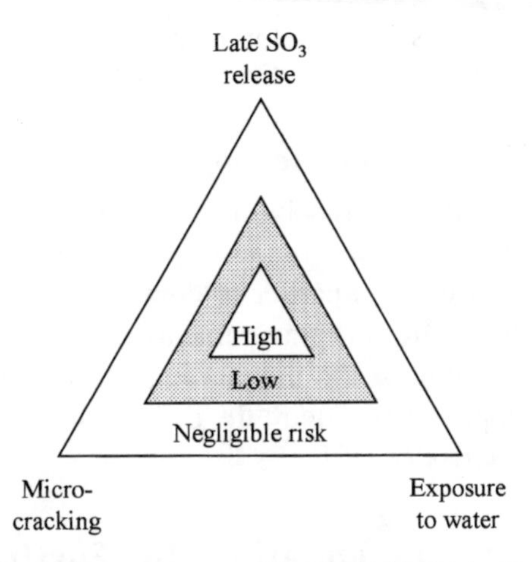

Figure 8. Holistic model of DEF (Collepardi 1997).

Holistic Approach to DEF Control

Collepardi (1997) recently presented a holistic approach to explain DEF and avoid its occurrence in concrete construction (see Fig. 8). Unfortunately, Collepardi's model considered only three elements as having a controlling influence on DEF: late sulfate release from the cement, microcracking, and the availability of moisture. On the basis of this model, Collepardi concluded that the only practical preventive measures are to limit the sulfate content of portland cement clinker or to reduce microcracking by modifying the prestressing process. This approach essentially disregards many other important factors that have been shown to have significant impact on expansion due to DEF, and thus falls a long way short of embracing the true doctrine of holism. A truly holistic approach would consider all of the influencing parameters, such as those listed at the beginning of this section (compare Figs. 7 and 8).

The advantage of developing guidelines based on such a universal approach is that it provides the manufacturer with choices. For example, a very crude guideline might provide a manufacturer with just two choices: control heat-curing or use only proven DEF-resistant cements. This could be extended to allow a third option: use a minimum of 25% fly ash or 40% slag. A more complex guideline might be based on a risk analysis, where all of the factors (processing, materials, environment) are taken into account to calculate an overall risk of DEF, the level of acceptable risk being either fixed or selected on the basis of the owner's requirements. This sort of approach is currently being developed for controlling the risk of damage due to alkali-silica reaction in Canada. To achieve a similar goal for DEF requires the development of a more extensive database on the interaction between damage, heat curing, materials, and exposure conditions.

Performance Testing

Another approach for minimizing the risk of damage due to DEF is for producers to carry out testing to determine what curing conditions can be safely used with particular combinations of their materials. Such testing is different from testing cements or blended cements to evaluate their DEF susceptibility; the distinctions are discussed below. However, both of these approaches require the development of suitable test methods for evaluating the risk of DEF.

Recommendations for Further Research

Test Methods

Many test methods have been used by different investigators in the study of DEF. Common features among most of these tests are the use of linear expansion as a means to quantify damage and, of course, the use of some excursion to elevated temperature to promote DEF. Differences between expansion tests include:

- Type of cements (e.g., laboratory- or industry-produced clinkers or cements).
- Specimen type (e.g., cement paste, mortar, or concrete).
- Type and grading of aggregates (e.g., siliceous, carbonates).
- Mixture proportions (e.g., aggregate/cement, water/cement, air entrainment).
- Specimen dimensions.
- Heat curing (preset time, heating rate, maximum temperature, holding time at maximum temperature, cooling rate, condition of specimen during heat cycle).
- Use of drying cycle.
- Post–heat cure exposure (e.g., wet/dry cycling, immersion in water or solution, storage in air at high relative humidity, temperature of storage).

All of these factors may effect the rate and extent of damage due to DEF, consequently it is not possible to directly compare the data between studies that have used different testing procedures.

There is a need to standardize testing procedures for DEF both for the evaluation of cements (and cementitious systems) and of material/manufacturing systems. A test aimed a determining the DEF susceptibility of a cementitious system would require all other materials and curing conditions to be standardized, although it is desirable that such a test would include heat curing at a range of elevated temperatures to establish the particular sensitivity of a cement to temperature. A test for a material/manufacturing system would evaluate the susceptibility of a particular combination of materials and processing conditions intended for use by a manufacturer. These tests provide different information and would probably be very different in methodology. The cement test, for instance, could probably be achieved using a simple mortar bar expansion test such as that used at Blue Circle Laboratories in the United Kingdom (Kelham 1996). An accelerated test involving a oven-drying cycle at 85°C immediately after moist heat curing has also been proposed (Fu 1996). Such tests take anywhere between 2 months and 2 years to produce usable information. Testing an actual production system would probably require the testing of concrete rather than mortar specimens. Expansion due to DEF may take many years to manifest in concrete specimens that are continuously soaked or stored at constant relative humidity after heat curing. A test method, known as the Duggan test, involving a number of cycles of dry heat at 82°C and soaking in water at 21°C, has been proposed by Grabowski et al. (1992). This test purportedly accelerates the deterioration of concrete due to DEF and may give useful information regarding the susceptibility of a particular concrete system in 90 days or less.

Despite the abundance of published data from various test methods, there has been little attempt that this author is aware of to correlate the test results with field performance data or long-term tests on concrete. This is clearly needed if laboratory tests are to be used with any confidence. To achieve this will require a concerted effort from the concrete industry, and the following list indicates some of the tasks that may assist such an exercise (some of the tasks may have already been undertaken):

- Initiate a comprehensive long-term testing program for concrete specimens stored under field conditions. This will require the establishment of a small-scale casting and heat-curing facility together with an outdoor exposure site. Specimens should be of sufficient size to represent real concrete elements (e.g., railway sleepers) and allow for instrumentation, nondestructive evaluation, and the periodic extraction of core samples for laboratory investigation.

- Collect a wide range of cements, including, where possible, those that have been implicated in field cases of DEF and those with known satisfactory field performance in heat-cured concrete. The samples collected should represent the range in cement and clinker composition likely to be encountered in North America, with the primary focus being Type III cements.
- Concrete specimens should be cast using the different cements and should be subjected to a range of curing conditions prior to long-term exposure. The cements should also be tested using various accelerated test methods, which should be calibrated by reference to the long-term concrete performance.

Such a study represents a substantial investment for the industry in terms of both time and money. It may take many years (e.g., >5–10 years) for useful information to be forthcoming from these tests. However, such an approach is necessary if nonarbitrary test methods and limits are to be established.

High–Early Strength Systems Optimized for High Temperatures

It is unfortunate that types of cement and processing conditions most favorable to the rapid production of precast concrete are also those most likely to give rise to damaging DEF. Ideally, precast operations require high–early strength cements that are insensitive to high-temperature curing. It may be difficult to achieve both aims with traditional portland cements, but alternatives based on either nonportland or blended portland cements may be suitable. For instance, blends of portland cement with pozzolans or slag have been shown to offer increased resistance to DEF expansion following heat curing. Extremely fine pozzolans such as silica fume or metakaolin can be used with Type III cements at moderate replacement levels without jeopardizing the early strength properties of the concrete. Problems may be encountered with fly ash or slag because of their relatively slow reactivity. However, high temperatures greatly accelerate the reactivity of both materials, and the initial early strength losses associated with both materials are much less marked when they are cured at elevated temperatures. Furthermore, early age strength losses may be compensated for through the judicious use of chemical admixtures, such as accelerators or superplasticizers.

Studies are currently underway at the University of Toronto to develop high–early strength concretes that are not susceptible to DEF damage even when cured at excessive temperature (e.g., 90–100°C).

Role of Clinker Sulfates: Ambient-Temperature DEF

The role of clinker sulfates and the potential for ambient-temperature DEF due to the late release of such sulfates was discussed in earlier in this paper. The balance of scientific evidence indicates that damage due to DEF can occur only in mortar or concrete that has been exposed to elevated temperature (generally >70°C) and that sulfates in the clinker phases behave in much the same manner as sulfates (gypsum) interground with the clinker to produce the finished cement. However, sufficient concerns regarding clinker sulfates have been expressed by eminent and respected scientists (e.g., Diamond 1996; Collepardi 1997) to warrant further investigation. To the author's knowledge, the studies of clinker sulfate distribution and release rates carried out by PCA in the United States and Aalborg Portland in Denmark are being extended to include expansion testing of mortar bars cured at ambient and elevated temperature (Tennis et al. 1997; Hertford et al. 1997). This should provide useful evidence to help the existing controversy. In the author's opinion, it would also be helpful to include tests on high-sulfate clinkers (with high $[SO_3]/[Na_2O_e]$) in the long-term concrete testing proposed above.

The evidence in support of ambient temperature DEF comes from field studies of deteriorated cast-in-place or non-heat-cured precast concrete structures. These cases should be adequately documented and further researched to confirm the actual role of DEF in the deterioration, estimate the potential autogenous temperature rise due to hydration, and, where possible, determine the composition of the cement, clinker, and other materials used for construction.

It may also be a worthwhile exercise to identify North American cement plants that produce high-SO_3 clinkers and try to establish field performance histories for the cements. This would involve a study of both cast-in-place and precast concrete elements produced using the cement.

Role of ASR and Other Damage Mechanisms

The associations between DEF and ASR and between DEF and freeze-thaw damage were discussed above. A number of issues remain unresolved, including whether mild ASR can promote DEF expansion that would not otherwise have occurred; whether DEF or secondary ettringite formation can compromise the air-void system of concrete, leading to freeze-thaw damage; and whether sufficient levels of air entrainment can effectively suppress DEF expansion. It is suggested that some of these issues be addressed in the long-term testing program proposed above. This could be achieved by using marginally reactive aggregate and nonreactive aggregate to produce otherwise identical concrete specimens and casting a series of concretes with a range of air contents.

Studies of prematurely failed concrete pavements in the midwestern United States have indicated that failure may be due to ettringite blocking air voids and lowering the freeze-thaw resistance of concrete. It is not clear from such postmortems whether the ettringite infiltrated the voids before damage occurred or whether the presence of ettringite in the void is merely the consequence of secondary ettringite precipitation after the concrete had cracked. As suggested above, a systematic study of pavements with both good and bad field performance is required to resolve these issues.

Review the Effectiveness of European Recommendations

Recommendations for heat curing concrete were first introduced in Germany in 1977. Since that time other European countries have adopted the same or similar practices. Although it has been suggested that such practices have been successful with regard to reducing the occurrence of DEF (Skalny and Locher 1997), there is insufficient documentation to confirm that improved performance is attributed solely to control of the heat-curing cycle. For instance, other changes in materials or production methods may have contributed. As mentioned above, a detailed review of the European experiences is required if lessons are to be learned from it.

References

Brown, P.W., and J.V. Bothe. 1993. "The stability of ettringite." *Adv. Cem. Res.,* **5** [18] 47–63.

Collepardi, M. 1997. "A holistic approach to concrete damage induced by delayed ettringite formation"; pp. 373–396 in *Proceedings of the 5th CANMET/ACI International Conference on Superplasticizers and Other Chemical Admixtures in Concrete,* ACI SP-173. American Concrete Institute, Detroit.

Day, R.L. 1992. "The effect of secondary ettringite formation on the durability of concrete: A literature analysis." Research and Development Bulletin RD108T. Portland Cement Association, Skokie, Illinois.

Deng, M., and M. Tang. 1994. "Formation and expansion of ettringite crystals," *Cem. Concr. Res.,* **24** [1] 119–126.

Detwiler, R.J., and L.J. Powers-Couche. 1997. "Effect of sulfates in concrete on its resistance to freezing and thawing." PCA R&D Serial No. 2090. Portland Cement Association, Skokie, Illinois.

Diamond, S., and S. Ong. 1994. "Combined effects of alkali silica reaction and secondary ettringite deposition in steam cured mortars"; pp. 79–90 in *Cement Technology,* Ceramic Transactions Vol. 40. Edited by E.M. Gartner and H. Uchikawa. American Ceramic Society, Westerville, Ohio.

Diamond, S. 1996. "Delayed ettringite formation — Processes and problems," *Cem. Concr. Compos.,* **18**, 205–215.

Fu, Y. 1996. "Delayed Ettringite Formation in Portland Cement Products," Ph.D. Thesis, University of Ottawa.

Fu, Y., P. Xie, P. Gu, and J.J. Beaudoin. 1994. "Significance of pre-existing cracks on nucleation of secondary ettringite in steam cured cement paste," *Cem. Concr. Res.,* **24** [6] 1015–1024.

Fu, Y., and J.J. Beaudoin. 1996. "Microcracking as a precursor to delayed ettringite formation in cement systems," *Cem. Concr. Res.,* **26** [10] 1493–1498.

Ghorab, H.Y., D. Heinz, U. Ludwig, T. Meskendahl, and A. Wolter. 1980. "On the stability of calcium aluminate sulphate hydrates in pure systems and in cements"; pp. 496–503 in *Proceedings of the 7th International Congress on the Chemistry of Cement,* Vol. 4. Paris.

Glasser, F.P. 1996. "The role of sulfate mineralogy and cure temperature in delayed ettringite formation," *Cem. Concr. Compos.,* **18**, 187–193.

Glasser, F.P., D. Damidot, and M. Atkins. 1995. "Phase development in cement in relation to the secondary ettringite problem," *Adv. Cem. Res.,* **7** [26] 57–68.

Grabowski, E., B. Czamecki, J.E. Gillott, C.R. Duggan, and J.F. Scott. 1992. "Rapid test of concrete expansivity due to internal sulfate attack," *ACI Mater. J.,* **89** [5] 469–480.

Heinz, D., and U. Ludwig. 1986. "Mechanism of subsequent ettringite formation in mortars and concrete after heat treatment"; pp. 189–194 in *Proceedings of the 8th International Congress on Chemistry of Cement,* Vol. 5. Rio de Janeiro.

Heinz, D., and U. Ludwig. 1987. "Mechanism of secondary ettringite formation in mortars and concretes subjected to heat treatment"; pp. 2059–2071 in *Concrete Durability: Katherine and Bryant Mather International Conference,* ACI SP-100, Vol. 2. Edited by J.M. Scanlon. American Concrete Institute, Detroit.

Heinz, D., U. Ludwig, and I. Rüdiger. 1989. "Delayed ettringite formation in heat treated mortars and concretes," *Concr. Precasting Plant Technol.,* **11**, 56–61.

Hempel, G., A. Böhmer, and M. Otte. 1992. "Investigation of the influence of material-technological parameters on the durability of heat cured concretes," *Concr. Precasting Plant Technol.,* **15**, 75–79.

Herfort, D., J. Soerensen, and E. Coulthard. 1997. "Mineralogy of sulphate rich clinker and the potential for internal sulphate attack"; presented at the ACI Spring Convention, Seattle, April.

Hime, W.G. 1996a. "Clinker sulfate: A cause for distress and a need for specification"; pp. 387–395 in *Concrete for Environment Enhancement and Protection*. Edited by R.K. Dhir and T.D. Dyer. E&FN Spon, London.

Hime, W.G. 1996b. "Delayed ettringite formation — A concern for precast concrete?" *PCI J.,* July-August, 26–30.

Hobbs, D.W. 1997. "Expansion and cracking in concrete attributed to 'Delayed Ettringite Formation'"; presented at the ACI Spring Convention, Seattle, April.

Johansen, V., N. Thaulow, and J. Skalny. 1993. "Simultaneous presence of alkali-silica gel and ettringite in concrete," *Adv. Cem. Res.,* **5** [17] 23–29.

Jones, T.N., and A.B. Poole. 1987. "Alkali silica reactions in several U.K. concretes: The effect of temperature and humidity on expansion and the significance of ettringite development"; pp. 446–450 in *Proceedings of the 7th International Conference on Alkali Aggregate Reactions.* Edited by P.E. Grattan-Bellew. Noyes Publications, Park Ridge, New Jersey.

Kalde, M., and U. Ludwig. 1997. "Some remarks on the late formation of ettringite in mortar and concrete"; presented at the ASTM Symposium on Internal Sulfate Attack on Cementitious Systems: Implications for Standards Development, San Diego, December.

Kelham, S. 1996. "The effect of cement composition and fineness on expansion associated with delayed ettringite formation," *Cem. Concr. Compos.,* **18**, 171–179.

Kelham, S. 1997. "Effects of cement composition and hydration temperature on volume stability of mortar"; paper 4iv060 in *Proceedings of the 10th International Congress on the Chemistry of Cement,* Göteborg, Sweden, Vol. 4. Edited by H. Justnes.

Klemm, W.A., and F.M. Miller. 1997. "Plausibility of delayed ettringite formation as a distress mechanism — Considerations at ambient and elevated temperature"; paper 4iv059 in *Proceedings of the 10th International Congress on the Chemistry of Cement,* Göteborg, Sweden, Vol. 4. Edited by H. Justnes.

Lagerblad, B., and P. Utkin. 1994. "The effect of water-cement ratio, sulphonated plasticizer and silica fume on delayed ettringite formation"; pp. 154–173 in *Proceedings of the RILEM International Workshop on the Durability of High Performance Concrete.* Edited by H. Sommer. RILEM, Cachan Cedex, France.

Larive, C., and N. Louarn. 1992. "Diagnosis of alkali aggregate reaction and sulphate reaction in French structures"; pp. 587–598 in *Proceedings of the 8th International Conference on Alkali Aggregate Reactions,* London, Vol. 2. Concrete Society, Wexham. U.K.

Lawrence, B.L., J.J. Meyers, and R.L. Carrasquillo. 1997a. "Premature concrete deterioration in Texas Department of Transportation precast elements"; presented at the ACI Spring Convention, Seattle, April.

Lawrence, B.L., J.J. Meyers, and R.L. Carrasquillo. 1997b. "Internal sulfate attack in Texas Department of Transportation precast elements"; presented at the ASTM Symposium on Internal Sulfate Attack on Cementitious Systems: Implications for Standards Development, San Diego, December.

Lawrence, C.D. 1993. "Laboratory Studies of Concrete Expansion Arising from Delayed Ettringite Formation." Technical Report C/16. British Cement Association.

Lawrence, C.D. 1995a. "Delayed ettringite formation: An issue?"; pp. 113–154 in *Materials Science of Concrete IV.* Edited by J. Skalny and S. Mindess. American Ceramic Society, Westerville, Ohio.

Lawrence, C.D. 1995b. "Mortar expansions due to delayed ettringite formation. Effects of curing period and temperature," *Cem. Concr. Res.,* **25** [4] 903–914.

Lawrence, C.D., J.A. Dalziel, and D.W. Hobbs. 1990. "Sulphate attack arising from delayed ettringite formation." BCA Interim Technical Note 12. British Cement Association, Wexham, U.K.

Lewis, M.C. 1996. "Heat Curing and Delayed Ettringite Formation in Concretes," Ph.D. Thesis, University of London (Imperial College).

Lewis, M.C., K.L. Scrivener, and S. Kelham. 1995. "Heat curing and delayed ettringite formation"; pp. 67–76 in *MRS Symposia Proceedings,* Vol. 370. Materials Research Society, Pittsburgh.

Marks, V.J., and W.G. Dubberke. 1996. "A different perspective for investigation of portland cement concrete deterioration." Transportation Research Record No. 1525.

Marusin, S.L. 1993a. "SEM studies of DEF in hardened concrete"; pp. 289–299 in *Proceedings of the 15th International Conference on Cement Microscopy,* Dallas, Texas, March/April.

Marusin, S.L. 1993b. "SEM studies of concrete failures caused by delayed ettringite formation"; in *Proceedings of the 4th Euroseminar on Microscopy Applied to Building Materials,* Visby, Sweden, June. Edited by J.E. Lindqvist and B. Nitz.

Marusin, S.L. 1994a. "Concrete failure caused by delayed ettringite formation. Case study." Third CANMET/ACI International Conference on Durability of Concrete — Supplementary Papers, Nice, France.

Marusin, S.L. 1994b. "A simple treatment to distinguish alkali-silica gel from delayed ettringite formations in concrete," *Mag. Concr. Res.,* **46** [168] 163–166.

Marusin, S.L., and W.G. Hime. 1997. "Concrete and delayed ettringite formation — Causes, development and case studies"; pp. 1–12 in *Supplementary Papers of the 4th CANMET/ACI International Conference on Durability of Concrete,* Sydney, Australia.

Meland, I., H. Justnes, and J. Lindgård. 1997. "Durability problems related to delayed ettringite formation and/or alkali aggregate reactions"; paper 4iv064 in *Proceedings of the 10th International Congress on the Chemistry of Cement,* Vol. 2, Göteborg, Sweden. Edited by H. Justnes.

Michaud, V., and R. Suderman. 1997a. "The solubility of sulfates in high-SO_3 clinkers"; paper 2ii011 in *Proceedings of the 10th International Congress on the Chemistry of Cement,* Vol. 2, Göteborg, Sweden. Edited by H. Justnes.

Michaud, V., and R. Suderman. 1997b. "Anhydrite in high SO_3/alkali clinkers: Dissolution kinetics and influence on concrete durability"; presented at the ASTM Symposium on Internal Sulfate Attack on Cementitious Systems: Implications for Standards Development, San Diego, December.

Michaud, V., A. Nonat, and D. Sorrentino. 1997. "Experimental simulation of the stability of ettringite in alkali silica solutions, produced by alkali-silica reaction, in concrete"; paper 4iv65 in *Proceedings of the 10th International Congress on the Chemistry of Cement,* Vol. 4, Göteborg, Sweden. Edited by H. Justnes.

Mielenz, R.C., S. Marusin, W.G. Hime, and Z.T. Zugovic. 1995. "Prestressed concrete railway tie distress: Alkali-silica reaction or delayed ettringite formation," *Concr. Int.,* **17** [12] 62–68.

Miller, F.M., and F.J. Tang. 1996. "The distribution of sulfur in present-day clinkers of variable sulfur content," *Cem. Concr. Res.,* **26** [12] 1821–1829.

Oberholster, R.E., H. Maree, and J.H.B. Brand. 1992. "Cracked prestressed concrete railway sleepers: Alkali-silica reaction or delayed ettringite formation"; pp. 739–749 in *Proceedings of the 9th International Conference on Alkali-Aggregate Reaction in Concrete,* London.

Odler, I., and Y. Chen. 1995. "Effect of cement composition on the expansion of heat-cured cement pastes," *Cem. Concr. Res.,* **25** [4] 853–862.

Odler, I., and Y. Chen. 1996. "On the delayed expansion of heat cured Portland cement pastes and concretes," *Cem. Concr. Compos.,* **18**, 181–185.

Ouyang, C., and O.J. Lane. 1997. "Effect of infilling air voids by ettringite on freeze-thaw durability of concrete"; presented at the ACI Spring Convention, Seattle, April.

Pettifer, K., and P.J. Nixon. 1980. "Alkali metal sulphate — A factor common to both alkali aggregate reaction and sulfate attack on concrete," *Cem. Concr. Res.,* **10** [2] 173–181.

Regourd, M., H. Hornain, and P. Poitevin. 1981. "Alkali reaction — Concrete microstructural evolutions"; paper S252/35 in *Proceedings of the 5th International Conference on Alkali-Aggregate Reactions,* Cape Town.

Saloman, N., J. Claude, and L. Hasni. 1992. "Diagnosis of concrete structures affected by alkali aggregate reaction"; pp. 902–915 in *Proceedings of the 8th International Conference on Alkali Aggregate Reactions,* London. Concrete Society, Wexham, U.K.

Schrama, H., M. Blumenthal, and E.C. Weatherhead, 1995. "A survey of tire burning technology for the cement industry"; pp. 285–306 in *IEEE Cement Industry Technical Conference, XXXVII Conference Record.* The Institute of Electrical and Electronics Engineers, Piscataway, New Jersey.

Scrivener, K.L. 1996. "Delayed ettringite formation and concrete railway ties"; pp. 375–377 in *Proceedings of the 18th International Conference on Cement Microscopy,* Houston.

Scrivener, K.L., D. Damidot, and C. Famy. 1997. "Possible mechanisms of expansion of concrete exposed to elevated temperatures during curing (a.k.a. DEF) and implications for avoidance of field problems"; presented at the ASTM Symposium on Internal Sulfate Attack on Cementitious Systems: Implications for Standards Development, San Diego, December.

Scrivener, K.L., V. Johansen, and J. Skalny. 1998. "Investigation of prestressed concrete railway tie distress." To be published.

Scrivener, K.L., and M. Lewis. 1997. "A microstructural and microanalytical study of heat cured mortars and delayed ettringite formation"; paper 4iv061 in *Proceedings of the 10th International Congress on the Chemistry of Cement,* Vol. 4, Göteborg, Sweden. Edited by H. Justnes.

Scrivener, K.L., and H.F.W. Taylor. 1993. "Delayed ettringite formation: a microstructural and microanalytical study." *Adv. Cem. Res.,* **5** [20] 139–146.

Shayan, A., and I. Ivanusec. 1996. "An experimental clarification of the association of delayed ettringite formation with alkali-aggregate reaction," *Cem. Concr. Compos.,* **18**, 161–170.

Shayan, A., and G.W. Quick. 1991/2. "Relative importance of deleterious reactions in concrete: Formation of AAR products and secondary ettringite," *Adv. Cem. Res.,* **4** [16] 149–157.

Shayan, A., and G.W. Quick. 1992. "Microscopic features of cracked and uncracked concrete railway sleepers," *ACI Mater.,* **89** [4] 348–361.

Shayan, A., and G.W. Quick. 1994. "Alkali aggregate reaction in concrete railway sleepers from Finland"; pp. 69–79 in Proceedings of the 16th International Conference on Cement Microscopy. Edited by G.R. Gouda et al. ICMA, Duncanville, Texas.

Shayan, A., G.W. Quick, and C.J. Lancucki. 1993. "Morphological, mineralogical and chemical features of steam-cured concrete containing densified silica fume and various alkali levels," *Adv. Cem. Res.,* **5** [20] 151–162.

Skalny, J., V. Johansen, N. Thaulow, and A. Palomo. 1996. "DEF: As a form of sulfate attack," *Materiales de Construcción,* **46** [244] 5–29.
Skalny, J., and F.W. Locher. 1997. "Curing practices and internal sulfate attack — The European experience"; presented at the ASTM Symposium on Internal Sulfate Attack on Cementitious Systems: Implications for Standards Development, San Diego, December.
Skalny, J., V. Johansen, and F.M. Miller. 1997. "Sulfates in cement clinker and their effects on concrete durability"; pp. 625–631 in *Proceedings of the Third International Symposium on Advances in Concrete Technology,* ACI SP-171, Vol. 2. Edited by V.M. Malhotra. American Concrete Institute, Detroit.
Stark, J., and K. Bollmann. 1997. "Ettringite formation — A durability problem of concrete pavements"; paper 4iv062 in *Proceedings of the 10th International Congress on the Chemistry of Cement,* Vol. 4, Göteborg, Sweden. Edited by H. Justnes.
Stark, J., K. Bollmann, and K. Seyfarth. 1998. "Ettringite — Cause of damage, damage intensifier or uninvolved third party," *ZKG Int.,* **5** [51] 280–292.
Sylla, H.M. 1988. "Reactions in cement stone due to heat treatment," *Beton,* **38** [11] 449–454. In German.
Taylor, H.F.W. 1990. *Cement Chemistry.* Academic Press, London.
Taylor, H.F.W. 1994a. "Delayed ettringite formation"; pp. 122-131 in *Advances in Cement and Concrete.* Edited by M.W. Grutzeck and S.L. Sarkar. American Society of Civil Engineers, New York.
Taylor, H.F.W. 1994b. "Sulfate reactions in concrete — Microstructural and chemical aspects"; pp. 61–78 in *Cement Technology,* Ceramic Transactions, Vol. 40. Edited by E.M. Gartner and H. Uchikawa. American Ceramic Society, Westerville, Ohio.
Taylor, H.F.W. 1996. "Ettringite in cement paste and concrete"; presented at RILEM Conference on Concrete: From Material to Structure in Tribute to M. Moranville-Regourd, Arles, France, September.
Tennis, P.D., S. Bhattacharja, W.A. Klemm, and F.M. Miller. 1997. "Assessing the distribution of sulfate in portland cement and clinker and its influence on expansion in mortar"; Presented at the ASTM Symposium on Internal Sulfate Attack on Cementitious Systems: Implications for Standards Development, San Diego, December.
Tepponen, P., and B. Eriksson. 1987. "Damages in concrete railway sleepers in Finland." *Nordic Concr. Res.,* [6] 199–209.
Vitousova, L. 1991. "Concrete sleepers in CSD tracks"; pp. 253–264 in *Proceedings of the International Symposium on Precast Concrete Sleepers,* Madrid, April.
Wieker, W., and R. Herr. 1989a. "On some problems of the chemistry of Portland cement," *Zeitschrift fur Chemie,* 29, 321–327. In German.
Wieker, W., and R. Herr. 1989b. "Sulphate ion equilibria in pore solutions of heat-treated mortars of Portland cement with respect to the expansion reaction by secondary ettringite formation"; pp. 58–66 in *Proceedings of the Second International Symposium on Cement.*
Wolter, S. *Ettringite: Cancer of Concrete.* Burgess Publishing, 1997.
Yang, R., C.D. Lawrence, and J.H. Sharp. 1996. "Delayed ettringite formation in 4-year old cement pastes," *Cem. Concr. Res.,* **26**, 1649–1659.

Use of Durability Indexes to Achieve Durable Cover Concrete in Reinforced Concrete Structures

M.G. Alexander and J.R. Mackechnie
Department of Civil Engineering, University of Cape Town, Rondebosh, South Africa

Y. Ballim
Department of Civil Engineering, University of the Witwatersrand, Johannesburg, South Africa

> *"In the long run men hit only what they aim at. Therefore, though they should fail immediately, they had better aim at something high."*
> — Henry David Thoreau

This chapter presents a practical engineering approach to achieving durability in reinforced concrete structures in the current context of pervasive steel corrosion problems. The chief issue is identified as the need to be able to characterize the cover layer to steel, using reproducible engineering measures of the concrete microstructure, and to use these measures to control quality through rational performance-based specifications. Three durability index texts — related to oxygen permeability, water sorptivity, and chloride conductivity — that can provide the required quantification of covercrete quality are described. These durability index tests have a sound theoretical basis, are easily and quickly performed, and have low statistical variability. The tests can also be used as a basis for prediction models of long-term performance related to control of steel corrosion.

Introduction

The durability performance of construction materials has long been a concern for engineers. Only in recent years, however, has the actual deterioration been accurately quantified and the extent of the problem been recognized. Durability may be defined as the ability of a material or structure to withstand the service conditions for which it was designed over a prolonged period without significant deterioration. Concrete has generally been regarded as having chemical and dimensional stability in most environments, thereby possessing inherently durable characteristics. This perception is also associated with reinforced concrete structures, which are expected to be relatively maintenance-free during their service life. These assumptions must be questioned given the weight of evidence of premature

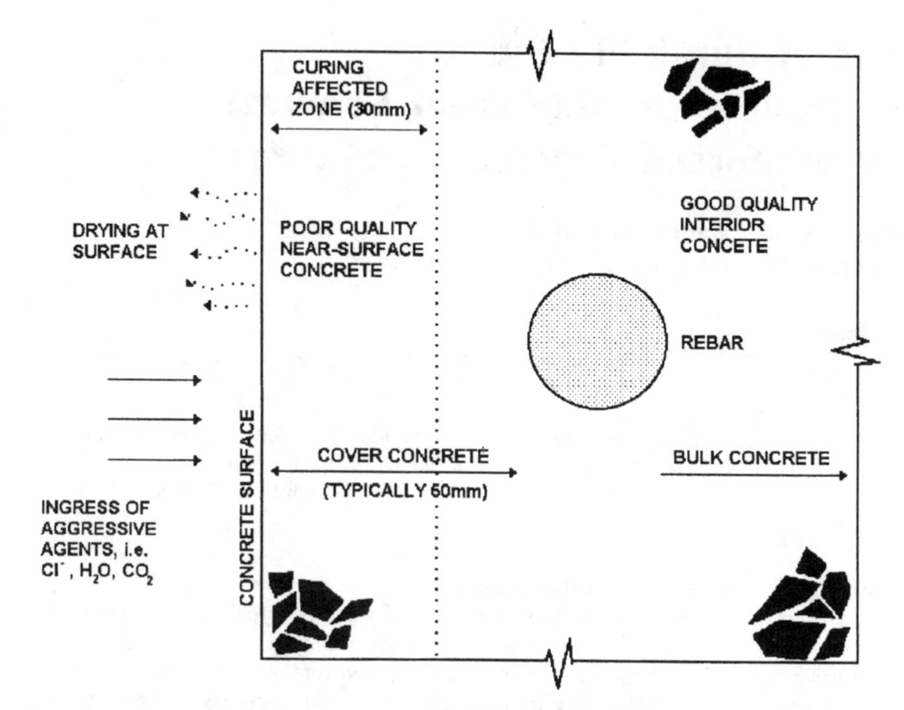

Figure 1. Schematic diagram of concrete protection of reinforcement.

deterioration of concrete structures. Many modern concrete structures need substantial repairs and maintenance during their service life with the resultant costs to the economy reaching 3–5% of GNP in some countries.[1]

Neville[2] suggests that the reasons for the widespread lack of durability include poor understanding of deterioration processes by designers, inadequate acceptance criteria of concrete on site, and changes in cement properties and construction practices.

There is an increasing awareness that concrete is a complex composite material, that environmental and service conditions vary widely, and that deterioration mechanisms interact dynamically with material and structural influences. Deterioration of concrete begins almost immediately after casting as the hardened properties are influenced by phenomena that occur at an early stage, such as plastic cracking, bleeding, segregation, and thermal effects. In the hardened state, concrete may be affected by a variety of internal and external factors that cause damage by physical and/or chemical mechanisms. Deterioration is often associated with ingress of aggressive agents from the exterior such that the near-surface concrete quality largely controls durability. The interaction between the various material and environmental elements influencing durability is shown in Fig. 1.

Many durability problems concern the corrosion of reinforcing steel rather than deterioration of the concrete fabric itself. The problem is then cast in terms of the protection to steel offered by the concrete cover layer, which is subjected to the action of aggressive agents such as chloride ions or agents of acidification arising from the surrounding environment. This chapter will focus on the issue of the quantification of the quality of the cover layer in reinforced concrete, in the context of the need to provide adequate protection to the reinforcing steel.

Modern design and construction practice has led to improvements such as the use of more consistent-quality cement, higher allowable stresses, faster concrete casting and setting times, and greater variety of binder types and admixtures. While these advances have improved concrete productivity, they have made concrete more sensitive to abuse that has contributed to the premature deterioration of modern concrete structures. The increasing number of concrete structures exhibiting unacceptable levels of deterioration has resulted in more stringent specifications for concrete construction. Unfortunately, the durability performance of concrete structures has not always shown a corresponding improvement despite these specifications. This appears to be due to a lack of understanding of what is required to ensure durability as well as inadequate means of enforcing or guaranteeing compliance with specifications during construction.

Most national codes and specifications are of the "recipe" type, setting limits on w/c ratios, cement contents, cover, and so on, but without really addressing the issue of achieving adequate quality of the concrete cover. Furthermore, it is difficult, if not impossible, to ensure compliance with these specifications on site, since they generally comprise difficult-to-measure aspects of construction. (The one notable exception is, of course, checking concrete cover to steel. Many have expressed the opinion that enforcing this one simple expedient would cure 90% of current durability problems.) It is clear that current durability specifications are not always very effective.

In response to this situation, research in South Africa has focused on the development of a durability index approach that seeks to characterize the potential durability of new concrete. This chapter outlines the philosophy of the durability index approach, reviews the current index tests, and indicates how the approach may used in practice. It is hoped that it will stimulate debate on how best to move forward in achieving better durability in concrete construction.

Philosophy of the Durability Index Approach

A plethora of durability tests have been developed to measure fluid transport rates by various mechanisms through concrete. Sophisticated equipment, complex monitoring, and lengthy testing periods are generally required to accurately model these mechanisms. While such techniques provide useful research information, they have limited practical value for site concrete given the rigorous constraints of the test methods. In response to the need for more practical durability tests, the philosophy of durability index testing of concrete was formulated, and is outlined below.

Improved durability will not be achieved unless some relevant durability parameter(s) can be unambiguously measured. This is where the crux of the problem lies. The issue is compounded by the fact that concrete is highly complex, changes with time, and is affected by a multitude of processes. However, engineers have grappled with this successfully in the past in respect of, for example, concrete strength, by adopting a simple quality control test: the cube or cylinder compression test. The test itself bears little resemblance to the conditions existing in a real structure. Nevertheless, experience has permitted the correlation of the results of the compression test with structural performance, so that structures may be designed for different levels of stress. In essence, the test can be thought of as an index test, which characterizes the intrinsic potential of the material to resist applied stresses. By experience, this index has come to be associated with acceptable performance and has generally worked very well.

It is this idea of indexing of the material that needs to be applied to the achievement of adequate protection of reinforcing steel in concrete. What is required is a means of characterizing the quality of the cover or surface layer, using parameters that influence the deterioration processes acting on the concrete. The use of strength parameters is not sufficient for this, since they merely measure the overall bulk response of the material to stress. It is the surface layer that is most affected by curing initially, and subsequently by external deterioration processes. These processes are linked with transport mechanisms, such as gaseous and ionic diffusion, water absorption, and so on. Thus a series of index tests is needed to cover the broad range of durability problems, each index test being linked to a transport mechanism relevant to that particular process.

Ultimately, the usefulness of index tests will be assessed only by reference to actual durability performance of structures built using the indexes for quality control purposes. This is a long-term undertaking. A framework for durability studies is therefore necessary, incorporating early age material indexing, direct durability testing, and observations of long-term durability performance.[3]

Material indexing requires quantifiable physical or engineering parameters to characterize the concrete at early ages. Such indexes (e.g., permeability, water sorptivity) must be sensitive to important material, processing, and environmental factors such as cement type, water/binder ratio, type and degree of curing, and so on. The purpose of material indexing is to provide a reproducible engineering measure of microstructure and properties of importance to concrete durability at a relatively early age (e.g., 28 days). The material indexes allow the material to be placed in an overall matrix of possible material values, these values depending on the important factors given above. Thus, it should be possible to produce concretes of similar durability by a number of different routes: additional curing, lower w/c ratio, choice of a different cement or extender, and so on.

Direct durability testing should comprise a suite of tests suited to a range of durability problems. Such tests would embrace accelerated as well as long-term evaluations. The need for accelerated testing is obvious in view of the long time periods involved in concrete deterioration. Nevertheless, it is usually necessary to also undertake long-term testing, since mechanisms dominant in accelerated tests may be different from those in the normal environment. Observations of long-term structural behavior are also necessary in order to provide a quantifiable estimate of durability performance.

Two other major issues exist with direct durability testing:

1. The definition of a suitable measure of deterioration, and a threshold or limiting value of deterioration (e.g., should the extent of sulfate attack be characterized by measuring expansion, mass loss, change in strength and stiffness, etc.?)
2. A lack of standard test methods, which prevents useful comparisons between reported results and hampers understanding of mechanisms.

Correlations are required between indexes and durability results, and between these two and actual structural performance, such that the index tests can ultimately be used as follows:

- As a means of controlling a particular property of concrete, or the quality of a particular zone of a concrete element, for example, the surface layer. (This control would typically be reflected by a suitable construction specification, in which limits to index values at 28 days would be specified.)

- As a means of assessing the quality of construction for compliance with a set of criteria.
- As a basis for fair payment for the achievement of concrete quality.
- As a means of predicting the performance of concrete in the design environment, on an empirical basis.

Index properties fulfill the requirements of a measurable property that can be specified.

Durability, viewed in this case as adequate protection of the reinforcing steel, should be viewed as a property that must be paid for, just as strength is at present. This is consistent with the idea of concrete as an engineered material, in which the requisite properties are provided by intelligent design. Each desired property must be specified and appropriately paid for. This is not to imply that the cost of concrete structures is likely to escalate dramatically due to such payment for durability. Life-cycle costing almost always proves that paying for durability initially is an investment that is best in the long run.

The criteria for suitable index tests require, among other things, that the tests:

- Be site or laboratory applicable (site applicable could involve retrieval of small core specimens from the structure for laboratory testing).
- Be linked to important fluid and ionic transport mechanisms and have a reasonable theoretical basis.
- Be quickly and easily performed with minimum demands on operator skill.
- Have sufficiently low statistical variability.
- Involve a minimum of specimen preparation, with uniform preconditioning to ensure standardized testing.
- Be conducted at a relatively early age (typically 28 days).

Durability Index Tests

Three durability index tests have been developed to characterize concrete according to transport mechanisms: oxygen permeability for permeation, water sorptivity for absorption, and chloride conductivity for diffusion. The development, results, and applications of these tests are discussed below. (Additional details on the philosophy and testing procedure of durability index tests are given in Refs. 4 and 5).

Oxygen Permeability Test

Permeation describes the process of movement of fluids through the pore structure under an externally applied pressure while the pores are saturated with the particular fluid. Permeability is therefore a measure of the capacity for concrete to transfer fluids by permeation. The permeability of concrete is dependent on the concrete microstructure, the moisture condition of the material, and the characteristics of the permeating fluid.

Permeability test methods that have been developed for concrete include throughflow, inflow, and falling-head types, while permeating fluids are either gases or liquids. Throughflow permeability tests attempt to determine the Darcy coefficient of permeability by measuring pressure gradient or flow rate through concrete under a sustained pressure head. Measuring permeability is intrinsically slow and often impractical for dense concretes. More empirical inflow permeability tests were therefore developed that measure the depth or amount of fluid penetration after a period of applied pressure. Falling-head permeameters apply an initial pressure to concrete and allow the pressure to decay as permeation proceeds. This approach allows for ease of testing while maintaining a high level of accuracy since pressure may be reliably monitored with time.

The falling-head gas permeameter developed at the University of the Witwatersrand by Ballim[6] is shown in Figure 2. The permeability of oven-dried concrete core samples to oxygen gas is determined by measuring the pressure decay with time (from the initial value of 100 kPa). The pressure decay curve measured either directly from gauges or using data logging from transducers is converted to a linear relationship by plotting the logarithm of the ratio of pressure heads versus time.

From the slope of the straight line produced by this plot, the coefficient of permeability may be determined as follows (see Appendix A):

$$k = \frac{\omega V g d}{R A \theta t} \ln \frac{P_0}{P} \tag{1}$$

where k is the coefficient of permeability (m/s), ω is the molecular mass of permeating gas (kg/mol), V is the volume of the pressure cylinder (m^3), g is the acceleration due to gravity (m/s^2), d is the sample thickness (m), R is the universal gas constant (Nm/K mol), A is the cross-sectional area of specimen (m^2), θ is the absolute temperature (K), t is the time (s), P_0 is the pressure at the start of the test (kPa), and P is the pressure at time t (kPa).

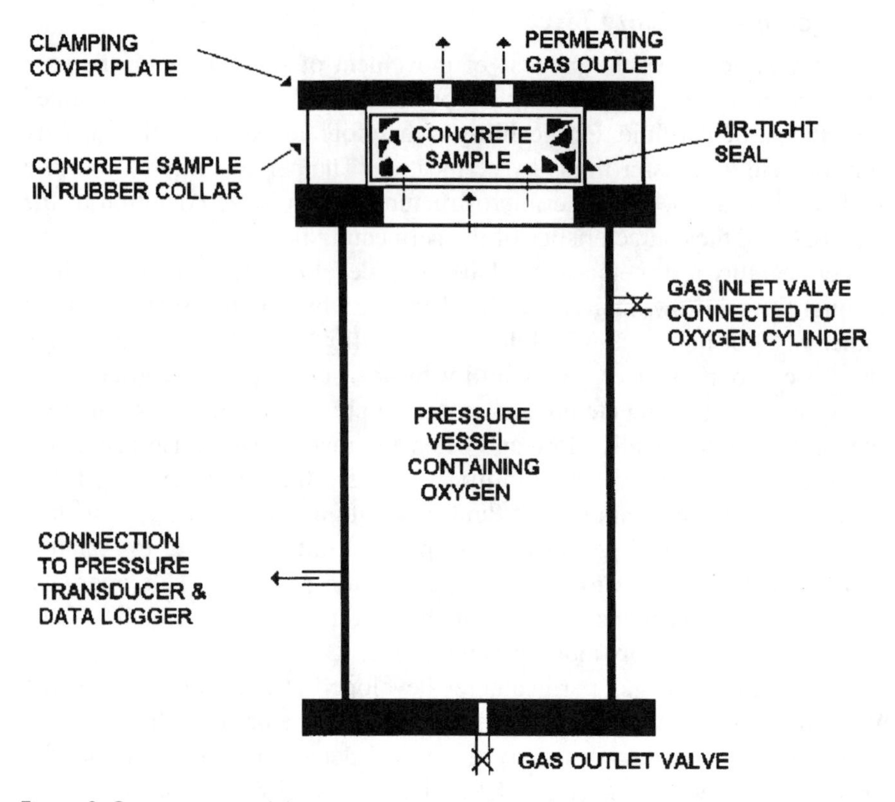

Figure 2. Oxygen permeability apparatus.

The coefficient of permeability is an unwieldy exponential number and it was therefore simplified by defining the permeability index (OPI) as the negative logarithm of the coefficient of permeability, that is:

$$\mathrm{OPI} = -\log_{10} k \tag{2}$$

Typical experimental plots from the oxygen permeability test are shown in Fig. 3 for Grade 40 OPC concrete tested at 28 days.[7] Figure 4 shows OPI results for three different grades of OPC concrete. Oxygen permeability results may be used to characterize young concrete for influences such as concrete grade, binder type, initial curing, and construction effects (such as compaction). Oxygen permeability may also be used at later ages to assess the deterioration of concrete but careful interpretation is required when assessing results.

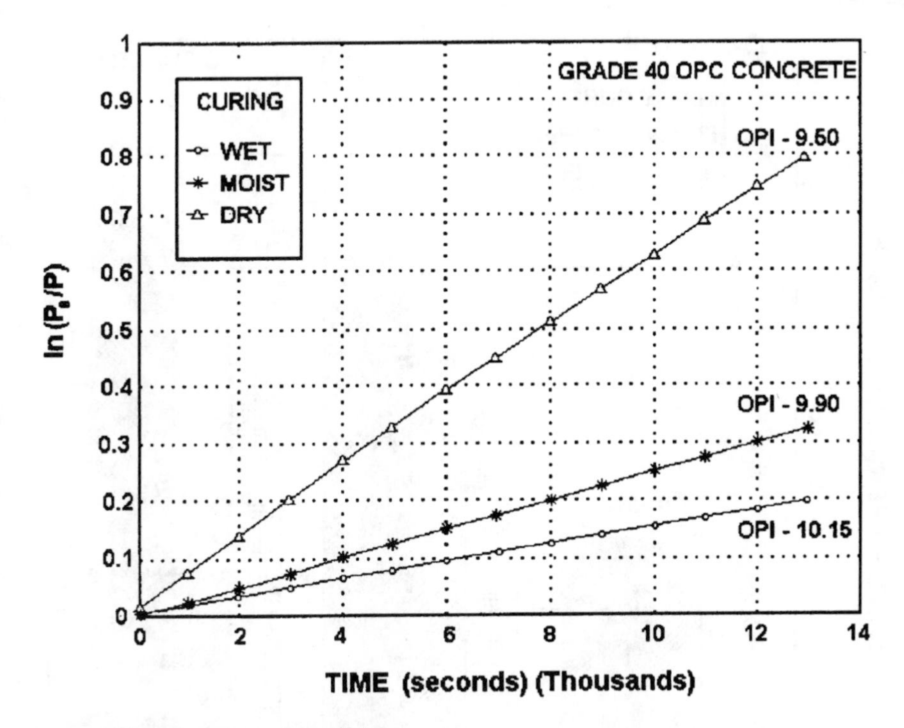

Figure 3. Typical oxygen permeability results.

Water Sorptivity Test

Absorption is the process whereby fluid is drawn into a porous, unsaturated material under the action of capillary forces. The capillary suction is dependent on the pore geometry and the saturation level of concrete. Water absorption caused by wetting and drying at the concrete surface is an important transport mechanism near the surface, but becomes less significant with depth. The rate of movement of a wetting front through a porous material under the action of capillary forces is defined as sorptivity.

Several general absorption tests have been developed for concrete in which concrete is immersed in water and the total mass of water absorbed is used as a measure of the absorption of the material. These tests merely measure the porosity of the concrete but cannot quantify the rate of absorption and do not distinguish between surface and bulk effects. Essentially such tests measure porosity, which may not be sensitive to the transport mechanisms influencing concrete durability.

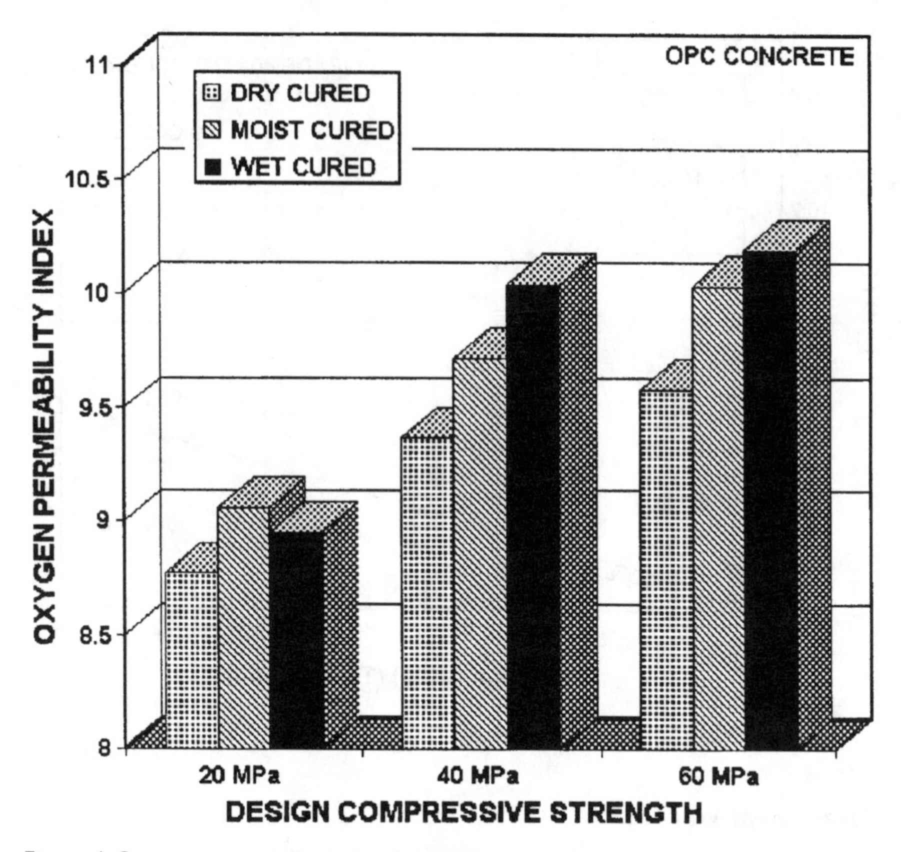

Figure 4. Oxygen permeability index for OPC concretes with varying degrees of moist curing.

A modified version of Kelham's sorptivity test was chosen as a compromise between acccuracy and ease of use.[8,9] Concrete samples (usually cores 68 mm diameter, 25 mm thick) are initially preconditioned at 50°C to ensure uniformly low moisture contents at the start of the test. The circular edges of the core are sealed with either epoxy or tape to ensure unidirectional absorption. The concrete specimens are then exposed to a few millimeters of water with the test surface facing downward, as shown in Fig. 5. At regular time intervals, the specimens are removed from the water and the mass of water absorbed is determined using an electronic balance. Measurements are stopped before saturation is reached and the concrete is then vacuum saturated in water to determine the effective porosity.

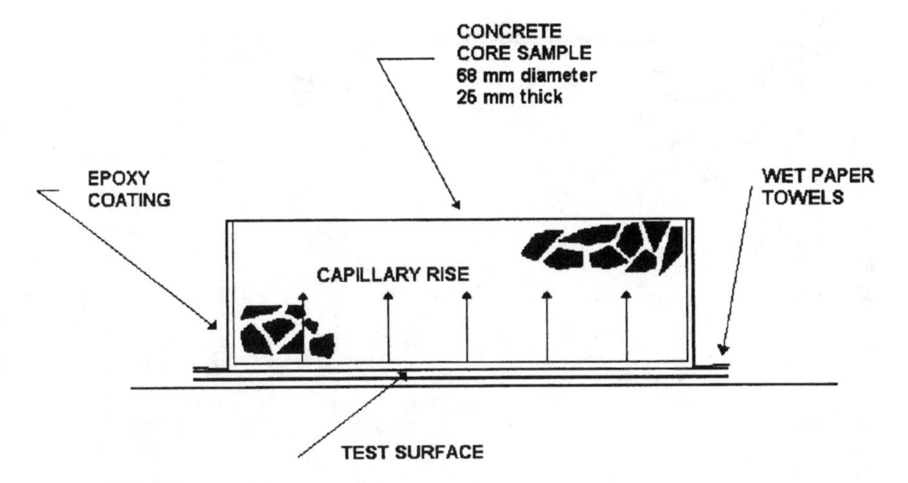

Figure 5. Water sorptivity test.

A linear relationship is observed when the mass of water absorbed is plotted against the square root of time. The sorptivity, S, of the concrete is determined from the slope of the straight line produced (see Appendix B), such that:

$$S = \frac{\Delta M_t}{t^{1/2}}\left(\frac{d}{M_{sat} - M_0}\right) \tag{3}$$

where ΔM_t is the change of mass with respect to the dry mass (g), M_{sat} is the saturated mass of concrete (g), M_0 is the dry mass of concrete (g), d is the sample thickness (mm), and t is the period of absorption (h).

The sorptivity test is able to characterize concrete for much the same influences as the permeability test, but with emphasis more on near-surface effects such as curing. Typical experimental data are shown in Figure 6. [Note that the early (1 min) rapid increase in mass is omitted in the calculations.]

Figure 7 shows water sorptivity indexes for three different grades of OPC concrete.

The degree of initial curing affects the quality of the concrete near surface, which in turn influences the sorptivity of the material. Sorptivity testing is best done on young concretes, as older concretes may be contaminated with salts or be carbonated, which alters the absorption of water into the pore structure.

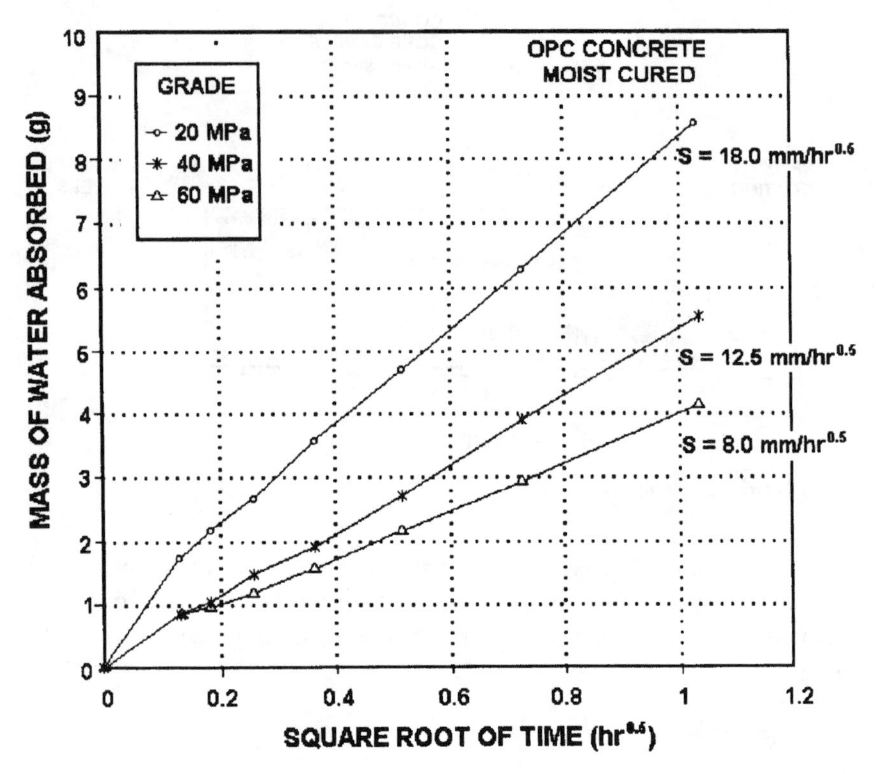

Figure 6. Water sorptivity results.[7]

Chloride Conductivity Test

Diffusion is the process by which liquid, gas, or ions move through a porous material under the action of a concentration gradient. Diffusion occurs in partially or fully saturated concrete and is an important internal transport mechanism for most concrete structures exposed to salts. High surface salt concentrations are initially developed by absorption, and the salt migrates by diffusion toward the low concentrations of the internal material. Diffusion rates are dependent on temperature, moisture content of the concrete, type of diffusant, and the inherent diffusibility of the material. Diffusion into concrete is complicated by chemical interactions, partially saturated conditions, defects such as cracks and voids, and electrochemical effects due to steel corrosion and stray currents. In marine or de-icing salt environments, diffusion of chloride ions is of particular importance due to the depassivating effect of chlorides on embedded steel, which may ultimately lead to corrosion.

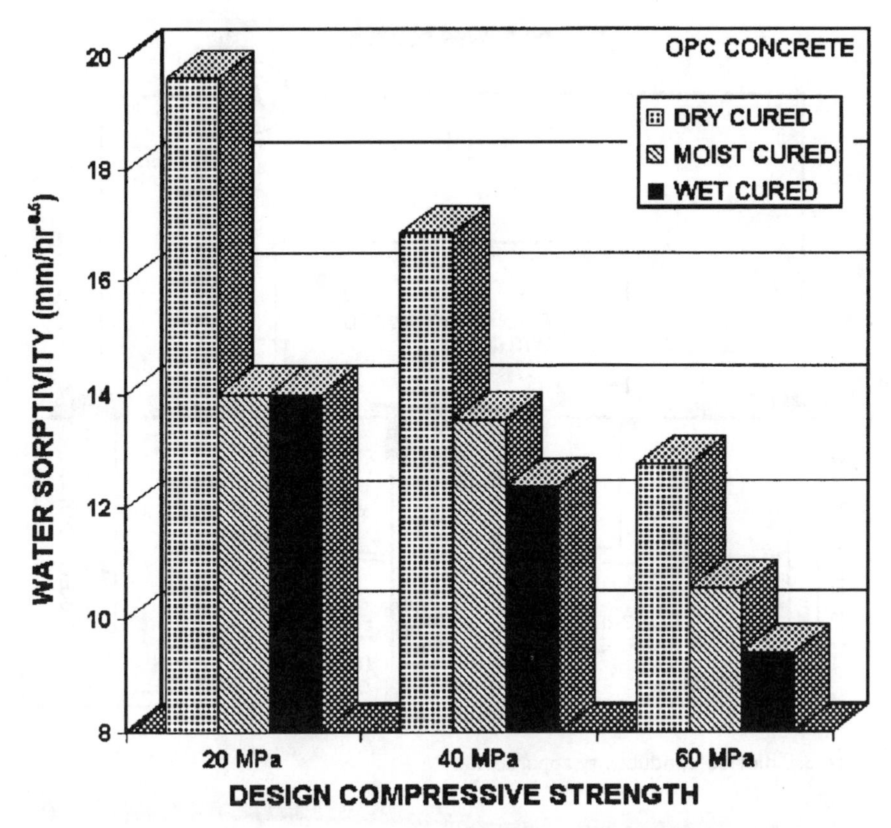

Figure 7. Water sorptivity indexes for OPC concretes with varying degrees of moist curing.

Intrinsic diffusion tests, where concrete is exposed to high and low concentrations of the diffusing species on opposite faces, have been successfully used to measure the coefficient of diffusion. Diffusion is a slow process, even when using high concentration gradients, and these tests may take several months to reach equilibrium. Accelerated diffusion tests using an applied potential difference have therefore been developed to obtain more rapid results in the laboratory. A fundamental weakness of some of these rapid chloride migration tests is that the increased ionic flux is caused by both diffusion and conduction, and some tests lack a sound theoretical basis.

Streicher has developed a rapid chloride conductivity test at UCT in which the ionic flux occurs by conduction due to a 10 V potential difference.[10] The apparatus, shown in Fig. 8, consists of a two-cell conduction rig in which concrete core samples are exposed on either face to 5 M NaCl

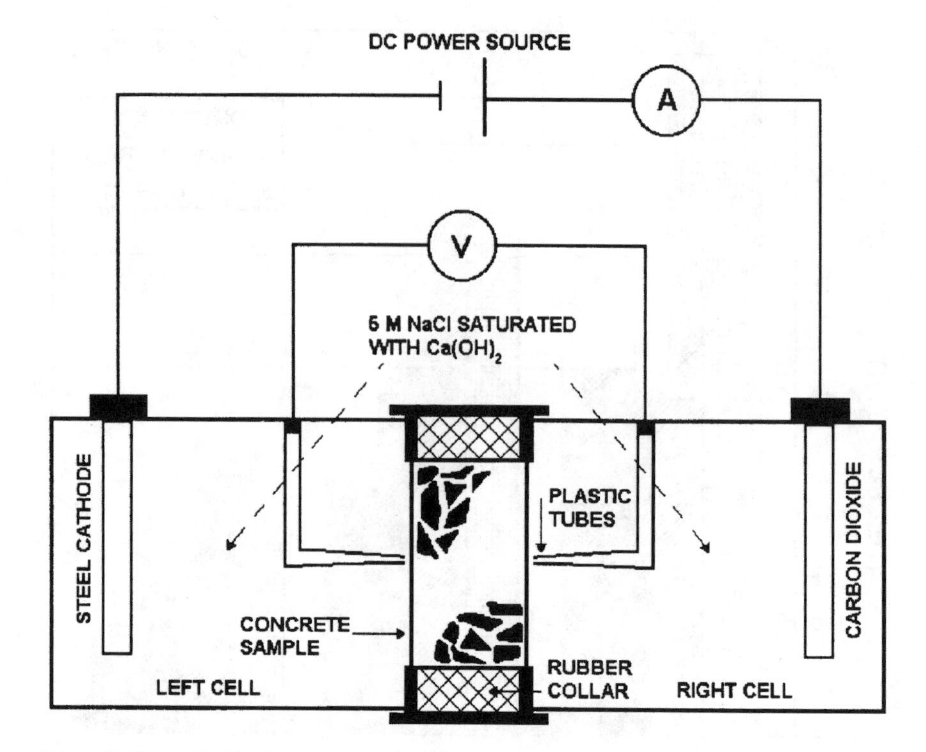

Figure 8. Chloride conductivity apparatus.

solution. The core samples are preconditioned before testing to standardize the pore water solution (oven-dried at 50°C followed by 24 h vacuum saturation in a 5 M NaCl solution). The movement of chlorides is accelerated by applying a 10 V potential difference and the chloride conductivity is determined by measuring the current flowing through the concrete specimen. The apparatus allows for rapid testing (virtually instantaneous readings) under controlled laboratory conditions.

The chloride conductivity of concrete may be defined as follows (see Appendix C):

$$\sigma = \frac{it}{VA} \tag{4}$$

where σ is the chloride conductivity (mS/cm), i is the current (mA), V is the voltage (V), t is the specimen thickness (cm), and A is the cross-sectional area (cm^2).

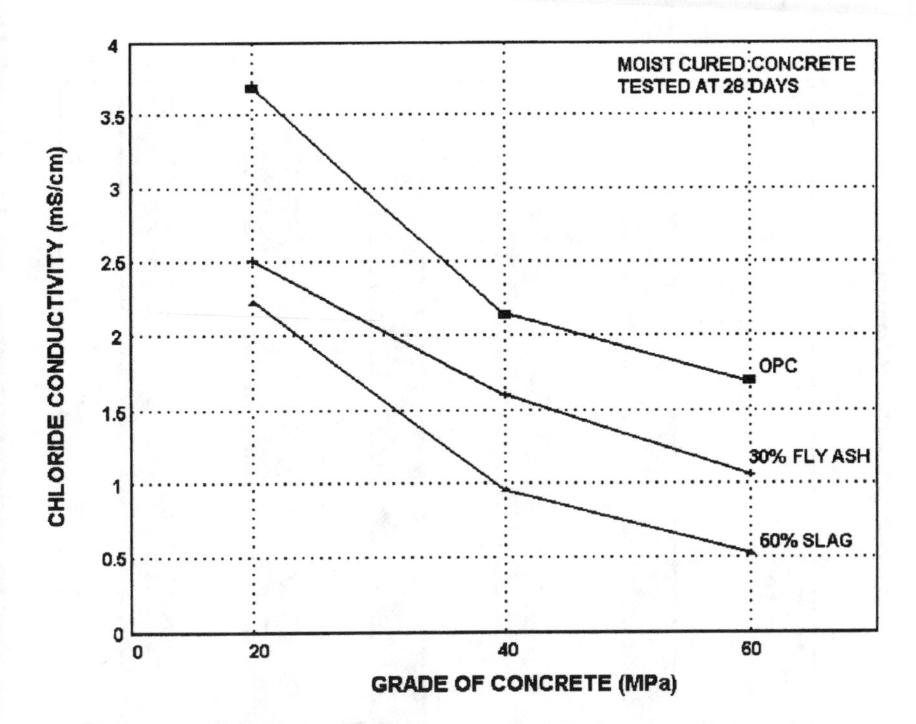

Figure 9. Chloride conductivity results.[7]

Chloride conductivity decreases with the addition of fly ash, slag, and silica fume in concrete; extended moist curing; and increasing grade of concrete. While the test is sensitive to construction and material effects that are known to influence durability, results are specifically related to chloride ingress into concrete. Figure 9 shows typical results measured at 28 days.

Use of Durability Index Tests

A suite of three durability index tests — oxygen permeability, water sorptivity, and chloride conductivity — has been shown to be sensitive to the important material, environmental, and constructional factors known to influence concrete durability. The index tests may be used in the following applications.

Quality Control of Site Concrete

The sensitivity of the index tests to material and construction effects makes them suitable tools for site quality control. Since the different tests measure

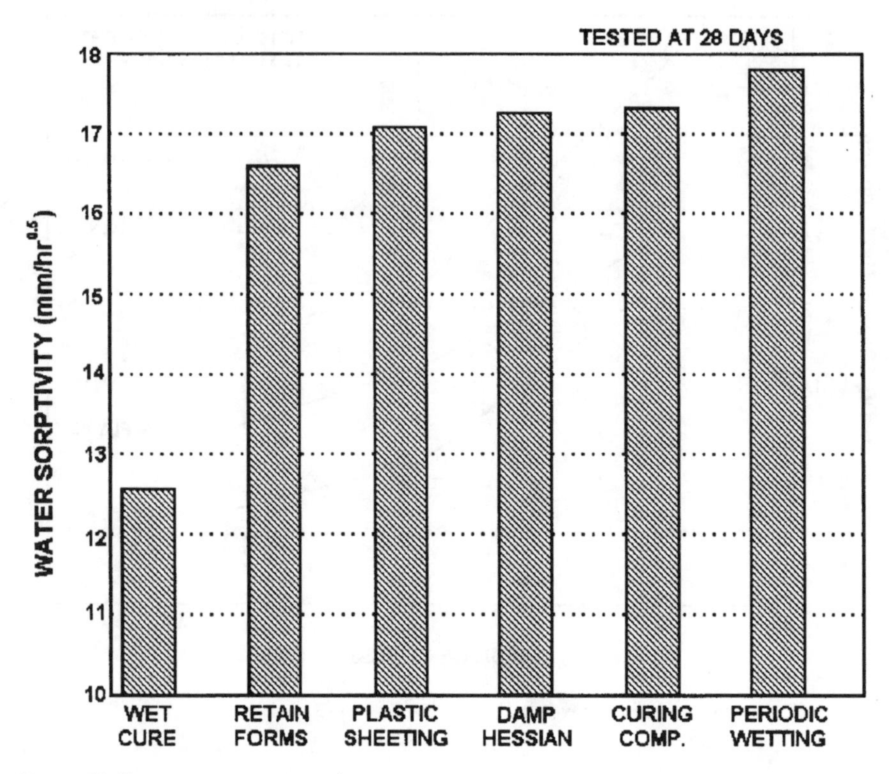

Figure 10. Sorptivity versus initial curing.

distinct transport mechanisms, their suitability depends on the property being considered. Permeability is best suited for assessing compaction since it is particularly sensitive to changes in the coarse pore fraction, sorptivity is sensitive to surface phenomena such as the effects of curing, while chloride conductivity provides good characterization of marine concretes or concretes exposed to de-icing salt conditions.

The sensitivity of sorptivity testing to curing is illustrated from a laboratory study where concrete panels were exposed to a variety of curing systems.[11] The concrete panels, made of grade 40 OPC concrete, were initially cured in a variety of ways for 7 days before being exposed to dry summer conditions until 28 days. Control concrete cured in water for 28 days had significantly lower sorptivity than any of the other curing systems, as shown in Fig. 10.

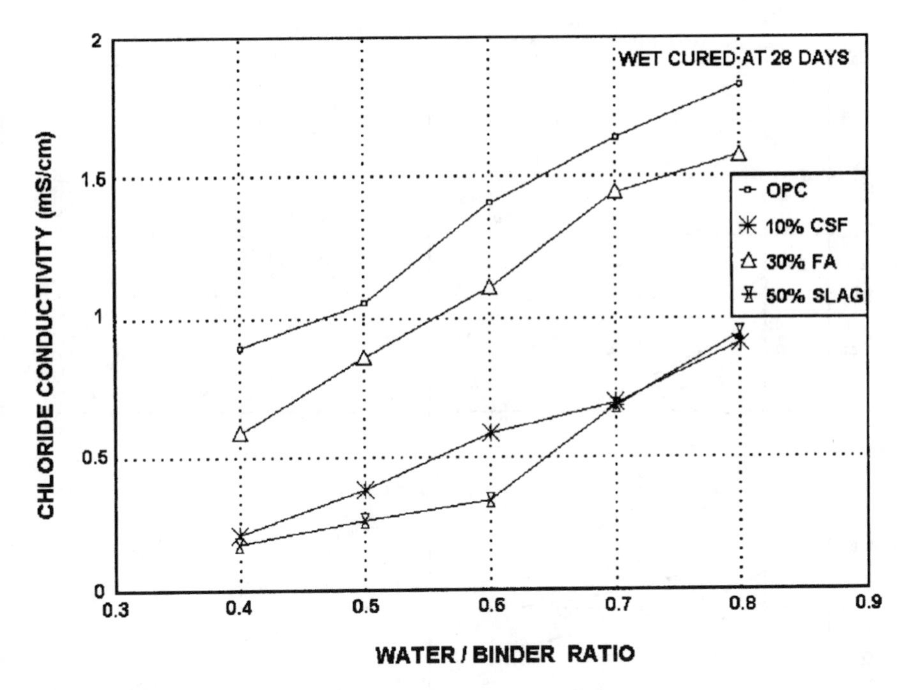

Figure 11. Chloride conductivity versus water/binder ratio.

Concrete Mix Optimization for Durability

Durability index testing may be used to optimize materials and construction processes where specific performance criteria are required. At the design stage the influence of a range of parameters such as materials and construction sytems may be evaluated in terms of their impact on the quality of the concrete, in particular the cover layer. A cost-effective solution for ensuring durability may in this way be assessed using a rational testing strategy.

When designing reinforced concrete structures for use in marine or de-icing salt environments, chloride resistance is critical for durability. The effect of various binder systems on the chloride resistance of concrete can be rapidly assessed using the chloride conductivity test. Figure 11 shows chloride conductivity results for a range of concrete types plotted against water-to-binder ratio.[12]

For mix design purposes, it is necessary to be able to choose appropriate proportions and materials to satisfy required index values. Figures 12–14

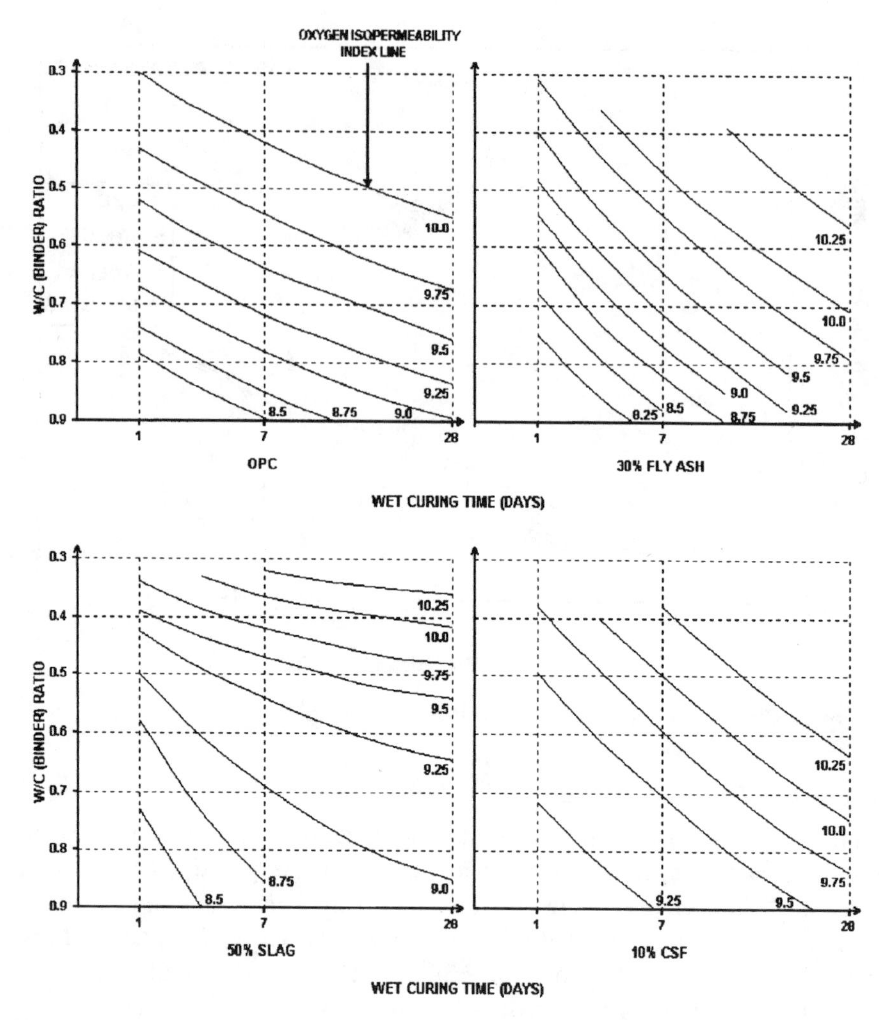

Figure 12. Isopermeability curves for concretes of different cement types.

provide guides for such selections for the three durability indexes and concretes based on OPC, or blends of OPC with FA (30%), GGBS (50%), or CSF (10%). These figures should help provide first estimates for design; it will usually be necessary to check by conducting laboratory or site trials with actual mixes.

Performance-Based Specifications

From controlled laboratory studies and site data, a matrix of durability index parameters is being developed that could be used to produce a set of

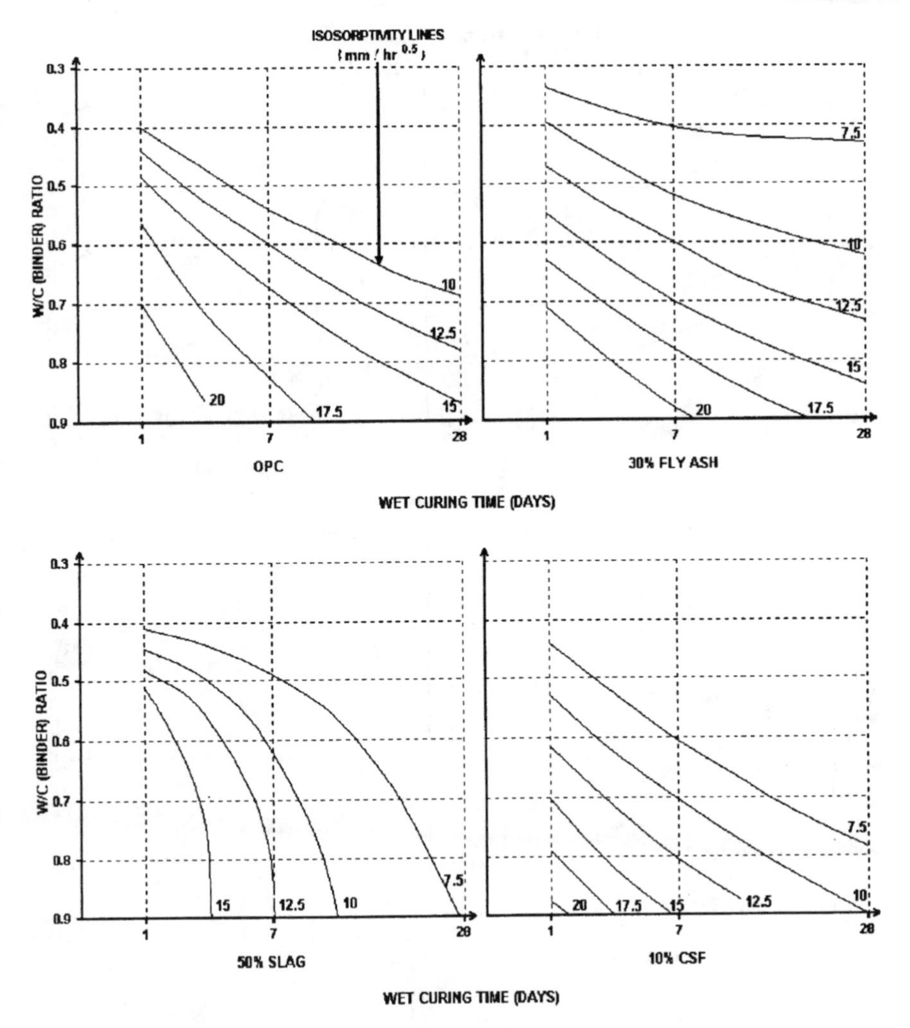

Figure 13. Isosorptivity curves for concretes of different cement types.

acceptance and rejection criteria for performance-based specifications. The approach needs to be refined and launched on an incremental basis but could have major benefits for all parties involved in construction. The current prescriptive approach to durability specifications is not only vague and sometimes inappropriate, but it is often inflexible. Performance-based specifications allow contractors more leeway in deciding how best to achieve the durability requirements while still having sufficient control to ensure satisfactory compliance. Suggested ranges for the durability classification

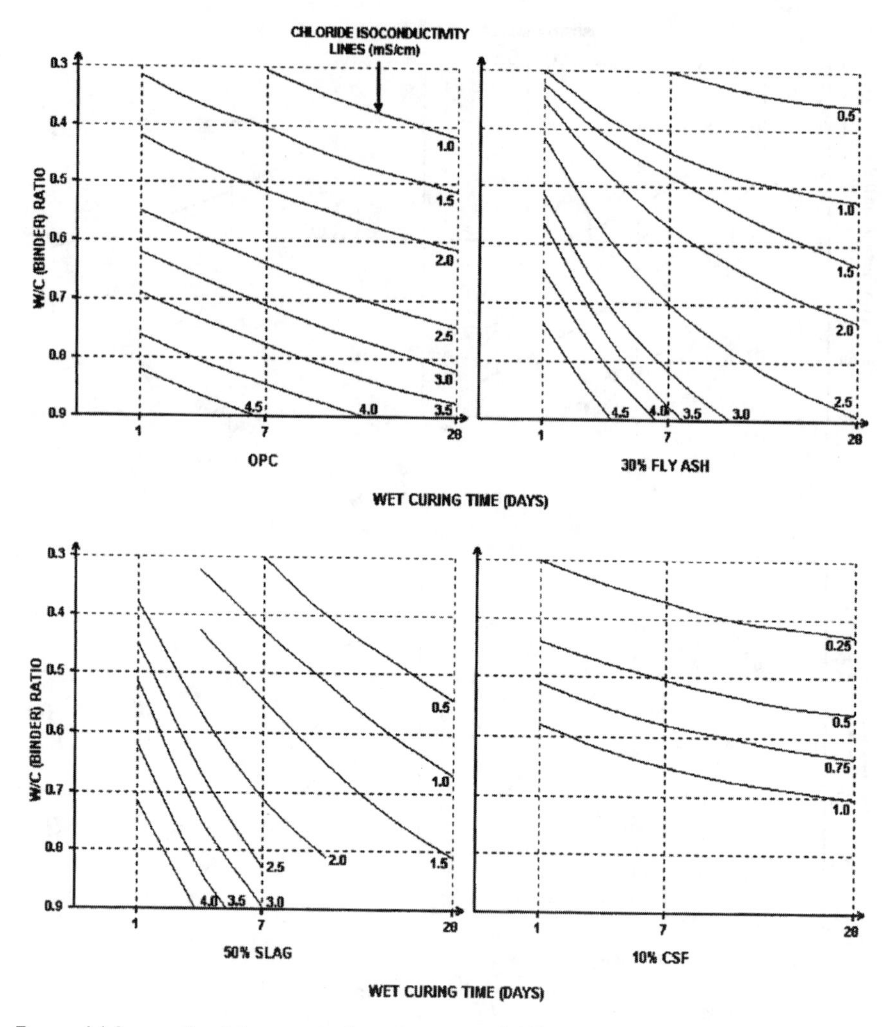

Figure 14. Isoconductivity curves for concretes of different cement types.

of concretes for the three index tests, based on site and laboratory data and in the context of quality of the cover layer, are shown in Table I.

Predictions of Long-Term Performance

Service life predictions of reinforced concrete structures are affected by a large number of variables that prevent precise estimates of durability performance. The durability prospects of concrete structures can be improved

Table I. Suggested ranges for durability classification (in terms of quality of the cover concrete) using index values

Durability class	OPI (log scale)	Sorptivity (mm/h$^{0.5}$)	Conductivity (mS/cm)
Excellent	>10	<6	<0.75
Good	9.5–10	6–10	0.75–1.50
Poor	9.0–9.5	10–15	1.50–2.50
Very poor	<9.0	>15	>2.50

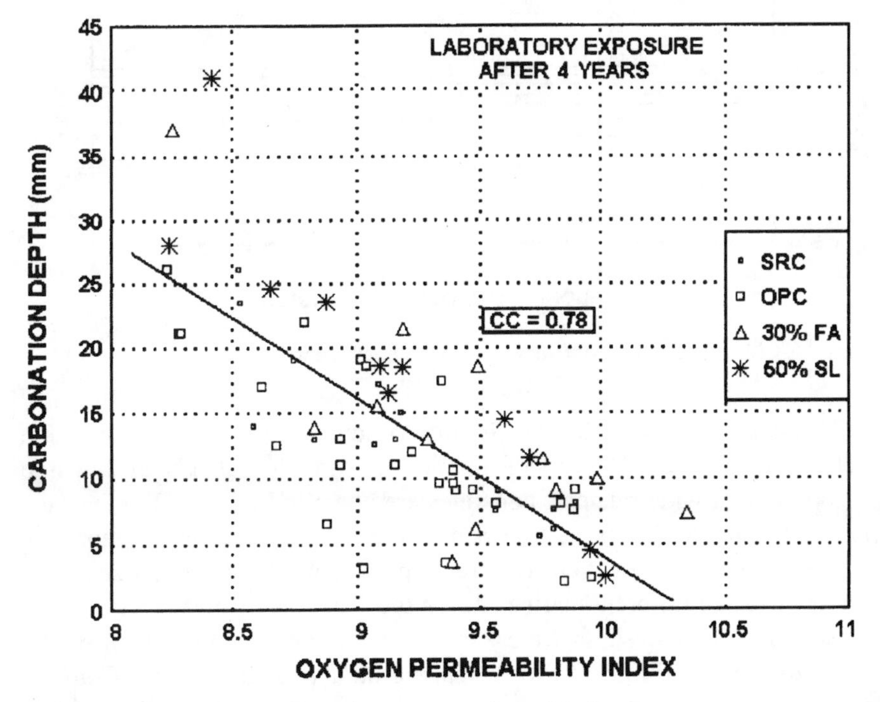

Figure 15. Oxygen permeability index versus carbonation depth.

by having a broad framework that allows for a system of rational designs, practical specifications, and a means of ensuring satisfactory compliance with specifications on site. Since durability index tests are based on transport mechanisms associated with deterioration, it is not surprising that these indexes are able to be used for durability predictions. Correlations between oxygen permeability and carbonation depth have been found to be good and are shown in Fig. 15.[12]

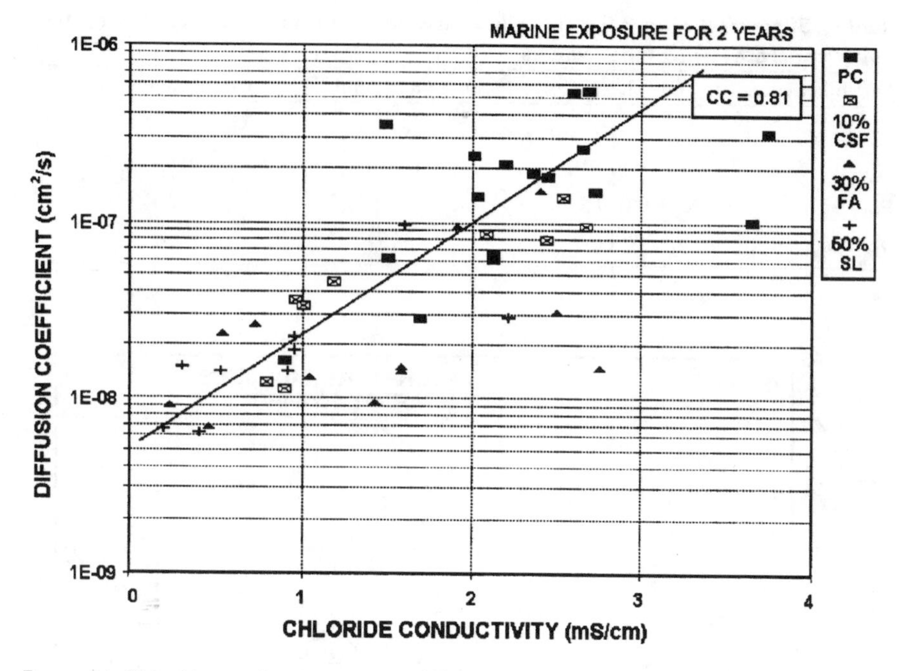

Figure 16. Chloride conductivity versus diffusion coefficient.

Correlations between 28-day chloride conductivity results and diffusion coefficients after several years of marine exposure have also been shown to be good over a wide range of concretes as shown in Fig. 16.[12] Some of the variability observed may be ascribed to the fact that the results represent three separate studies where concrete specimens of differing sizes were exposed to three different marine sites in the Cape Peninsula.

Durability index tests that measure early age concrete properties should not be used indiscriminately when making durability predictions. Comparisons between different concretes in particular may be misleading unless long-term effects are considered. Durability predictions should be made only once the relationship between early age properties and medium- or long-term durability performance has been established. Further details about durability predictions for marine concrete structures are given in Ref. 13.

Conclusion

A new approach is required to solve the problem of lack of durability in reinforced concrete structures, in the context of the need to provide adequate protection to the reinforcing steel. This approach must take a broad view, incorporating a proper definition of the environment, characterization of the material, and long-term observations of durability performance. By integrating these various aspects, performance specifications can ultimately be produced that require certain durability criteria to be met, while at the same time allowing contractors opportunity to provide innovative and cost-effective solutions.

The durability index approach discussed in this chapter is premised upon the need to provide quantifiable physical and engineering parameters to characterize concrete, particularly the cover layer, at relatively early ages. These parameters can be expressed as durability indexes, which must be sensitive to the important material, environmental, and construction factors that influence the quality of the cover layer. The purpose of such material indexing is to provide a reproducible engineering measure of microstructure, reflecting the transport properties of importance to concrete durability.

Three durability index tests have been proposed: an oxygen permeability index test, a water sorptivity test, and a chloride conductivity test. Each of these measures distinctly different transport properties of the cover layer of concrete. As such they span a range of transport processes that influence concrete durability, that is, gas permeation, water absorption, and chloride diffusion. The advantages of the tests are that they have a sound theoretical basis, are site and lab applicable, are quickly and easily performed, and have sufficiently low statistical variability. The properties of oxygen permeability, water sorptivity, and chloride conductivity are not measured for their own sake, but rather as material indexes, that is, reproducible engineering measures of the ability of the concrete surface layer to resist certain durability-related processes.

It has been shown that the index tests can be used for a number of engineering purposes:

- For quality control of site concrete.
- For concrete mix optimization for durability.
- As a basis for performance specifications.
- For predictions of long-term durability performance.

There is a need to exercise due care in interpreting the results of the index tests, particularly if the samples are retrieved from site concrete. All factors such as the type and nature of the concrete, the likely or actual environment, and possible deterioration must be taken into account. It must also be remembered that the purpose of the approach is to characterize potential durability in terms of relatively early age properties. It appears that this indexing method of improving concrete durability holds promise for the future.

Acknowledgments

This chapter is based on an internally published monograph covering work done by the authors in the Department of Civil Engineering, University of Cape Town, covering work done by the authors and postgraduate students in the Industry/NRF joint research program on concrete durability at the University of Cape Town and the University of the Witwatersrand. Particular acknowledgment is given to PPC Cement and Alpha Ltd, who sponsored specific research projects from which much of the data has been generated.[14,15] Reference 16 was also used extensively in compiling the monograph, and the Concrete Society of Southern Africa is thanked for permission to use this paper.

References

1. J.P. Broomfield, *Corrosion of Steel in Concrete.* Chapman and Hall, 1997.
2. A.M. Neville, "Why we have concrete durability problems"; pp. 21–48 in *Katherine and Bryant Mather International Conference on Concrete Durability,* ACI SP-100. ACI, Detroit, 1987.
3. M.G. Alexander and Y. Ballim, "Experiences with durability testing of concrete: A suggested framework incorporating index parameters and results from accelerated durability tests"; pp. 248–263 in *Proceedings of the 3rd Canadian Symposium on Cement and Concrete,* Ottawa, August 1993. Nat. Res. Council, Ottawa, Canada, 1993.
4. M.G. Alexander, P.E. Streicher, and J.R. Mackechnie, *Rapid Chloride Conductivity Testing of Concrete,* Research Monograph No. 3. University of Cape Town, 1999.
5. M.G. Alexander, Y. Ballim, and J.R. Mackechnie, *Concrete Durability Index Testing Manual,* Research Monograph No. 4. University of Cape Town, 1999.
6. Y. Ballim, "A low cost falling head permeameter for measuring concrete gas permeability," *Concr. Beton,* **61**, 13–18 (1991).
7. J.R. Mackechnie, "Predictions of reinforced concrete durability in the marine environment," Ph.D. Thesis, University of Cape Town, 1996.
8. S. Kelham, "A water absorption test for concrete," *Mag. Concr. Res.,* **40** [143] 106–110 (1988).
9. Y. Ballim, "Curing and the durability of OPC, fly ash and blast-furnace slag concretes," *Mater. Struct.,* **26** [158] 238–244 (1993).
10. P.E. Streicher and M.G. Alexander, "A chloride conduction test for concrete," *Cem. Concr. Res.,* **25** [6] 1284–1294 (1995).
11. A. Krook and M.G. Alexander, "An investigation of concrete curing practice in the Cape Town area," *SAICE J.,* **38** [1] 1–4 (1996).

12. J.R. Mackechnie, unpublished laboratory results, University of Cape Town, 1998.
13. J.R. Mackechnie, *Predictions of Reinforced Concrete Durability in the Marine Environment,* Research Monograph. University of Cape Town, 1997.
14. P.E. Streicher, "PPC/UCT Research Project: Durability of Marine Cements," Final Project Report. Dept. of Civil Engineering, University of Cape Town, August 1996.
15. B. Magee, "Performance of Silica Fume Concrete," Research Report 1/98. Concrete Materials Research Group, Department of Civil Engineering, University of Cape Town, June 1998.
16. M.G. Alexander, "An indexing approach to achieving durability in concrete structures"; pp. 571–576 in *FIP '97 Symposium: The Concrete Way to Development,* Johannesburg, March 1997. Concrete Society of Southern Africa, 1997.

Appendix A: Permeability Theory

The D'Arcy equation for permeation can be expressed as:

$$\frac{\partial m}{\partial t} = \frac{-k}{g}\left(\frac{\partial P}{\partial z}\right) \tag{A1}$$

where $\partial m/\partial t$ is the rate of mass flow per unit cross-sectional area, $\partial P/\partial z$ is the pressure gradient in the direction of flow, k is the coefficient of permeability, and g is the acceleration due to gravity.

For a gas, mass is related to volume V and pressure P by the equation:

$$m = \frac{\omega VP}{R\theta} \tag{A2}$$

where ω is the molecular mass of permeating gas, θ is absolute temperature, and R is the universal gas constant.

The change in total mass of stored gas in time ∂t is:

$$\frac{\partial m}{\partial t} = \frac{\omega V}{R\theta}\left(\frac{\partial P}{\partial t}\right)$$

For a test specimen of cross-sectional area A, measured normal to the direction of flow, and thickness d; Eq. A1 can be rewritten as:

$$\frac{\partial m}{\partial t} = \frac{k}{g}\left(\frac{PA}{d}\right)$$

Hence,

$$\frac{\omega V}{R\theta}\left(\frac{\partial P}{\partial t}\right)=\frac{-k}{g}\left(\frac{PA}{d}\right)$$

Rearranging,

$$\frac{-\omega Vgd}{R\theta kA}\left(\frac{\partial P}{P}\right)=\partial t$$

Integrating,

$$\frac{-\omega Vgd}{R\theta kA}\ln(P)=t+\text{constant} \tag{A3}$$

At $t = 0$, $P = P_0$, therefore

$$\text{constant}=\frac{\omega Vgd}{R\theta kA}\ln(P_0)$$

Substituting into Eq. A3,

$$t=\frac{\omega Vgd}{R\theta kA}\ln\frac{P_0}{P}$$

Rearranging,

$$k=\frac{\omega Vgd}{RA\theta t}\ln\frac{P_0}{P} \tag{A4}$$

Appendix B: Absorption Theory

Using the one-dimensional case of water absorption with defined boundary conditions, it is possible to define sorptivity in terms of the extended D'Arcy equation such that

$$i=St^{1/2} \tag{B1}$$

where i is the cumulative water absorption per unit area, S is sorptivity, and t is time.

Using Eq. B1, the change of mass with time after water infiltration can be described by the equation:

$$\Delta M_t = An\rho_w St^{1/2} \tag{B2}$$

where ΔM_t is the change of mass with respect to the dry mass, A is the cross-sectional area, n is the effective porosity of the concrete, and ρ_w is the density of water.

The effective porosity is determined as follows:

$$n = \frac{M_{sat} - M_0}{Ad\rho_w} \tag{B3}$$

where M_{sat} is the saturated mass of concrete, M_0 is the dry mass of concrete, and d is the sample thickness.

Substituting into Eq. B2,

$$\Delta M_t = \frac{M_{sat} - M_0}{d} St^{1/2}$$

Thus

$$S = \frac{\Delta M_t}{t^{1/2}} \left(\frac{d}{M_{sat} - M_0} \right) \tag{B4}$$

where $\Delta M_t / t^{0.5}$ is the slope of the straight line produced when the mass of water absorbed is plotted against the square root of time.

Appendix C: Diffusion Theory

The diffusion process is described by Fick's first law of diffusion for steady state conditions, such that

$$J_x = -D_f \frac{\partial C}{\partial x} \tag{C1}$$

where J_x is the flux per unit cross-sectional area of the liquid, D_f is the diffusion coefficient, and C is the concentration of the liquid.

When the liquid is constrained in a pore structure with longer diffusion paths than those of the free liquid phase, a diffusion coefficient D_p may be defined with respect to the pore liquid of the material such that

$$J_x = -D_p \frac{\partial C}{\partial x} \quad \text{(C2)}$$

The configuration of the pore structure with longer diffusion paths and localized constrictions produces lower diffusion coefficients, D_p compared to D_f for the free liquid phase. This is due to the constrictivity δ and tortuosity τ of the pore system, which effectively restricts diffusion such that

$$D_p = D_f \frac{\delta}{\tau^2} \quad \text{(C3)}$$

The flux may be expressed in terms of the medium rather than the liquid to produce the following equations, where $< >$ refers to the medium generally and not the pore liquid.

$$< J_x > = -D_i \frac{\partial C}{\partial x}$$

and

$$D_i = D_f \frac{\varepsilon\delta}{\tau^2}$$

where ε is the volume fraction of porosity and D_i is the intrinsic diffusion coefficient.

The quantity $\varepsilon\delta/\tau^2$ is a material property that characterizes the pore structure and is referred to as the diffusibility Q of the material.

Diffusibility is an intrinsic material property defined by the concrete pore structure while diffusivity D defines the material resistance to diffusion being dependent on material properties and the concentration and mobility of diffusing ions.

$$Q = \frac{D}{D_0} = \frac{\sigma}{\sigma_0} \quad \text{(C4)}$$

where D is the diffusivity of the ions through the material, D_0 is the diffusivity of the ions in the pore solution, σ is the conductivity of the material, and σ_0 is the conductivity of the pore solution.

The diffusion of chloride ions through concrete under non-steady-state conditions may be described by Fick's second law of diffusion as follows:

$$\frac{\partial C}{\partial t} = D_c \frac{\partial^2 C}{\partial x^2} \qquad \text{(C5)}$$

where D_c is the apparent diffusion coefficient and x is the depth into the material.

The partial differential equation C5 may be solved by applying the following boundary conditions and assumptions:

- The surface concentration C_s reaches a constant value almost immediately.
- The internal concentration C_x is zero at some point internally.
- The diffusivity of the material is constant with depth and time.
- The concrete is saturated throughout the diffusion process.
- There is no significant interaction between diffusant and concrete.

The solution may be written as follows

$$C_x = C_s \left[1 - erf\left(\frac{x}{2\sqrt{D_c t}} \right) \right] \qquad \text{(C6)}$$

where *erf* is the mathematical error function.

Despite the lack of compliance with the above conditions, the error function solution of Fick's Law is widely used. The equation has been found to be useful for characterizing chloride ingress into concrete, but may be misleading when used to predict long-term chloride concentrations in concrete.

Microfiller Partial Substitution for Cement

Moncef Nehdi
University of Western Ontario, London, Ontario, Canada

Sidney Mindess
University of British Columbia, Vancouver, British Columbia, Canada

"All truth goes through three stages: First it is ridiculed, then it is violently opposed, finally it is accepted as self-evident."
— Schöpenhauer

The world's cement production is expected to increase from 1.4 billion tons in 1995 to about 2 billion tons in 2010. This implies that by the year 2010, 2 billion tons of CO_2 per year might be released into the atmosphere due to cement production. The World Earth Summits of 1992 and 1997 in Rio de Janeiro and Kyoto, respectively, strongly urged industries to reduce their CO_2 emissions. At the same time, on a planetary scale, the generation of various powdered industrial wastes continues daily. It is believed that when a real science based on a knowledge of the particulate nature of these wastes and their effects on hydration reactions, strength development, and durability of concrete is developed, a number of these wastes could be partially substituted for cement. This could lead to the manufacture of microfiller-blended cements, providing economic benefits and environmental relief. It is therefore of vital importance to develop a scientific understanding of the microfiller effect on the rheology, microstructure, strength, and durability of cement-based materials. This chapter is a contribution toward this understanding, with special focus on limestone microfillers.

Environmental Considerations

The annual global production of concrete is about 5 billion tons. Mankind consumes only water in larger quantities. In the United States, the use of concrete averages 2 metric tons per person per year. The world's cement production is expected to increase from 1.4 billion tons in 1995 to almost 2 billion tons by the year 2010. The production of one ton of cement releases about one ton of CO_2 into the atmosphere. The level of CO_2 in the atmosphere has increased significantly during the last two centuries due to the industrial revolution. Carbon dioxide belongs to the so-called greenhouse gases, which allow high-frequency heat waves to penetrate the atmosphere and reach the earth, but do not allow the same type of radiation to escape from the earth's surface back into space. This contributes to what is now commonly known as global warming.

The effect of greenhouse gases in the atmosphere on the global temperature is nonlinear; small increases can cause exaggerated temperature rises because the higher temperatures cause more water to evaporate. The World Earth Summits in Rio de Janeiro in 1992 and Kyoto in 1997 strongly urged industries to decrease their CO_2 emissions. Developed countries are considering imposing mandatory quotas on greenhouse-gas emissions. Pressure from public opinion and legislatures in this regard is expected to rise in the near future.

The burning of cement clinker at about 1400°C is also costly in terms of fossil fuels and coal, and consumes irreplaceable raw materials. After aluminum and steel, cement is the most energy-intensive construction material. In Japan, it is estimated that limestone sources could be exhausted in 50 years. In view of a potential energy crisis in this century, the increased demand for cement from a steadily growing world population, and the threat of global warming, cement production and use must become more rational. In particular, the extra 100–200 kg of cement per cubic meter of high-performance concrete (HPC) compared to conventional concrete must be seriously reconsidered. The very low water content of HPC results in only a partial hydration of the cement grains. Thus, a significant proportion of the cement acts merely as microfiller and its potential hydraulic activity is not fully utilized.

In view of the above concerns, the concrete industry has resorted to the use of several supplementary cementing materials as partial replacements for cement. Most of these are recycled by-products or unprocessed materials that are not only ecological and economic, but also impart to the concrete superior workability, strength, and durability. However, the mechanisms underlying these benefits are still a matter of considerable controversy. Whether the positive effects of mineral additions are due primarily to their pozzolanic activity or to a physical filler effect is still in dispute.

Limitations of Supplementary Cementing Materials

Table I lists the amounts of the major siliceous mineral admixtures produced as byproducts in various countries, and the level of utilization of these resources. Clearly, there is still a great potential for the implementation of these materials in the concrete industry. In recent years, new research has greatly expanded the use of mineral admixtures in concrete. For instance, up to 40% low-calcium fly ash can now be incorporated into superplasticized concrete to produce high-strength concrete.[1] Also, ultra-low-heat cements made of proportions of slag, fly ash, and portland cement

Table I. Annual production and utilization rates of siliceous by-products (t), 1988*,[3]

	Fly ash × 10^6		Blast furnace slag × 10^6		Condensed silica fume × 10^3	
Country	Production	Utilization	Production	Utilization	Production	Utilization
Australia	3.5	0.25	4.7	0.12	60	20
Canada	3.3	0.8	2.9	0.2	23	11
China	35	7.2	22	16	0	0
Denmark	1	0.45	0	0	0	0
France	5.1	1.5	10.4	1.9	60	0
Federal Republic of Germany	2.6	2.0	15	2.8	25	0
India	19	0.5	7.8	2.8	0	0
Japan	3.7	0.5	24	8.2	25	0
Netherlands	0.5	0.3	1.1	1	0	0
Norway	0	0	0.1	0	140	40
South Africa	12.9	0.1	1.5	0.6	43	0
Sweden	0.1	0.02	0.1	0.03	10	1
United Kingdom	13.8	1.3	1.5	0.25	0	0
United States	47	5	13	1	100	2

*These figures have certainly changed since 1988. In particular, the use of silica fume has increased and a shortage of this material in the future is foreseen.

have been used in large-scale bridge construction in Japan.[2] Silica fume has became a familiar ingredient of HPC in leading North American cities where ready-mix HPC is commercialized.

Nonetheless, the use of supplementary cementing materials is associated with a few problems. The properties of these minerals are variable, and often little care is given to their control during production since they are only by-products. The quality control of concrete and its sensitivity to curing become more acute when mineral admixtures are used. The dosage of air-entraining admixtures to achieve a certain air content increases dramatically in silica fume, fly ash, and rice husk ash, and to a lesser extent in GBFS concretes. Furthermore, the early age development of strength is slower in slag/fly ash concrete. This can hinder their use in jobs that are restrained to a tight schedule for formwork removal, and can be particularly troublesome in winter concreting. In addition, silica fume and rice husk ash concretes exhibit an increased tendency toward plastic shrinkage.

In many countries, no standards have yet been established to regulate the use of mineral admixtures in concrete. Data regarding such materials are often not available, and there has not been sufficient effort to transfer the technology of using mineral admixtures to developing countries. Many cement plants still do not include provisions for the production of blended cements containing mineral admixtures. The cost of transportation associated with the use of such materials dissuades some cement plants from pursuing this venture. Also, mineral admixtures are not available in all places. Some by-products such as silica fume have become even more expensive than cement, and a future shortage of this product can be expected.

Potential of Inert Fillers and Triple-Blended Cements

Conventionally, mineral admixtures have been used to reduce the cost of concrete. However, because of new developments in chemical admixture technology, mineral additions are increasingly being used to improve the performance of concrete. In this respect, the strength is generally not the most critical issue, since the current strength values obtainable with HPC are beyond those that structural engineers require for the design of new structures. In many applications, large amounts of mineral admixtures are desired only as a fine powder (filler) rather than as a cementing material. With the increasing cost of labor, concrete is shifting toward a self-leveling material that consolidates under its own weight. Some heavily reinforced structures require high-fluidity concretes that can be placed easily without vibration. Unfortunately, material segregation and bleeding are characteristic of this kind of concrete. To cope with this situation, the water content should be reduced through the use of superplasticizers, and the fines in the concrete mixture should be increased. For this purpose slag and/or fly ash can be used, with the additional benefits of lower heat of hydration and higher long-term strength. Yet, these materials are sometimes not available, a high transportation cost is associated with their use in concrete, and their effect on early strength development can be detrimental. In these cases, inert fillers, such as limestone, which is available in all cement plants,* can provide the rheological benefits and enhance the early age strength. Furthermore, if the mechanisms of action of these microfillers are well understood, they can even improve the long-term strength of HPC.

Other opportunities for the concrete industry lie in the field of triple- and multi-blended cements. Often, single-mineral admixtures have drawbacks. However, when combined with other mineral admixtures, a countermeasure for the unwanted behavior can be obtained. Some simple examples can illustrate this concept. Suppose a slow-hydrating pozzolan must be used in a HPC mixture, yet removal of formwork is an issue and early strength is desired. The addition of a nonpozzolanic filler such as limestone powder, which can enhance the early strength, could be a solution. By the same token, suppose that a large proportion of an inert filler must be substituted for cement to reduce the heat of hydration, yet the strength should not be compromised. In this case, incorporating a very efficient supplementary cementing material such as silica fume[†] can solve the problem. An excellent practical example for this is the use of belite cement (in which the C_2S content is higher than 50%) in conjunction with slag and fly ash to produce an ultra-low-heat cement with an acceptable initial strength. This material has been used in Japan in submerged bridge abutments. The previous option for this application was to use normal cement with fly ash and slag, which also provides low heat, though neutralization of the material occurred quickly and durability problems could be expected.

Historical Case: Limestone Filler

In the wake of the oil shortage of 1974, the U.S. Congress passed a law mandating cement producers to reduce their energy consumption. By 1985, 92% of the clinker produced in the United States was burned using coal or petroleum coke, as compared to only 36% in 1972. The energy crisis also stimulated a worldwide interest in adding fillers to portland cement as a means of saving energy. Since limestone is available at all cement plants and thus has no additional transportation costs associated with its possible use as a filler in cement, this subject has gained increased interest. The question of whether limestone additions should be permissible has stimulated a great deal of debate and research.

Various national standards have adopted different positions. The French cement standards have allowed up to 35% filler additions to portland cement since 1979. The filler cements (known as CPJ) have developed con-

*In fact, not all limestone powder can be used for this purpose. Requirements regarding fineness, calcite content, presence of clay or organic matter in the limestone filler, and so on should be satisfied.

[†]Silica fume has an efficiency factor of about 3–4 when used in small percentages as a replacement for cement; that is, 1 kg of silica fume can replace 3–4 kg of OPC and still result in a similar strength of concrete.

siderably since then, and represent more than 60% of the total cement production in France.[4] The European pre-standard ENV 197 permits portland cement to contain 5% of additional constituents such as limestone, and defines a portland-filler cement that may contain up to 20% of high-purity limestone. It was suggested that this upper limit be increased to 30%[5] before the standard was actually adopted in 1992.

The Canadian Standards Association has resolved the issue of limestone additions to cement in CSA-A5-M83, where it recognizes the existence of an optimum carbonate addition to cement and allows a maximum of 5% limestone of specified quality to be added to Type 10 (normal) portland cement. This was later extended to Type 30 (high–early strength) cement. Since not all cements can contain carbonate additions, the option of limestone addition is not used by all cement plants. In the United States, however, a 1985 proposal before the ASTM C-1 committee to allow up to 5% limestone to be ground with the clinker has resulted in considerable controversy. The proponents suggest that the cost and energy effectiveness can be improved without a degradation in the overall quality of cement, and even report some improvements in cement and concrete properties. The opponents claim that ground limestone acts merely as an adulterant and that its addition to cement should be rejected on ethical grounds.[6,7]

Since then, some new issues have arisen that have not yet received proper analysis. For instance, the chemical effects on the hydration reactions occurring upon the addition of limestone to cement and the subsequent effects on the rheology of fresh concrete have been extensively studied. However, the same effects in a low w/c ratio concrete in the presence of a superplasticizer are still not clearly understood. In addition, the use of carbonate additions to produce cements of different strength grades is well documented. Yet, the possible use of ultrafine limestone to produce fluid HPC, increase the early age strength, or/and alter the porosity and the interfacial properties in high-strength mixtures has not been fully investigated.

Limestone Substitution for Gypsum as a Set Regulator

Among the early attempts to use limestone in cement was limestone substitution for gypsum as a set regulator. Some cement plants experience seasonal problems related to the dehydration of gypsum due to high grinding mill temperatures, with subsequent false set tendencies. Since limestone is often mined at costs around 10% the cost of mining gypsum,[8] the financial impact of such substitution is quite clear, especially in areas of the world where natural gypsum deposits are rare, which results in high transportation costs as well.

Regulations regarding the addition of gypsum and the permissible SO_3 content in cement go as far back as the 1890s. Yet, as a result of the energy crisis, the cement industry resorted to fuels having higher sulfur contents, which brought about new concerns regarding this issue. The gypsum additions should be kept to a minimum to account for the additional clinker SO_3 content and to keep the SO_3 level within specification limits. Many attempts have been made to find an adequate replacement for gypsum,[9] among which limestone has attracted the most interest.

Bobrowski et al.[8] found that with the correct limestone/gypsum ratio for a given clinker, the autoclave expansion and the setting time could be maintained at acceptable values, the concrete strengths were favorable except at early ages, and the severe false setting systems were altered dramatically. Later work by Negro et al.[9] and by Bensted[10,11] confirmed that a partial substitution of limestone for gypsum of up to 50% was possible without deleterious effects to the cement performance (Fig. 1). However, it was also noted that each clinker-limestone-gypsum system should be individually investigated to ascertain the optimum replacement rate. It was also observed[12] that the optimum SO_3 decreases exponentially with an increase in the limestone content. In this regard, it would be of interest to investigate the set regulation effect of ultrafine limestone in modern low–w/c ratio superplasticized concrete. In such a system, the rapid loss of workability due to the delicate equilibrium between the reactivity of C_3A and the solubility of the sulfates is a major issue.

Reactivity and Effect on Hydration Reactions

Limestone has traditionally been considered as an inert additive in a hydrating cement paste. However, although there is agreement that limestone is not pozzolanic, several studies have shown that it has significant reactivity. As early as 1938, Bessey[13] suggested the existence of calcium carboaluminates, which result from the reaction of carbonates and calcium aluminate according to the chemical reaction:

$$C_3A + CaCO_3 + 11\,H_2O \Rightarrow C_3A \cdot CaCO_3 \cdot 11\,H_2O \qquad (1)$$

Manabe et al.[14] observed that calcium monocarboaluminate hydrate ($C_3A \cdot CaCO_3 \cdot 11\,H_2O$) was formed in an equimolar mixture of C_3A and $CaCO_3$ and water. They also observed that in the presence of sulfate and

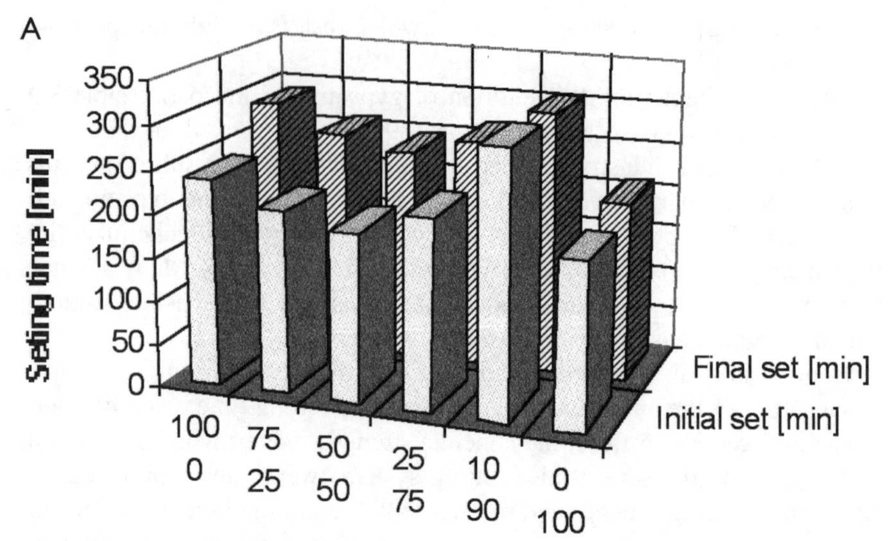

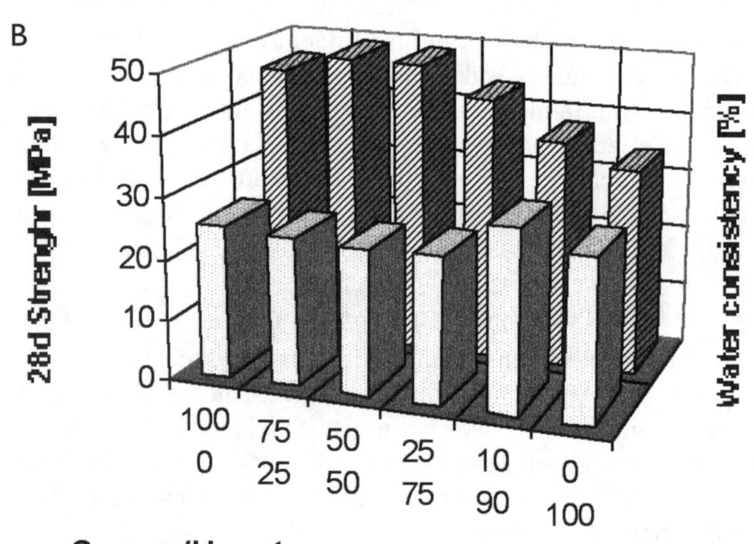

Figure 1. Effect of limestone substitution for gypsum on (a) initial and final set, and (b) water consistency and 28-day compressive strength.[11]

carbonate salts, both trisulfoaluminate and monocarboaluminate were formed. Feldman et al.[15] studied the influence of $CaCO_3$ on the hydration of C_3A. Mixtures of different proportions were observed using differential thermal analysis, X-ray diffraction, and infrared absorption. They concluded that the hydration of C_3A was suppressed by $CaCO_3$ additions due to the formation of calcium carboaluminate on the C_3A grains. This effect was also observed by Ramachandran and Zhang,[16] who noticed further that ettringite formation and its conversion to monosulfoaluminate were accelerated in the presence of $CaCO_3$. Conversely, Vernet[17] has suggested that the early formation of ettringite is similar to the condition in which no limestone is added. But when the gypsum is exhausted, the limestone reaction dominates, and after C_3A is consumed the stable compounds are ettringite and monocarboaluminate.

Ramachandran and Zhang[18] suggested that in the presence of $CaCO_3$, the hydration of C_3S was accelerated as the particle size became finer and the amount of $CaCO_3$ was increased. There was some evidence of partial incorporation of $CaCO_3$ into the C-S-H phase, a slight modification in the C/S ratio of C-S-H and an increase in the microhardness and the density of the hardened cement paste (HCP). Husson et al.[19] and Barker and Cory[20] reported similar trends where limestone addition to portland cement enhanced the hydration of C_3S and the formation of calcium hydroxide, probably because it offered nucleation sites for its growth. On the other hand, Ushiyama et al.[21] suggested that the addition of small amounts of carbonates retards the early hydration of alite, while the addition of large amounts accelerates its hydration. More recently, work by Evrard et al.[22] has highlighted the reactivity of calcite in an alkaline environment and has confirmed the hypothesis of Ramachandran and Zhang[18] that some reaction occurs between calcite and C-S-H. Several other investigations dealing with the reactivity of limestone and the formation of carboaluminates have also been reported.[23–26]

Other evidence of the reactivity of limestone comes from the use of limestone aggregates in concrete. Farran[27] studied the effect of the mineralogical nature of the aggregate on its bond with cement paste. He concluded that calcareous aggregates formed more intimate bonds compared to other minerals, and that calcite cannot be considered as truly inert in concrete. Buck and Dolch[28] examined the bond of different aggregates to cement paste in concrete; limestone aggregate offered the best bonding strength. It was observed more recently by Aïtcin and Mehta[29] that high-strength concrete made with limestone aggregate showed higher strength and elastic modulus than gravel concrete, due to interfacial reactions. Grandet and Ollivier[30] explained the lower calcite orientation at the cement paste/carbonate aggregate interface by the reactivity of limestone and the presence of monocarboaluminates in the interfacial zone. On the other hand, Monteiro and Mehta[31] explained the same reactivity between carbonate rock and cement paste by a possible reaction between calcite and calcium hydroxide, which resulted in the formation of a basic calcium carbonate hydrate.

Although the data on the reactivity of limestone are extensive, it is not clear how important this reactivity is, whether it occurs under field conditions, and how it varies in different concrete mixtures and in the presence of different admixtures. An important aspect of the effect of the reactivity of limestone filler in hydrated cement paste is its effect on early age properties. This aspect will be discussed in more detail later in this chapter.

Microfiller Effect on the Rheology of Cement Paste and Concrete

Generally, fine fillers complement the deficiency in fine particles of the particle size distribution of cement ($CaCO_3$ having mean particle sizes of 5 μm[4,32] and 3 μm[33] have been used), which can enhance the stability of fresh concrete and mortar. Ultrafine grains can fill in between the relatively coarser cement grains. Beyond the lubricating role they can play, the filler particles can reduce the room available for water, decreasing the water demand and obstructing the capillary pores at later ages.[4,32,34]

Bombled[32] noted that when the effect of limestone filler on the rheology of fresh concrete is significant, it is usually in the sense of a higher cohesion, better plasticity, and in some instances a pseudo-acceleration of the setting. Neto and Campitelli[35] used a two-point test to characterize the rheology of limestone filler cements. They observed that there was a reduction in the yield stress as the limestone addition increased, and also an increase of the plastic viscosity beyond a certain fineness/limestone relationship (Fig. 2). Brookbanks[36] presented the results of a comprehensive study on the effect of limestone additions to cement in the range of 5–28% on the properties of fresh concrete. He observed that the setting times were marginally reduced as the limestone addition increased. For the materials used, there was no discernible difference in the water demand at 5% replacement,

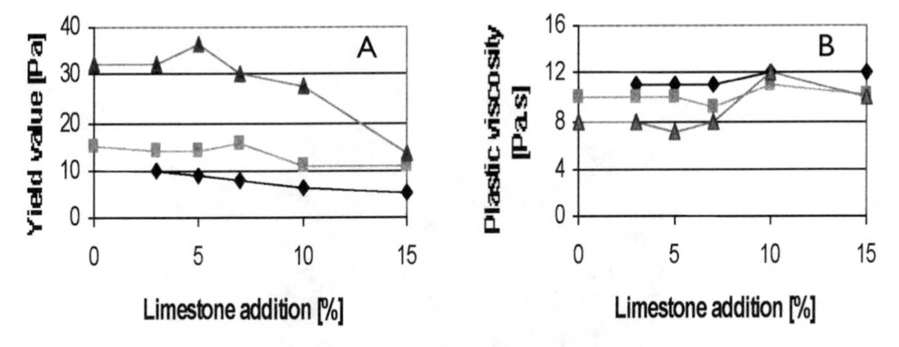

Figure 2. Effect of limestone addition on (a) yield value and (b) plastic viscosity (after Neto and Campitelli[35]). Surface area of cement: C1: 300 m^2/kg; C2: 400 m^2/kg; C3: 500 m^2/kg.

while a marginal reduction was observed at higher filler levels. The limestone fillers greatly reduced the bleed water and seemed to have no effect on the air-entraining properties.

The mechanisms underlying the effect of limestone filler on the rheology of cement paste and concrete seem to be controlled by the filler particle size and its flocculation capacity. In most of the above studies, limestone was interground with clinker, which increases the overall fineness due to the lower grinding energy of limestone. Whether the use of blended ultrafine limestone when deflocculated with a superplasticizer can play a different role, similar to the filler effect of silica fume on the rheological properties of cement paste, has not yet been fully investigated. Being a grinding mill product with a nonspherical shape and unsaturated surface electrical charges may limit the effect of ultrafine limestone on the rheological properties. Yet, it was noticed that the lubricating effect of limestone filler was dominant for the lower w/c ratio concretes.[4,37]

Study on Cement Pastes and Mortars

Nehdi [38] conducted statistical modeling of the microfiller effect on the rheology of cement pastes and mortars using 0–25% limestone replacement in silica fume and non–silica fume systems. The w/b ratio was systematically varied between 0.3 and 0.4. Ordinary ASTM Type I cement (OPC), high-purity limestone microfiller (LF) having a 3-µm mean particle size, and silica fume (SF) were used. A naphthalene-sulfonate superplasticizer and tap water were employed for the mixing. The different mortars were made with standard Ottawa sand (ASTM C778-91).

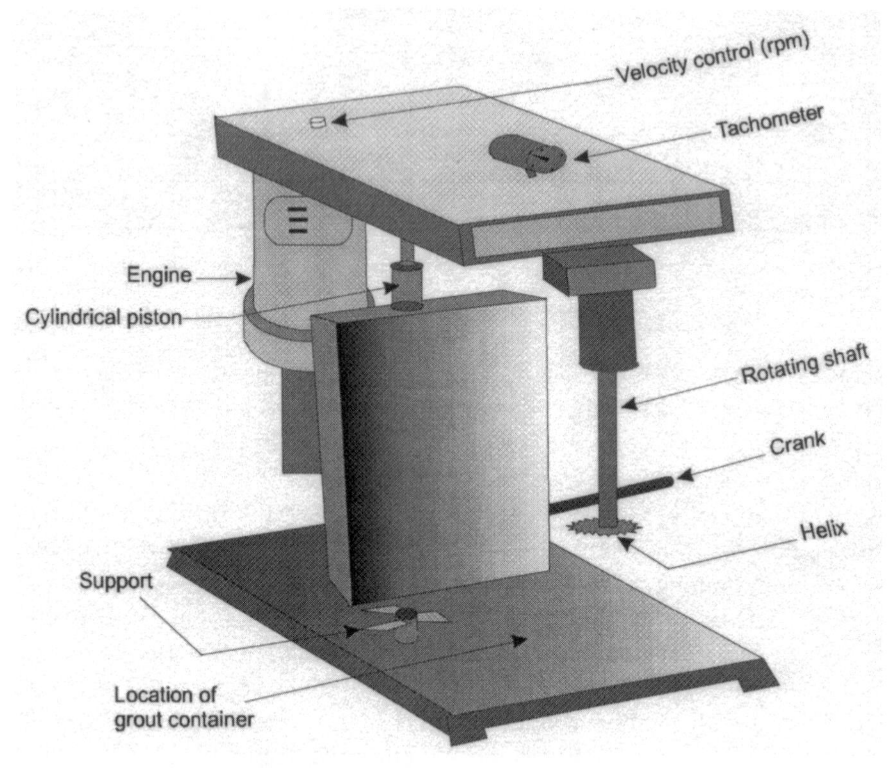

Figure 3. Schematic of the helical mixer.

The mixing procedure was identical for all the cement pastes. First, the superplasticizer was added to the mixing water, which was at a constant temperature of 17 ± 1 °C. Then the binder was added over a period of 1 min while mixing the grout at a constant speed of 3500 rpm for 3 min using a helical mixer (Fig. 3). A rest period of 1 min followed, during which the inner sides of the mixer pan were scraped down. Finally, the grout was mixed at the same velocity for an additional 1 min, and its temperature was measured at the end of the mixing.

The flow time was measured just after the mixing using a modified Marsh cone with a 5-mm outlet. A volume of 1.1 L of grout was placed in the cone while locking the outlet with the index. A chronometer was switched on while removing the index and the time was measured for each 100-mL flow of grout up to 1 L.

For each mix, the percentage of superplasticizer was systematically varied to obtain the saturation dosage; that is, the superplasticizer dosage beyond which no improvements in the fluidity of the grout were observed. For the superplasticizer dosage that achieved a flow time of 110 ± 10 s at 5 min, the grout was further characterized using a mini slump test, a rotational viscometer, and an induced bleeding test.

The yield stress and plastic viscosity were measured from the flow curve established using a coaxial-cylinders rotational viscometer (smooth cylinders, no serration). The test grout was contained within the annular space between the two cylinders. The outer cylinder (rotor, radius = 1.8415 cm) was rotated at 12 controlled speeds varying from 1 to 600 rpm, and the viscous drag exerted by the grout created a torque on the inner cylinder (bob, radius = 1.7245 cm, height = 3.80 cm). The torque was transmitted to a precision torsion spring whose deflection was measured and related to shear stress (Fig. 4) using the following formulae, where θ is the viscometer reading, k_1 is the torsion constant of spring per unit deflection (N-cm/degree), k_2 is the shear stress constant for the effective bob surface (cm^{-3}), and k_3 is the shear rate constant (s^{-1}/rpm):

$$\text{Shear stress} = k_1 \, k_2 \, \theta \tag{2}$$

$$\text{Shear rate} = k_3 \, N \tag{3}$$

The yield stress and plastic viscosity were obtained from linear regression of the shear stress versus shear rate data thus obtained. The slump of the cement pastes was measured using a miniature slump test as described in detail by Kantro.[39] The induced bleeding test consisted of pouring a representative sample of 200 mL of grout into an airtight vessel equipped with a Baroid-type filter and a paper filter. After closing the cell, a graduated cylinder was placed under the outlet of the cell. A pressurization system was then connected to the cell to apply a constant nitrogen gas pressure of 0.55 MPa. The volume of induced bleeding was recorded at 15 and 30 s, and then at every minute up to 10 min.[38]

A mortar corresponding to each grout was made using standard Ottawa sand. The mixing procedure for the mortars followed that of ASTM C305-94 (Standard Practice for Mechanical Mixing of Hydraulic Cement Pastes and Mortars of Plastic Consistency). A proportion of sand of 2.3 times the mass of the binder was generally found suitable to achieve a flow of 100 ± 10% on a flow table for the designed mortars. The flow measure of the mortars was carried out using ASTM C230-90 (Standard Specification for Flow Table for Use in Tests of Hydraulic Cement). The mortar-flow results were

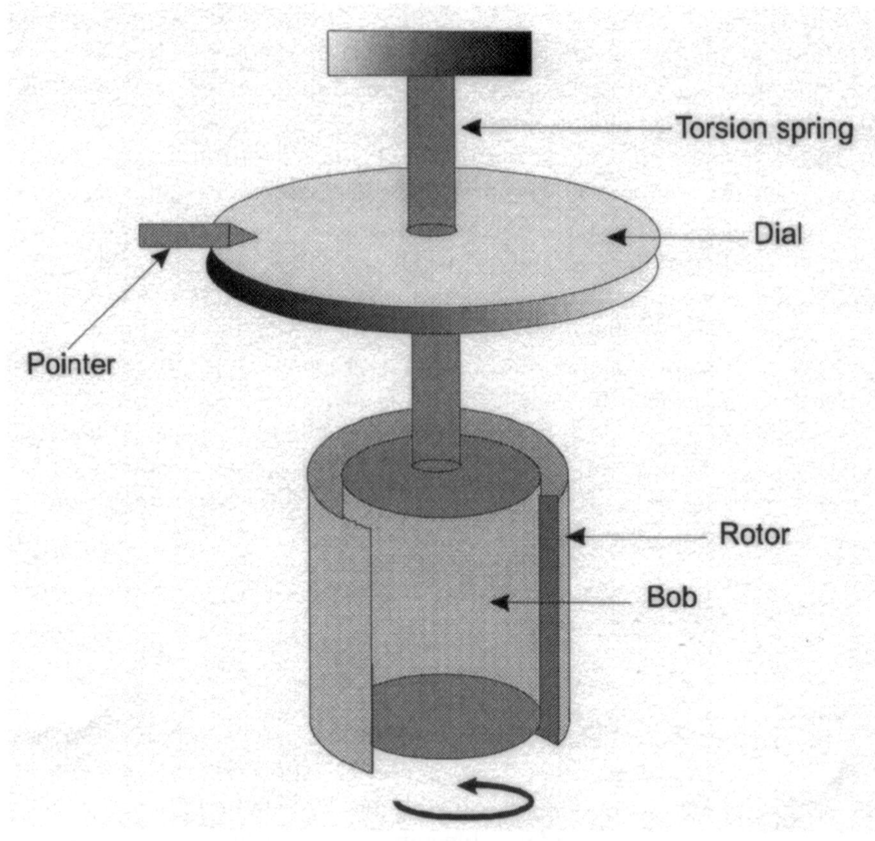

Figure 4. Schematic of the coaxial-cylinders viscometer.

calculated as $[(d - d_0) / d_0 \times 100]$, where d_0 is the initial diameter and d is the diameter after spreading.

The experiments were designed according to a two-level uniform-precision factorial plan.[40] This approach was selected to limit the number of cement pastes to be investigated, while first- and second-order models could be used to fit the data. In addition, this method highlights the significance of the effect of the experimental variables and their interactions, and has a predictive capability for the response of other experimental points located within the experimental domain. The data corresponding to the various responses that resulted from the designed experimental program were analyzed and plotted using a statistics software package.[41]

The superplasticizer efficiency was improved by the LF replacement of

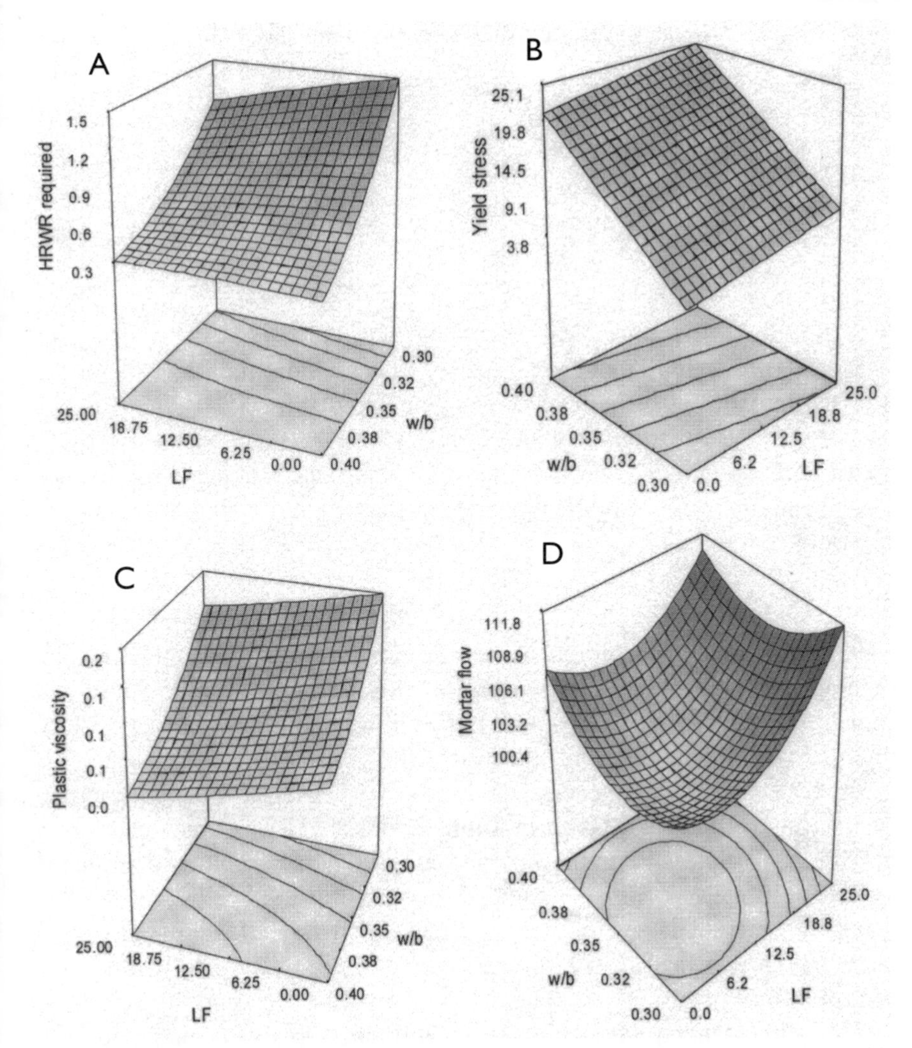

Figure 5. Microfiller effect on rheological properties of cement pastes and mortars: (a) superplasticizer required (%); (b) yield stress (Pa); (c) plastic viscosity (Pa·s); (d) mortar flow (%).

cement, and this contribution offset the negative effect due to the consequent increase in the surface area [Fig. 5(a)]. The LF replacement of cement slightly increased the yield stress of cement paste [Fig. 5(b)] and decreased its plastic viscosity [Fig 5(c)], which implies a better stability and flowability of the cement paste.

Table II. Fineness and average particle size of the cement and fillers

	BET surface area (m^2/kg)	Average particle size (μm)
OPC	350	25.6
SF	18 000	0.26
LF1	2300	2.9
LF2	10 000	0.7
GS	1250	13.8

Mortars made with cement pastes of constant flow time had increased flow as the LF replacement of cement increased [Fig. 5(d)]. This implies that besides the effect on the cement paste yield stress and plastic viscosity, the LF effect on the reduction of the friction and interlocking of aggregates can be important. This lubricating aspect may be more significant in low w/b ratio concretes.

Study on Concrete

Nehdi[38] conducted a comprehensive study of the microfiller effect on the rheological properties of fresh concrete. Ordinary ASTM Type I portland cement (OPC) was used. Two high-purity limestone fillers (LF1 and LF2), finely ground silica (GS), and silica fume (SF) were used as partial replacements for cement. The fineness and average particle size of the cement and various microfillers are given in Table II. Washed gravel having 10 mm maximum particle size was employed as coarse aggregate, and siliceous sand having a fineness modulus of 2.3 was used as the fine aggregate. Tap water was employed for the mixing. A naphthalene sulfonate superplasticizer was used as the superplasticizer.

Proportions of 0, 5, 10, 15, and 20% of LF1, LF2, GS, and SF were individually dry-blended with the cement. HPC mixtures were prepared with these composite cements at a w/b of 0.33 and a constant slump of 200 ± 20 mm. The mixing was carried out mostly using a pan mixer, but some of the mixtures were also conducted using a much more efficient Omni-Mixer. The rheometer test was carried out for these concretes at 15, 30, 45, 60, and 90 min, and the slump test (ASTM C143-90) was carried out at 15, 30, and 60 min. In addition, self-leveling HPC mixtures having a w/b of 0.25 and incorporating 15% of various microfillers were prepared and tested in the rheometer at the same time intervals as above. The 15% cement replacement level was selected because it seemed optimal[42] in maintaining high compressive strengths.

The rheometer used in this study was developed at the University of British Columbia.[43] A computer drives a motor from rest to the desired speed and then back to rest, which causes a planetary motion of a four-finger impeller. A tachometer is used to measure the impeller speed, and a torque-measuring device equipped with four strain gauges measures the torque from the deflection of a small beam in bending. The impeller speed and torque data are acquired by a data acquisition system. A computer program allows the user to customize test parameters, such as the number of readings, the speed increment, the speed decrement, the sampling interval between readings, and so on. In this study the rheometric test consisted of an impeller speed loop starting from rest and going up to about 1.08 rev/s in 10 increments. The speed was then gradually reduced to zero in 30 decrements. For each step, a total of 50 measurements of speed and torque were made in approximately 0.06 s followed by a waiting period of 1.2 s. The test duration was 2.8 min, a compromise between accuracy and the segregation due to longer tests. The rheometer is shown in Fig. 6.

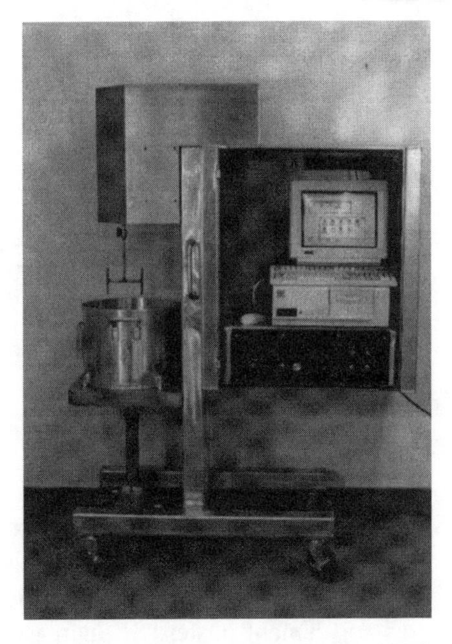

Figure 6. Illustration of the UBC rheometer.

Superplasticizer Requirement

The superplasticizer requirement to achieve a slump of 200 ± 20 mm for the 0.33 w/b concrete mixtures is shown in Fig. 7. Partial substitution of GS and LF for cement slightly reduced the superplasticizer requirement. However, SF replacement of cement increased the superplasticizer demand to achieve a constant workability. It is interesting to note that 20% replacement of cement by LF2 (mean particle size = 0.7 μm) slightly reduced the superplasticizer requirement, while a 5% replacement by SF (mean particle size = 0.26 μm) increased the superplasticizer requirement. This implies that the high surface area of SF may not be the sole factor affecting the increase in the superplasticizer demand for SF mixtures, and supports the

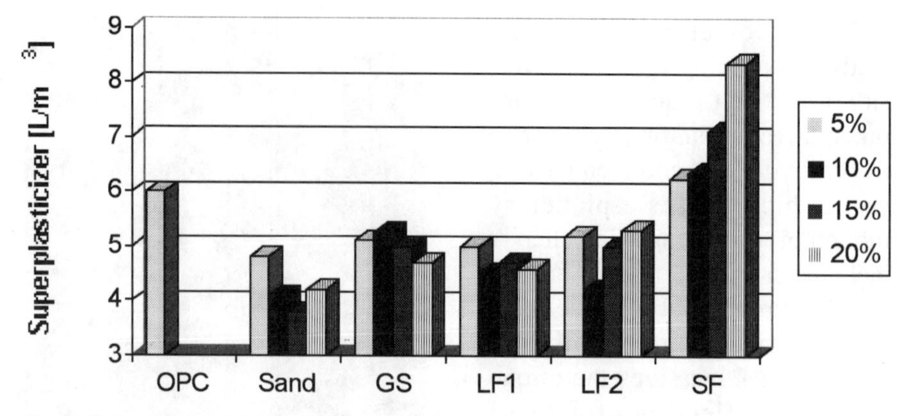

Figure 7. Superplasticizer requirement for a constant workability.

idea that SF may have an affinity for multilayer adsorption of superplasticizer molecules. Previous work[44] also has shown that while using a 3-µm LF tended to reduce the superplasticizer requirement to achieve a constant flow for cement pastes, SF increased the superplasticizer requirement significantly.

Flow Resistance

The flow resistance, *g,* at various ages after mixing for different levels of partial replacement of cement by GS, LF1, LF2, and SF microfillers is illustrated in Fig. 8. For fluid HPC mixtures, *g* tended to increase with time, reflecting the stiffening behavior of concrete with aging. Partially substituting various fine fillers for cement reduced *g,* although the surface area, and thus the wettable surface, were increased. The improved gradation of the binder and the lubricating effect imparted by the fine particles may have reduced the aggregate interlocking, and may have reduced *g* as a result.

Torque Viscosity

The torque viscosity, *h,* for various mixtures at different ages is illustrated in Fig. 9. Generally, the finer the microfiller, the more *h* was reduced.[‡] This was true for both fluid and self-leveling mixtures. Generally, higher proportions of filler replacement of cement were more effective in reducing the

[‡]It should be remembered that the mixtures under investigation contain fair amounts of superplasticizer.

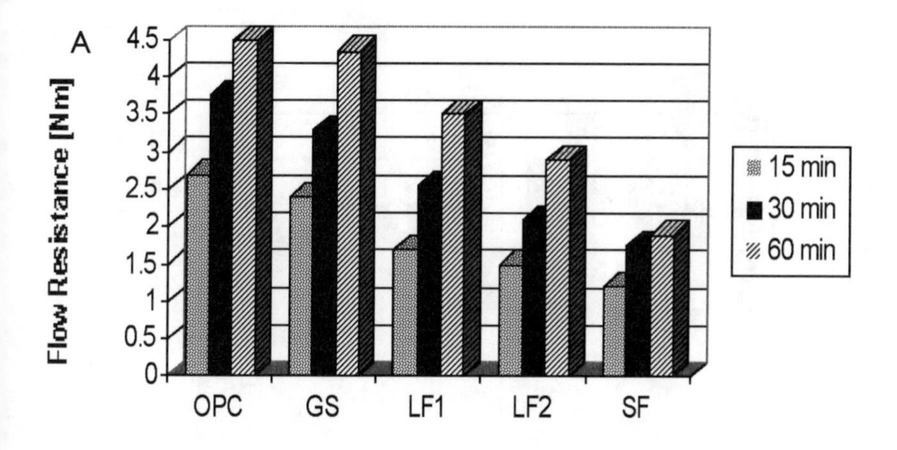

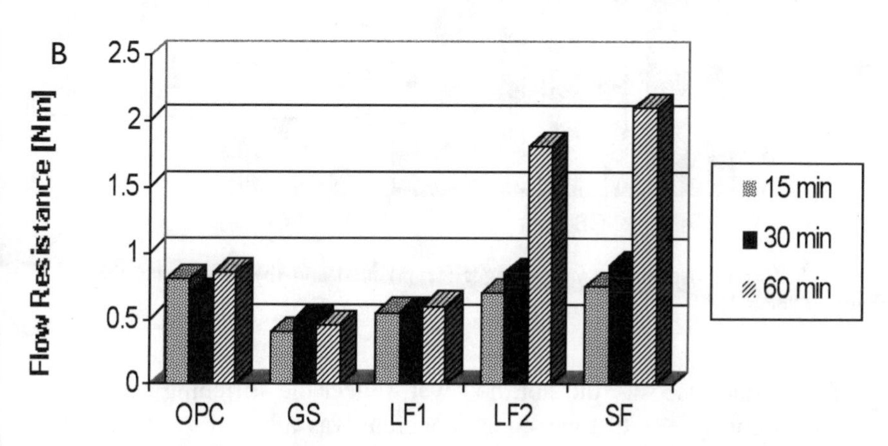

Figure 8. Flow resistance of (a) fluid and (b) self-leveling HPC mixtures (filler = 15%).

torque viscosity. It was observed earlier that higher filler proportions reduced the plastic viscosity of cement paste in both silica fume and non–silica fume systems.[44] Yet, this effect was more significant in concrete than in cement paste, probably because the ultrafine particles played a more important lubricating role via a reduction of the aggregate interlocking. This will be illustrated more clearly later, in the discussion of the results of the statistical modeling of the filler effect on the torque viscosity of HPC made with triple-blended composite cements.

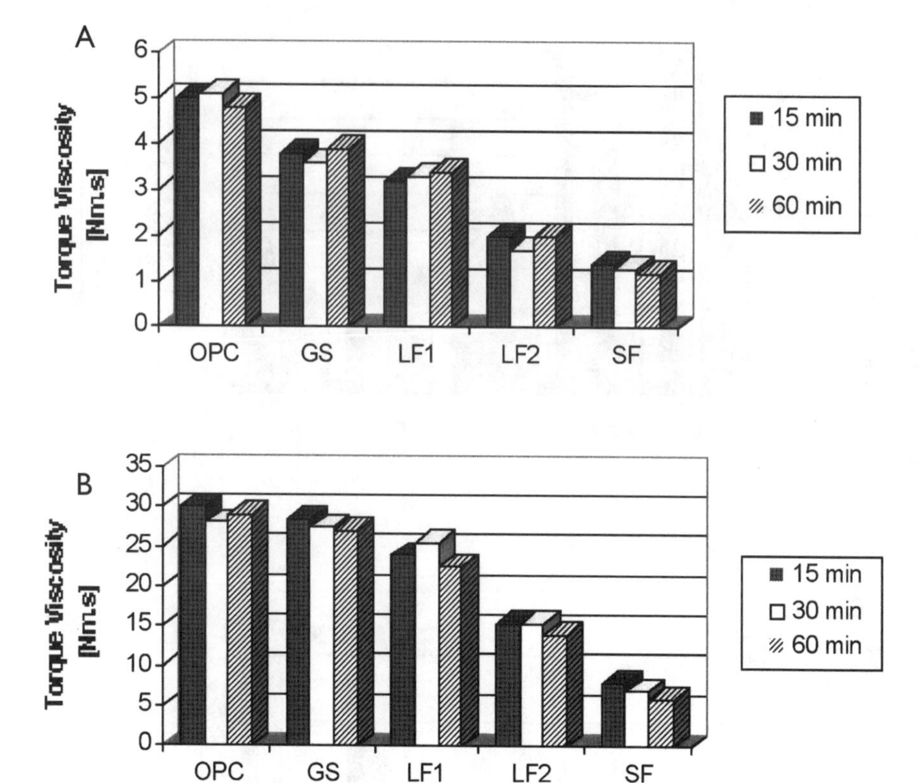

Figure 9. Torque viscosity at various ages for (a) fluid and (b) self-leveling HPC mixtures (filler = 15%).

The torque viscosity did not, however, reflect the stiffening behavior of the fresh concrete with time. Since concrete was allowed to rest between consecutive rheological tests, the torque requirement for the low impeller speeds increased more than that for the high impeller speeds, probably due to flocculation of binder particles, growth of hydration products, and exhaustion of superplasticizer molecules in the complex chemical processes. This resulted in a reduction of the slope of the shear stress versus shear rate curve, which may explain why the torque viscosity did not increase with time and tended to decrease slightly in some cases. Induced segregation may also contribute to this behavior. It should be remembered that the torque viscosity is a dynamic value, as opposed to the flow resistance and slump tests, which can be regarded as static values.

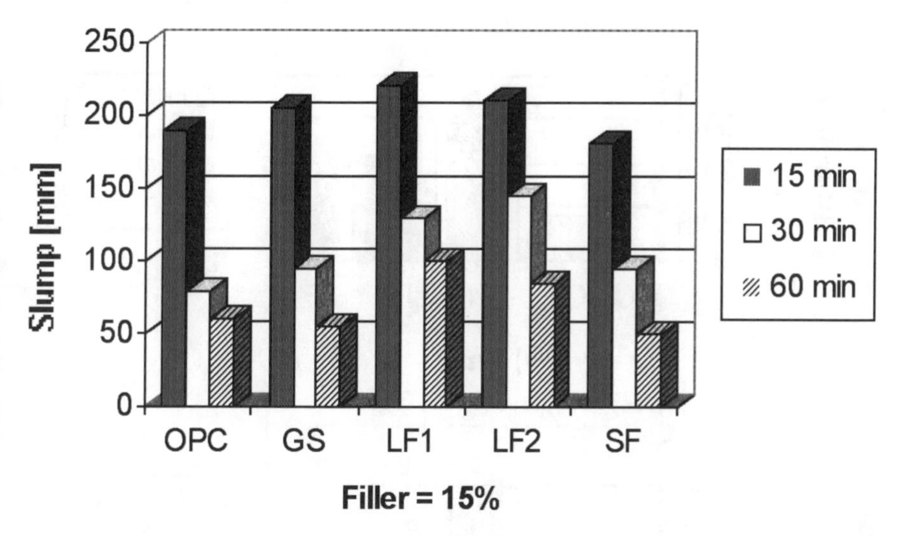

Figure 10. Slump variation of concretes made with 15% of various fillers.

Slump Loss

Although superplasticizers are effective in dispersing the cement particles through a deflocculation process, this action is time dependent, and very low w/b concrete usually undergoes a rapid slump loss. There are two concerns regarding this issue: the cement should have the lowest rheological reactivity (that is, the amount of water fixed immediately after mixing should be minimal) and the superplasticizer should not compete with the calcium sulfate to neutralize the C_3A.[45] The changes of the slump loss behavior occurring upon partial replacement of cement by microfillers of various mineralogies and mean particle sizes is not yet clear. Figure 10 illustrates the slump variations of 0.33 w/b concretes at 15, 30, and 60 min. Results are shown for 15% replacement of cement by GS, LF1, LF2, and SF. Starting at slumps of 200 ± 20 mm, all the mixtures had slumps higher than 50 mm after 60 min.

The blended composite cements generally showed performance equal to or better than pure OPC. Generally, concretes that had the highest slumps at 15 min kept higher slumps at 60 min. It should be remembered that only the silica fume mixtures had higher superplasticizer dosages. Therefore, the effect of the difference in the superplasticizer dosage on the slump loss is limited for the rest of the mixtures.

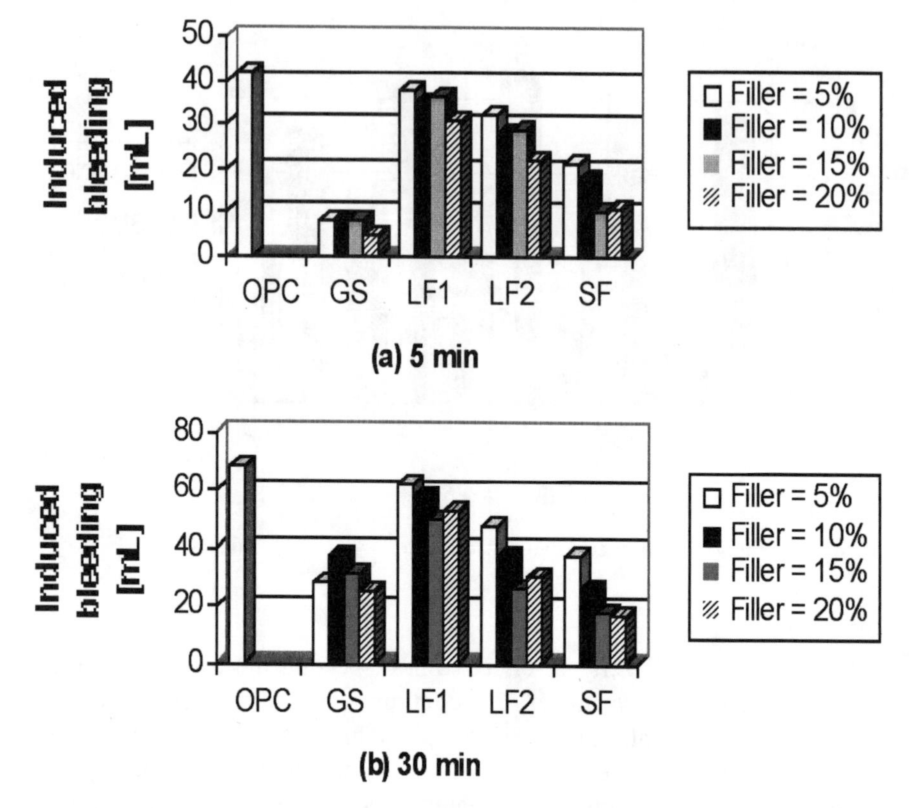

Figure 11. Microfiller effect on induced bleeding of fresh high-strength concrete.

Induced Bleeding

It has long been known that the bleeding rate and bleeding capacity of cement paste are strongly dependent on the water content and the specific surface area of the cement.[46] Thus, it can be reasonably expected that partial replacement of cement by finer microfillers would reduce the induced bleeding of cement paste and concrete. The water thus retained may participate in lubricating the concrete mixture and improving the rheology. The induced bleeding of the microfiller concrete mixtures at 5 and 30 min is illustrated in Fig. 11.

Replacing some of the cement by finer fillers reduced the bleeding at both 5 and 30 min. The finer the microfiller, the more the bleeding was inhibited, except for GS, which, though being the coarsest microfiller used, was the most effective in reducing the bleeding at 5 min. The reason for this behavior is not clear. It may be however that the GS changed the electrolytic environment, which in turn increased the interparticle attraction. This is known to reduce the bleeding capacity while generally having little influence on the bleeding rate.[46]

Rheology of HPC Made with Triple-Blended Cements

The iso-response curves resulting from a factorial experimental plan[47] for the various rheological characteristics under investigation are illustrated in Fig. 12. The superplasticizer demand to achieve a 200 ± 20 mm slump [Fig. 12(a)] was not markedly affected by the LF proportion at low SF rates. At high SF levels, the superplasticizer demand was reduced as the LF proportion increased. A previous investigation[44] also showed that increased LF levels reduced the superplasticizer amount required to achieve a constant flow time of low w/b ratio silica fume cement paste.

The slump loss of the fresh concrete mixtures is illustrated in Fig. 12(b). Increased levels of LF or SF reduced the slump loss of the concrete. A combination of moderate levels of LF and SF reduced the slump loss, while high levels seemed to enhance the slump loss, although a 10% LF–10% SF binder is still advantageous compared to an OPC. However, the ability of the slump test to describe properly the effect of time on the rheology of this kind of concrete is questionable; the flow resistance and torque viscosity are more relevant.

The flow resistance, g, at 15 min decreased with higher SF levels, and increased with higher LF levels up to about 6%. Higher LF levels seemed to reduce g slightly. A combination of moderate levels of LF and SF reduced g, while high levels seemed to increase g. At 60 min, g increased with higher LF levels, yet a combination of high LF-SF levels did not seem to be deleterious as was suggested by the slump test.

The torque viscosity, h, was reduced as the LF and SF levels were increased, at both 15 and 60 min. The overall effect of replacing cement by a LF-SF combination on the workability was positive. It was rather surprising that h seemed to decrease from 15 to 60 min. It seemed that the increase of g with time and the torque requirement for the lower impeller speeds was not accompanied by an equal increase of the torque at higher impeller speeds. This resulted in a reduction of the slope of the torque/impeller-speed curve, and an artificial decrease of viscosity as a consequence. It is possible that induced segregation may also have contributed to

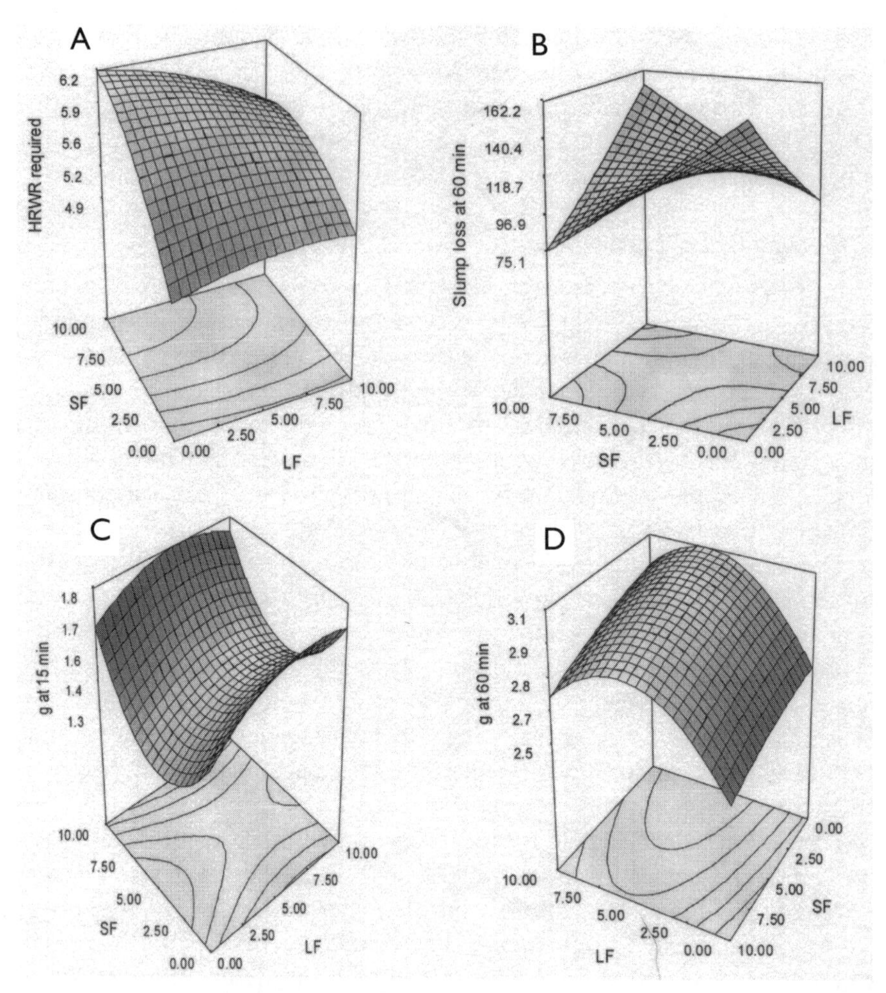

Figure 12. Statistical modeling of various rheological responses for OPC-LF2-SF triple-blended composite cements: (a) HRWR required (%); (b) Slump loss at 60 min (mm); (c) g at 15 min (Nm); (d) g at 60 min (Nm).

this behavior. The results obtained above for concrete support similar results of previous work,[44] where it was shown that increased LF levels resulted in an increased yield stress and a reduced viscosity of low w/b ratio cement pastes, in both SF and non-SF systems.

Ultrafine Particles and Rheology of Cement Suspensions

Various kinds of forces coexist in a cement suspension. First, there is the Brownian randomizing force, which influences the spatial orientation and arrangement of particles. This force is strongly size dependent and has a major influence below a particle size of 1 μm. The effect of microfiller partial substitution of cement on this kind of force is beyond the scope of this investigation.

Second, there are forces of colloidal origin, which arise from mutual interactions between particles and are affected by the polarizability of water. When the van der Waals attraction between cement grains and the electrostatic attraction between unlike charges on the surface of particles are dominant, the net result is an attraction and the particles tend to flocculate. However, in the presence of polymeric or surfactant materials on the surface of cement grains, the net result may be a repulsion and the particles remain separate. In this respect, the filler material can influence the electrostatic forces, depending on its mineralogical nature and the state of its particle surface charges. Since colloidal forces depend also on the average distance between nearest neighbor particles, the interposition of finer filler grains between cement particles may affect their electrostatic attraction and thus their flocculated structure. Likewise, replacing cement by a material of different specific surface area would change the wettable surface area and the amount of water adsorbed. Some fillers having a certain solubility in water may modify the electrolyte solution and thus the electrostatic forces.

Third, there are viscous forces, which are proportional to the local velocity differences between a cement particle and the surrounding water, and between an aggregate and the surrounding cement paste. Since cement-based materials fall in the range of dense suspensions, particles have to move out of the way of each other, especially when flocs are formed. The filler effect on the rheology will depend on its fineness, its particle size distribution, and its particle shape. Broader particle-size ranges have higher maximum particle packing because the finer particles fit into the gaps between the coarser particles. The viscosity of suspensions usually increases as the deviation from ideal grading increases. For a certain volume of water, the viscosity reaches a minimum at the most compact arrangement of particles.[48] In addition, any deviation from a spherical shape implies an increase in viscosity for the same phase volume. Thus, in the presence of

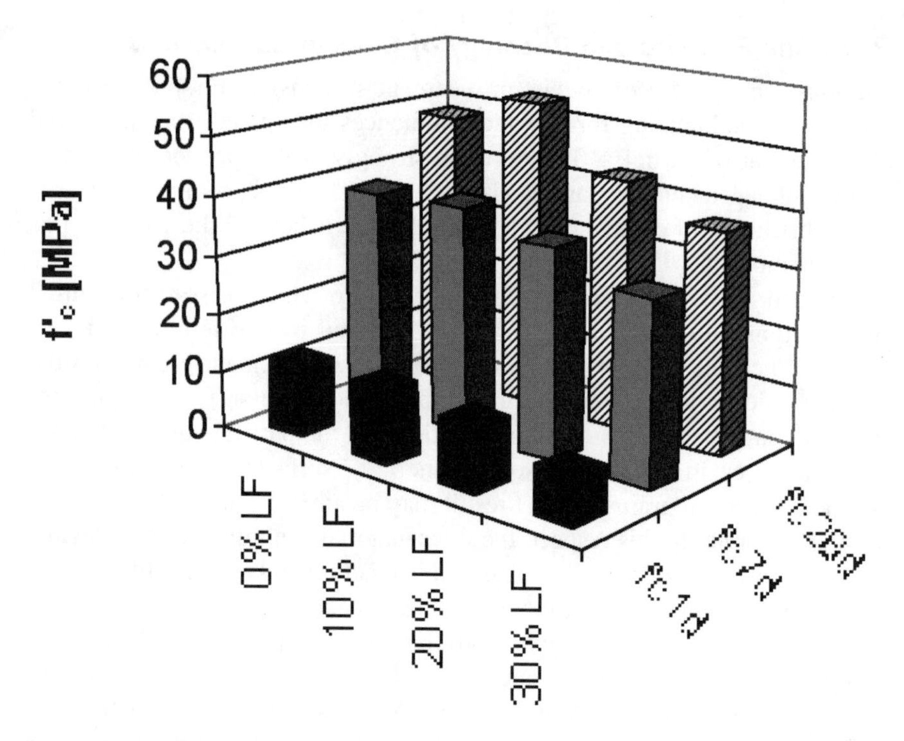

Figure 13. Effect of limestone partial replacement of cement on compressive strength.[50]

superplasticizers, the finer and the more spherical the filler, the better the rheological properties. Filler materials can also have different efficiencies in the adsorption of superplasticizers. They can, if soluble, introduce certain ions, which may affect the kinetics of the hydration reactions and the nucleation of hydration products.

Microfiller Effect on Mechanical Properties

There is some agreement that at up to 5% limestone replacement by mass of cement, the compressive strength of concrete is not affected. There are some reports that claim that up to 15% of replacement[49,50] and up to 20% of addition[7,51,52] the mechanical performance of cement paste is equivalent to pure OPC and even more favorable for some ages (Fig. 13). Livesey[53] reported on the strength development characteristics of British cements containing limestone filler. A 5% limestone addition seemed not to affect either the compressive or flexural strengths, while a 25% addition reduced the strength class of the cement. It was also observed that at a constant strength, increasing the amount of limestone filler increased the modulus of elasticity.[49] A calcite having a disorganized crystalline network was shown to enhance the mechanical performance in a 75% OPC and 25% limestone cement better than either a well-crystallized calcite or a quartz filler of the same fineness.[54]

The European practice for dealing with the reduction of the compressive strength at high levels of cement replacement by carbonates is to increase the fineness of the cement and/or to increase its C_3A content so that the formation of carboaluminates compensates a bit for the loss of strength. In view of its fluidifying effect,[4,32,34,35] its possible reduction of the transition zone thickness,[4] and its refinement of the grain size of the hydration products due to a possible nucleation effect, ultrafine limestone may play a different role in HPC. There is an increasing tendency to use multiblended binders having an optimal particle packing for the development of high-strength materials.[55,56] In these systems, a portion of even the cement plays a filler role due to incomplete hydration.

Study on High-Strength Concrete

ASTM Type I cement along with the microfillers shown in Table II were used in this investigation. A total of 78 concrete mixtures were examined. HPC mixtures were made at a w/b ratio of 0.33. Levels of 0, 5, 10, 15, and 20% filler replacement by volume of cement were used. In addition, a uniform precision factorial plan was designed to optimize OPC-LF2-SF triple-blended cements. The two experimental variables were the proportion of LF2 and the proportion of SF replacement of cement. Likewise, in order to investigate the influence of the w/b ratio on the filler effect, concrete mixtures made with binary- and ternary-blended cements were made with a 0.25 w/b ratio. In all of these mixtures, a pan mixer was used. Since the filler effect depends on the degree of dispersion of the ultrafine particles, several concrete mixtures were produced using a much more efficient OMNI mixer (Fig. 14).

For all mixtures, tap water, washed 10 mm maximum particle size gravel, silica sand having a fineness modulus of 2.3, and a naphthalene sulfonate superplasticizer having 42% solid content were employed. Cylinders (75 × 150 mm) and beams (100 × 100 × 350 mm) were cast using a vibrating table. Specimens were demolded at 24 h after casting (except for the 12-h compressive strength tests) and cured in lime-saturated water until

Figure 14. The OMNI mixer.

testing. The compressive strength was measured at 12 h and 1, 3, 7, 28, and 91 days as per ASTM C39 using 75 × 150 mm cylinders with mechanically ground ends. For each age, three specimens were tested.

The modulus of elasticity was measured at 1, 3, 7, and 28 days according to ASTM C469 on 75 × 150 mm cylinders. The axial load was measured using a load cell, and the displacement was measured by means of two LVDTs placed vertically on each side of the specimen (using the middle 100 mm of the specimen as the gauge length (Fig. 15). Each specimen was loaded and unloaded at a cross-head speed of 0.5 mm/min up to about 40% of the average compressive strength of the concrete mixture. The load cell and LVDT signals were recorded using a digital data acquisition system. The modulus of elasticity, E, was calculated from the stress (σ) vs. strain (ε) curve obtained from the load cell and the averaged LVDT signals. The first loading/unloading cycle was ignored, and the modulus of elasticity was taken as the mean of the two values obtained from the two subsequent loading cycles.

The flexural strength test was carried out at 1 and 28 days on 100 × 100 × 350 mm beam specimens by means of a third point loading test according

Figure 15. Experimental setup to measure the modulus of elasticity of concrete specimens.

to ASTM C78-84. A hydraulic digital Instron machine with a 150 kN capacity was used with a cross-head speed of 0.05 MPa/s. A load cell and two LVDTs continuously measured the load and the corresponding deflection of the beam specimen, and the signals were acquired by a data acquisition system. The flexural strength represents the average value obtained on three specimens.

Filler Effect on Compressive Strength of HPC

The compressive strength values for the 0.33 w/b HPC mixtures at ages varying from 12 h to 91 days are illustrated in Fig. 16. Results are shown for 0–20% replacement of cement by various microfillers.

At 12 h, the compressive strength depended strongly on the type of microfiller added to the cement. The pure OPC mixture had very low strength. The addition of 5–20% of ground silica did not seem to affect the 12-h strength significantly. The addition of silica fume increased the 12-h strength; this effect increased as replacement rates increased. The 3-μm mean particle size limestone filler was not effective in increasing the 12-h strength at a 5% replacement rate, but had a significantly higher effect at increased replacement rates. The much finer 0.7-μm mean particle size

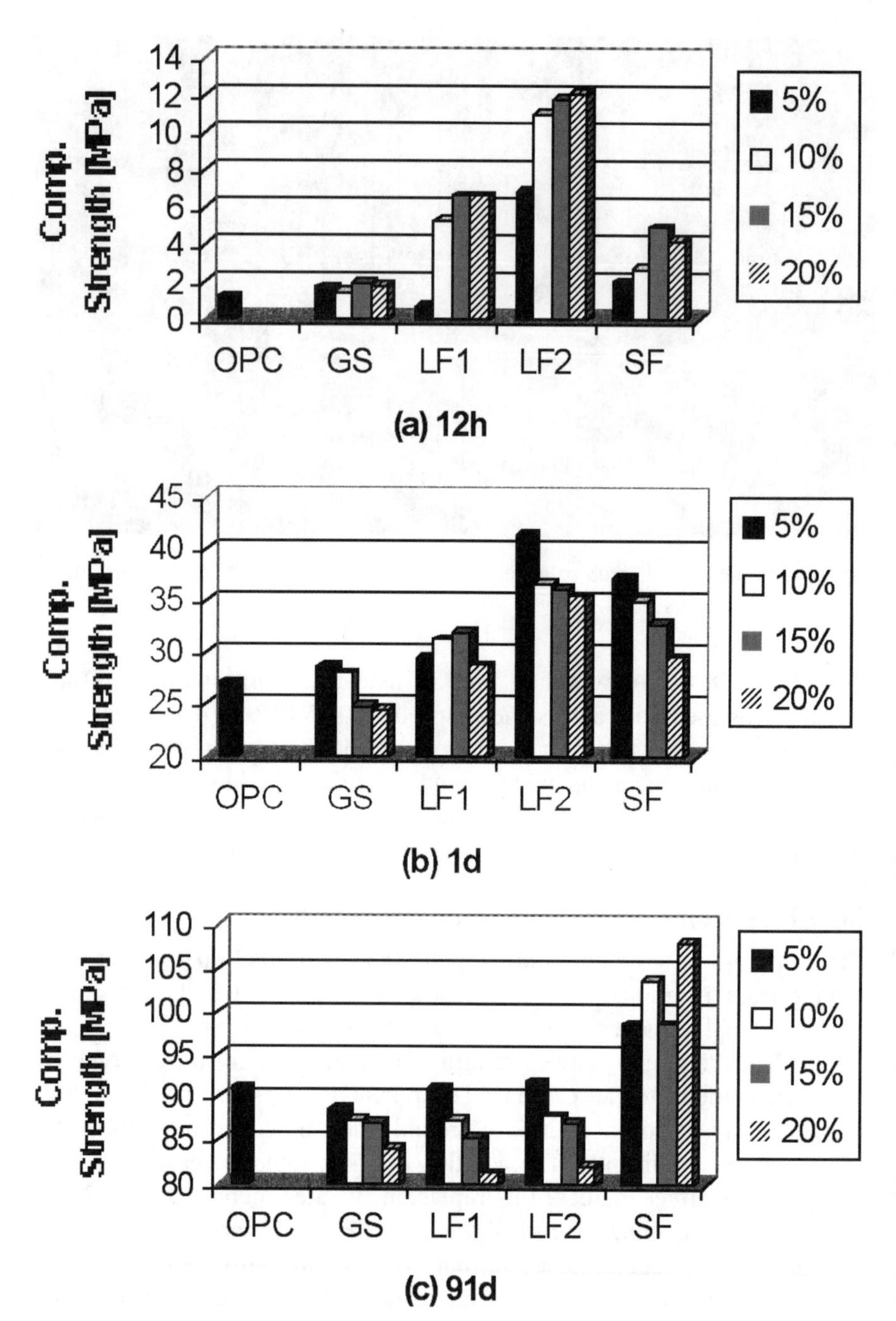

Figure 16. Filler effect on compressive strength (w/b = 0.33).

limestone filler had a greater influence and increased the very early strength by an order of magnitude at a 15% replacement rate. The limestone fillers had the greatest effect on the 12-h compressive strength; this was more significant as the fineness and the replacement rate increased. The explanation of this behavior will be addressed later on in this chapter.

At 1 day the compressive strength of the limestone filler mixtures was still advantageous compared to a pure OPC mixture. At this age, the silica fume concrete started to outperform the rest of the mixtures except those containing the ultrafine limestone. Only the ground silica mixtures had lower strengths than the OPC mixtures at this age.

At ages greater than 3 days, the OPC mixtures gradually reached the compressive strength of the limestone filler mixtures or outperformed them, especially at high replacement rates. Contrary to the early age behavior, there was not a major difference between the strengths of the concretes made with the various inert fillers. The silica fume mixtures achieved distinctly higher long-term strengths. It is worth mentioning that 80 MPa concretes with good rheological properties were achieved using as much as 20% replacement of cement by inert fillers in 0.33 w/b mixtures.

Dependence of the Filler Effect on w/b Ratio

Figure 17 shows the compressive strength ratio (strength of filler cement concrete divided by strength of OPC concrete) at different ages for mixtures with various proportions of fillers to a reference OPC mixture. Results are illustrated for mixtures incorporating 15% of microfillers at w/b ratios of 0.33 and 0.25.

The silica fume mixtures had compressive strength ratios higher than 1 at both w/b ratios. However, for a w/b ratio of 0.33, the limestone filler mixtures had compressive strength ratios higher than 1 only at early ages; these values were lower than 1 at later ages. At a w/b ratio of 0.25, the limestone filler mixtures had compressive strength ratios higher than 1, even at 91 days. The practical implication of these results is that HPC mixtures can be made with as much as 15% of a grinding mill filler with improved strength both at early and later ages. The increased efficiency of the microfiller effect at very low w/b ratios has been discussed from a microstructural point of view elsewhere.[38] Briefly, it is thought that at a very low w/b ratio, a significant part of the cement plays a filler role due to the limited hydration. Partial replacement of cement with a finer filler may improve the particle packing of the system and reduce its porosity without much loss in the bond, thereby increasing the strength.

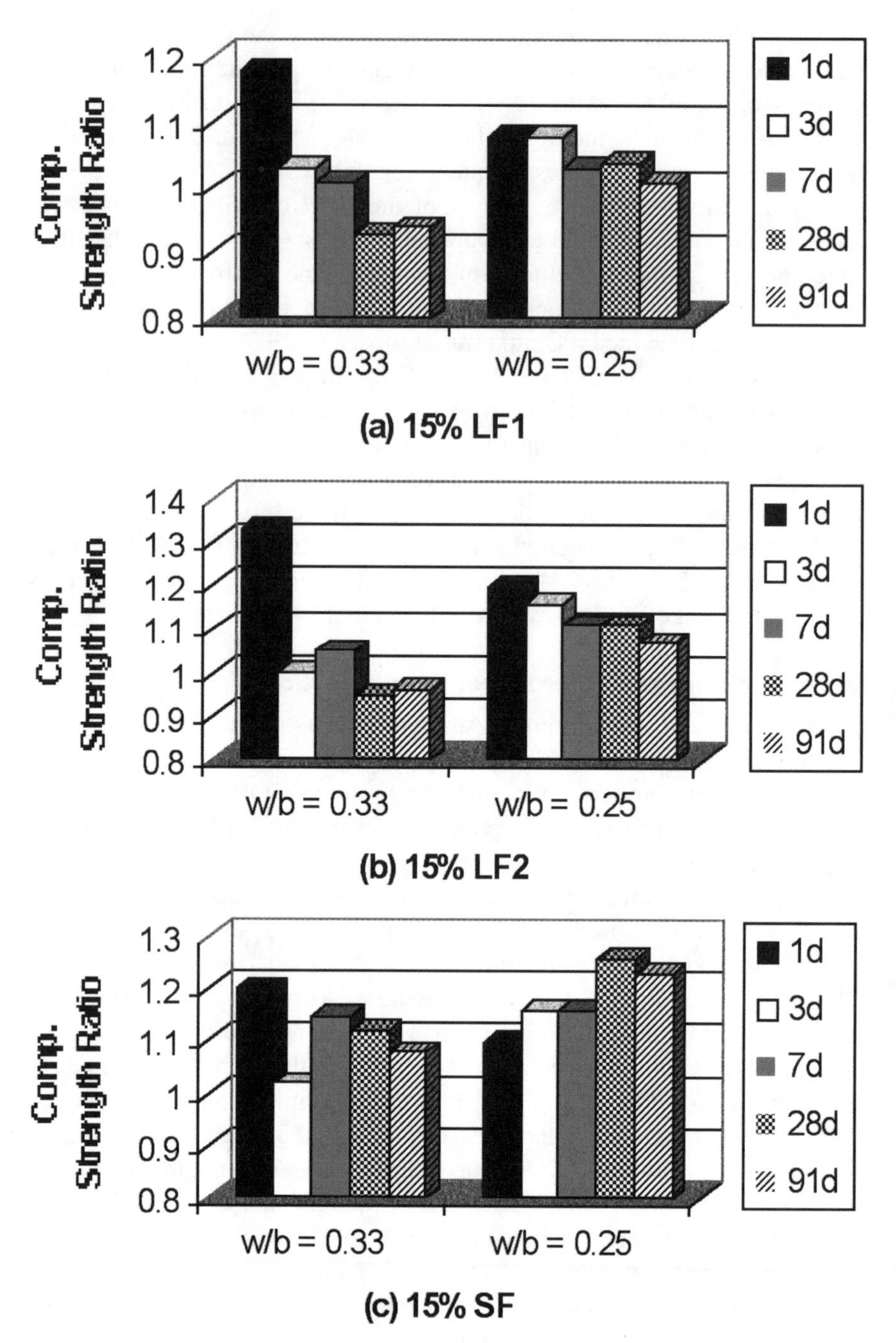

Figure 17. Dependence of filler effect on the w/b ratio.

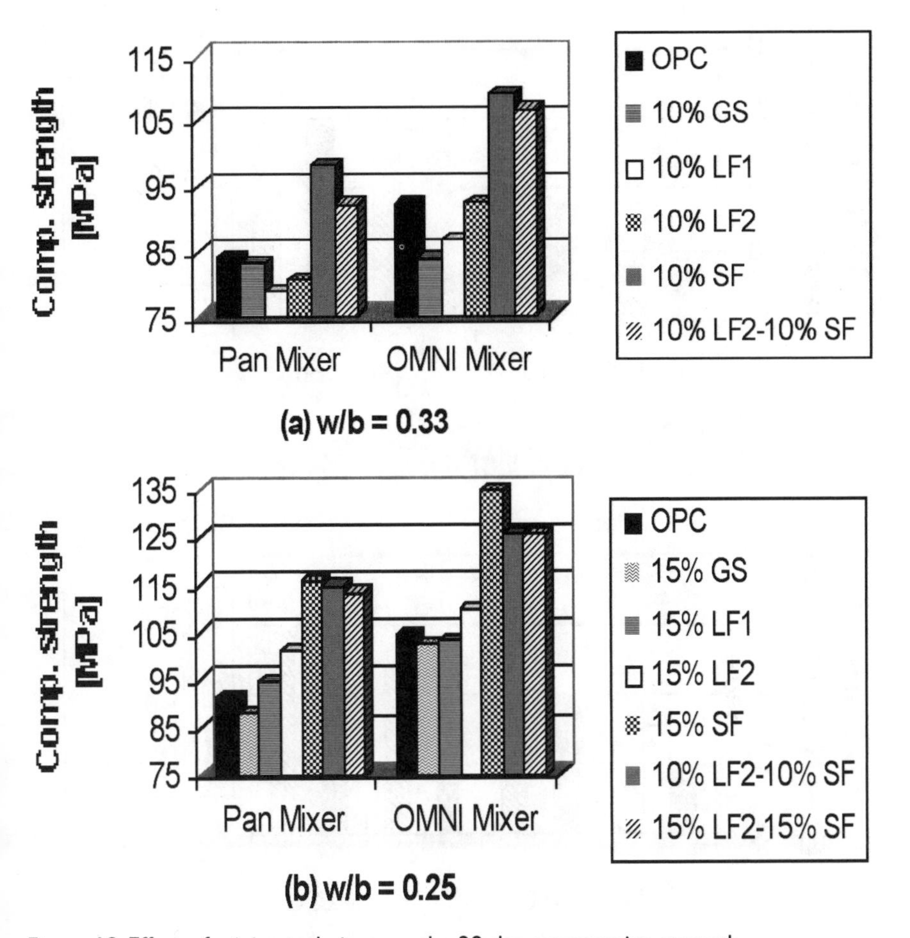

Figure 18. Effect of mixing technique on the 28-day compressive strength.

Effect of Mixing Technique on Compressive Strength

Figure 18 compares compressive strength results obtained with HPC mixtures made using a regular pan mixer versus the OMNI mixer. Results are shown for mixtures incorporating various microfillers at two different w/b ratios. It was observed that the compressive strength generally increased when the OMNI mixer was used, and this was more significant in mixtures containing silica fume. Of particular interest in this figure is that the replacement of 30% of the cement by a combination of 15% limestone filler and 15% silica fume yielded compressive strength values 20% higher than an OPC mixture at 28 days. This triple-blended binder also achieved

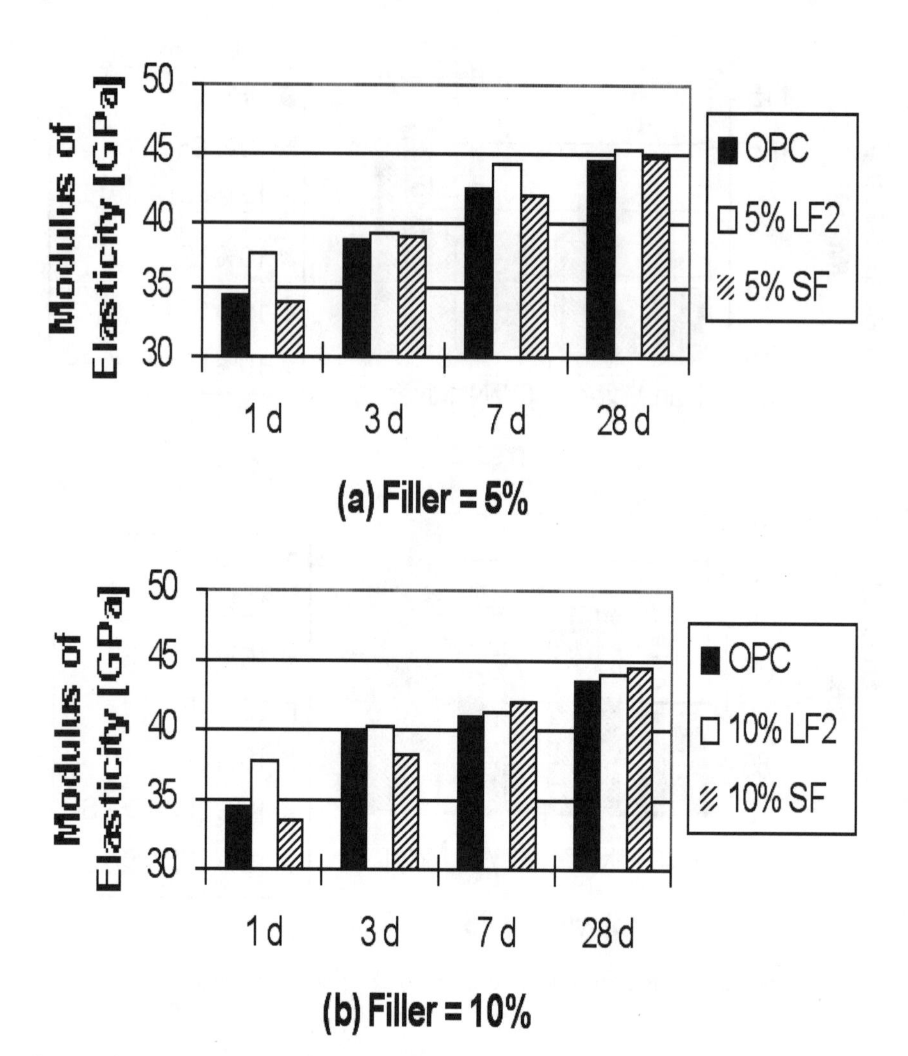

Figure 19. Filler effect on modulus of elasticity at various ages.

strengths that outperformed (at early age) and were comparable to (at later age) an OPC plus 15% silica fume binder.

Filler Effect on Modulus of Elasticity

The modulus of elasticity values measured at various ages for concrete mixtures incorporating various proportions of microfillers are illustrated in Fig. 19. At early ages, the modulus of elasticity values for limestone filler mixtures were higher than for the reference OPC mixtures. This was main-

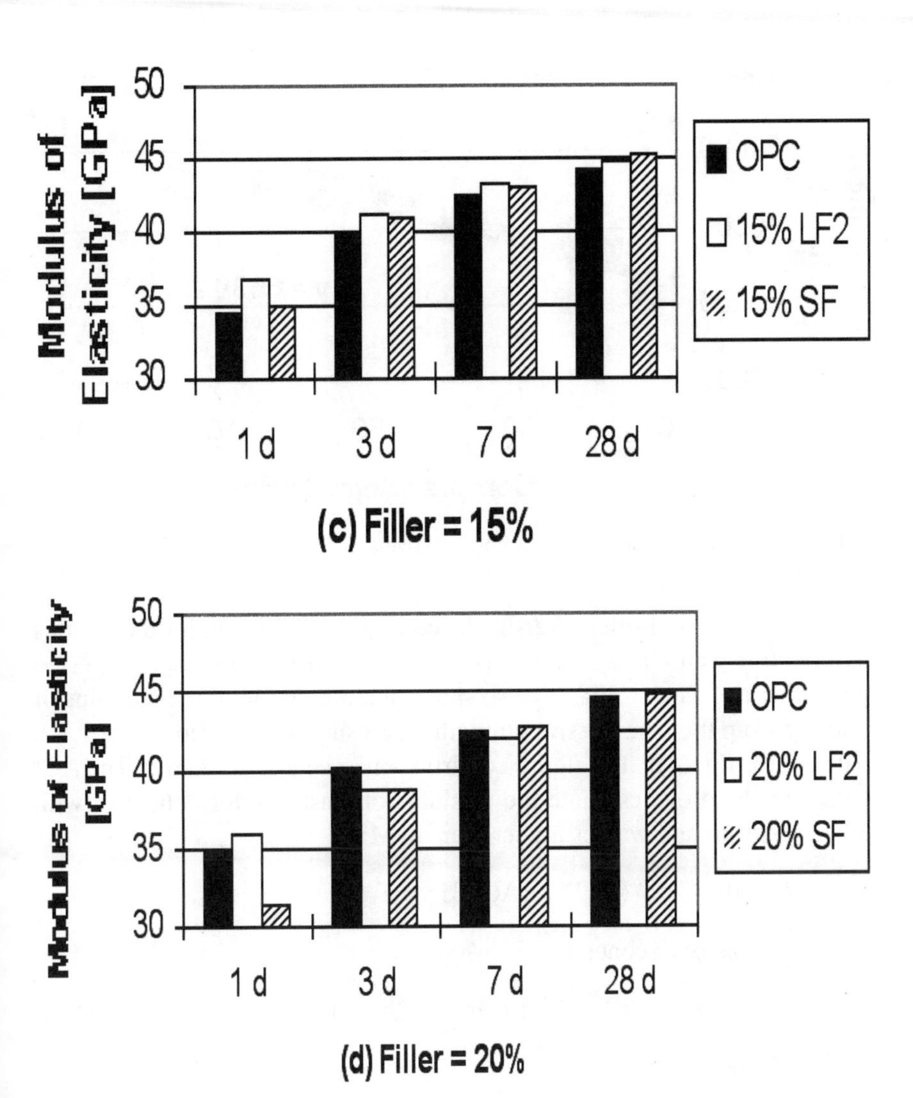

Figure 19, continued.

tained up to 28 days for filler replacement rates lower than 15%. For 20% filler replacement, the modulus of elasticity was lower than that of a reference OPC mixture. On the other hand, the silica fume mixtures generally had a lower elastic modulus at early ages. After 7 days, these mixtures often had higher modulus values than either the OPC or limestone filler mixtures.

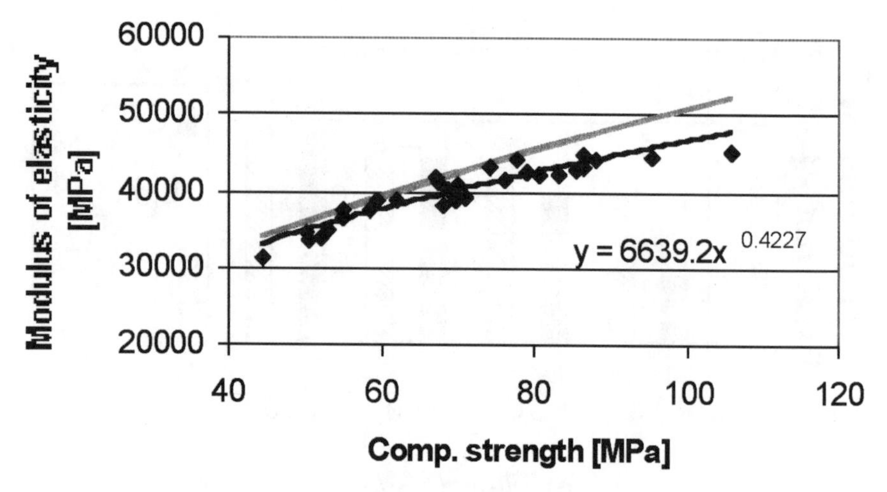

Figure 20. Modulus of elasticity versus compressive strength for various HPC mixtures.

The modulus of elasticity versus the compressive strength for the various mixtures followed a power model (Fig. 20) similar to the model suggested by Gardner and Zhao.[57] Fig. 20 also shows the estimation of the modulus of elasticity from the compressive strength data using the ACI Committee 318 relationship [Eq. (4)]. The figure confirms what we already know: The ACI method tends to overestimate the modulus of elasticity for concretes with compressive strength values higher than 40 MPa.

$$E_c = 0.043\,\rho^{1.5}\,(f_c')^{0.5} \quad \text{(ACI 318)} \tag{4}$$

where ρ = density of concrete.

$$E_c = 9000\,(f_c')^{0.33} \quad \text{(Gardner and Zhao, 1991)} \tag{5}$$

for $f_c' > 27$ MPa.

$$E_c = 6640\,(f_c')^{0.42} \quad \text{(from results above)} \tag{6}$$

The elastic modulus mainly depends on the nature and fraction of aggregates per unit volume, which was maintained constant in all the mixtures. However, the various mineral admixtures affected not only the modulus of the cement paste, but also that of the aggregate-cement paste transition zone. This transition zone adds up to a considerable volume[58]: about one-third to one-half of the hardened cement paste. It is generally observed that a layer of oriented crystalline $Ca(OH)_2$, about 0.5 μm thick, covers the sur-

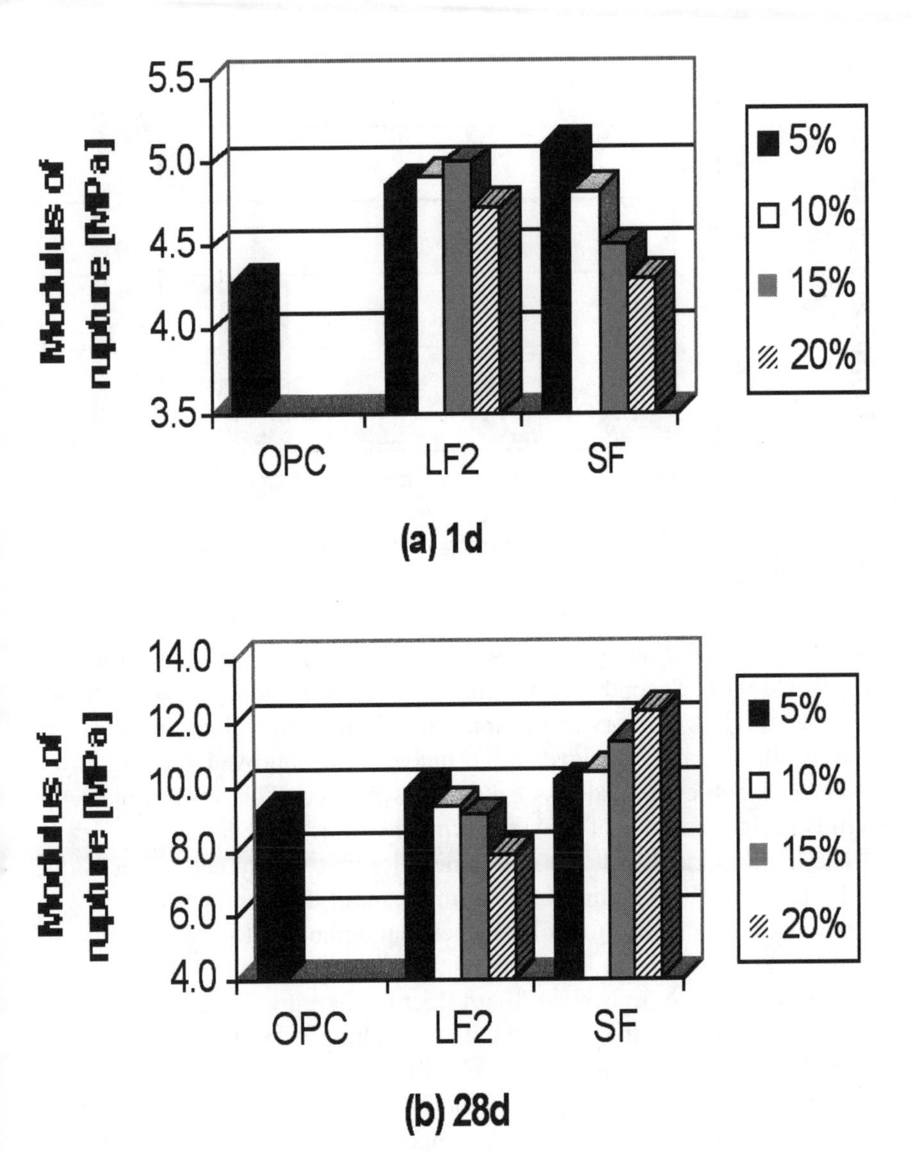

Figure 21. Filler effect on flexural strength.

face of the aggregate, behind which there is a calcium silicate hydrate layer, also about 0.5 μm thick. The two layers constitute what is known as the duplex film. The main interfacial zone is the next 50 μm farther away from the aggregate, which contains larger $Ca(OH)_2$ crystals and a low unhydrated cement content. The significance of this is twofold. First, the absence of

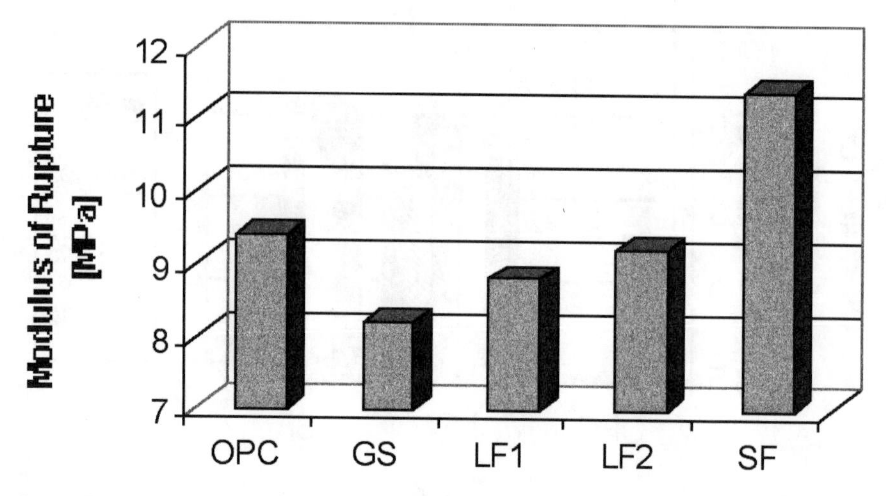

Figure 22. Dependence of the 28-day flexural strength on the fineness of the filler (w/b = 0.33, filler = 15%).

unhydrated cement suggests a lower initial cement content and a locally higher w/c ratio. Second, the presence of large calcium hydroxide crystals implies a higher porosity at the interface. If fine pozzolanic microfillers such as silica fume are added in the mixture, an improved packing at the interface is achieved, and less $Ca(OH)_2$ is formed. This may explain the slightly higher modulus of the silica fume mixtures at 28 days.

The higher modulus of the limestone filler mixtures at early ages is due to the accelerated hydration and the more rapid development of strength. These mixtures also had modulus values that compared to and even outperformed OPC mixtures at later ages. It seems that the improved packing at the interface due to a physical filler effect of the ultrafine limestone particles compensated for the reduced C-S-H resulting from the partial replacement of the cement by a nonpozzolanic filler.

A model was proposed by Lutz and Monteiro[59] in which the aggregate particles are modeled as spheres surrounded by a radially nonhomogeneous matrix. A constant and a term that decays with the radius following a power law represent the modulus of the matrix. It was found that the modulus of the transition zone was 15–50% lower than that of the bulk cement paste, which may explain why concrete falls below the Hashin-Shtrikman bounds for a two-component material.

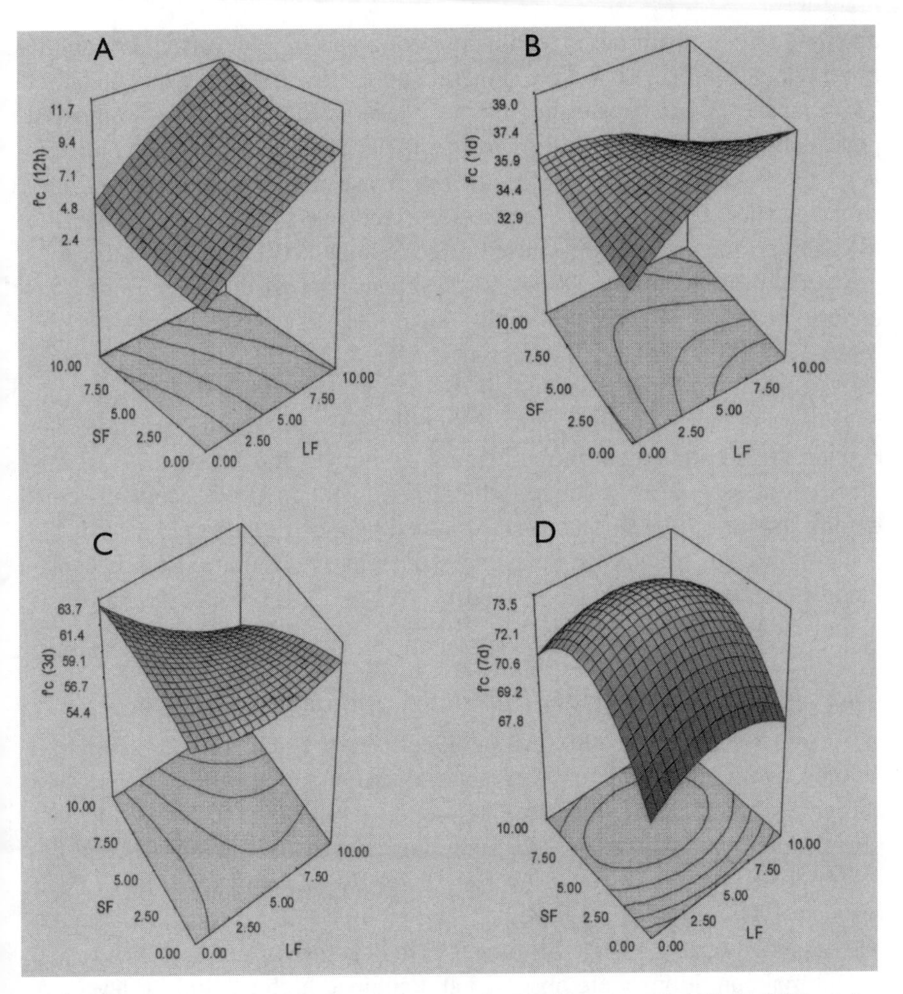

Figure 23. Iso-strength curves at various ages for HPC made with LF2-SF-OPC triple-blended cements: (a) at 12 h (MPa); (b) at 1 day (MPa); (c) at 7 days (MPa); (d) at 91 days (MPa).

Filler Effect on Flexural Strength

The modulus of rupture at 1 and 28 days of HPC mixtures made with various proportions of limestone microfiller and silica fume is shown in Fig. 21. At 1 day, the limestone filler mixtures had higher flexure strengths than a reference OPC mixture, perhaps due to the more advanced hydration and better development of bond. At this age, the silica fume mixtures had decreased flexural strengths with increased replacement levels, although the mixtures outperformed an OPC concrete up to 20% silica fume content. At 28 days, the flexural resistance of the limestone filler mixtures compared positively to an OPC mixture with up to 15% replacement. Beyond this level the flexural resistance tended to decrease with the addition of limestone powder. Escadeillas[4] has also observed that at 28 and 90 days, the filler replacement of cement caused a decrease in the flexural strength.

On the other hand, the 28-day flexural resistance of the silica fume concrete increased with increased replacement rates. The paste-aggregate bond represents the major factor affecting the resistance of concrete to tensile stresses. Thus, the densification of the transition zone by the silica fume particles and the consumption of portlandite in the pozzolanic reaction may explain this improvement in the flexural strength. It has been observed earlier[4] that at early ages, a higher adhesion was obtained with cements containing the finest fillers. This was explained by a decreased thickness of the paste-aggregate transition zone and a decreased orientation of the portlandite crystals. This was confirmed by Nehdi[38]; the flexural strength was higher the finer the microfiller (Fig. 22).

Microfiller Effect in Triple-Blended Composite Cements

The iso-strength curves at various ages over the selected experimental domain, which resulted from the factorial experimental program described earlier, are illustrated in Fig. 23. The proportion of LF in the triple-blended binder dominated the very early strength development [Fig. 23(a)]. The 12-h compressive strength in a 0% LF binder was around 1 MPa, while it reached 9 MPa in a 10% LF binder. The effect of SF on the 12-h strength was less significant. A 10% SF binder could achieve around 4.5 MPa at 12 h. Optimal conditions were observed at combined high LF and SF levels. A 10% LF–10% SF triple-blended binder could achieve 12 MPa at 12 h; about one order of magnitude higher than a pure OPC.

Kessal et al.[60] obtained similar results, though they used fundamentally different materials. A low-heat cement having lower C_3A and C_3S contents and lower fineness was employed with a coarser limestone filler (mean particle size = 3 μm). A melamine-based superplasticizer was selected to reduce the retarding effect. It was possible to develop a high-early-strength concrete without increasing the heat of hydration, which is of interest for mass concrete applications.

The 1-day compressive strength was also significantly influenced by the LF proportion in the triple-blended composite cement (Fig. 23). The high-

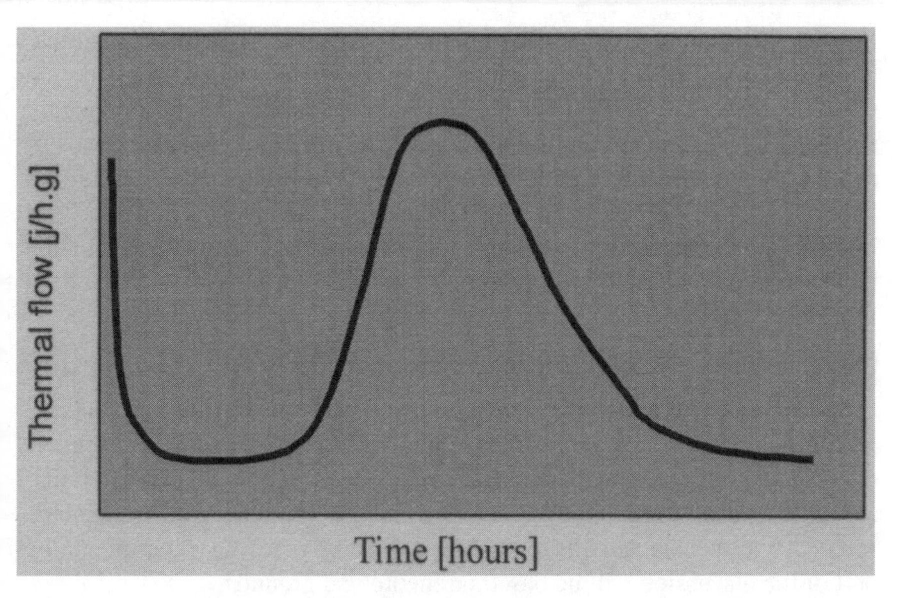

Figure 24. Typical isothermal calorimetric curve for the hydration of portland cement.

est 1-day strength was obtained in a 6–10% LF binder with no SF. As the SF proportion increased, the 1-day strength at high LF proportions became lower, probably because of the low cement content and because much of the hydration of the SF had not occurred yet.

At 3 days, the LF had only a limited effect on the strength for SF contents lower than 5%. For higher SF levels, the strength decreased as the LF proportion increased [Fig. 23(c)]. At 7 days, the presence of SF was a dominant factor, and the compressive strength increased with increased SF proportions regardless of the LF levels [Fig. 23(d)]. The LF proportion did not seem to have a significant effect on the strength at this age. Optimal strength was obtained using a 4–7.5% SF / 2.5–7.5% LF triple-blended binder. Combined high levels of SF and LF were advantageous at this age as opposed to the age of 3 days.

The 28-day compressive strength increased with higher SF proportions and decreased with higher LF levels. A similar trend was observed at 91 days. It is interesting to point out that the 28-day strength obtained using an OPC can be outperformed using 20% replacement of the cement by a combination of 10% LF–10% SF. Figure 18 shows that a 10% LF2–10% SF binder would have a compressive strength 10% higher than that of a pure OPC binder at 28 days, using a pan mixer.

If an OMNI mixer with higher fine-particle dispersive capability is used, the strength of the composite binder can reach 16% higher than the pure OPC. At a lower w/b ratio of 0.25, a 15% LF2–15% SF binder would achieve a 20% higher strength than a pure OPC at 28 days. These kinds of triple-blended cements can achieve increased strength at both early and later ages, while saving energy and raw materials, reducing toxic emissions into the atmosphere, and improving the rheological characteristics of the HPC mixtures.

Mechanisms of Microfiller Effect on Early Age Properties

It has been shown above that the presence of very fine calcite in concrete mixtures increases early-age strength development. This section discusses the possible mechanisms that could explain this effect. Some of this discussion is based on experimental results from the literature. However, since some of the mechanisms have not been directly observed experimentally, part of the discussion will be based on theoretical grounds.

Impermeable C-S-H Layer Theory

To better understand why a limestone microfiller seems to accelerate the early age strength development of cement-based materials, it is pertinent to approach this topic from the point of view of the main theories for the early hydration of cement. Due to the complexity of the reactions between cement and water, most research in this field has been carried out on dilute suspensions of alite. Nonetheless, it has been observed that the mechanisms underlying the induction period and the onset of hydration are similar in alite and cement systems.

When cement comes into contact with water, a reaction that is characterized by a short and intense heat release occurs (Fig. 24). An induction period that extends up to several hours follows, in which low heat release and hydration activity take place. An acceleration phase involving high hydration activity is then initiated. Subsequent events are of little relevance for early-age behavior.

Stein and Stevels[61] among others have suggested that a layer of C-S-H forms on the surface of C_3S particles in the early stages of hydration. The permeability of this layer is assumed to be sufficiently low to inhibit further hydration, which explains the onset of the induction period. Later on, it is thought that this layer would convert through a recrystallization process into a more permeable structure. Thus, the rate of hydration would increase, which explains the initiation of the acceleration phase.

Several theories have been put forward to explain the transformation of the impermeable C-S-H layer into a more permeable one. For instance, Damidot and Nonat[62] believe that this transformation occurs when the lime concentration in the solution reaches a certain level (22 mM), and that the lime concentration is the most crucial parameter in the C_3S hydration. On the other hand, Gartner and Jennings[63] support the idea that the conversion of this impermeable layer proceeds via a solid state reaction, but depends essentially on the concentration of CaO and SiO_2 in the solution. Experimental results have shown that the initial nucleation of CH and the end of the induction period coincide with the maximum Ca^{2+} concentration. This observation has led Young et al.[64] among others to theorize that the induction period will end when CH starts to grow from the solution. This would occur when the solution is saturated in Ca^{2+} and OH^- ions. The removal of these ions from the solution by means of CH nucleation would reactivate the C_3S hydration, thus ending the induction period. There is no consensus for this explanation. Others have found that crystalline CH formed even before the end of the induction period and the Ca^{2+} concentration peak.[65,66] Skalny and Young[67] explained this CH formation previous to the Ca^{2+} peak by a possible nucleation of CH in the vicinity of the C_3S particles where the Ca^{2+} concentration is locally higher than in the rest of the solution.

Osmotic Membrane Theory

Powers[68] and Double et al.[69] proposed the osmotic membrane theory to explain the occurrence of the induction period. It is believed that water can diffuse through the primary membrane that forms around C_3S particles at the very early stages of the hydration of cement, and dissolves calcium and silicate ions. By virtue of their smaller size and better mobility, the calcium ions can diffuse out into the bulk solution, while the silicate ions are restrained by the hydration membrane. This theory is experimentally supported by the high concentration of calcium ions and the very low concentration of silicate ions in the bulk solution measured during the early stages of hydration. This differential diffusion process causes a concentration gradient between the solution inside versus the solution outside the primary hydration layer. This generates osmotic pressures that would reach, by the end of the induction period, a level capable of bursting this hydration layer. The previously blocked hydrosilicate solution combines with the calcium ions to produce C-S-H. Thus, the rupture of the protective layer marks the end of the induction period and the onset of the acceleration phase.

Others[65] believe that crystal defects generated in the processing stage of the anhydrous cement manufacture determine to a large extent the subsequent hydration characteristics. These crystal defects constitute active sites for the growth of hydration products. Once the initial hydration nuclei reach a certain critical size, they start to grow rapidly, which marks the end of the induction phase. However, recent environmental electron microscope investigations[70] confirm the previous theories, which argue that a certain layer of hydration products surrounds the C_3S particles at the early stage of the hydration reaction.

Nucleation Mechanism Theory

Three mechanisms are believed to be involved in the process of hydration: a diffusion mechanism through the hydration layer surrounding the anhydrous grains, a phase-boundary interaction, and a nucleation and growth mechanism. It is conceivable that ultrafine particles may act as preferential substrates for the initiation of hydration nuclei. The nucleation and growth process might in fact be a predominant factor in determining the rate of hydration and strength development at early ages. At first it would seem that ultrafine particles provide more sites for the growth of hydration products, thus increasing the early age strength. Previous results in this chapter showed that limestone microfiller enhanced the early age strength more than silica fume, though the latter is finer and would provide more numerous fine sites for nucleation.

Jiang et al.[71] observed that limestone, titanium dioxide, and barium carbonate fillers accelerated the cement hydration, while quartz and alumina retarded the hydration. They suggested that the filler acceleration effect depends on the number and nature of interparticle contacts achieved in a cement-filler system. Only fillers with an acceleration effect would have coagulation contacts on which nuclei of hydration products will initiate.

However, rheological results presented earlier, in addition to zeta-potential measurements by Kjellsen and Lagerblad,[72] provide no evidence for such coagulation. On the contrary, the rheology of cement suspensions was improved when ultrafine limestone particles were added to the mixture, which contradicts any extra coagulation of the system due to calcite.

Likewise, Beedle et al.[73] have observed that graphite and α-alumina did not affect the rate of hydration, while clays and γ-alumina had a significant acceleratory effect. Being a surface mechanism, the nucleation will thus depend on the surface characteristics of the filler, namely its chemical composition, atomic structure, and surface morphology. However, calcite does not have a strong ionic nature, and would thus not precipitate ions through particularly high interatomic forces. Rather, it would constitute a preferential substrate for the germination and growth of hydration products, thus accelerating the hydration process.

Nucleation of C-S-H or Nucleation of CH?

Although there is abundant literature suggesting the nucleation mechanism as the primary cause for the microfiller acceleration of the early age hydration, there is a certain confusion as to what is nucleating: C-S-H or portlandite or both, and in what sequence?

For instance, Ramachandran and Zhang[18] suggested that calcite not only accelerates the hydration of C_3S, but that a certain percentage of calcite is consumed before 1 day in this process. They explained the increased hydration rate by the nucleation of C-S-H around calcite particles, which would incorporate a part of the calcite in some kind of a composite. They supported this idea by 1-day SEM micrographs illustrating the growth of C-S-H around calcite grains. Kjellsen and Lagerblad[72] argued that the acceleration of the hydration is initiated during the induction period, before any notable long-range nucleation of C-S-H can be expected. Therefore they questioned the above explanation.

A rational explanation for the accelerating effect of microfillers should account for the fact that the mechanism responsible for such an acceleration is initiated within the induction period, and be compatible with the theories of hydration mentioned earlier. During the induction period, the solution is increasing in its concentration of calcium and hydroxyl ions. Calcite does not have a strong ionic character, and would not exhibit particularly strong ionic attractions to precipitate calcium and hydroxyl ions on its surface. But let us assume that calcite has an affinity to grow CH on its surface, or, in other words, CH energetically prefers to germinate on calcite as a substrate. Thus, in the presence of calcite, the removal of calcium ions from the solution would start earlier, and would enhance further dissolution of calcium ions. By the same token, there will be a stronger concentration gradient for the migration of calcium ions through the hydration membrane that engulfs the cement particles.

The continued germination of CH on calcite implies an increased dissolution of C_3S since the system will tend to restore the equilibrium between the surrounding solution and the hydration membrane. Furthermore, more

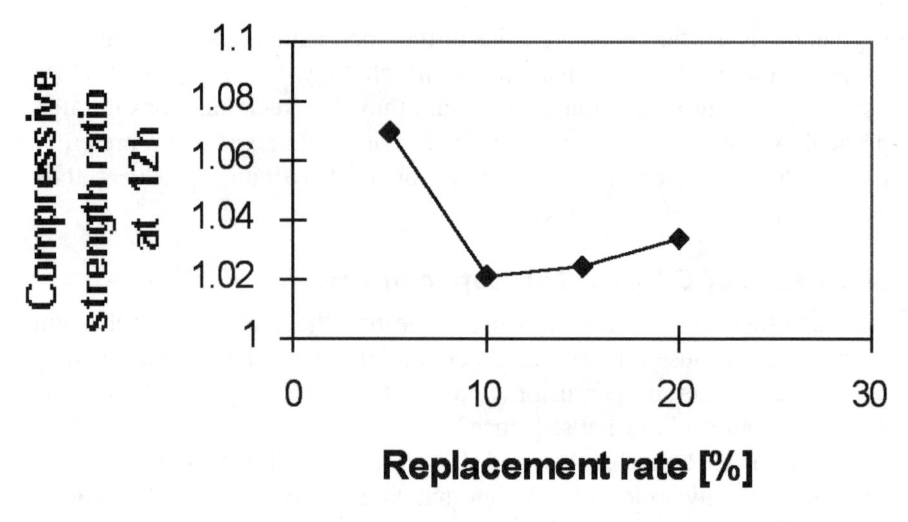

Figure 25. Compressive strength ratio at 12 h of limestone to hydrated lime HPC mixtures.

osmotic pressure would have built up since more silicate ions are now trapped inside the hydration shell. This promotes the conditions for rupture of this shell, and thus an earlier end of the induction period. By the same token, this would stimulate the appropriate thermodynamic conditions for a recrystallization of the hydration shell around the C_3S particles into a more permeable structure. This also signals an earlier end of the induction period. This explanation thus seems to be compatible with the hydration theories stated earlier. During the acceleration period, calcite may continue to be a preferential site for CH and C-S-H growth, which is compatible with the observations of Ramachandran and Zhang[18] and others.

It is conceivable that calcite can be a preferential substrate for the nucleation and growth of seemingly compatible calcium-bearing CH and other hydration products. However, why would other noncarbonate microfillers also have an acceleration effect on the hydration of cement, although not to the same extent? The results discussed earlier illustrated that silica fume also has a significant acceleration effect. As mentioned earlier, Beedle et al.[73] reported that microfillers of α-alumina and graphite had no effect on the hydration rate and titania had little effect, while clays and γ-alumina had a significant effect. Their explanation for the acceleration of the rate of hydration is compatible with the CH nucleation theory above. Amorphous silica and clays would reduce the calcium ion potential in the solution

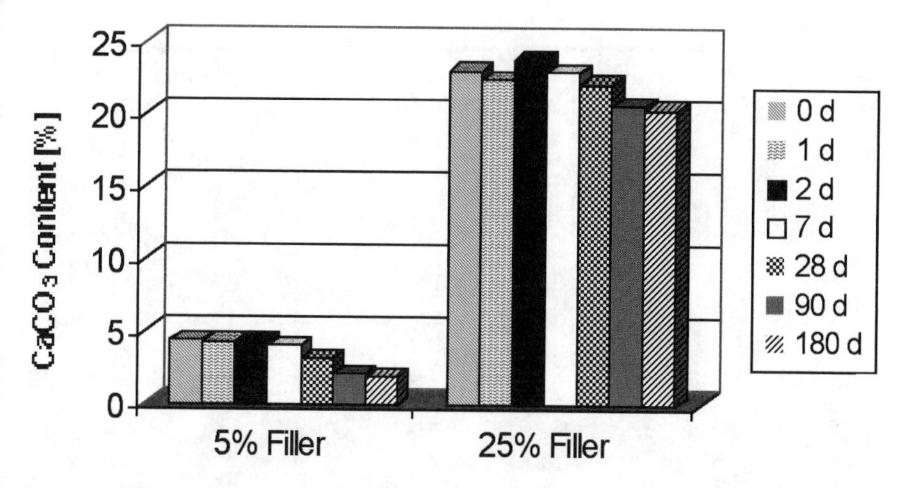

Figure 26. Illustration of consumption of $CaCO_3$ in the hydration of cement paste.[74]

through the introduction of silicate ions into the system. This is similar to the reduction of this potential by removal of calcium ions during CH germination and growth. Both mechanisms catalyze further dissolution of C_3S because the system will tend to shift toward equilibrium. Hence, the acceleration of cement hydration due to some noncarbonate microfillers is compatible with the explanation based on the nucleation and growth of CH discussed earlier.

Effect of Calcium Ions Dissolved from Calcite

It might also be that calcite would dissolve in the limestone-cement-water system, providing extra Ca^{2+} ions and thus modifying the kinetics of the hydration reactions. Because of the low solubility of calcite in alkaline systems, this theory does not seem to be thermodynamically plausible. To investigate this aspect further, HPC mixtures were prepared with 5, 10, 15, and 20% of limestone filler and hydrated lime powder. The hydrated lime is much more soluble in water than limestone powder and would provide many more Ca^{2+} ions in the solution, thus further increasing the rate of hydration. Fig. 25 shows the compressive strength ratio of limestone powder to hydrated lime mixtures at 12 h. The limestone powder slightly outperformed the hydrated lime at all replacement levels. However, since the limestone powder had higher fineness, and since the affinity to CH growth of the two microfillers might be different, the real contribution from the calcium ions dissolved from the calcite remains uncertain.

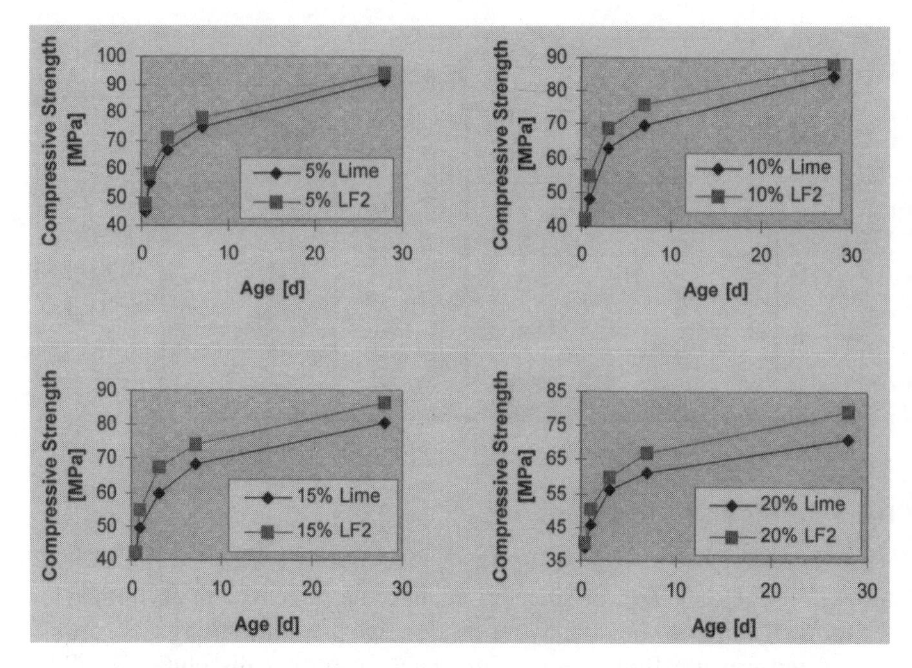

Figure 27. Comparison of the effect of limestone and hydrated lime partial replacement of cement on compressive strength (w/b = 0.33).

Effect of Reaction between Calcite and C_3A

The potential reaction between calcite and calcium aluminate in the presence of water to produce calcium carboaluminates ($C_3A \cdot CaCO_3 \cdot 11H_2O$) was discussed above. The question is, could this reaction be responsible for the reduction of the induction period and the increase of the rate of the early hydration? Or is it possible that the formation of carboaluminates enhances the early age strength in the same way as sulfoaluminates do?

The attempt to substitute limestone for gypsum as a set regulator rather supports the idea that this reaction would form carboaluminates around the C_3A grains, thus preventing their further hydration. This is analogous to the set control action of gypsum through the formation of ettringite. Also, based on the observation that calcite accelerates the hydration of pure C_3S (no potential formation of carboaluminates), the hypothesis that the reaction between calcite and calcium aluminate is responsible for the acceleration of the hydration reactions is unlikely.

Using X-ray diffraction of limestone filler cement pastes, Escadeillas[4] observed that the percentage of calcite started to decrease after the first few

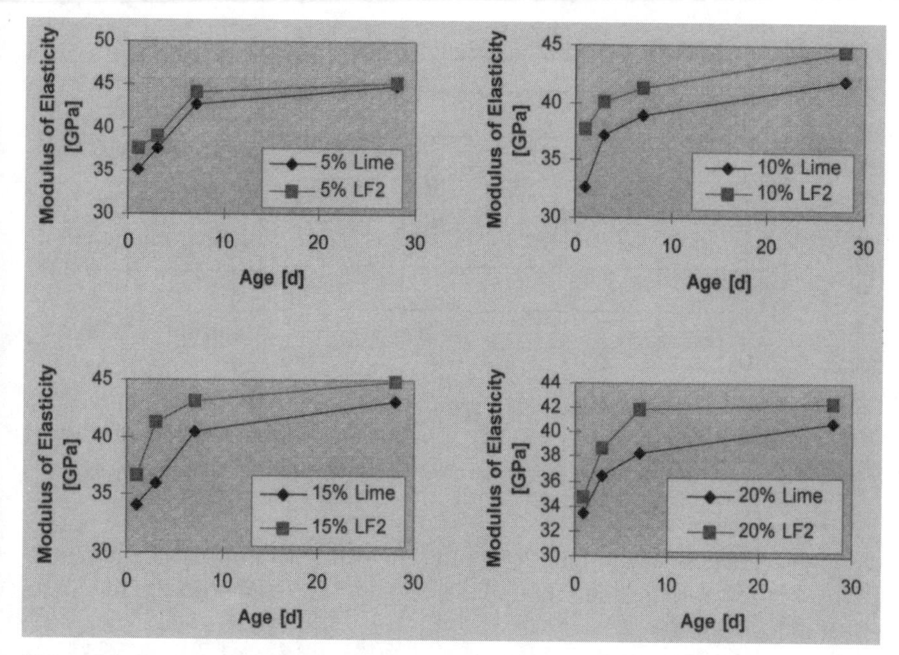

Figure 28. Comparison of the effect of limestone and hydrated lime partial replacement of cement on modulus of elasticity (w/b = 0.33).

hours of hydration. However, carboaluminates were not detected before 7 days, either because they were too small to be detected by X-ray diffraction, or because the crystallization of carboaluminates is only possible at high carbonate levels in the solution. It was claimed that the transformation of aluminates to carboaluminates is practically complete at around 9 months. Fig. 26 shows the calcium carbonate variation up to the age of 180 days in and 25% limestone filler cement pastes. Since most of the consumption of $CaCO_3$ occurred at later ages, it is unlikely that the production of calcium carboaluminates is responsible for the very early age strength enhancement.

Figures 27 and 28 compare the compressive strength and modulus of elasticity values measured on HPC mixtures made with various proportions of limestone or hydrated lime powder. At early ages, hydrated lime and limestone provided comparable effects on the rate of hydration and enhanced the early strength development. At later ages, the limestone microfiller led to higher strength and modulus values. This difference tended to be more significant as the replacement rate increased. It might be that

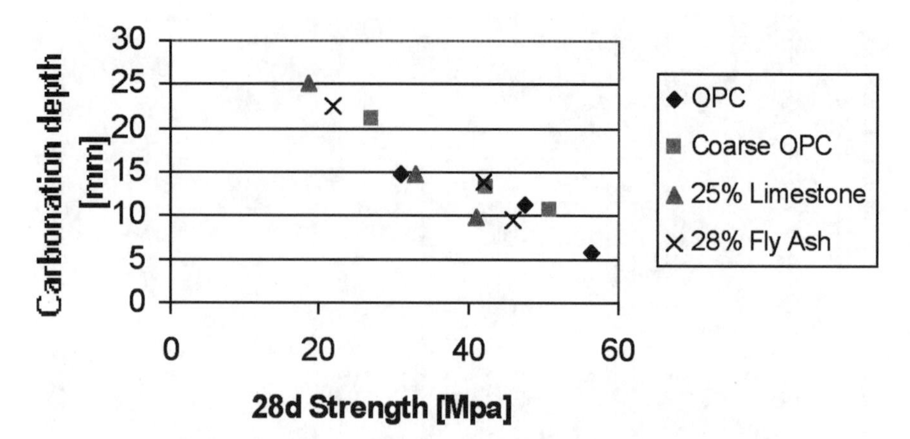

Figure 29. Relationship between 5-year carbonation depth and 28-day strength.[79]

the higher fineness of the limestone filler yielded a better filler effect. But it is also conceivable that carboaluminates could have enhanced the limetone effect at later ages.

Limestone Microfiller Effect on Durability of Concrete

Carbonation Potential in Limestone Filler Cement

The presence of $Ca(OH)_2$ is primarily responsible for maintaining the high alkalinity level of concrete. As CO_2 diffuses into the concrete and reacts with the $Ca(OH)_2$ to form calcium carbonate, the pH may be reduced to a level at which the passive oxide layer on the reinforcing steel becomes unstable. Therefore, oxidation of the steel can proceed. To what extent does the addition of carbonates to cement affect the carbonation depth and carbonation rate of concrete? Ingram and Daugherty[7] stated that "one of the most interesting side effects of carbonate additions to cement is that atmospheric carbonation damage might be reduced." They proposed two mechanisms for this possible improvement: The first is that the CO_3^{2-} and CO_2 react in hydrating cement to produce the bicarbonate ion HCO_3^-. In such a scenario, two moles of CO_2 are consumed by one mole of incorporated carbonate before attack of the calcium hydroxide proceeds. The second mechanism is that CO_2 does attack the $Ca(OH)_2$ normally, but the system stays basic for a longer time owing to the presence of calcium carbonate.

Closer examination of the published data regarding this issue reveals that such a statement is very optimistic. Research carried out by the Building

Research Establishment[47] at three independent laboratories resulted only in two widely held views: The carbonation depth increases proportionally to the square root of time, and the carbonation rate is inversely proportional to the compressive strength of concrete (Fig. 29). There was no evidence that the type of cement had any significant effect on carbonation. This trend applies also to the previous data acquired during the relatively long French experience with the CPJ cements.[37]

Potential of Steel Corrosion in Limestone Filler Cement

There are certain requirements for the corrosion of steel embedded in concrete to proceed. The first of these requirements is inherent to the steel itself and to the presence of microstructural and stress gradients within it. These are beyond the scope of this discussion. The second requirement is the reduction of the alkalinity of the concrete (usually around 12.5–13) to a critical value (usually below 11), which causes the passive oxide layer coating the steel rebars to become unstable and eventually be destroyed. The major reason for this reduction is the carbonation reaction discussed above. Based on the carbonation criterion alone, there is no evidence to believe that a drastic change in the corrosion behavior should be expected in limestone filler cements.

Nevertheless, chloride ions have the special ability to destroy the passive oxide film of steel even if high alkalinity conditions prevail in the concrete. The attack of this passive layer can create an electrochemical process in which the unprotected part of the steel can act as an anode (oxidation of iron) and the remaining steel can act as a cathode (reduction of water). For such an electrochemical process to be maintained, there is need for an electrolyte, which requires the presence of both oxygen and moisture.

The difference in the corrosion potential of embedded steel in concrete made with OPC or limestone filler cement will be directly related to the resistance of the "skin" of the concrete to the mechanisms of mass transfer, that is, the diffusion of chlorides, oxygen, and moisture. The mechanism of chloride incorporation in the HCP (which delays the further ingress of chlorides) is mainly related to the formation of Friedel salt, which in turn depends on the C_3A content of the cement. This will not be discussed further here.

The French experience with the CPJ cements[76] has demonstrated that the diffusion of the chloride ions is primarily dependent on the strength grade of the cement, and is not related to the amount of carbonate that is incorporated. Research carried out at the Building Research Establishment[75] has

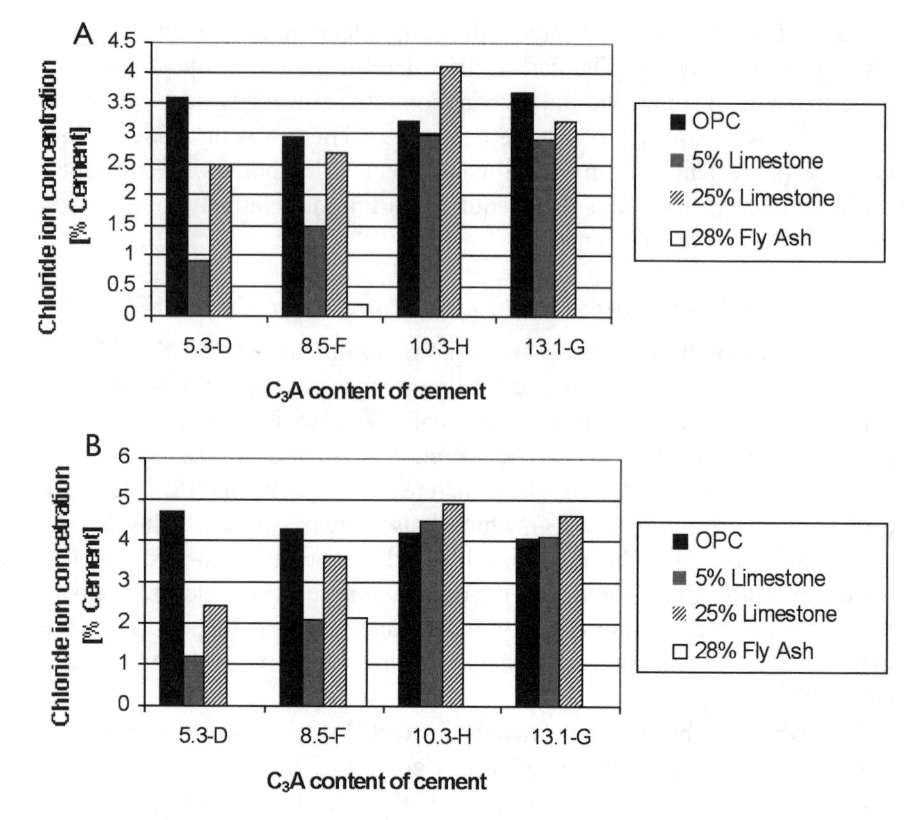

Figure 30. Five-year chloride concentrations: (a) interval 6–11 mm and (b) interval 26–31 mm.[79]

shown no clear trends in the chloride diffusion of cements with or without carbonate additions (Fig. 30). Although the 25% filler cement showed higher diffusion rates, these results are comparable to some OPC controls in the study. More recent work[77] showed that the addition of limestone filler reduces the diffusion coefficient of chloride ions. This was attributed to the effect of limestone filler particles on the tortuosity of the system.

Relatively lean concrete prisms were also placed in a tidal zone exposure site.[75] The 25% limestone cements appeared to allow for greater corrosion rates than OPC. No difference was observed with respect to the 5% filler cements. Expressing the chloride concentrations relative to the weight of the OPC, rather than the total weight of the binder, leads to the 25% filler cement lying on the same curve describing the behavior of the other

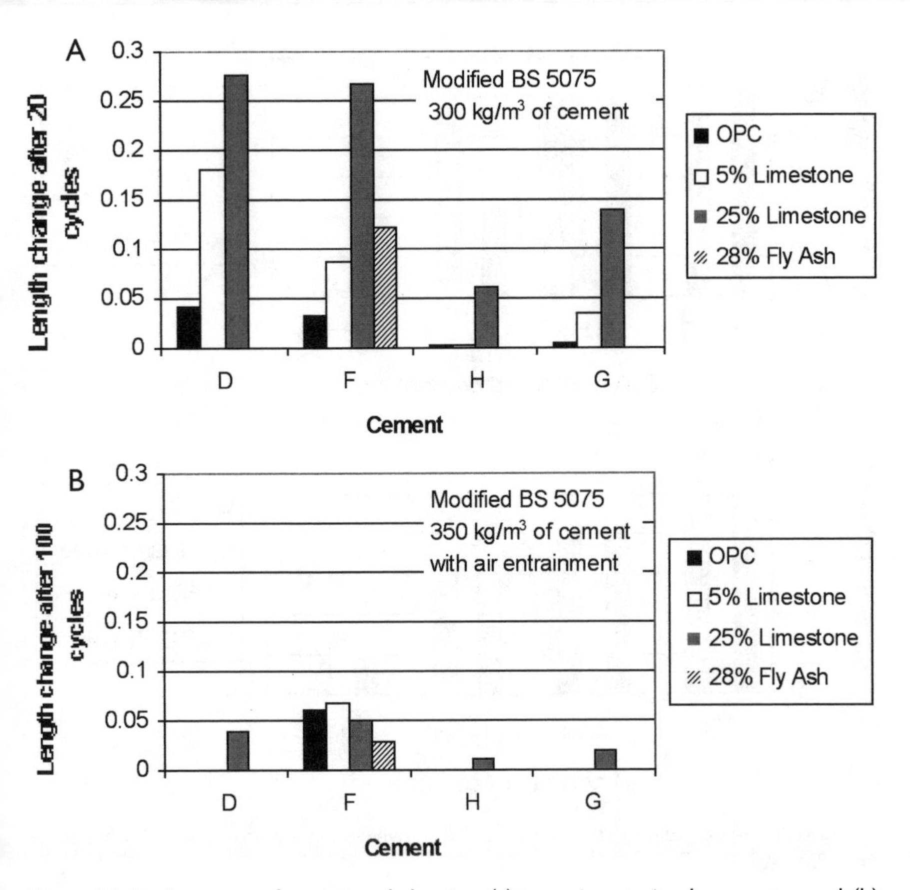

Figure 31. Resistance to freezing and thawing: (a) non-air-entrained concretes and (b) air-entrained concretes.[79]

cements. This indicates that the strength grade of concrete and its porosity, rather than its carbonate content, controls the diffusion of chlorides. The same research program showed that the filler cements tended to have a somewhat lower permeability than oxygen. Their porosity was shown to increase with the filler content, but this effect was very slight.[53,75]

Freeze-Thaw Performance

It was shown[78] that for a constant water/binder ratio, the critical average spacing of air bubbles decreased with an increase of the carbonate addition to cement. If the comparison was carried out on the basis of the same w/c

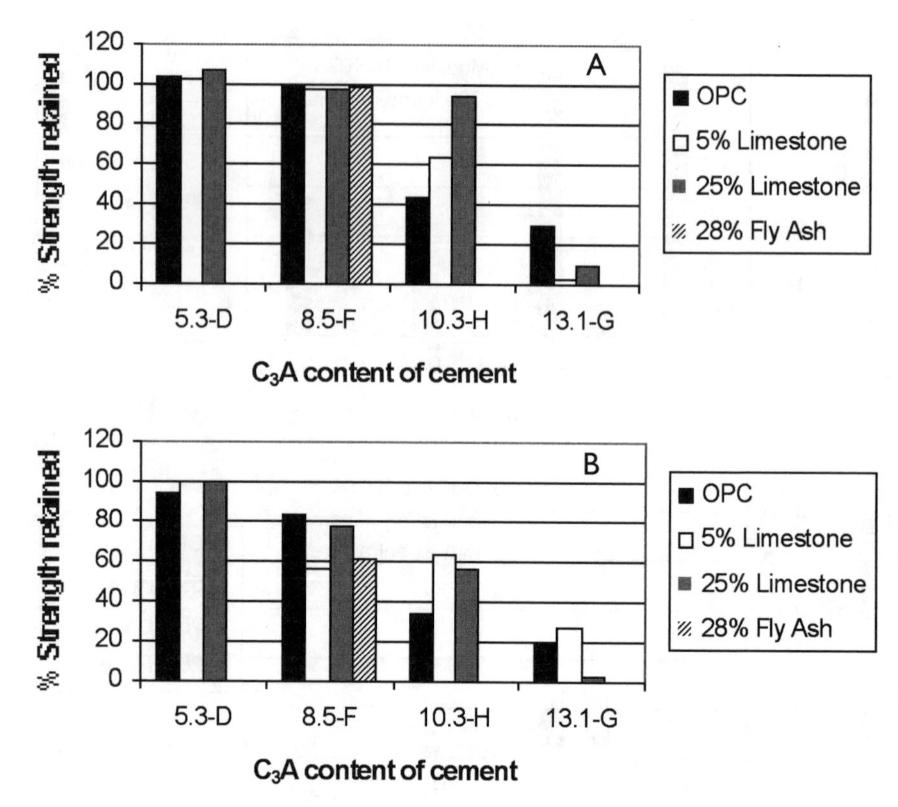

Figure 32. Two-year sulfate resistance: (a) 1.5% magnesium solution and (b) 1.5% sodium sulfate solution.[79]

ratio instead of the w/b ratio, the behavior of the filler cement was similar to the OPC. Livesey[53] reported that the performance of cements in freeze/thaw tests was improved as the filler limestone replacement level was reduced, although the performance of a 25% limestone replacement was comparable to that of fly ash cements. It was also shown that below a replacement threshold of 15%, the freeze/thaw resistance of cements was not an issue.[37] Matthews[79] has reported the results of a study on the frost performance of limestone filler cements using ASTM C-666 (Standard Test Method for Resistance of Concrete to Rapid Freezing and Thawing) and an adaptation of BS 5075 (Part 2, Appendix C, Test for Resistance to Freezing and Thawing). It was observed that the freeze/thaw resistance became poorer as the level of limestone addition increased, although some 5% limestone cements gave better results than the poorer control cements. When an appropriate level of air was entrained in the concrete mixtures, the performance of the limestone filler concrete was satisfactory (Fig. 31).

Sulfate Attack: A New Concern

The sulfate resistance of cements made with finely ground carbonate additions has been studied by several investigators. Hooton[80] carried out a study on three pairs of cements produced with limestone additions and having medium and high contents of C_3A, using both ASTM C452 (Test for Potential Expansion of Portland Cement Mortars Exposed to Sulfate) and ASTM C1012 (Test for Length Change of Hydraulic Cement Mortars Exposed to Sulfate Solution). Whether the specimens passed or failed for moderate sulfate resistance limits was unaffected by the carbonate additions. The French experience with CPJ cements supports the widely held opinion that limestone additions to cement increase its sulfate resistance, and that this is more noticeable in cements rich in C_3A.[37] Ingram and Daugherty,[7] Livesey,[53] and Matthews[79] reported the same effect (Fig. 32). Only one study[81] was available in which cement pastes containing large proportions of limestone showed decreased resistance to sulfate attack as measured by expansion. The positive effect on the sulfate resistance of limestone filler cement is so well-documented that a German patent was actually issued for sulfate-resistant portland cement containing finely ground limestone.[82]

The chemical reactions involved in the sulfate attack that have been identified are the transformation of portlandite [$Ca(OH)_2$] to gypsum and calcium aluminate hydrate to ettringite ($3CaO \cdot Al_2O_3 \cdot 3CaSO_4 \cdot 32H_2O$). To account for these mechanisms, sulfate-resisting portland cement (SRPC) having a low level of tricalcium aluminate was formulated. SRPC has been used satisfactorily where sulfate resistance has been specified, though to be effective w/c ratios must also be held to less than 0.50 for moderate levels of sulfates, and to less than 0.45 (even 0.40) for severe sulfate concentrations. In recent years, however, cases of severe sulfate attack have been reported in concrete foundations specifically designed to provide good sulfate-resisting properties.[83–85] A third chemical reaction was identified as responsible for these degradations. Where a wet and cold environment prevails and a constant source of sulfate and carbonate ions is provided, C-S-H reacts to form the mineral thaumasite ($CaSiO_3 \cdot CaCO_3 \cdot CaSO_4 \cdot 15H_2O$). In this regard, SRPC is as vulnerable as normal cement.

Thaumasite formation is not simply an expansive reaction leading to the cracking of concrete; it affects in a more fundamental way the binding capabilities of the hydrated cement paste. The concrete undergoes a considerable softening, which in structural terms means that the concrete element loses its load-bearing capacity. Comparing the conventional expansive reaction due to ettringite formation to the thaumasite form of sulfate attack, Taylor[86] states that "the quantity of ettringite that can be formed from a cement is limited by the amount of Al_2O_3 available, but assuming a continuing source of SO_4^{2-}, thaumasite formation is limited only by the available CaO and SiO_2. It can therefore form in large quantities, reducing a mortar into a mush."

A detailed list of the published papers and technical reports regarding thaumasite[87] and its occurrence in modern structures[88] and a review of the analytical methods used to investigate its presence[89] are available. The thaumasite form of sulfate attack in concretes containing ground calcium carbonate was reported only for very limited cases and the extent of the problem in practice is not yet clear.[85] However, in most reported cases of thaumasite attack, an available carbonate source was identified. After the abundant literature on the beneficial effects of carbonate additions to cement on its sulfate resistance, the thaumasite reaction might be a surprise.[§] To date, only a few precautions against this form of sulfate attack have been recommended: one for concretes buried in sulfate soils in the UK[91]; the other for concreting in cold regions in the Canadian Arctic.[84] Other cold and wet parts of the world with sulfate soils, including the Prairie Provinces of Canada and parts of the western United States, present a possible scenario for the thaumasite reaction. Since the potential of this reaction with local cements, especially when carbonates are added, is not fully understood, more research is required.

Concluding Remarks

More than ever before, the cement and concrete industry is compelled to preserve resources, reduce waste, and become more environmentally efficient. With the advent of the century of the environment and the infrastructure and housing needs of a steadily growing world population, better use of existing resources, development of new resources, and implementation of better technologies are needed if this industry is to satisfy the goals of sustainable development.

The construction industry is still far from achieving optimal use of cement. Superplasticizers can decrease the water demand, and therefore the

[§]In Greek, "thaumasite" literally means to be surprised.[90]

amount of cement needed for equal strength. However, they are not used in all concrete construction projects. In high-strength concrete, a significant part of the cement acts merely as a microfiller due to the limited hydration (low w/c ratio). Several supplementary cementing materials are not fully utilized due to their transportation costs or simply because of the lack of implementation of technology related to their use. Technology can also be used to increase the strength of clinker, which theoretically permits the replacement of part of the cement with an inert filler while maintaining equal strength. A pertinent example in this regard is the use of mineralized clinker.[92] The introduction of controlled levels of fluoride and sulfate in high-C_3S clinkers has been shown to markedly enhance the early strength potential of cement. The industrial use of such options is however limited. Often, simply moving away from closed-circuit grinding of cement, which results in fairly uniform particle sizes, toward methods that result in a wider range of particle sizes would allow a better use of cement.

Some inert microfillers, including a large variety of industrial wastes, can be partially substituted for cement at rates of 5–35% to produce cements of different strength classes. This can provide economic benefits and environmental relief. Furthermore, if a proper knowledge of the effect of these wastes on the particle-packing density of cement blends, hydration reactions, microstructure, strength development, and durability of concrete is developed, they can even increase the mechanical strength. However, the development and implementation of microfiller-blended cements faces controversy in the technical community and divergence of national standards.

The purpose of this chapter has been to demonstrate (using limestone powder as an example) that it is possible to develop microfiller-blended cements that provide enhanced rheological properties and in some cases better mechanical properties. Increasing amounts of data in the literature support this idea. What is needed now are forward-looking cement producers that will undertake the production of microfiller-blended cements, and appropriate adjustments within the construction industry to deliver the right cement for the right job.

References

1. V.M. Malhotra, "Fly ash, slag, silica fume, and rice-husk ash in concrete: A review," *Concr. Int.,* **15** [4] 23–28 (1993).
2. S. Nagataki, "Mineral admixtures in concrete: State of the art and trends"; p. 447 in *Concrete Technology: Past, Present, and Future.* V.M. Malhotra Symposium, ACI SP-144. Edited by P.K. Mehta. 1994.

3. RILEM Technical Report, 73-SBC RILEM Committee, "Final Report on Siliceous Byproducts for Use in Concrete," *Mater. Constr.*, **21** [121] 69–80 (1988).
4. G. Escadeillas, "Les ciments aux fillers calcaires: Contribution à leur optimisation par l'étude des propriétés mécaniques et physiques des bétons fillérisés," Doctoral Thesis in Civil Engineering, Université Paul-Sabatier, Toulouse, France, 1988.
5. A.T. Corish, "European cement standards"; pp. 2.1–2.25 in *Proc.: Performance of Limestone Filled Cements,* BRE/BCA Garston Seminar. 1989.
6. L.L. Mayfield, "Limestone additions to cement—An old controversy revisited," *Cem. Concr. Aggregates,* **10** [1] 3–8 (1988).
7. K.D. Ingram and K.E. Daugherty, "A review of limestone additions to portland cement and concrete," *Cem. Concr. Comp.,* **13** [3] 165–170 (1991).
8. G.S. Bobrowski, J.L. Wilson, and K.E. Daugherty, "Limestone substitutes for gypsum as a cement ingredient," *Rock Prod.,* **80** [2] 404–410 (1977).
9. A. Negro, G. Abbiati, and L. Cussino, "Calcium carbonate substitute for gypsum as a set regulator," *Il Cemento,* **4**, 537–544 (1986).
10. J. Bensted, "Some hydration investigations involving portland cement — Effect of calcium carbonate substitution of gypsum," *World Cem. Technol.,* **11** [8] 395–406 (1980).
11. J. Bensted, "Further hydration investigations involving portland cement and the substitution of limestone for gypsum," *World Cem. Technol.,* **14**, 383–392 (1983).
12. V.C. Campitelli and M.C. Florindo, "The influence of limestone additions on optimum sulfur trioxide content in portland cements"; pp. 30–40 in *Carbonate Additions to Cement,* ASTM STP 1064. 1990.
13. G.E. Bessey, "Calcium aluminate and calcium silicate hydrate"; pp. 178–215 in *Proceedings of the Symposium on the Chemistry of Cements, Stockholm,* Vol. I. 1938.
14. T. Manabe, N. Kawada, and M. Nishiyama, "Studies on $3CaO{\cdot}Al_2O_3{\cdot}CaCO_3{\cdot}nH_2O$ (Calcium monocarboaluminate hydrate)," *CAJ Rev.,* pp. 48–55 (1961).
15. R.F. Feldman, V.S. Ramachandran, and J.P. Sereda, "Influence of $CaCO_3$ on the hydration of $3CaO{\cdot}Al_2O_3$," *J. Am. Ceram. Soc.,* **48** [1] 25–30 (1965).
16. V.S. Ramachandran and C.-M. Zhang, "Hydration and microstructural development in the $3CaO{\cdot}Al_2O_3$–$CaCO_3$-H_2O system," Materials and Structures, Vol. 19, No. 114, pp. 437-444 (1986).
17. C. Vernet and Noworyta, "Mechanisms of limestone filler reactions in the system {C_3A-CSH_2-CH-CC-H}"; pp. 430–436 in *9th International Congress on the Chemistry of Cements, New Delhi,* Vol. IV. 1992.
18. V.S. Ramachandran and C.-M. Zhang, "Dependence of fineness of calcium carbonate on the hydration behavior of tricalcium silicate," *Durabil. Building Mater.,* **4**, 45–66 (1986).
19. S. Husson, B. Gulhot, and J. Pera, "Influence of different fillers on the hydration of C_3S"; pp. 83–89 in *9th International Congress on the Chemistry of Cements, New Delhi,* Vol. IV, III-A.013. 1992.
20. A.P. Barker and H. Cory, "The early hydration of limestone filled cements"; pp. 107–124 in *Proc.: Blended Cements in Construction.* Edited by R.N. Swamy. University of Sheffield, 1991.
21. H. Ushiyama et al., "Effect of carbonate on early stage of hydration of alite"; pp. 154–159 in *8th International Congress on the Chemistry of Cements, Rio de Janeiro,* Vol. II. 1986.

22. O. Evrard and M. Chloup, "Réactivité chimique des calcaires en milieu basique: application aux ciments et bétons," *Annales ITBTP,* N. 529, Série 316, 83–87 (1994).
23. K.E. Ingram, "Limestone additions to cement: Uptake, chemistry and effects"; pp. 180–186 in *9th International Congress on the Chemistry of Cements, New Delhi,* Vol. III. 1992.
24. C. Vernet, "Séquence et cinétique des réactions d'hydratation de l'aluminate tricalcique en présence de gypse, de chaux et de fillers calcaires"; pp. 70–74 in *8th International Congress on the Chemistry of Cements, Rio de Janeiro,* Vol. III. 1986.
25. J. Jambor, "Influence of $3CaO{\cdot}Al_2O_3{\cdot}CaO_3{\cdot}nH_2O$ on the structure of cement paste"; pp. 487–492 in *Proceedings of the 7th International Congress on the Chemistry of Cement, Paris,* Vol. IV. 1980.
26. W.A. Klemm and L.D. Adams, "An investigation of the formation of carboaluminates"; pp. 60–72 in *Carbonate Additions to Cement,* ASTM STP 1064. 1990.
27. J. Farran, "Contribution minéralogique à l'étude de l'adhérence entre les constituants hydratés des ciments et les matériaux enrobés," *Revue des Matériaux et Constructions,* No. 490–491, 155–172 and 191–209 (1956).
28. A.D. Buck and W.L. Dolch, "Investigation of a reaction involving non-dolomitic limestone aggregate in concrete," *ACI J.,* **63** [7] 755–763 (1966).
29. P.C. Aïtcin and P.K. Mehta, "Effect of coarse-aggregate characteristics on mechanical properties of high-strength concrete," *ACI Mater. J.,* **87** [2] 103–107 (1990).
30. J. Grandet and J.P. Ollivier, "Etude de la formation du monocarboaluminate de calcium hydraté au contact d'un granulat calcaire dans une pate de ciment portland," *Cem. Concr. Res.,* **10**, 759–770 (1980).
31. P.J.M. Monteiro and P.K. Mehta, "Reaction between carbonate rock and cement paste," *Cem. Concr. Res.,* **16**, 127–134 (1986).
32. J.P. Bombled, "Rhéologie du béton frais: Influence de l'ajout de fillers aux ciments"; pp. 190–196 in *8th International Congress on the Chemistry of Cements, Rio de Janeiro,* Vol. IV. 1986.
33. M. Kessal, "Développement d'un système cimentaire à haute résistance initiale à base de ciment de Type 20M," M.A.Sc. Thesis, Université de Sherbrooke, 1995.
34. S. Sprung and E. Siebel, "Assessment of the suitability of limestone for producing portland limestone cement (PKZ)," *Zement Kalk Gips,* **1**, 1–11 (1991).
35. S.N. Neto and V.C. Campitelli, "The influence of limestone additions on the rheological properties and water retention value of portland cement slurries"; pp. 24–29 in *Carbonate Additions to Cement,* ASTM STP 1064. Edited by Klieger and Hooton. 1990.
36. P. Brookbanks, "Properties of fresh concrete"; pp. 4.1–4.15 in *Performance of Limestone Filled Cements,* Proc. BRE/BCA seminar, Garston. 1989.
37. G. Cochet and F. Sorrentino, "Limestone filled cements: Properties and uses"; pp. 266–295 in *Mineral Admixtures in Cement and Concrete,* vol. 4. Edited by S.N. Ghosh, S.L. Sarkar, and S. Harsh. ABI Book Pvt. Ltd., New Delhi, India, 1993.
38. M. Nehdi, "Microfiller Effect on Rheology, Microstructure, and Mechanical Properties of High Performance Concrete," Ph.D. Thesis, University of British Columbia, 1998.
39. D.L. Kantro, "Influence of water-reducing admixtures on properties of cement paste — A miniature slump test," *Cem. Concr. Aggregates,* **2** [2] 95–102 (1981).
40. D.C. Montgomery, *Design and Analysis of Experiments.* John Wiley and Sons, New York, 1984. Pp. 261–292 and 460-470.

41. Design-Expert, Version 5.0.3, STAT-EASE Inc., 2021 East Hennepin Avenue, Suite 191, Minneapolis, Minnesota, 55413.
42. M. Nehdi, S. Mindess, and P.C. Aïtcin, "Optimization of high-strength limestone filler cement mortars," *Cem. Concr. Res.,* **26** [6] 883–893 (1996).
43. D. Beaupré, "Rheology of High Performance Shotcrete," Ph.D. Thesis, University of British Columbia, 1994.
44. M. Nehdi, S. Mindess, and P.C. Aïtcin, "Statistical modeling of the microfiller effect on the rheology of composite cement pastes," *Adv. Cem. Res.,* **9** [33] 37–46 (1997).
45. P.C. Aïtcin, "The use of superplasticizers in high performance concrete"; p. 14 in *High Performance Concrete: From Material to Structure.* Edited by Yves Malier. E & FN Spon, 1992.
46. T.C. Powers, *The Properties of Fresh Concrete.* John Wiley & Sons Inc., 1968. P. 553.
47. M. Nehdi, S. Mindess, and P.C. Aïtcin, "Optimization of triple-blended composite cements for making high-strength concrete," *World Cem. R&D,* **27** [6] 69–73 (1996).
48. H.A. Barnes, J.F. Hutton, and K. Walters, *An Introduction to Rheology.* Elsevier, 1989. P. 115.
49. P. Gegout et al., "Texture et performance des ciments fillerisés"; pp. 197–203 in *8th International Congress on the Chemistry of Cements, Rio de Janeiro,* Vol. I. 1986.
50. CRIC, "Les ciments portland au filler"; pp. 8–28 in *Centre National de Recherches Scientifiques et Techniques pour l'Industrie Cimentaire, Report RA-f-1986/87.* Belgium, 1987.
51. I. Soroka and N. Setter, "The effect of fillers on strength of cement mortars," *Cem. Concr. Res.,* **7** [4] 449–456 (1977).
52. I. Soroka and N. Stern, "Calcareous fillers and the compressive strength of portland cement," *Cem. Concr. Res.,* **6** [3] 367–376 (1976).
53. P. Livesey, "Performance of limestone filled cements"; pp. 1–15 in *Blended Cements in Construction.* Edited by R.N. Swamy. Elsevier, 1991.
54. M. Regourd, "Ciments spéciaux et ciments avec addition: Caractéristiques et activation des produits d'addition"; pp. 119–229 in *8th International Congress on the Chemistry of Cements, Rio de Janeiro,* Vol. I. 1986.
55. P. Richard and M.H. Cheyrezy, "Reactive powder concrete with high ductility and 200–800 MPa compressive strength"; pp. 507–518 in *Concrete Technology: Past, Present, and Future,* ACI SP-144. 1994.
56. P. Fidjestol and J. Frearson, "High performance concrete using blended and tripleblended binders"; pp. 135–157 in *International Conference on High Performance Concrete, Singapore,* ACI SP-149. Edited by V. M. Malhotra. 1994.
57. N.J. Gardner and J.-W. Zhao, "Mechanical properties of concrete for calculating long term deformations"; pp. 150–159 in *Proceedings, Second Canadian Symposium on Cement and Concrete, Vancouver.* 1991.
58. P.J. Monteiro, J.C. Maso, and J.P. Ollivier, "The aggregate mortar interface," *Cem. Concr. Res.,* **15** [6] 953–958 (1985).
59. M.P. Lutz and P.J.M. Monteiro, "Effect of the transition zone on the bulk modulus of concrete"; pp. 413–418 in *Microstructure of Cement-Based Systems/Bonding and Interfaces in Cementitious Materials,* MRS Sym. Proceedings Vol. 370. 1995.
60. M. Kessal, M. Edwards-Lajnef, A. Tagnit-Hamou, and P.C. Aïtcin, "L'optimization de la résistance à court terme des bétons fabriqués avec un ciment de Type 20M," *Can. J. Civ. Eng.,* **23** [3] (1996).

61. H.N. Stein and J.M. Stevels, "Influence of silica on the hydration of 3CaO-SiO_2," *J. Appl. Chem.*, **14**, 338–346 (1964).
62. D. Damidot and A. Nonat, "Investigations of the C_3S hydration process during the first hours of hydration"; pp. 23–34 in *Proceedings: International RILEM Workshop on Hydration and Setting of Cements, Dijon.* 1991.
63. E.M. Gartner and H.M. Jennings, "Thermodynamics of calcium silicate hydrates and their solutions," *J. Am. Ceram. Soc.*, **70** [10] 743–749 (1987).
64. J.F. Young, H.S. Tong, and R.L. Berger, "Compositions of solutions in contact with hydrating tricalcium silicate pastes," *J. Am. Ceram. Soc.*, **60** [5-6] 193–198 (1977).
65. P. Fierens and J.P. Verhagen, "Hydration of tricalcium silicate in paste kinetics of calcium ions dissolution in the aqueous phase," *Cem. Concr. Res.*, **6** [3] 337–342 (1976).
66. P.A. Slegers and P.G. Rouxhet, "The hydration of tricalcium silicate: calcium concentration and portlandite formation," *Cem. Concr. Res.*, **7** [1] 31–38 (1977).
67. J. Skalny and J.F. Young, "Mechanisms of portland cement hydration"; in *Proceedings of the 7th International Congress on the Chemistry of Cement, Paris,* Vol. 1. 1980.
68. T.C. Powers, "Some physical aspects of the hydration of portland cement," *J. Portland Cem. Assoc.*, **3** [1] 47–56 (1961).
69. D.D. Double, A. Hellawell, and S.J. Perry, "The hydration of portland cement," *Proc. R. Soc. London,* **A359**, 435–451 (1978).
70. K. Sujata, T.B. Bergström, and H.M. Jennings, "Preliminary studies of wet cement pastes by an environmental scanning electron microscope," *Microbeam Anal.*, **1**, 195–198 (1991).
71. S.P. Jiang, J.C. Mutin, and A. Nonat, "Effect of fillers (fine particles) on the kinetics of cement hydration"; pp. 126–131 in *Proceedings of the 3rd International Symposium on Cement and Concrete, Beijing,* Vol. 3. 1993.
72. K.O. Kjellsen and B. Lagerblad, "Influence of natural minerals in the filler fraction on hydration and properties of mortars," Swedish Cement and Concrete Research Institute, Report S-100 44. Stockholm, 1995.
73. S.S. Beedle, G.W. Groves, and S.A. Rodger, "The effect of fine pozzolanic and other particles on the hydration of C_3S," *Adv. Cem. Res.*, **2** [5] 126–131 (1989).
74. A.P. Parker and H.P. Cory, "The early hydration of limestone filled cements"; pp. 107–124 in *Blended Cements in Construction.* Edited by R.N. Swamy. Elsevier Science Publishers, 1991.
75. G.K. Moir and S. Kelham, "Durability"; pp. 7.1–7.67 in *Performance of Limestone Filled Cements,* BRE/BCA Garston Seminar. 1989.
76. G. Cochet and B. Jesus, "Diffusion of chloride ions in portland cement-filler mortars"; pp. 365–376 in *International Conference on Blended Cements in Construction.* Edited by R.N. Swamy. University of Sheffield, 1991.
77. H. Hornain, J. Marchand, V. Duhot, and M. Moranville-Regourd, "Diffusion of chloride ions in limestone filler blended cement pastes and mortars," *Cem. Concr. Res.*, **25** [8] 1667–1678 (1995).
78. P. Gegout, H. Hornain, B. Thuret, and M. Regourd, "Résistance au gel des ciments aux fillers calcaires"; pp. 47–52 in *8th International Congress on the Chemistry of Cements, Rio de Janeiro,* Vol. VI. 1986.
79. J.D. Matthews, "Sulfate and freeze thaw resistance"; pp. 8.1–8.16 in *Performance of Limestone Filled Cements,* BRE/BCA Garston Seminar. 1989.

80. R.D. Hooton, "Effects of carbonate additions on heat of hydration and sulfate resistance"; pp. 73–81 in *Carbonate Additions to Cement,* ASTM STP 1064. 1990.
81. B.K. Marsh and R.C. Joshi, "Sulfate and acid resistance of cement paste containing pulverized limestone and fly ash," *Durabil. Building Mater.,* **4**, 67–80 (1986).
82. Portland-Zementwerke Heiderberg, A.G., "Increasing sulfate resistance of portland cement," German Patent 1 646 910, May 1972.
83. N.J. Crammond and P.J. Nixon, "Deterioration of concrete foundation piles as a result of thaumasite formation"; pp. 295–305 in *6th International Conference on Durability of Building Materials, Japan.* 1993.
84. J.A. Bickley et al., "Thaumasite related deterioration of concrete structures"; pp. 159–175 in *Concrete Technology: Past, Present and Future,* ACI SP: 144. Edited by P.K. Mehta. 1994.
85. N.J. Crammond and M.A. Halliwell, "The thaumasite form of sulfate attack in concretes containing a source of carbonate ions: a microstructural overview"; in *Advances in Concrete Technology,* 2nd CANMET-ACI Symposium. Edited by V.M. Malhotra. 1995.
86. H.F.W. Taylor, *Cement Chemistry.* Academic Press, 1990. Pp. 401–402.
87. N.J. Crammond, "Thaumasite: A detailed list of published papers and technical reports," BRE Internal Note N. 148/91, 1991.
88. N.J. Crammond, "The occurrence of thaumasite in modern constructions," BRE Internal Note N. 147/91, 1991.
89. L. Hjorth, "Thaumasite: Review of analytical methods," EUREKA, Project EU-672, 1991.
90. M. Berra and G. Baronio, "Thaumasite in deteriorated concretes in the presence of sulfates"; pp. 2073–2089 in *Concrete Durability,* Vol. II, ACI SP-100. 1987.
91. "Sulfate and acid resistance of concrete in the ground," BRE Digest 363. 1991
92. S. Kelham, J.S. Damtoft, and B.L.O. Talling, "The influence of high early strength mineralized clinker on the strength development of blended cements containing fly ash, slag and ground limestone"; pp. 229–247 in *Proceedings, 5th International Conference on Fly Ash, Silica Fume, Slag and Natural Pozzolana in Concrete,* ACI SP 153. Edited by V.M. Malhotra. 1995.

Credits

Hearn and Figg

Figure 1.
Reprinted with permission from "Results of Experiements Made to Determine the Permeability of Cements and Cement Mortars," R.F.M Bakker, *J. Franklin Inst. Philadelphia*, 199–207 (1889).

Figure 4.
Reprinted with permission from "A Model for Hydrated Portland Cement Paste as Deducted from Sorption-Length Change and Mechanical Properties," R.F. Feldman and P.J. Sereda, *Materials and Structures*, RILEM, **1** [6] 509–520 (1968).

Figure 7.
Reprinted with permission from "Studies of Physical Preoperties of Hardened Portland Cement Paste," *R&D Bulletin* 22. Portland Cemenet Association, 1948.

Figure 8.
Reprinted with permission from "In-situ Permeability Testing—A Basis for Service Life Prediction," A.E. Long, P.A.M. Basheer, and F.R. Montgomery, pp.651–670 in *Advances in Concrete Technology,* Proceedings of the 3rd CANMET/ACI International Concference, Auckland, New Zealand, (1997).

Figure 10.
Reprinted with permission from "Capillary Continuity or Discontinuity in Cement Pastes," *J. PCA R&D Labs.*, **1** [2] 38–48 (1959).

Figure 16.
Reprinted with permission from "Effect of Cracking on Drying Permeability and Diffusivity of Concrete," Z.P. Bazant, S. Senere, J.K. Kim, *ACI Materials Journal*, **9–10** 351–357 (1987).

Figures 18 and 23.
Reprinted with permission from "Effect of Shrinkage and Load-Induced Cracking on Water Permeability of Concrete," N. Hearn, *ACI Materials Journal*, **96** [2] 234–241 (1999).

Figures 19 and 24.
Reprinted with permission from “Influence of Microcracking on the Mass Transport Properties of Concrete,” H. Rusch and K.C. Hover, *ACI Materials Journal*, **89** [4] 416–424 (1992).

Figure 21.
Reprinted with permission from "Fracture mechanism of Concrete under Compressive Loads," S.D. Santiago and H.K. Hilsdorf, *Cement and Concrete Research* **3** 363–388 (1973).

Figure 22.
Reprinted with permission from “Research toward a General Flexural Theory for Structural Concrete,” H. Rusch, *ACI Materials Journal*, pp. 1–29 (1961).

Figure 25.
Reprinted with permission from “A Method of Measuring Permeability of Mortar under Uniaxial Compression,”N. Hearn and G. Lok, *ACI Materials Journal*, **95** [6] 691–694, 1998.

Figure 26.
Reprinted with permission from “Permeability of Cracked Concrete,” C.M. Aldea, S.P. Shah, and A. Karr, *Materials and Structures*, pending publication.

Figure 29.
Reprinted with permission from “Laboratory Studies and Calculations on the Influence of Crack Width on Chloride-Induced Corrosion of Steel in Concrete,” P. Schiessl and M. Raupach, *ACI Materials Journal*, **94** [1] 56–62 (1997).

Figure 30.
Reprinted with permission from “Corrosion of Reinforced Steel in Concrete: Effects of Materials, Mix Composition and Cracking,” T. Lorentz and C. French, ACI Materials Journal, 28 [March-April] 181–190 (1995).

Hall and Bosbach

Figure 7
Reprinted from *Geochimica et Socmochimica Acta*, Vol. 57, A.J. Gratz, P.E. Hillner, and P.K. Hansma, "Step Dynamics and Spiral Growth on Calcite," pp. 491–495, copyright 1993 with permission from Elsevier Science.

Figure 10.
Reproduced from "Scanning Force Microscopy of Gypsum Dissolution and Crystal Growth,"
42, 232–238 (1996), with permission of the American Institute of Chemical Engineers. Copyright 1996 AIChE. All rights reserved.

Berliner

Figure 3.
Reprinted with permission from Institut Laue-Langevin.

Figure 4.
Reprinted with permission from International Union of Crystallography.

Figure 5.
Reprinted with permission from "Rate of Reactions between D_2O and $Ca_xAL_yO_z$," *J. Solid State Chem.*, 51, 196–204 (1984).

Figures 6 and 7.
Reprinted with permission from *Acta Chemica Scandinavica.*

Figure 8.
Reprinted from *Cement and Concrete Research* 27, R. Berliner, C. Ball, and P.B. West, "Neutron Powder Diffraction Investigation of Model Cement Compounds," 551–575, copyright 1997, with permission from Elesevier Science.

Figures 11 and 16.
Reprinted from *Cement and Concrete Research* 28, R. Berliner, M. Popovici, K.W. Herwig, M. Berliner, H.M. Jenning, and J.J. Thomas, "Quasielastic Neutron Scattering Study of the Effect of Water to Cement Ratio on the Hydration Kinetics of Tricalcium Silicate," 231–243, copyright 1998, with permission from Elesevier Science.

Figures 14 and 15.
Reprinted with permission from "An in-situ Quasielastic Neutron Scattering Study of hte Hydration Reaction in Tricalcium Silicate," S.A. FitzGerald, D.A. Neumann, J.J. Rush, D.P. Bentz, and R.A. Livingston, *Chem. Mater.* **10** 397, Copyright 1998 American Chemical Society.

Figure 20.
Reprinted from "Development of the Fine Porosity and Gel Structure of Hydrating Cement Systems," by A.J. Allen et al., *Hil. Mag.* B 56 263–288 (1987) with permission from Taylor & Francis.

Author and Keyword Index